Reliability Technology for Integrated Circuit Packaging

Bin Zhou · Yunfei En · Si Chen

Reliability Technology for Integrated Circuit Packaging

中国工信出版集团
電子工業出版社
PUBLISHING HOUSE OF ELECTRONICS INDUSTRY
http://www.phei.com.cn
Springer

Bin Zhou
China Electronic Product Reliability
and Environmental Testing Research
Institute
Guangzhou, Guangdong, China

Yunfei En
China Electronic Product Reliability
and Environmental Testing Research
Institute
Guangzhou, Guangdong, China

Si Chen
China Electronic Product Reliability
and Environmental Testing Research
Institute
Guangzhou, Guangdong, China

ISBN 978-981-95-3884-3 ISBN 978-981-95-3885-0 (eBook)
https://doi.org/10.1007/978-981-95-3885-0

Jointly published with Publishing House of Electronics Industry
The print edition is not for sale in China (Mainland). Customers from China (Mainland) please order the print book from: Publishing House of Electronics Industry.
ISBN of the Co-Publisher's edition: 978-7-121-46151-4

Translation from the English language edition: "集成电路封装可靠性" by Bin Zhou et al., © Publishing House of Electronics Industry Co., Ltd 2023. Published by Publishing House of Electronics Industry Co., Ltd. All Rights Reserved.

This Springer imprint is published by the registered company Springer Nature Singapore Pte Ltd.
The registered company address is: 152 Beach Road, #21-01/04 Gateway East, Singapore 189721, Singapore

Preface by the Editor-in-Chief of the *Integrated Circuit Series*

The Soil of Cultivation · The Spring of Seedling · The Key to Enlightenment · The Foundation of a Strong Nation

In his poem "Butterflies in Love with Flowers," Wang Guowei wrote: "Things hardest to keep in this human world—Rosy cheeks which fade in the mirror and flowers that fall from trees." This appears to be an unchangeable law of nature. Nevertheless, through various means and using diverse media, people still retain memories of time and express hopes for the future.

Books, particularly paper ones, stand as the most numerous and enduring medium for recording thoughts and knowledge. From the "Book of Songs," we experience the stirrings of youth; through the "Records of the Grand Historian," we hear the thundering hooves of war horses; reading "The Analects," we delve into philosophical reasoning; appreciating Tang poetry, we grasp the essence of humanities and customs.

Despite the ability to store dynamic audiovisual images in films, tapes, and chips, books still possess a unique charm. They delineate the overarching development, chronicle the fortunes of victory and defeat, exhibit the passion of sensibility, express the assertiveness of reason, encapsulate the allure of color, resonate with the clang of musical notes, gallop with the fervor of the soul, and sparkle with the brilliance of wisdom.

In the "Cihai" (The Collection of Words), books, periodicals, albums, pictures, and other publications are collectively termed as "books." Through this vast array of "books," we come to understand the inventions of electronic tubes, transistors, and integrated circuits, as well as the growth trajectory and developmental patterns of integrated circuit science, technology, market, and applications. Building upon this foundation, pioneers of integrated circuit technology and industry in China embarked on a journey of exploration from the 1950s onwards. Since the turn of the twenty-first century, China's integrated circuit industry has entered a period of rapid development, achieving significant progress in various domains such as basic research, design, manufacturing, packaging, equipment, and materials, with some accomplishments earning recognition on the world stage.

To summarize the past experience, depict the present landscape and envision the future dream, the idea of compiling *The Integrated Circuit Series* (hereinafter called *The Series*) has become a shared aspiration among the many workers in China's integrated circuit science, technology, and industry.

In 2016, the editorial committee for *The Series* was established, bringing together nearly 500 authors from across China to begin drafting the first volume of *Handbook of Integrated Circuit Industry* (hereinafter called *Handbook*). On September 12, 2018, the launch ceremony for the *Handbook* was held at the Great Hall of the People in Beijing, officially introducing the *Handbook* to readers, where it received enthusiastic praise and wide approval from the education, research and industry sectors. Subsequently, the English version of the *Handbook* started to be compiled with joint publication by the Publishing House of Electronics Industry and Springer, one of the largest scientific publisher in the world. Due to the limitation of its volume, the *Handbook* adopts concise "Term" format to describe aspects such as products, production, economics, and markets in the integrated circuit industry, with each term consisting of around 1000 words. The advantage of the format is its brevity and clarity, making it easy to consult and be referred to. However, it also has limitations, which lies in that the *Handbook* does not allow comprehensive and detailed coverage of any one specialized field. On the other hand, each volume in *The Series* is not constrained by its size and can offer in-depth and broad discussions, illustrated and detailed, on specific specialized fields. *The Series* and the *Handbook* complement each other, satisfying different readers' needs and enhancing each other's value. To better organize the compilation of *The Series*, the editorial committee established 12 sub-committees, each responsible for the following volumes:

- Integrated Circuit Series: The History of Integrated Circuit Development
- Integrated Circuit Series: Economics of the Integrated Circuit Industry
- Integrated Circuit Series: Management of the Integrated Circuit Industry
- Integrated Circuit Series: The Integrated Circuit Industry, Education, and Talents
- Integrated Circuit Series: Frontiers of Integrated Circuit Development and Basic Research
- Integrated Circuit Series: Integrated Circuit Products and Markets
- Integrated Circuit Series: Integrated Circuit Design
- Integrated Circuit Series: Integrated Circuit Manufacturing

VI

- Integrated Circuit Series: Integrated Circuit Packaging and Testing
- Integrated Circuit Series: Specialized Equipment for the Integrated Circuit Industry
- Integrated Circuit Series: Specialized Materials for the Integrated Circuit Industry
- Integrated Circuit Series: Research and Applications of Compound Semiconductors
- Integrated Circuit Series: Integrated Micro-Nano Systems
- Integrated Circuit Series: Electronic Design Automation.

Through the concerted efforts of industry colleagues in 2021, approximately 10 volumes in *The Series* were published and released to commemorate the centenary of the Chinese Communist Party. As the beginning of a continuing endeavor, from 2021 onwards, new works will be included in *The Series* annually, consistently contributing to the development of the integrated circuit industry in our nation. By 2035, our vision is to see the number of new or republished volumes reach nearly a hundred, creating a flourishing *Series* that showcases a diversity of ideas and innovations.

As integrated circuits are revolutionizing both production and lifestyles today, they have emerged as pivotal assets in the global competition among major nations. On the path to the rejuvenation of the Chinese nation, integrated circuits are shouldering a new and arduous historical mission. We firmly believe that, whether serving as textbooks for the primary discipline of "Integrated Circuit Science and Engineering" or as reference materials for frontline researchers and industry professionals, *The Series* will serve as timely assistance and vital resources to meet the urgent need for talent cultivation and accelerate industrial development.

The development of science, technology, and industry knows no bounds. When China achieves its second centenary goal in 2049, future generations may discover various shortcomings in *The Series* authored in the 2020s. However, they will still see the clear footprints of those predecessors who contributed to the self-reliance and strength of the Chinese nation within the rigorous words of *The Series*. They will feel the fervent passion that surges through their earnest expressions and hear the enduring heartbeat of their dreams that continues to resonate.

Books serve as invaluable mentors for acquiring knowledge, essential tools for spreading ideas, and crucial vessels for preserving culture. They are significant markers of human progress and civilization. May *The Series* perpetually act as the fertile soil that nurtures the development of integrated circuit science and technology in our nation. May it become the nourishing spring that propels the growth of China's integrated circuit industry. May it serve as the golden key that unlocks the intellectual potential of our integrated circuit talent. Ultimately, may it be the genetic cornerstone that fulfills the dream of establishing China as a global leader in the integrated circuit industry.

Compiling *The Series* is a monumental endeavor, involving an extensive collection of works. Looking at classic texts from history, it is not uncommon to find those that took over a decade to complete or covered a wide array of disciplines, sometimes spanning over a hundred categories.

Shi Ji (Records of the Historian), written by Sima Qian in the Western Han Dynasty, comprises 130 volumes with over 526,500 characters, and was completed over a span of 14 years.

Zizhi Tongjian (Comprehensive Mirror in Aid in Government), written by Sima Guang in the Northern Song Dynasty, consists of 294 volumes and took 19 years to compile.

The Siku Quanshu (The Complete Library of the Four Treasuries), encompassing 36,300 volumes and approximately 800 million characters, was compiled by 360 scholars and transcribed by 3826 individuals during the Qing Dynasty, taking 13 years to complete.

Mengxi Bitan (Dream Pool Essays), composed by Shen Kuo in the Northern Song Dynasty, includes 30 volumes, 17 categories, and 609 essays. It covers a range of disciplines such as astronomy, mathematics, physics, chemistry, and biology, and is acclaimed as a "milestone in the history of Chinese science."

Tiangong Kaiwu (The Exploitation of the Works of Nature), authored by Song Yingxing in the Ming Dynasty, is the world's first comprehensive work on agricultural and industrial production. It is divided into 3 volumes with 18 chapters and includes 123 illustrations, earning it the title of the "Encyclopedia of Chinese Technology in the 17th Century."

These classics embody the academic spirit of "true learning comes from heartfelt understanding" and the scholarly attitude of "persistence leads to excellence." These are precisely the exceptional traditions that our generation should perpetuate and carry forward in the process of compiling *The Series*. We hope that all the editors of *The Series* will take inspiration from the exemplary standards set by our predecessors, persistently striving for perfection in their work. By doing so, they will continually contribute to the enrichment of our nation's repository of knowledge on integrated circuits. May the spirit of learning, seeking truth, exploring, and innovating as encapsulated in *The Series* be passed down from generation to generation.

Yan Garden, Beijing, China
July 2021

Preface

Reliability is one of the key indicators for evaluating product quality and is an interdisciplinary field encompassing materials science, electronics, thermodynamics, and mechanics. The core of reliability engineering lies in ensuring that a product's functional performance is fully realized under actual operating conditions. With the implementation of the national strategy to strengthen China through quality, reliability has received unprecedented attention. Quality initiatives focused on improving reliability have become a central theme in industry development and represent a necessary path for achieving high-quality growth in China's manufacturing sector.

With the rapid advancement of semiconductor process technologies, advanced semiconductor nodes have entered the 7 nm and 5 nm eras and are now progressing toward the 3 nm and 2 nm generations. As process nodes approach atomic dimensions, further scaling of feature sizes is increasingly constrained by physical limits. To sustain the development of electronic products toward higher performance, smaller form factors, and lower power consumption, both industry and academia have proposed parallel solutions under the paradigms of "More Moore" and "More than Moore," among which advanced packaging plays a pivotal role. In recent years, advanced packaging technologies such as chip-on-wafer-on-substrate (CoWoS), fan-out packaging (InFO), chiplet architectures in 2.5D packaging, through-silicon via (TSV)-based 3D packaging, and microsystem integration have evolved rapidly. These technologies have driven disruptive improvements in the performance of electronic products such as smartphones and unmanned aerial vehicles. As advanced packaging continues to evolve, the boundary between integrated circuit (IC) chips and IC packaging is becoming increasingly blurred, giving rise to a new trend of integrated development.

The reliability of integrated circuit (IC) packaging is a systematic engineering endeavor involving multiple aspects such as materials, processes, structures, and operating environments. With the widespread adoption of new materials, technologies, and processes in IC packaging, some advanced packaging technologies are encountering common industry challenges, including unclear failure mechanisms, increased difficulty in defect localization, and a lack of targeted reliability assessment methods. It is essential to investigate the failure modes and mechanisms of advanced packaging structures—such as micro-bumps, through-silicon vias (TSVs), and redistribution layers (RDLs)—under the combined effects of thermal , electrical, and mechanical stresses. Furthermore, there is a need to develop new techniques for defect localization and failure analysis in complex packaging configurations, including wafer-level chip packaging on substrates and 3D wafer-level packaging. Advancing targeted, physics-of-failure-based reliability assessment methods for microsystem integrated packaging will play a crucial role in enhancing both the functionality and reliability of next-generation electronic products and in supporting the development of China's electronic information industry.

Seizing the strategic opportunities of the "Post-Moore era," and leveraging China's comprehensive advantages—including a complete range of foundational industrial sectors, the rapid rise of packaging and testing technologies, and a significantly heightened awareness of reliability—this book, *Reliability Technology for Integrated Circuit Packaging*, has been compiled under the guidance and coordination of the Editorial Committee of the *Integrated Circuit Series: Integrated Circuit Packaging and Testing*. The book is authored by the Fifth Electronics Research Institute of the Ministry of Industry and Information Technology. It represents not only the Committee's recognition of our work but also a mission entrusted to us by the times—and, more importantly, a responsibility borne by history.

The Fifth Electronics Research Institute of the Ministry of Industry and Information Technology is one of the earliest authoritative institutions in China engaged in research on electronic product quality and reliability and has pioneered a number of national "firsts" and "onlys" in the field of quality and reliability. Since the official launch of the compilation work by the Editorial Committee of the *Integrated Circuit Series: Integrated Circuit Packaging and Testing*, we immediately established a writing team, set clear objectives, defined the tasks, and carried out multiple rounds of discussion, adjustment, and revision throughout the writing process. This book consists of ten chapters. Based on an introduction to the classification of IC packaging technologies and packaging reliability characterization techniques, it describes the typical failure modes, failure mechanisms, and physical characteristics of IC packaging from both the product dimension—namely plastic packaging and hermetic packaging—and the stress dimension—specifically thermal and mechanical stresses. It then presents failure analysis techniques and quality reliability evaluation methods related to packaging, taking into account the structural characteristics of advanced packaging technologies. Finally, it discusses board-level assembly reliability of integrated circuits from the perspectives of materials, structure, and stress. The primary authors of this book are core members of the national key laboratory team specializing in advanced packaging and microsystem reliability technologies, with many years of experience in packaging reliability research and IC failure analysis. During the writing process, we consulted a large volume of literature, technical resources, and prior research achievements of the writing team in related fields. The authorship and division of work are as follows: Chap. 1 was written by En Yunfei, He Xiaoqi, and Zhou Bin; Chap. 2 by He Xiaoqi, En Yunfei, and Yang Shaohua; Chap. 3 by Yang Shaohua, En Yunfei, and Yang Xiaofeng; Chap. 4 by Chen Si, Zhou Bin, and

Chen Yuan; Chap. 5 by Yang Xiaofeng, Shi Yijun, and Fu Xing; Chap. 6 by Zhou Bin, Fu Zhiwei, and Fu Xing; Chap. 7 by Su Wei, Chen Si, and Fu Xing; Chap. 8 by Chen Yuan, Yang Shaohua, and Fu Xing; Chap. 9 by Wang Hongyue, En Yunfei, and Wu Huiwei; and Chap. 10 by Fu Zhiwei, Zhou Bin, and Yao Bin. En Yunfei and Zhou Bin were responsible for the overall organization, planning, compilation, and final review of the book, while the other contributing authors were responsible for the self-review and cross-review of their respective chapters.

After completing internal proofreading and review, this book was submitted to the Editorial Committee of the *Integrated Circuit Series: Integrated Circuit Packaging and Testing*, and was reviewed by Prof. Qin Fei of Beijing University of Technology, Prof. Yu Daquan of Xiamen University, and Deputy Director Ming Xuefei of the 58th Research Institute of China Electronics Technology Group Corporation, as organized by the Committee. Upon completion of the review process, revisions were made based on the reviewers' comments and suggestions.

We would like to express our sincere gratitude to Mr. Bi Keyun, Honorary Chairman of the Packaging Branch of the China Semiconductor Industry Association, and Mr. Wang Xinchao, Chairman of Xinchao Group, for their strong support. We also thank Mr. Shen Yang and Mr. Zhou Jian of the Editorial Committee of the *Integrated Circuit Series: Integrated Circuit Packaging and Testing* for their enthusiastic assistance during the compilation of this book. Our appreciation also goes to colleagues from the Fifth Electronics Research Institute of the Ministry of Industry and Information Technology—Zhang Zhangang, Wang Lei, Niu Hao, Liu Yuebo—for providing technical materials; to Zhu Jianyuan, Hong Danni, and others at the same institute; to Liang Zhentang, a graduate student at South China University of Technology; to Zhang Yiming, a graduate student at Xidian University; and to Zhou Xiangsheng, a graduate student at Xiamen University of Technology—for their diligent efforts in literature research, editing, and typesetting. During the writing process, we also consulted materials and references from several industry experts and colleagues, including those from the Reliability Research and Analysis Center, Key Laboratory, and Component Testing Center of the Fifth Electronics Research Institute of the Ministry of Industry and Information Technology. We express our heartfelt thanks to them all.

Given the wide scope and rapidly evolving nature of advanced packaging reliability, continued in-depth research and exploration are necessary. Due to the limitations in our understanding and knowledge, errors and omissions may still be present in this book. We sincerely invite readers to offer their valuable feedback so that we may make improvements in future revisions.

This book is respectfully dedicated to all professionals striving for the enhancement of electronic product reliability and the development of the integrated circuit packaging industry in China.

Guangzhou, China

Bin Zhou
Yunfei En
Si Chen

Introduction

Integrated circuits (ICs), often referred to as the "heart" of electronic products, are the core of all information technology industries. IC packaging technology, which involves the "packaging" of integrated circuits, has become a crucial technological approach in the post-Moore era. Reliability technology for IC packaging forms the foundation and core of the reliability of integrated circuits and entire electronic systems. Approximately half of IC failures are caused by packaging failures, making packaging reliability a widely focused issue. Based on an introduction to the classification of IC packaging technologies and reliability characterization methods, this book describes the typical failure modes, failure mechanisms, and physical characteristics of IC packaging from the product dimension—namely plastic packaging and hermetic packaging--and the stress dimension, including thermal and mechanical stresses. In combination with the structural features of advanced packaging, it presents failure analysis techniques and quality reliability evaluation methods related to packaging. Additionally, from the perspectives of materials, structure, and stress, the book addresses board-level assembly reliability of integrated circuits. This book aims to open a window of communication for those interested in packaging reliability technology and to build a bridge between IC reliability and electronic product reliability.

This book is primarily intended for engineers and managers engaged in electronic components, electronic packaging, as well as research, design, manufacturing, testing, and experimentation related to electronic system products. It can also serve as a reference textbook for relevant majors in various higher education institutions.

Contents

About the Authors

Bin Zhou, Ph.D. is Senior Engineer, Deputy Chief Engineer, and Academic Leader of the National Key Laboratory at the Fifth Electronics Research Institute of the Ministry of Industry and Information Technology. He is recognized as an outstanding young and middle-aged expert at the provincial and ministerial level, an Excellent Science and Technology Worker of the Chinese Institute of Electronics, a representative of the Guangzhou Municipal People's Congress, a member of the Guangzhou Municipal People's Congress Budget Committee, and a high-end domestic talent of Zengcheng District, Guangzhou. He is an expert registered with the Ministry of Science and Technology and the Ministry of Industry and Information Technology, a member of the Microsystems Special Group under the Military Electronic Components Standardization Technical Committee, a member of the Integrated Microsystem Modeling and Simulation Committee of the Chinese Society of Simulation, a member of the Chinese Materials and Testing Group Standards Committee, and a member of the Degree Evaluation Subcommittee of the Guangzhou Research Institute of Xidian University. Dr. Zhou has led nearly 30 national and provincial research projects; he has been awarded one first-class and four second-class provincial and ministerial Science and Technology Progress Awards; he also received the National Defense Science and Technology Innovation Team Award and the China Electronic Information Science and Technology Innovation Team Award. He has published 51 SCI/EI-indexed papers, holds 22 authorized invention patents, authored 3 monographs, contributed to 5 standards, and owns 3 software copyrights.

Yunfei En is a Senior Engineer, serves as the Party Committee Secretary and Deputy Director of the Fifth Electronics Research Institute of the Ministry of Industry and Information Technology, and is an expert with special government allowances from the State Council. She is also a fellow of the China Institute of Electronics. With a long-term focus on the reliability of electronic components, she has led and undertaken numerous national and provincial-level scientific research projects, and has published several monographs including Reliability Physics.

Si Chen is a researcher at the National Key Laboratory of Reliability Physics and Application Technology of Electronic Components, The Fifth Electronics Research Institute of the Ministry of Industry and Information Technology (MIIT), China. She earned her bachelor's degree in Materials Science and Engineering from Hohai University and her Ph.D. in Mechanics from Beijing University of Technology, and subsequently completed postdoctoral research at the MIIT Fifth Electronics Research Institute. Her work focuses on reliability physics, failure analysis, and lifetime prediction for advanced semiconductor packaging and 3D integration, with particular emphasis on TSV-related dielectric leakage and TDDB behavior, electromigration and microstructural evolution, characterization of interfacial residual stress, and electro-thermo-mechanical multi-field coupled reliability modeling; in recent years, her research has further expanded to the failure mechanisms and lifetime modeling of Cu/SiO_2 hybrid bonding in HBM modules.

Chapter 1
Overview of Integrated Circuit Packaging Technology and Reliability

Integrated Circuit (IC) consists of internal chips with specific functions and external packages that support and protect these chips. Tummala [1] from the Georgia Institute of Technology provides a definition of IC packaging: it involves mounting functional IC chips into corresponding carriers using various methods, which provides a stable and reliable operating environment for the chips, effectively dissipates the heat generated by the internal chips to the outside, and ensures the normal functionality of the IC. Packaging, as an essential component of an integrated circuit, acts as the bridge for electrical connections between the internal chip and the external environment. It also serves to secure, seal, and protect the chip while ensuring its thermal, electrical, mechanical performance, and reliability.

With the development of society and the continuous advancement of microelectronic device technology, there is a growing demand for smaller sizes and a greater number of leads to meet the increasing requirements of the IC integration. At the same time, packaging reliability must satisfy higher standards for thermal dissipation, enhanced mechanical performance, and improved protective capabilities. This chapter introduces the types and technological developments in IC packaging, examines the relationship between packaging technology and reliability, and outlines the current state of packaging reliability technologies.

1.1 Overview of Packaging Technology Development

1.1.1 Functions of Integrated Circuit Packaging

Due to the brittleness of semiconductor materials and the susceptibility of metal wiring to corrosion in Integrated Circuit (IC) chips, it is essential to protect the chip through packaging to prevent harmful gases and external impurities from eroding the internal circuitry. This ensures that the chip can operate reliably within its specified

B. Zhou et al., *Reliability Technology for Integrated Circuit Packaging*,
https://doi.org/10.1007/978-981-95-3885-0_1

lifespan. The purpose of IC packaging is to protect the chip while enabling electrical connections between the internal chip's input/output (I/O) leads and external circuits. Depending on the application, the structure, packaging materials, and the number of I/O leads can vary significantly. However, the basic functions of the packaging remain largely the same, including power distribution, signal distribution, thermal dissipation, mechanical support, and environmental protection [2].

(1) Power Distribution: Packaging must ensure effective communication between the internal chip and external circuits and meet the power distribution requirements within different sections of the package.
(2) Signal Distribution: The packaging interconnections must minimize signal loss and delay, and prevent external signal interference and internal signal crosstalk. With the improvement of the packaging integration, Electromagnetic Compatibility (EMC) of IC becomes increasingly important, it often serves as a source of radiation in electronic systems. Effective Electromagnetic Interference (EMI) shielding can be achieved through the design of package structure.
(3) Thermal Dissipation: Packaging provides a low thermal resistance path for heat dissipation from the chip, efficiently expelling the heat generated from the chip. When chips operate, power consumption causes them to heat up, leading to temperature rise, which is more pronounced in high-power chips. To prevent beyond-tolerance electrical parameter drift due to high temperature, it is necessary to dissipate the heat generated in the active region of the chip as quickly as possible through packaging, ensuring stable functional performance and long-term reliability.
(4) Mechanical Support: The external packaging structure provides mechanical support for the internal chip. IC semiconductor materials, such as silicon (Si) and gallium arsenide (GaAs), are very thin and brittle, requiring substrates with greater mechanical strength and protective encapsulating materials to meet assembly process requirements and ensure long-term use.
(5) Environmental Protection: Packaging acts as a barrier against moisture and other harmful gases, minimizing impact of external environmental factors on the internal chip's performance. For example, aluminum metallization wiring bonding windows on the chip surface without passivation layer protection are prone to contamination and corrosion, leading to open circuits. Therefore, appropriate packaging structures and materials are necessary to prevent moisture intrusion.

In addition, IC packaging must meet the mounting process requirements of Printed-Circuit Boards (PCBs) by designing appropriate I/O leads. Based on application needs, IC chips must adopt packaging forms with varying lead pitches and structures to ensure compatibility with PCB circuit design and assembly processes.

Statistics indicate that for the manufacturing of electronic components, including ICs [3], 70–90% of the volume of electronic components is determined by the packaging. More than 50% of signal delays, over 55% of increased resistance, over 60% of thermal performance anomalies, and over 50% of component failures are also related to the packaging. Additionally, the cost of packaging accounts for 30–80%

of the total component cost. Therefore, IC chips without packaging are like a heart with no strength, unable to function properly. Similarly, ICs without reliable packaging are bound to have strength that doesn't follow the heart's command, and fail prematurely. As the applications of integrated circuits expand, the technology and reliability of their packaging continue to evolve. There is now a demand not only for more pins and higher density but also for greater reliability, robustness against harsh environments, and longer lifespans.

1.1.2 Common Types of Integrated Circuit Packaging

IC packaging can be classified into four categories according to packaging carrier materials: plastic packaging, ceramic packaging, metal packaging, and glass packaging [2]. It is the most widely used packaging type, offering advantages such as low cost and light weight. Plastic packaging is extensively utilized in various consumer electronics, automotive electronics, medical electronics, and even aerospace and equipments. Ceramic packaging features excellent hermeticity, insulation, stability and high structural strength. Some ceramics also have superior thermal conductivity, making ceramic packaging ideal for high-reliability applications such as aerospace devices. Metal packaging is known for its outstanding thermal conductivity and mechanical properties. It is also commonly employed in high-reliability fields. Glass packaging, with its exceptional high-temperature resistance, chemical resistance, electrical insulation and hermeticity, is widely used in MEMS sensors, photovoltaic cells, LEDs, etc.

Based on hermeticity, IC packaging can be classified into two types: hermetic packaging and non-hermetic packaging. Plastic packaging is a common type of non-hermetic packaging, while ceramic, metal and glass packaging are usually hermetic packaging. By packaging form, IC packaging is mainly divided into traditional packaging and advanced packaging. The following section will provide a brief introduction to the types within these two categories.

1. Traditional Packaging

Traditional packaging primarily includes Single In-line Package (SIP), Dual In-line Package (DIP), Transistor Outline Package (TOP) (including Small Outline Package (SOP) and Small Outline J-leaded Package (SOJ)), Quad Flat Package (QFP), Quad Flat No-lead Package (QFN), Ball Grid Array (BGA), Ceramic Pin Grid Array (CPGA), and Ceramic Column Grid Array (CCGA), etc. [2].

1) Single In-line Package (SIP)

The Single In-line Package features leads extending from one side of the package body in a single row, as shown in Fig. 1.1. It represents a basic form of through-hole packaging. Typically made of plastic, SIP is predominantly used for custom products. Its standard specifications include a lead pitch of 2.54 mm and the number of leads ranging from 2 to 23.

Fig. 1.1 Single in-line packaging (SIP)

Fig. 1.2 Dual in-line packaging (DIP)

2) Dual In-line Package (DIP)

The Dual In-line Package is another form of through-hole packaging, as illustrated in Fig. 1.2. Its leads extend from both sides of the package body and are bent vertically downward. DIP is the most commonly used type of through-hole package. It is typically employed for small to medium-scale integrated circuits and is available in plastic or ceramic materials. Standard specifications include 6–64 leads, a lead pitch of 2.54 mm, and a package body width of approximately 15 mm.

3) Small Out-line Package (SOP)

The Small Out-line Package (SOP) is a type of surface-mounted packaging, also known as Small Out-line Integrated Circuit (SOIC), as shown in Fig. 1.3. The leads extend from both sides of the package body and are bent into a shape of a seagull wing (L shape). The packaging materials are typically plastic or ceramic, The number of leads ranges from 8 to 44, with a lead pitch of approximately 1.27 mm. SOPs with a package height of less than 1.27 mm are referred to as TSOP (Thin Small Out-line Package). SOPs with a lead pitch smaller than 1.27 mm for high-density leads are referred to as SSOP (Shrink Small Out-line Package). SOPs combining both thin-body and high-density lead features are referred to as TSSOP (Thin Shrink Small Out-line Package).

Fig. 1.3 Small out-line package (SOP)

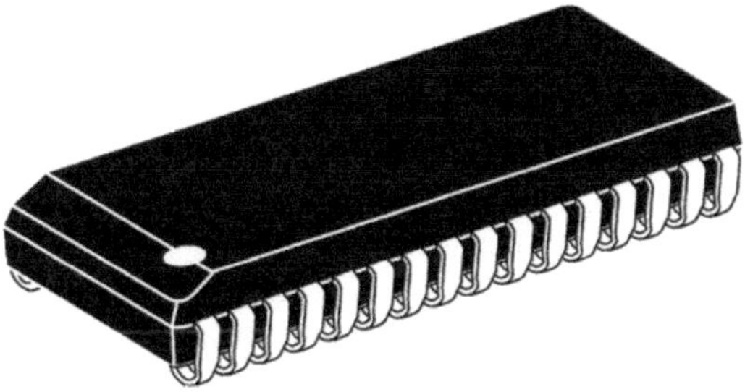

Fig. 1.4 Small out-line J-leaded package

4) Small Out-line J-leaded (SOJ) Packaging

SOJ is a type of surface-mount package. The leads are bent vertically from the sides of the package body into a J-shape. The packaging material is usually plastic. Common dimensions and lead counts: lead pitch is 1.27 mm, with 20–40 leads (Fig. 1.4).

5) Quad Flat Package (QFP)

The Quad Flat Package (QFP) is a common form of surface-mounted packaging, as shown in Fig. 1.5. The leads extend from all four sides of the package body and bend vertically to form a shape of a seagull wing (L shape). The packaging materials are typically plastic and ceramic, with plastic packaging being more widely used. QFP is an early form of high-density packaging, generally featuring more than 100 leads, with a lead pitch ranging from 0.3 mm to 1.0 mm.

6) Quad Flat No-lead (QFN)

The Quad Flat No-lead (QFN) Package is a widely used type of surface-mounted packaging. Leads extend from four sides of the package body, forming contact areas for the electrodes of the IC output terminals. To provide higher device assembly density, this type of packaging has no leads, known as a Leadless Chip Carrier (LCC). The packaging materials include plastic and ceramic. Figure 1.6 illustrates

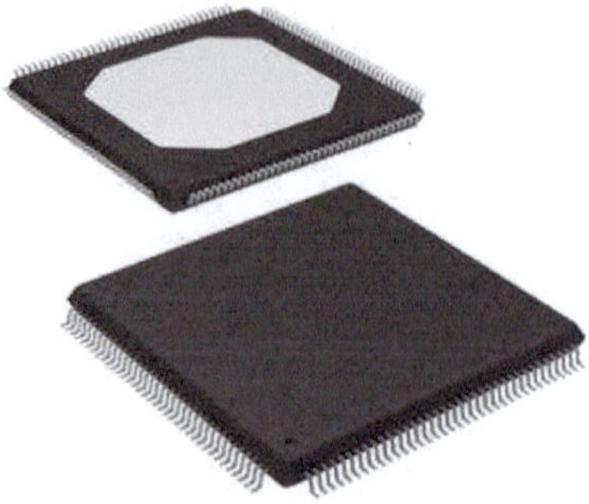

Fig. 1.5 Quad flat package

a Quad Flat No-lead Package. Typical sizes and lead numbers are as follows: lead pitch include 1.27 mm, 0.65 mm, and 0.5 mm, with lead numbers ranging from 14 to 100.

Ceramic Leaded Chip Carrier (CLCC), as shown in Fig. 1.7, is a type of packaging in which leads extend from all sides of the package body and bend downward and inward to form a J-shape. The packaging material is typically ceramic. Typical sizes and the number of leads are as follows: a lead pitch of 1.27 mm, with the number of leads ranging from 18 to 84.

Fig. 1.6 Quad-flat no lead (QFN) package

Fig. 1.7 Ceramic leaded chip carrier (CLCC)

Fig. 1.8 Plastic BGA package

7) Ball Grid Array (BGA)

Ball Grid Array (BGA) Package is a recently developed structure for surface-mounted devices. The substrate is typically an organic multi-layer board with 2–4 layers, and the interconnection between the chip output terminals and the substrate circuits is primarily achieved through either wire bonding or flip-chip bonding. The packaging carrier materials include plastic, ceramic, and metal. Figure 1.8 illustrates a plastic BGA package.

8) Ceramic Pin Grid Array (CPGA)

Ceramic Pin Grid Array (CPGA) packaging is a type of package designed specifically for CPGA plug-in applications, allowing easy installation and removal, as shown in Fig. 1.9. CPGA packaging uses round or square pins arranged in a square grid layout at equal intervals around the chip. The pins can form two or more rows, enabling direct use by inserting into sockets with reliable contact. The packaging carrier materials may include plastic, but ceramics are predominantly used. The number of pins in the grid array typically exceeds 100, often ranging from 250 to 528, with lead pitch specifications primarily at 2.54 mm and 1.27 mm.

9) Ceramic Column Grid Array (CCGA)

Fig. 1.9 CPGA package

(a) Front View of a Typical CCGA Package (b) Back View of a Typical CCGA Package

Fig. 1.10 Typical view of CCGA package

Ceramic Column Grid Array (CCGA) packaging is a form of Column Grid Array (CGA). The CGA device packaging structure consists of three parts: the IC chip package at the top, the ceramic substrate package in the middle, and the high-lead solder column array package at the bottom. A typical CCGA package is shown in Fig. 1.10.

In the CCGA packaging form, the increased space between solder columns and the substrate allows for the more effective heat dissipation and helps buffer thermal stress between the solder column and ceramic die as well as the ceramic substrate, which helps to extend the thermal fatigue life. Generally, the solder columns in CGA packaging exhibit higher resistance to thermal mismatch and better thermal fatigue life compared to the solder balls in BGA packaging, although their resistance to mechanical shock is lower than that of BGA solder balls.

packaging form, the stability of a packaging technology is primarily determined by its control over packaging efficiency—the ratio of packaging area to chip area. Additional considerations include the impact of packaging leads on signal transmission delay and the effect of packaging materials on heat dissipation. The basic requirements are as follows:

(1) Improve packaging efficiency by ensuring the ratio of packaging area to chip area approaches 1:1 as closely as possible.
(2) Minimize the impact of leads on signal transmission delays and avoid mutual interference; keep leads as short as possible and maximize the space between them.
(3) Ensure good thermal performance. Select packaging materials with high thermal conductivity that meet the requirements, and aim for the thinnest possible package.

2. Advanced Packaging

As Moore's Law approaches its physical limitation, further reducing transistor size to increase the density of functional units in IC has become more challenging. Advanced packaging technology is widely acknowledged as one of the most critical approaches for continuously promoting IC performance. Consequently, semiconductor manufacturers worldwide are intensifying their investments in the research and development, as well as production, of advanced packaging technologies. New packaging formats, such as Wafer Level Package (WLP), 2.5D package, 3D package and System in package (SiP), are typically categorized under advanced packaging [2].

1) Wafer Level Package (WLP)

The previously introduced packaging methods are designed for individual chips, whereas Wafer Level Package (WLP) is a one-time packaging process applied to all chips on a wafer. Subsequent aging and testing are conducted simultaneously, and the wafer is eventually diced into individual devices for mounting on substrates or PCBs. The advantages of WLP include high packaging efficiency, compact size, and low cost. It allows for the simultaneous consideration of chip design and WLP design, thereby improving product design efficiency and shortening development cycles. WLP is predominantly applied to integrated circuits such as Analog ICs, PA/RF, and CMOS image sensors.

WLP technology includes Fan-In WLP and Fan-Out WLP. Fan-In WLP involves packaging chips before dicing the wafer into individual devices. A typical Fan-In WLP structure is illustrated in Fig. 1.11. The size of the packaged device after dicing is almost the same as the size of the chip inside the package, with the packaging area to chip area ratio being approximately 1:1. Fan-Out WLP, a newly developed packaging technology in recent years, addresses the soldering challenges caused by the continuous reduction in chip size. Fan-Out WLP uses wafer reconstruction technology to attach multiple known-good-die (KGD) chips to a wafer carrier. The distance between the chips determines the area of the Fan-Out package. During packaging, mould materials fill the gaps between the KGD chips and the surrounding space. After separating the wafer carrier, processes for the Re-Distribution Layer (RDL) and BGA solder balls for terminals are performed. Finally, the wafer is diced into individual Fan-Out WLP devices [4]. A typical Fan-Out WLP structure is illustrated in Fig. 1.12.

2) 2.5D Package

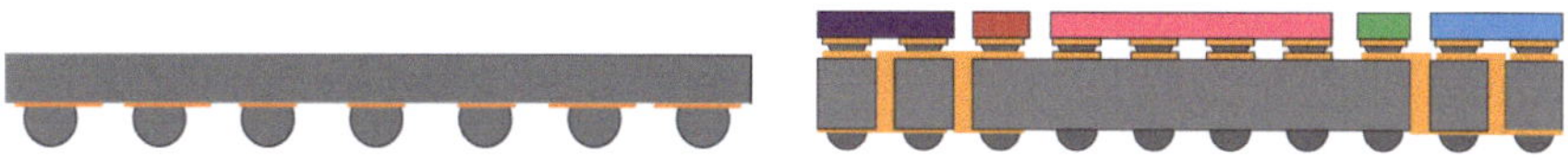

(a) Single-chip Fan-In WLP (b) Multi-chip Fan-In WLP

Fig. 1.11 Typical fan-in WLP structure

(a) Single-chip Fan-Out WLP (b) Multi-chip Fan-Out WLP

Fig. 1.12 Typical fan-out WLP structure

Fig. 1.13 Typical 2.5D package structure

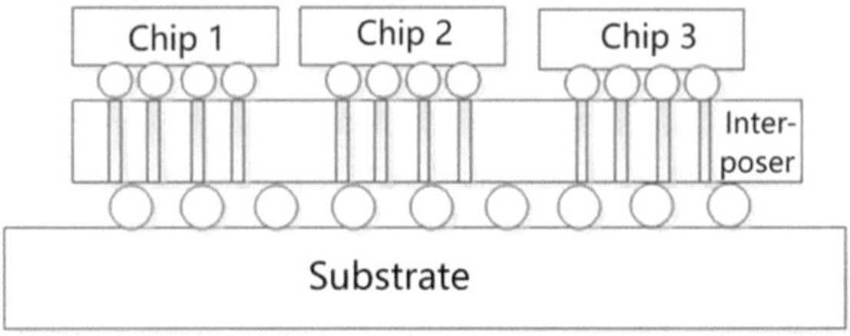

2.5D package builds upon 2D package by introducing a (silicon) interposer between the chip and the package carrier. The interposer typically employs Through Silicon Via (TSV) technology to connect the metallized wiring on its upper and lower layers, thereby significantly enhancing the density and performance of planar 2D package. A typical 2.5D package structure is illustrated in Fig. 1.13.

Intel, the world's largest integrated device manufacturer, and TSMC, the largest wafer foundry, have focused heavily on developing advanced packaging technologies in recent years alongside the introduction of new-generation process nodes. These advancements prominently include Intel's Embedded Multi-die Interconnect Bridge (EMIB) technology, TSMC's Chip on Wafer on Substrate (CoWoS) technology, and Chiplet technology. A shared characteristic of these advanced packaging technologies is their departure from the traditional 2D planar integration on packaging substrates. Instead, they incorporate chip packaging and integration in the vertical dimension. Among these, CoWoS and Chiplet packaging structures represent current wafer-level system packaging solutions. CoWoS technology allows for combining small chips of different types and process nodes into a 2.5D configuration, resulting in a structure similar to a system-on-chip (SoC).

CoWoS is a technology that integrates both chips and substrates into a single package. It first installs semiconductor chips onto a wafer through the Chip on Wafer (CoW) process and subsequently connects and integrates the CoW chips with the substrate [5]. CoWoS is classified as a type of 2.5D packaging technology. A typical CoWoS package structure is shown in Fig. 1.14.

Chiplet is a packaging model that integrates chips in a modular way, where multiple chips are interconnected internally and packaged together with a base chip, forming a multifunctional heterogeneous System in Package (SiP) module. So far, many semiconductor manufacturers have constructed their own Chiplet technology chains, such as Marvell's MoChi technology and Intel's EMIB technology. A typical Chiplet package structure is shown in Fig. 1.15. Currently, 2.5D packaging is the primary method for the Chiplet architecture. In the future, 3D packaging technology will also be used for Chiplets to achieve chip stacking and high-density interconnection.

3) 3D Package

3D package achieves vertical stacking and interconnection of chips using technologies such as wire bonding, flip-chip, or Through Silicon Via (TSV). This approach further reduces product size, shortens interconnect distances, minimizes signal delay,

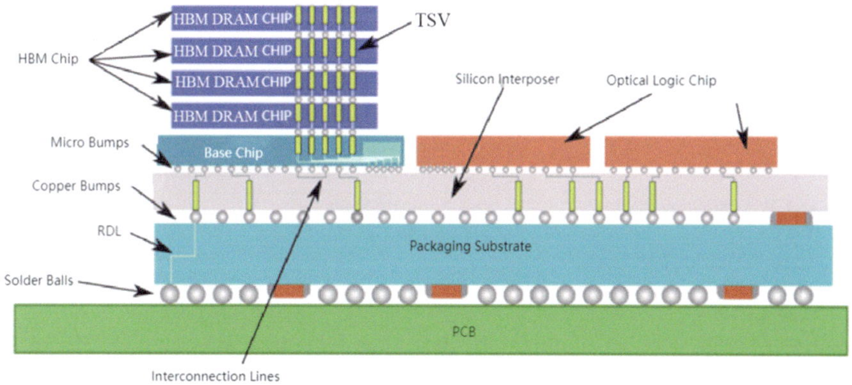

Fig. 1.14 Typical CoWoS package structure

Fig. 1.15 Typical Chiplet package structure

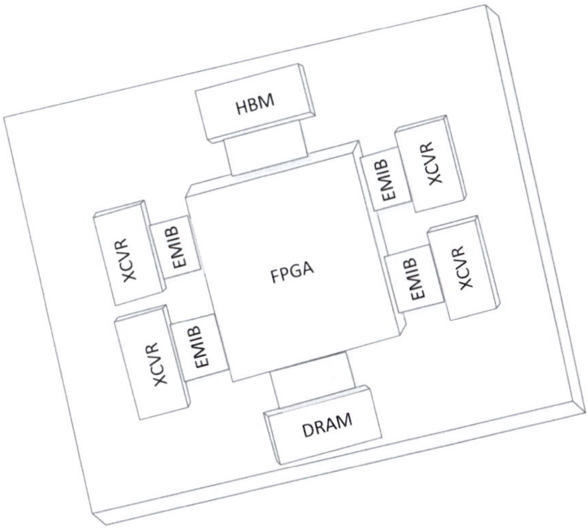

and enhances performance or capacity. Wire-bonded stacked 3D packaging typically features structures like pyramid, cantilever, vertical, and side-by-side arrangements. Flip-chip stacking involves creating micro-bumps directly on the chip pads, then flipping the chip upside down with the micro-bumps facing downward to achieve vertical stacking with the lower substrate or chip. TSV stacking creates vertical vias directly on silicon chips and achieves electrical connections between the upper and lower chips through these vias. It represents the highest interconnect density among 3D packaging methods. However, high cost, poor heat dissipation, and filler layer defects are the main factors limiting its further development and application. Leading memory manufacturers Samsung, Hynix, and Micron have begun mass production of 3D dynamic random-access memory (DRAM) chips based on TSV interconnected

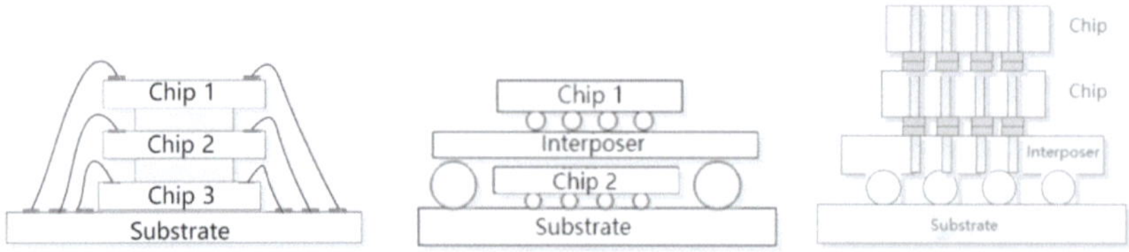

(a) Wire Bonding Stacking (b) Flip-chip Stacking (c) TSV Stacking

Fig. 1.16 3D packaging structure

multilayer chip stacking. Figure 1.16 illustrates schematic structures of 3D packaging based on wire bonding, flip-chip, and TSV stacking. Chapter 5 will provide a detailed introduction to 3D packaging and its reliability.

4) System in Package (SiP)

System in Package (SiP) is a packaging format that integrates microelectronics, optoelectronics and sensor devices into a single independent package. It uses high-density integration packaging technologies such as WLP, CSP, 3D stacking, and passive device integration. SiP primarily includes four types: horizontally tiled SiP, vertically stacked SiP, package-stacked SiP, and embedded chip SiP [6]. Structures of horizontally tiled and vertically stacked SiP packaging are shown in Fig. 1.17 [7]; while the structures of package-stacked and embedded chip SiP packaging are shown in Fig. 1.18.

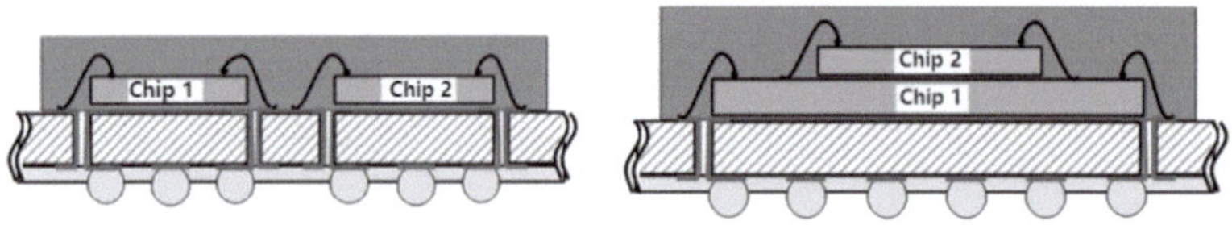

(a) Horizontally Tiled SiP (b) Vertically Stacked SiP

Fig. 1.17 Structure of horizontally tiled and vertically stacked package

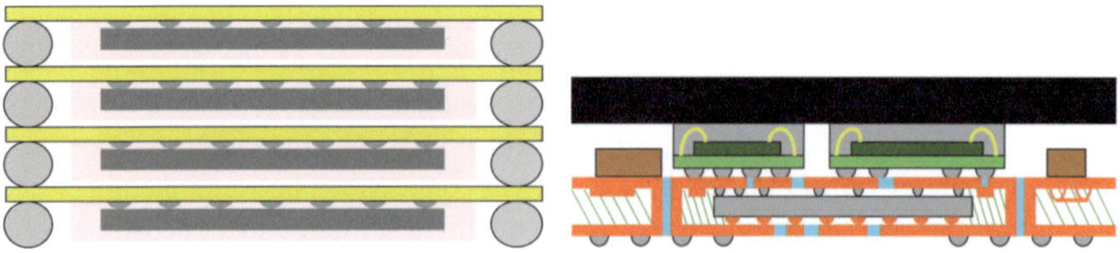

(a) Package-stacked SiP (b) Embedded Chip SiP

Fig. 1.18 Structure of package-stack and embedded chip SiP package

1.1.3 Development Trends in Integrated Circuit Packaging Technology

The core of packaging lies in the structure of the leads, the materials used, and the methods of chip packaging. Initially, IC packaging was developed to provide physical support, protection, and interconnection for chips. However, as chip manufacturing technology has progressed, integrated circuit packaging technology has also advanced to meet the requirements for high performance, multifunctionality, and miniaturization. The evolution of lead structures, packaging materials, and chip packaging methods, demonstrates that integrated circuit packaging technology is continuously advancing towards higher density, a greater number of leads, lower latency, smaller sizes, and reduced costs. This evolution aims to align with the multifunctional and highly integrated characteristics of integrated circuit chips.

Since 1947, when Bell Labs in the United States invented the world's first germanium transistor, and in 1958, when Texas Instruments in the United States developed the world's first semiconductor integrated circuit, marking the beginning of semiconductor technology development, semiconductor integrated circuit packaging technology has undergone nearly 80 years of evolution. Over this period, significant advancements have been made in three key areas: lead structure, packaging materials, and chip packaging methods.

1. Changes in IC chip packaging methods

The packaging methods for integrated circuit chips have evolved from chip-on-board packaging to tape automated bonding packaging, then to flip-chip bonding packaging, and finally to stacked chip packaging. The driving force behind these changes in chip packaging is the need to reduce the transmission distance from the chip to the external pins and to improve the integration of the circuit.

Conventional chip packaging mainly involves wire bonding (WB), the most traditional IC packaging method. This technique uses wire bonding to create electrical connections between the chip's electrodes and external leads. The structural characteristics of wire bonding packaging include the chip being adhered to the package base or substrate, which provides heat dissipation channels, with internal wires bonded at both ends to the chip and the leadframe. Generally, the number of wire bonds on the chip equals the number of external pins. Conductive adhesive and solder are the primary materials used in chip packaging, with conductive adhesive used for low-power integrated circuits and solder for higher-power integrated circuits. The chip area required for wire bonding is about one-third larger than that for flip-chip packaging [8], because it is difficult to form wire bonds directly in the active area of the chip.Metallization is required to connect the active area to the periphery of the chip to form the electrodes.

Tape Automated Bonding (TAB) packaging is a chip packaging form that uses tape bonding technology. It is developed from wire bonding packaging technology and creates interconnections between the chip and the leadframe. The tape material is typically made of copper foil, which offers excellent electrical and thermal conductivity, high strength, and good ductility. It also adheres firmly to different substrates and is easy to electroplate. The substrate material is usually polyimide (PI), which provides good adhesion and thermal compatibility with copper foil, low shrinkage, stability, and strong corrosion resistance. The bump materials include gold, copper-gold, gold-tin, and lead-tin. The tape can be single-layer, double-layer, or triple-layer. Compared to wire bonding packaging, TAB packaging offers lighter, thinner, shorter, and smaller structures. The electrode size and the pitch between the electrodes and the soldering areas can be reduced. Additionally, the wire resistance, capacitance, and inductance are lower, the bonding strength is higher, and production is more easily automated. However, the drawbacks include poor thermal matching, dimensional stability, corrosion resistance, and mechanical strength of the TAB and chip bump materials.

Flip-Chip Bonding (FCB) packaging is an IC packaging technology in which the active area of the chip, equiped with bump electrodes, is flipped downward and bonded to the substrate's wiring layer for interconnection. Its features include bump pitches ranging from 4 to 14 mils, and ball diameters from 2.5 to 8 mils. After the chip is flipped onto the substrate, a filler is required between the chip and the substrate to protect the chip and reduce thermal mismatch that could damage the bumps. Compared to wire-bonded chips, the advantages of FCB packaging include smaller interconnect dimensions, higher performance, increased pin density per unit area, improved heat dissipation, and the ability to use heat sinks or other cooling mechanisms on the chip's backside. However, FCB packaging faces challenges due to the small bump pitches and ball diameters, which place higher demands on the bumping process, substrate technology, material compatibility, manufacturing processes, and defect detection techniques.

Stacked Die (SD) packaging is a 3D packaging technology developed to further enhance integration density. It improves both packaging efficiency and circuit operation speed, making it particularly suitable for packaging chips such as programmable logic circuits, processors, memory chips, and digital-to-analog converters. Its key feature is stacking multiple chips within a single package to form a three-dimensional structure. This approach offers benefits such as reduced size, shorter interconnect lines between chips, and higher packaging efficiency.

2. Changes in the lead structure of IC packaging

The lead structure of IC packaging has undergone three major technological revolutions: from long-lead through-hole packaging to short-lead or leadless surface-mount packaging, and eventually to BGA surface-mount packaging. These changes aim to continuously reduce the package size to meet the requirements of device miniaturization.

In the 1960s, the first packaging revolution led to the development of through-hole packaging with long leads. This packaging featured long leads were directly inserted into the PCB, with the number of leads ranging from 6 to 64. The main forms of packaging includes Single In-line Package (SIP) and Dual In-line Package (DIP). The advantage of these long-lead packages was that they initiated the traditional plating through-hole (PTH) insertion technology; however, the downside was the large package size, which occupied a significant area on the PCB, making it difficult to further increase the assembly density and operating frequency of the PCB, and also challenging to improve the efficiency of automated production.

In the 1980s, the second packaging technology revolution shifted from through-hole to surface-mount packaging. The characteristic of this packaging was that the short leads or leadless packages were mounted directly onto the PCB, with the number of leads ranging from 14 to 100. The main forms of packaging included Small Out-line Package (SOP), Small Out-line J-leaded (SOJ), Quad Flat Package (QFP), and Quad Flat No-lead (QFN), as well as J-leaded chip carriers such as Plastic Leaded Chip Carrier (PLCC) and Ceramic Leaded Chip Carrier (CLCC). The advantage of these short-lead or leadless packages was that they were surface-mounted on the PCB, with smaller lead spacing and higher package density, which made them suitable for automated production. However, the downside was that the number of I/O leads and frequency could not keep up with the rapid development demands of ASICs and microprocessors.

In the mid-1990s, the third packaging technology revolution occurred, shifting from peripheral lead-out style to area array lead-out style on the backside [9–11]. The characteristic of this packaging is smaller lead space and higher packaging density, with the number of leads ranging from dozens to hundreds. The main forms of packaging include Pin Grid Array (PGA) and Ball Grid Array (BGA). The advantages of these array-style lead packaging are evident, as they can meet the high-density packaging requirements of ASICs and microprocessors; however, the downside is that BGA surface-mount soldering is difficult to inspect and repair after reflow soldering.

In the late 1990s, the rapid development of Chip Scale Package (CSP) technology marked the third major revolution in packaging, significantly enhancing the packaging efficiency of ICs and their integration density on PCBs. CSP technology reduced the size and weight of ICs, leading to improved thermal performance, increased product performance, reduced parasitic effects in portable communication products, and lower impedance, thus meeting the requirements for radio frequency (RF) performance. CSP involves initially partitioning the wafer into individual IC chips followed by post-packaging processes. The innovative Wafer Level Chip Scale Package (WLCSP) refers to the integrated completion of packaging on a wafer that has undergone all previous processes, which is then diced into separate, standalone devices. The local structure of WLCSP is shown in Fig. 1.19.

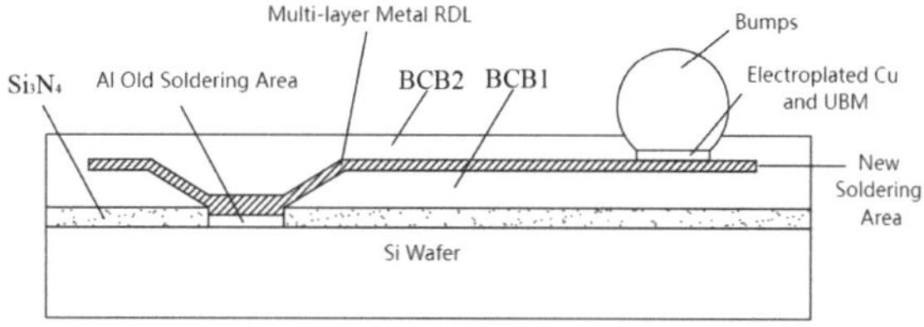

Fig. 1.19 Local structure of WLCSP

Compared to CSP, WLCSP reduces the number of required testing steps, This advantage has made it the mainstream technology for micro- and nano-scale system packaging [12]. The shortcomings of CSP and WLCSP applications include high manufacturing complexity and costs, as well as high performance requirements for packaging materials due to their small form factor. Additionally, the PCB traces used for CSP and WLCSP are narrow and require a high degree of flatness.

3. Changes in IC packaging materials

The transition in IC packaging from metal to ceramic, and then to plastic, aims to continuously reduce packaging costs, minimize package size, and decrease package weight, while enhancing circuit performance.

Metal packaging was the earliest form of IC packaging, initially featuring three-pin metal-to-glass encapsulation. The packaging materials included Kovar, carbon steel, copper, and others. However, with the development of ceramic and plastic packaging technologies, metal packaging is now mainly used for hybrid ICs and module circuits. The greatest advantage of metal packaging lies in its high thermal conductivity, high packaging strength, and the ability to achieve hermetic sealing. The main drawbacks include high costs, large volume, high density, and a large coefficient of thermal expansion leading to thermal mismatch [13]. As IC technology continues to evolve, traditional metal packaging materials can no longer meet the requirements. New types of metal packaging materials with appropriate coefficients of thermal expansion, lightweight and high-strength, and high thermal conductivity are being continuously explored and applied, including copper/carbon fiber, aluminum/silicon carbide alloys, and negative thermal expansion materials, etc. [14–16].

With the advancement of technology, ceramic packaging has become widely used in IC packaging. Ceramic packaging exhibits excellent overall thermal, electrical, mechanical, and dimensional properties. The main ceramic packaging materials are Al_2O_3, AlN, and BiO. Typical parameters of ceramic packaging [8] include the range of dielectric constants, thermal conductivity, and coefficient of thermal expansion. Ceramic packaging is currently one of the most reliable packaging forms, but due to its high costs, it is mainly used in high-end fields such as aerospace. The greatest advantage of ceramic packaging lies in its hermetic sealing and the stability of the its materials. However, ceramic packaging also has drawbacks, including high packaging process costs, brittleness of ceramic materials, and limited resistance of the ceramic body to mechanical shock.

With the further development of packaging technology, plastic packaging (EMC), which combines extremely low material and process costs, has increasingly replaced ceramic packaging. Plastic packaging has significantly reduced the packaging costs of ICs. Although it is not as reliable as metal or ceramic packaging, it is still widely used in consumer electronics, accounting for approximately 97% of the IC packaging market. The biggest advantage of plastic packaging are its low cost, small weight, and small packaging size. The mass of plastic-packaged devices is about half that of ceramic-packaged devices, and the smaller size greatly reduces signal delay. However, the shortcomings of plastic packaging include the generation of internal stress due to thermal mismatch, susceptibility to deformation at high temperatures, low thermal conductivity (only 1/50 of ceramic packaging), and poor moisture resistance.

Over the next decade, the packaging industry is expected to grow even more rapidly, with packaging materials continuing to evolve toward higher performance and low cost. In the civilian sector, new types of epoxy packaging materials, composite packaging materials, and environmentally friendly packaging materials that meet the requirements of new packaging forms will be developed. In the high-reliability sector, new materials such as aluminum nitride, silicon carbide-aluminum alloys, and silicon-aluminum alloys for high-density and high-heat dissipation areas, as well as nano silver and nano copper for high-power areas, will develop more quickly.

4. The development trend of IC packaging technology

Plastic packaging technology will remain the mainstream for IC packaging for a considerable time in the future [17]. The future development of metal packaging will focus on high performance, low cost, and high reliability. Materials such as silicon-aluminum alloys and silicon carbide-aluminum alloys, which are lightweight, have high thermal conductivity, and are thermally matched, will have a broad prospect in the future. The ceramic packaging technology will remain the mainstream in fields such as aerospace and high-end consumer electronics, with ceramic hermetic packaging developing towards more advanced integrated ceramic packaging, and low-temperature co-fired ceramics (LTCC) holding a promising future.

Advanced packaging technologies and trends in integrated circuits: From the through-hole packaging of the 1970s to today's wafer-level and chip-stacking packaging, density has greatly increased. Over the past 50 years, the semiconductor industry has followed Moore's Law, with transistors shrinking and packaging line widths decreasing from 1000 to 1 μm, even to sub-micron levels, enhancing capabilities by 1000 times. Trends in advanced packaging are shown in Fig. 1.20 [18]. As Moore's Law for semiconductors approaches its end, system-level packaging will extend it, pushing for higher integration and finer line widths in IC packaging.

According to Moore's Law, at a fixed cost, the integration level of ICs doubles approximately every 18 months, and performance also doubles. The integrated circuit industry, after only a few decades of development, has seen nodes shrink from 10 μm [15] to 14 nm, 7 nm, 5 nm, and even 3 nm [12, 14]. Although 2D integration has made significant progress, the improvement in cost-performance ratio of integrated circuits brought about by technological advancements is diminishing as integration and performance increase. The main reasons for this trend include the following aspects [12, 17].

(1) Limited by material properties, process capabilities and physical laws, the feature size of transistors is approaching atomic dimensions and process limits, leading to increasingly severe quantum effects and short-channel effects.
(2) In terms of reliability, the increase in power density makes device cooling difficult, and the thermal stress and strain from different process steps such as

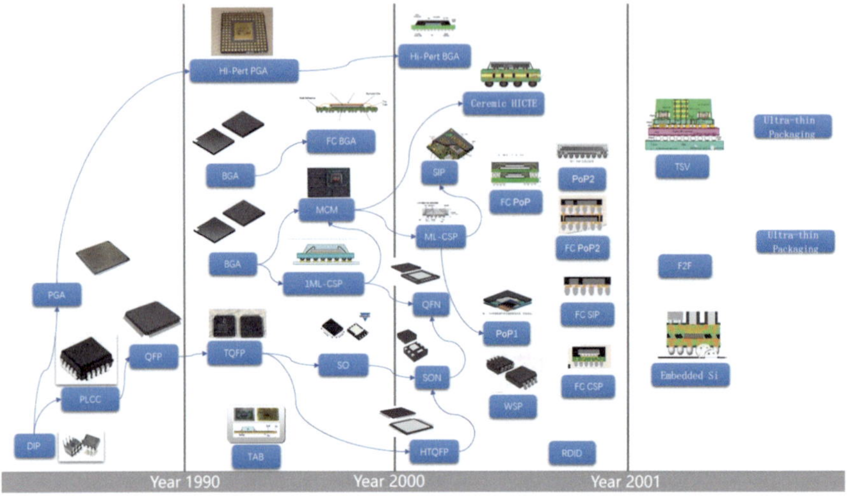

Fig. 1.20 Development trend of advanced packaging technology

annealing and thermal cycling in semiconductor manufacturing lead to more reliability issues for devices.

(3) Starting from the 180 nm process node, chip performance is more determined by the length of interconnects on the chip, and the performance gains from device scaling are not as significant as the delays caused by the increased length of interconnects [19].

Currently, various methods have been proposed to improve interconnect delay, such as inserting repeaters [20, 21] and using ultra-low-k dielectric materials [22], but the effects are far from meeting the demands. The interconnect delay issue has become one of the bottlenecks in the development of integrated circuits, and Moore's Law faces a severe challenge.

In 2009, the Semiconductor Industry Association proposed a "post-Moore's Law" in the International Technology Roadmap for Semiconductors (see Fig. 1.21) [23]. The "post-Moore's Law" shifts the focus towards integrated innovation driven by advanced packaging technologies, achieving 3D integration through vertical extension.

The concept of 3D integration was proposed as early as the 1960s [24], but due to thermal dissipation constraints, it did not reach practical application Only in recent years, with the development of ultra-low power technology, advancements in thermal management, and the encounter of technical bottlenecks in 2D integration, that 3D integration has ushered in a period of rapid development opportunities [25]. The advantages of adopting 3D integration are twofold: on one hand, compared to 2D integration, the significant reduction in global interconnect lengths directly decreases the parasitic capacitance of the interconnects, thereby enhancing performance in terms

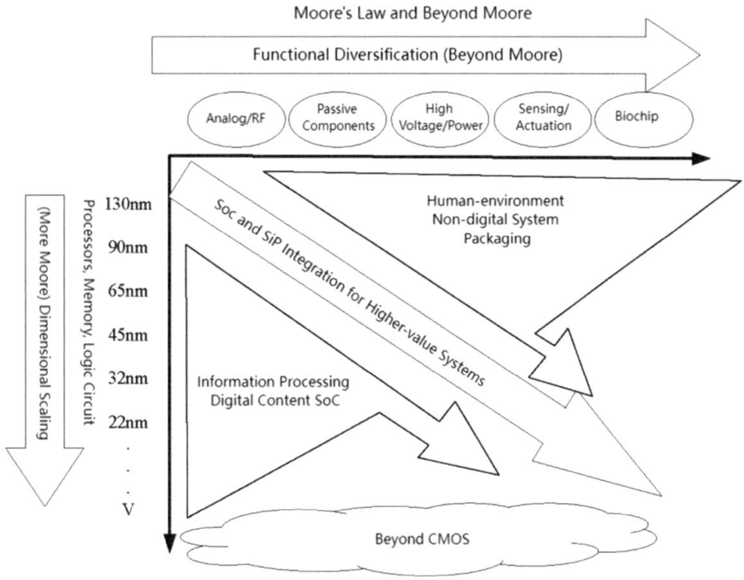

Fig. 1.21 International technology roadmap for semiconductors

of delay and power consumption; on the other hand, 3D integration enables heterogeneous integration using different substrate materials (such as GaAs, glass, etc.) and technical modules (such as digital circuits, memory, sensors, etc.), combining components that are completely incompatible with each other in terms of process technology.

Currently, there are primarily two forms of 3D integration: chip stacking and package stacking, with several distinct 3D packaging technologies corresponding to these forms [26]. As illustrated in Fig. 1.22, these technologies include wire bonding packaging, TSV packaging, and stacked packaging. Among them, TSV packaging provides the shortest interconnects for 3D integration and is currently favored by researchers for its advantages [27].

TSV primarily interconnects vertically stacked chips, serving as signal conductors, heat dissipators, and mechanical support [28]. Based on TSV technology, the 3D IC stacking integration is illustrated in Fig. 1.23. The application of TSV allows for circuits to be fabricated on both the front and back sides of each chip or interposer, providing the shortest interconnects between chips, the smallest pad size, and the tightest pitch. Currently, the applications of TSV are mainly divided into two categories [14, 29]: one is utilizing TSV and flip-chip microbump technology to stack chips, which is 3D IC integration; the other uses only TSV to stack wafers/chips, which is 3D Si integration without bump processes. Integrating wafers/chips into a 3D device through TSV can result in products with superior electrical performance, smaller form factors, reduced weight, and consequently, lower production costs.

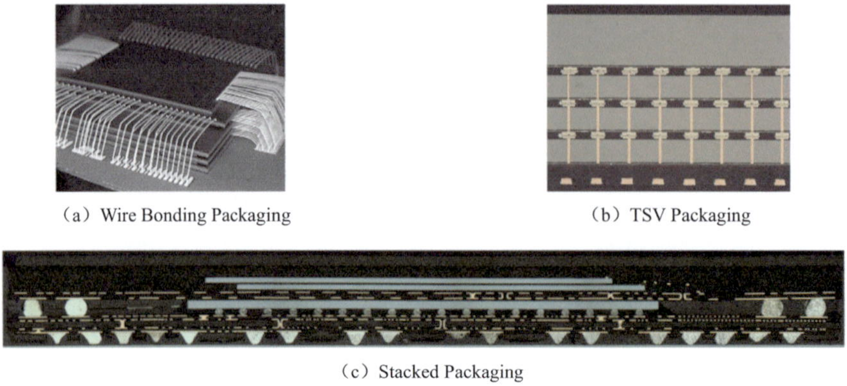

Fig. 1.22 3D packaging technology

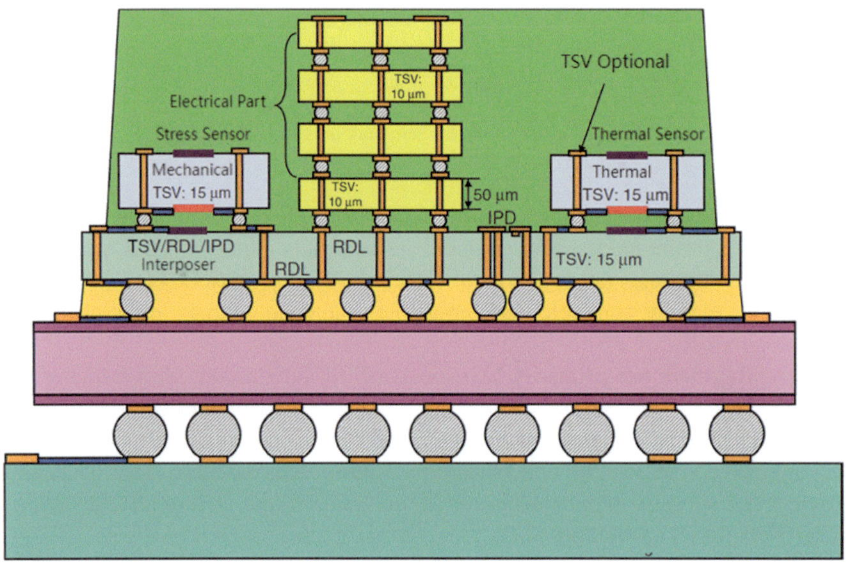

Fig. 1.23 3D IC stacking based on TSV technology

Currently, the known applications of TSV mainly include: ① 3D packaging of chips; ② 3D packaging of heterogeneous devices, such as integrating integrated circuits with MEMS, RF modules, and optoelectronic devices; ③ wafer-level 3D packaging; ④ silicon interposer packaging. There are already a large number of high-density packaging device products on the market that use TSV technology, such as stacked memory, logic 3D system-in-package products, high-brightness LED modules, MEMS, FPGAs, images and optical sensors, etc.

Figure 1.24 illustrates the 3D IC products available on the market that based on TSV technology. In 2006, Samsung developed a 16 GB NAND flash memory chip,

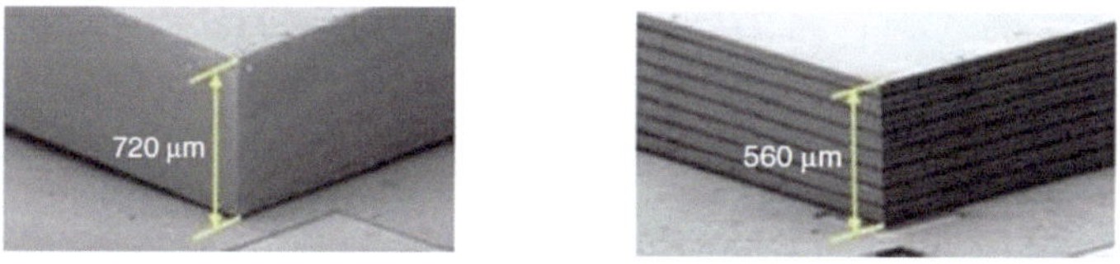

(a) Thickness of a Single Wafer (b) Samsung's 8-layer Stacked Packaging Product

Fig. 1.24 3D IC products based on TSV technology in the market

which was one of the earliest TSV applications on the market, based on wafer-level stacking package technology [30]. This chip stacked eight NAND flash memory chips, with a total height of 0.56 mm, reducing the thickness by 0.16 mm compared to individual chips using wire bonding technology. In 2011, IBM and Micron Technology achieved the manufacturing of Hybrid Memory Cube (HMC) DRAM chips based on TSV process technology [31]. The HMC data bandwidth is 14 times higher than that of existing memory chips, while the chip package size is reduced by 90%. To date, SK Hynix's world-first 128 GB Double Data Rate 4 (DDR4) RDIMM, which has been mass-produced, incorporates TSV packaging technology.

1.2 Relationship Between Packaging Technology and Reliability

Integrated Circuit (IC) packaging technology is closely related to the reliability of ICs, fundamentally due to the packaging structure, packaging materials, lead form, and the internal chip packaging method, all of which directly affect the thermal performance, mechanical performance, moisture resistance, radiation resistance, and electromagnetic interference (EMI) resistance of the ICs. The basic failure rate and inherent lifespan of ICs depend on their thermal performance, while their environmental adaptability depends on their mechanical performance, moisture resistance, radiation resistance, and EMI resistance.

1.2.1 Packaging Thermal Performance and Reliability

Long-term research has shown that for semiconductor devices, every 10°C increase in temperature can halve the device's lifespan. Therefore, temperature is a primary factor affecting the reliability of semiconductor devices, and the packaging structure, packaging materials, lead form, and internal chip packaging methods all significantly impact the heat dissipation performance of integrated circuits, which is crucial in determining the operating temperature of semiconductor chips.

The structure of the packaging I/O terminals has a significant impact on the operating temperature of IC chips, as it serves as a crucial thermal conduction path for heat dissipation. The heat dissipation efficiency of I/O terminals is as follows: Array I/O > Peripheral I/O > Dual-inline I/O.

Under normal circumstances, the thermal conductivity of various packaging materials is as follows: Metal > Ceramic > Plastic

The comparison of heat dissipation efficiency for various internal chip packaging methods is: Flip-chip bonding heat dissipation efficiency > Wire bonding heat dissipation efficiency > Stacked die heat dissipation efficiency.

The 10 °C rule, summarized from practical experience, demonstrates that for every 10 °C increase in the operating temperature of semiconductor devices, their lifespan is reduced by half. The MIL standards provide a model for the failure rate of semiconductor ICs, where the thermal performance of the package has a noticeable impact on the failure rate; the JEDEC standards offer life models for four typical failure mechanisms of semiconductor devices, all reflecting the impact of the device chip temperature on the lifespan of ICs, with the junction temperature of the device chips being related to the thermal resistance of the package.

1.2.2 Packaging Mechanical Properties and Mechanical Environmental Adaptability

The mechanical properties of IC packaging directly reflect the mechanical environmental adaptability of the ICs, which is related to the packaging structure, packaging materials, and lead form of ICs.

IC may inevitably encounter issues such as vibration, shock, inertial load during operation or transportation, which could cause damage like breaks in leads or cracks in the package body. Particularly in PCB electronic automated assembly, IC need to withhold certain mechanical grasping force when being picked up by mechanical arms and installed onto the PCB. Additionally, when the function unit of PCBs are installed in the integrated system, the surface-mounted ICs on the PCB may have to endure mechanical force caused by PCB bending and vibration, all of which will directly test the mechanical environmental adaptability of the IC packaging.

The mechanical environmental adaptability of integrated circuit packaging can be characterized by the package I/O terminals, internal wire bonding strength, chip adhesive shear strength, and the sealing strength of the package lid. Testing and evaluation are carried out according to relevant standards, such as GJB 548 and others.

1.2.3 Packaging Airtightness and Humidity Adaptability

The airtightness of integrated circuit (IC) packaging reflects the adaptability of the IC to work in atmospheric or humid environments. It is related to the packaging structure and materials used. For example, the airtightness and reliability of ICs with metal or ceramic airtight packaging have much better airtightness and reliability than those with plastic packaging.

For plastic packaging, this non-airtight packaging form has a major disadvantage compared to airtight metal and ceramic packaging: it has low resistance to humidity. When plastic-packaged ICs operate in a humid environment, the penetration of moisture can lead to corrosion of the IC chip, reduced reliability, and degraded electrical performance. The primary path for moisture to enter a plastic-packaged IC is through the gaps between the plastic packaging material and the lead frame. Moisture can enter through these gaps and corrode the metallization layer on the chip surface. Because water molecules are very small (with a diameter of about 2.5×10^{-8}cm), they have strong penetration and diffusion abilities. Thus, moisture can also seep into the packaging through the capillary pores and interstitial gaps in the plastic material. For example, in the PCB reflow soldering process, the "popcorn" cracking failure that occurs in plastic-packaged integrated circuits is caused by excessive moisture absorption, leading to process failure.

The assessment of the moisture resistance of ICs can be conducted by testing the moisture content and airtightness of sealed packaging devices, evaluating the airtightness of airtight packaging devices; performing moisture sensitivity tests to assess the reflow soldering process adaptability of plastic-packaged devices; and conducting high-temperature, high-humidity tests to evaluate the moisture resistance of sealed devices and plastic-packaged devices.

1.2.4 Packaging Materials and Electromagnetic Interference

The electromagnetic interference (EMI) of the integrated circuit (IC) packaging structure on the performance of the IC mainly manifests in two aspects: first, the mutual EMI caused by the multi-chip assembly structure; second, the impact of the coating materials inside the packaging on electromagnetic shielding effectiveness.

The source of the mutual electromagnetic interference caused by the multi-chip assembly structure lies in the transistor switching noise of the IC chips. The electromagnetic effects of integrated circuits are their inherent characteristics, and these switching noises in the IC can be summarized as dI/dt and dV/dt, which refer to the effects of transient current and transient voltage. As the scale of ICs increases, the number of internal transistors has reached the billion-level, and therefore, switching noise cannot be ignored. At the same time, in order to accelerate computational speed (which is related to clock frequency), the rising and falling edges, which determine the magnitude of dI/dt and dV/dt, become steeper, and their impact on the IC is also fundamental. With the advancement of Moore's Law, the integration of ICs continues to increase, and the problems caused by dI/dt and dV/dt are becoming more prominent and unavoidable. For SoC (System on Chip) and SiP (System in Package) designs, the chips include digital circuits, analog circuits, and radio frequency (RF) circuits, and the electromagnetic interference resistance of various chips in the circuit differs. Analog and RF circuits, in particular, are more susceptible to electromagnetic signals, and improper packaging design may cause the chip to fail to function properly. Therefore, the higher the circuit integration, the more severe the mutual interference between chips within the module. Internal chip assembly structure design is one way to solve EMI.

The shielding effectiveness of the coating materials inside the packaging determines the EMI resistance of the module's circuits. Various new electromagnetic shielding materials, including metal materials, inorganic non-metallic materials, polymer materials, and composite materials, are used in the packaging structure in the forms of coatings, thin films, sheets, and powders. For RF modules, the electromagnetic shielding materials belong to near-field electromagnetic shielding materials. A soft magnetic shielding material can be applied between the electromagnetic radiation source and the metal package body to prevent the impact of electromagnetic radiation on the metal packaging carrier. Electric field shielding refers to the suppression of electric field radiation interference using electric field shielding materials in near-field environments with high voltage and low current. Magnetic field shielding refers to the suppression of magnetic field radiation interference using magnetic field shielding materials in near-field environments with low voltage and high current. Electromagnetic field shielding refers to the suppression of electromagnetic field radiation interference using electromagnetic field shielding materials in near-field environments with complex electromagnetic fields. Therefore, there are three types of electromagnetic shielding materials designed for near-field application environments: electric field shielding materials, magnetic field shielding materials, and electromagnetic field shielding materials.

1.2.5 Packaging Materials and Radiation Resistance Performance

In various packaging materials of semiconductor devices, impurities such as uranium (U) and thorium (Th) exist. These impurities are naturally radioactive, and the α particles they emit have strong ionization ability. When these α particles pass through electronic devices, they can generate a large number of electron–hole pairs. These electron–hole pairs are collected by the device, leading to soft errors in the device [32]. Although soft errors can be corrected by various means, if they occur in critical locations (such as the instruction cache of a central processing unit), they may result in severe catastrophic consequences. An illustration of α-particle decay is shown in Fig. 1.25.

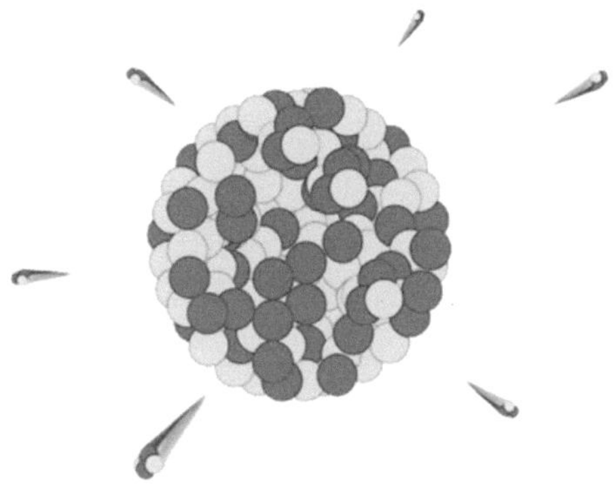

Fig. 1.25 α-particle decay

With the continuous development of IC technology, factors such as increased integration, reduced supply voltage, and smaller node capacitance have made soft errors caused by ionizing particles a significant threat [32, 33] to the high-reliability applications of ICs. In atmospheric environments, α-particles, high-energy neutrons, and thermal neutrons are the main sources of soft errors [34]. The soft error rate caused by α particles is closely related to various factors such as packaging material grade, chip process, and operating environment [30], and it constitutes an important part of the overall soft error rate.

Currently, the main sources of α particles in semiconductor devices are molding compounds, solder balls, and underfill materials, as shown in Fig. 1.26. The radiation levels are classified into three categories: ordinary alpha, low alpha (Low Alpha, LA), and ultra-low alpha (Ultra-Low Alpha, ULA). Under current packaging technology, using LA or ULA materials is an effective measure to reduce the radiation impact of packaging materials. Special purification techniques can significantly lower the surface α particle radiation rate of packaging materials by several orders of magnitude, with a minimum rate of less than 0.001 cph/cm^2. By testing the surface α-particle radiation rate of the materials and conducting device-level radiation tests, the soft error rate caused by α particle radiation in packaging materials can be determined.

1.3 Packaging Reliability Technology and Its Development

1.3.1 Integrated Circuit Packaging Reliability

The basic function of integrated circuit (IC) packaging is to provide physical support, protection, and interconnection for the IC chip. However, these protective and interconnecting functions may degrade or fail under the influence of thermal dissipation stress and environmental stress. Therefore, reliability design must be implemented to address potential failure issues that may occur under various.

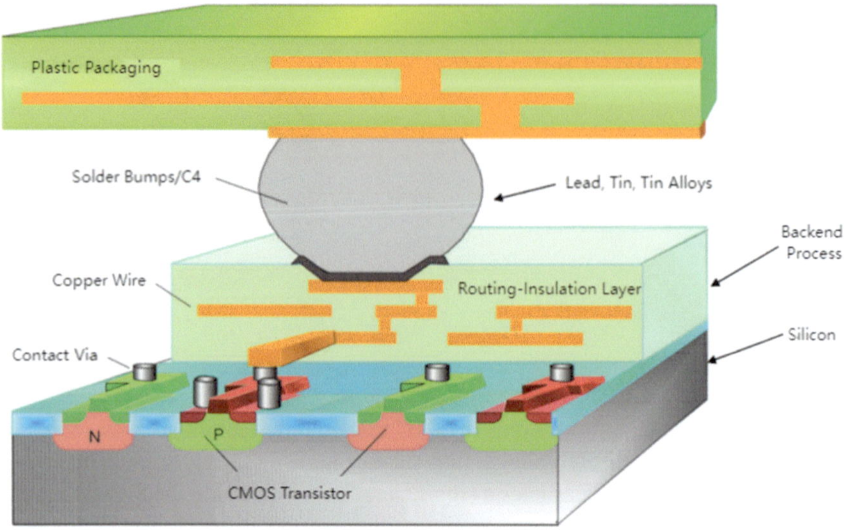

Fig. 1.26 α-particle sources in semiconductor devices (from: IMB)

The reliability of IC depends on two parts: the semiconductor chip reliability and the packaging reliability, covering failure rates, service life and environmental adaptability. The reliability of semiconductor chips is in the field of microelectronics and will not be discussed here. Packaging reliability involves components such as the chip bonding layer, bonding wires, substrates and interconnections, and the encapsulation materials or carriers. They all contribute to the failure rate, service life and environmental adaptability of ICs. The bathtub curve diagram of IC failure rate is shown in Fig. 1.27.

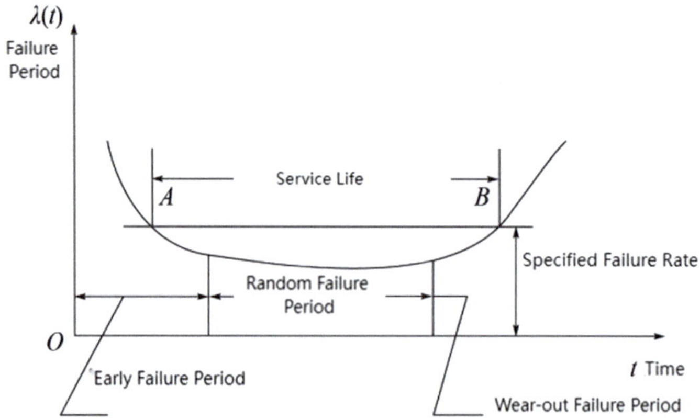

Fig. 1.27 The bathtub curve diagram of IC failure rate

To facilitate the improvement and enhancement of IC packaging reliability, the assessment of IC failure rates is divided into two parts: one is the chip failure rate, and the other is the packaging failure rate. For instance, the GJB 299 standard provides a predicted failure rate model for semiconductor ICs [21], which specifically accounts for the packaging complexity failure rate C_3; the basic failure rate prediction model for hybrid integrated circuits separately considers the process and packaging failure rate λ_{SF}; the FIDES standard's failure rate model for integrated circuits [35], separately considers the basic failure rate $\lambda_{0TCyCase}$ of the packaging carrier under thermal cycle stress.

The service life of IC packaging depends on the weak lines in packaging materials and structures, such as the degradation of life of Au-Al bonding and the fatigue life of packaging solder joints.

The environmental adaptability of IC packaging includes resistance to moisture, salt fog, vibration, and other environmental adaptability, and these tests can be conducted in accordance with standards like GJB 597B [20, 22].

1.3.2 Research on the Failure Mechanism of Integrated Circuit Packaging

From the perspective of products, the failure mechanism of integrated circuit (IC) mainly include two aspects: the failure mechanisms of IC semiconductor chips and the failure mechanism of IC packaging. Generally speaking, failure mechanisms related to electrical stress are primarily concentrated on the integrated circuit semiconductor chips, while those related to environmental stress are mainly focused on the IC packaging. However, regardless of which level the failure mechanism originates from, the ultimate failure mode of the IC is either electrical performance deviation or loss of function.

1. Failures related to packaging structure

Failures related to IC packaging structure mainly manifest as impacts on the physical properties of the IC, with changes in these properties leading to the failure of the IC.

For the thermal performance of ICs, regardless of the packaging structure designs, heat dissipation must be considered. The basic principle is to ensure that the junction temperature or channel temperature of the internal IC chip does not exceed the specified temperature. Once the junction temperature or channel temperature of the IC chip exceeds the specified temperature, the IC becomes unreliable or fails.

For the mechanical properties of ICs, the packaging structure design directly determines the rigidity of the integrated circuit, which in turn affects its mechanical strength and vibration resistance. The quality of the IC's packaging also influences its mechanical strength and vibration resistance. The basic principle is that the mechanical performance of the IC must meet the impact and vibration resistance requirements specified by the standards. If the mechanical performance of the IC is substandard, its failure may manifest as chip cracking or package cracking.

Regarding the moisture resistance of ICs, metal or ceramic hermetically sealed packaging structures are the best choice, with internal moisture content typically kept below 5000 ppm. If the internal moisture content exceeds this limit, it is considered a failure. Plastic packaging structures, which are non-hermetic, can also protect the IC chip. However, external moisture can penetrate through gaps between the plastic packaging material and the lead frame, reaching the surface of the chip. Over time, this accumulation can lead to corrosion and failure of the IC chip.

2. Failures related to packaging materials

Failures related to IC packaging materials mainly manifest in terms of moisture resistance, thermal cycle adaptability, and radiation resistance, all of which can lead to IC failure.

Metal and ceramic packaging materials have excellent moisture resistance and do not require consideration of moisture infiltration. However, plastic packaging materials, being organic polymer materials, inherently exhibit a certain level of moisture absorption, leading to two main issues: improper moisture sensitivity control in plastic-packaged devices during reflow soldering processes can cause them to explode (commonly referred to as "popcorn"); and during long-term storage of plastic-packaged devices, moisture infiltration can lead to chip corrosion, including infiltration at the interface between the plastic packaging and the metal frame.

For the thermal cycle adaptability of ICs, both metal and plastic packaging materials must be considered. The impact of metal packaging materials: For high-power circuit chips, if directly bonded to the metal casing, a significant mismatch in the thermal expansion coefficients of the silicon wafer and the metal packaging material can cause the chip to crack during switching operations. The impact of plastic packaging materials: Due to the mismatch in the thermal expansion coefficients between plastic packaging materials and metal bond wires, after prolonged use, the bond wires in plastic-packaged devices may become detached, leading to open circuits.

Regarding radiation resistance of ICs, charged particles and cosmic rays in space environments can alter the electrical properties of semiconductor materials, causing the integrated circuit to lose its intended functionality or suffer radiation damage [25]. Although the radiation protection provided by packaging materials is limited, ongoing research into new materials continues. Studies have shown that [26] specialized composite shielding materials can absorb space ionizing radiation, and employing sealing processes to reinforce memory devices can improve their resistance to electron source radiation by 1–2 orders of magnitude. If the purity of the IC plastic packaging material is insufficient, it may contain radioactive elements [28]. For instance, zirconium silicate filling materials in low-melting-point glass can emit α-particle radiation rates of 150–200 cph/cm^2. Additionally, underfill materials at the bottom of flip-chip packages can also emit α-particles. These α-particle radiations from packaging materials can impact the B10 element in the semiconductor chip, thereby inducing soft errors in the IC.

3. Failure related to lead terminal forms

Failures related to lead terminal forms mainly involve the effect of lead heat dissipation efficiency on the thermal performance of ICs, as well as the mechanical strength of the leads themselves. For example, mechanical shock or vibration may cause leads to fracture or detach.

After electrical assembly on the PCB, the lead terminals of the IC are subjected to additional forces under mechanical shock or vibration. Their resistance to shock strength and vibration capacity are crucial indicators for evaluating their environmental adaptability.

4. Failures related to chip mounting methods

Failures related to chip mounting methods are primarily due to thermal mismatch during packaging, leading to chip cracking, flip-chip solder joint cracking, bonding wire disconnection, TSV failure, etc.

Thermal mismatch causing chip cracking refers to the installation method of high-power chips soldered to the back of the chip. Due to the severe mismatch in thermal expansion between the silicon chip and the metal base, cracking may occur during switching.

Flip-chip solder joint cracking refers to the generation of cracks at the interface of the flip-chip bump after long-term operation, due to metal ion migration and thermal cycling forces, ultimately leading to interconnection failure.

Bonding wire disconnection is most typically observed in Au-Al aluminum bonding structures. After long-term operation, the Au-Al interface of the bonding wire degrades, leading to disconnection.

TSV failure includes issues such as TSV expansion under process stress, cracking under prolonged high temperature, thermal cycling, or mechanical stress, and time-dependent dielectric breakdown (TDDB) under electrical stress.

1.3.3 Development of Integrated Circuit Packaging Reliability Technology

The development of integrated circuit (IC) packaging technology is reflected in both packaging structure design and the selection of packaging materials, with the goal of continuously improving packaging density and performance. The evolution of IC packaging reliability technology is seen in the improvement of packaging reliability, environmental adaptability, and the advancement of reliability in advanced packaging.

1. Development of packaging reliability technology

With the continuous development of packaging structures, the lead pitch and packaging thickness have undergone significant changes. Lead pitch (the center-to-center distance between two adjacent leads of an IC package) is specified by international standards and regulations, with nominal values. Typical dual-inline packages have a lead pitch of 2.54 mm, and flat packages have 1.27 mm, both conforming to international standards. As packaging technology progresses and the demand for higher packaging density increases, the number of leads needs to be increased, which results in smaller lead pitches. Traditional lead pitches of 2.54 mm and 1.27 mm are gradually being replaced by lead pitches of 1.27 mm, 1.00 mm, 0.80 mm, and 0.65 mm, with the trend moving toward 0.30 mm, or even below 0.25 mm. To arrange a large number of leads on a single packaging substrate, three methods are commonly used: first is to increase the substrate area, although this contradicts the miniaturization requirement and can lead to higher lead inductance and resistance, thus degrading electrical performance. Second is to optimize the packaging structure, fully utilizing the sides or bottom of the packaging substrate, such as moving the lead exits from two sides to all four sides or the bottom. And the third method is to reduce the lead pitch. However, an excessively small lead pitch reduces mechanical strength, increases crosstalk between leads, and complicates the technology.

Over the past few decades, the feature sizes of ICs have progressed from 0.25 to 0.13 μm, 65 nm, 28 nm, 14 nm, 7 nm, etc. The advancement of IC microelectronics technology has maintained the pace set by Moore's Law. However, compared to microelectronics technology, the development of IC packaging technology has lagged behind. In many applications, the packaging density of ICs has become the bottleneck restricting performance improvement. 3D packaging reliability and wafer-level packaging reliability are critical issues in this regard. Ensuring the reliability of IC packaging, so that designed ICs can fully realize their functionality, has become a crucial task within the entire industry chain.

In the development of IC packaging reliability technology, the study of failure mechanisms in packaging has always been a hot research topic, serving as the core foundation for improving packaging reliability.

In terms of heat dissipation, IC packaging technology focuses on effectively dissipating heat while improving packaging density, i.e., thermal management technology, and addressing material thermal expansion caused by thermal loads. The mismatch of thermal expansion coefficients between different packaging materials can lead to localized over-stress and failure.

In terms of mechanical properties, IC packaging technology is more concerned with large-size packages, such as plastic and ceramic packages, and the elastic and plastic deformations under mechanical shock and vibration, which may cause damage. There is also particular concern regarding the impact resistance of 3D stacked packages.

Electrically, IC packaging technology emphasizes the insulation properties of high-density packages, including insulation between leads, between plastic packaging materials, and between base materials. In microwave applications, it also focuses on parameters such as inter-lead capacitance, load capacitance, and lead inductance.

Chemically, IC packaging technology continues to focus on failure issues caused by corrosion, oxidation, and dendritic growth of ions on the surface due to humid environments. Moisture ingress into plastic packaging is a major concern. Water vapor can extract ions from catalysts or other additives in the material, forming by-products that enter the chip surface or internal layers, causing parameter drift or failure of the IC.

Regarding radiation resistance, IC packaging technology is increasingly concerned with trace radioactive elements such as uranium and thorium in plastic packaging or filler materials, which can cause α-particle radiation. This is especially problematic for memory devices, leading to flip errors and other soft errors. One solution to mitigate this is the use of PI coatings or artificial composite fillers on the chip surface to shield against α-particle radiation.

2. Development of packaging reliability testing and evaluation technology

To ensure the long-term stability of packaging performance, reliability testing has become an essential method for guaranteeing packaging reliability. With the advancement of IC technology, the service life and adaptability of IC packaging to harsh environments have become key areas of focus. Therefore, accelerated life tests and High Accelerated Stress Tests (HAST) are critical for assessing the reliability of IC packaging.

Accelerated life testing is a method for evaluating the failure mechanisms of semiconductor devices under stress conditions. It involves applying higher stress to accelerate degradation, such as temperature-accelerated life tests or humidity-accelerated life tests. The key to this testing technique lies in the selection of temperature stress and monitoring of sensitive parameters. Although the object of evaluation is the IC product, the impact of packaging on failure is also considered. For example, in temperature-accelerated life tests, the junction temperature of the integrated circuit serves as the baseline, and the environmental temperature control is determined based on the thermal resistance of the IC junction-to-case. Therefore, the acceleration factor in any packaging form's accelerated life test is closely related to the thermal performance parameters of the package. Monitoring sensitive parameters needs to be determined based on the IC's technical specifications.

High Accelerated Stress Testing (HAST) is a method for evaluating the environmental adaptability of IC packaging and can quickly expose packaging defects and identify potential issues. To simulate real-life usage environments, integrated device-level comprehensive environmental tests, such as three-environmental tests (vibration, temperature cycling, humidity), are used. These tests can rigorously assess the environmental adaptability of IC packaging to meet the requirements for operating in harsh environments. Specifically, comprehensive environmental tests under electrical load are a key focus and direction for test technology development.

3. Development of packaging reliability simulation and evaluation technology

With the rapid development of technology, the reliability assessment, life prediction, and enhancement of ICs have become increasingly important. Traditional reliability testing methods often fail to meet the requirements of high-reliability, fast turnover, and short development cycles for next-generation electronic products. Reliability simulation methods can quickly identify the weak points and reliability levels of products during the design phase. These methods bridge the gap between digital design and performance testing. It creates virtual prototypes and test environments, enabling designers to assess product reliability using high-performance computing, finite element analysis, failure mechanism analysis, and reliability modeling. This helps quickly understand product performance and reliability metrics, guiding design improvements and enhancing inherent reliability. Reliability simulation in its full process generally involves digital prototype modeling, stress analysis based on finite elements, device-level reliability analysis based on failure physics, and reliability evaluation at the board/microsystem/unit levels.

In the United States, the CALCE research center at the University of Maryland has developed the CalcePWA software, and DfR Solutions in the United States has launched the commercial reliability simulation software Sherlock. And in China, the Fifth Research Institute of the Ministry of Industry and Information Technology, a professional research institution for reliability, has developed the RSE-PoF software, which is a reliability simulation software based on physics of failure, as shown in Fig. 1.28. RSE-PoF is a simulation and evaluation software for reliability physics that integrates multiple mechanisms and fusion. It uses finite element simulation of various physical fields such as thermal, mechanical, and electrical to locate weak links at the device, packaging, board, microsystem, and single machine levels, as well as to analyze potential failure causes and predict working life.

The RSE-PoF software, relying on the technical achievements and a rich library of proprietary failure physics models accumulated by the Fifth Research Institute of the Ministry of Industry and Information Technology over the years, can simulate damage models considering the stress of the entire lifecycle from process to use, as

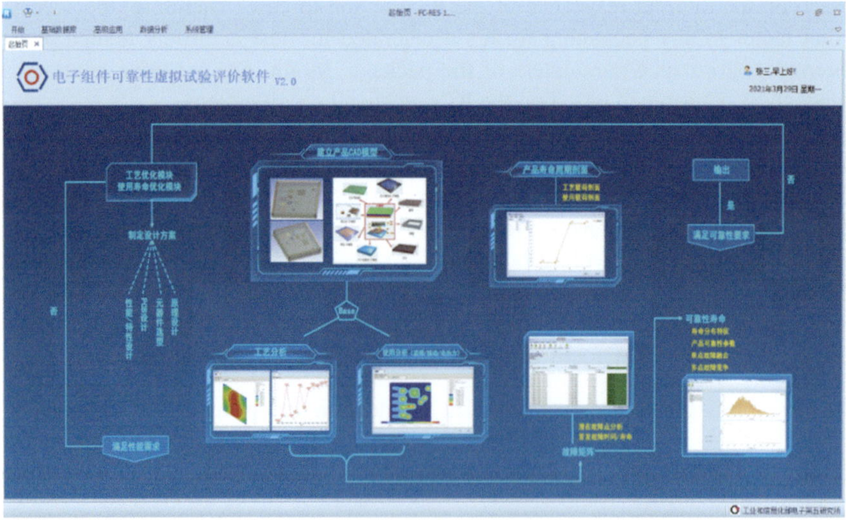

Fig. 1.28 A reliability simulation software RSE-PoF based on the physics of failure

well as the fine interconnection structure. It has a comprehensive reliability prediction function. In terms of board/microsystem-level reliability modeling, the software is based on fundamental reliability theory, combined with numerical algebra for solving nonlinear implicit complex functions, joint probability distribution solution technology, and artificial intelligence-based theoretical analysis technology. It has established a reliability algorithm library for devices, packaging, boards, microsystems, and single machine levels, covering analytical algorithms, iterative algorithms, data-driven algorithms, and value transfer algorithms. The software also provides process optimization models for reliability design, supporting the optimized design and process improvement of electronic equipment.

References

1. Tummala RR (2001) Fundamentals of microsystems packaging. McGraw-Hill Education
2. Wang Y (2018) Integrated circuit industry manual, vol II. Publishing House of Electronics Industry, Beijing
3. Liu J (2000) Manufacturing and packaging trend for Swedish microelectronics industry. J High Compon Fail Anal 15
4. Liu CC, Chen SM, Kuo FW et al (2012) High-performance integrated fan-out wafer level packaging (InFO-WLP): technology and system integration. In: 2012 International electron devices meeting. IEEE, 14.1.1–14.1.4
5. Hou SY, Chen WC, Hu C et al (2017) Wafer-level integration of an advanced logic-memory system through the second-generation CoWoS technology. IEEE Trans Electron Devices 64(10):4071–4077
6. ITRS (2015) International technology roadmap for semiconductors 2.0: executive report. Int Technol Roadmap Semicond 79
7. Committee AECCT (2017) Failure mechanism based stress test qualification for multi-chip modules (MCM) in automotive applications: AEC-Q104-Rev-September 14, 2017. The Automotive Electronics Council, USA
8. Tummala R, Rymaszwski E, Klopfenstein A (2001) Microelectronics packaging handbook. Electronics Industry Press, Beijing
9. Songchun ZHU (1998) New microelectronics packaging technology—BGA. Electron Technol 19(2):5
10. Mearig J, Goers B (1995) An overview of manufacturing BGA technology. In: Seventeenth IEEE/CPMT international electronics manufacturing technology symposium. 'Manufacturing technologies-present and future'. IEEE, pp 434–437
11. Marrs RC, Freman B, Martin J (2013) High density BGA technology. In: Proceedings of the second international conference and exhibition on multichip modules, pp 326–329
12. Jacobson DM, Sangha S (1998) Future trends in materials for lightweight microwave packaging. Microelectron Int 15(3):17–21
13. Pecht MG, Agarwal R, McCLUSKEV P et al (2017) Electronic packaging: materials and their properties. RC Press, Boca Raton
14. Pan MY, Gupta M, Tay A et al (2006) Development of bulk nanostructured copper with superior hardness for use as an interconnect material in electronic packaging. Microelectron Reliab 46(5/6):763–767
15. Chien C, Lee S, Lin J et al (2002) Effects of Sip size and volume fraction on properties of Al/Sip composites. Mater Lett 52(4–5):334–341
16. Gao S, Nan Z, Li Y et al (2020) Copper matrix thermal conductive composites with low thermal expansion for electronic packaging. Ceram Int 46(11):18019–18025

17. Tang T, Zhang X, Xu Z (2010) Current status and trends of electronic packaging materials research. J Nanjing Univ Technol (Nat Sci Ed) 32(4):6
18. Lyer M (2009) Advanced packaging technology development trends. Integr Circuit Appl 9:34–37
19. China Electronics Technology Standardization Institute (2003) Handbook of reliability prediction for electronic equipment: GJB/Z 299C-2006
20. China Aviation Society (2001) General requirements for equipment environmental engineering. In: 2001 China Aviation Society Environmental Engineering Academic Annual Meeting
21. JEDEC (2011) Failure mechanisms and models for semiconductor devices: JEDEC JEP122G-2011. In: Foreign-Foreign Academic Associations-(USA) Solid State Technology Association, EIA US-JEDEC
22. Qualified Manufacturers Certification General Specification for Semiconductor Integrated Circuits: GJB7400-2011 (2011)
23. De Sio C, Azimi S, Sterpone L (2019) On the evaluation of the PIPB effect within SRAM-based FPGAs. In: 2019 IEEE European test symposium (ETS). IEEE, pp 1–2
24. Karim N, Mao J, FAN J (2010) Improving electromagnetic compatibility performance of packages and SiP modules using a conformal shielding solution. In: 2010 Asia-Pacific international symposium on electromagnetic compatibility. IEEE, pp 56–59
25. China Electronics Technology Standardization Institute (2005) Test methods and procedures for microelectronic devices: GJB548B-2005
26. State Administration for Market Regulation (1993) Outline dimensions of semiconductor integrated circuits: GB/T 7092-1993. In: Domestic—National Standard—State Administration for Market Regulation of China (CN-GB)
27. Electronic Packaging Technology Editorial Board (1998) Handbook of integrated circuit packaging tests. Publishing House of Electronics Industry, Beijing
28. State Standardization Administration (1999) General Specification for Semiconductor Integrated Circuit Cases: GJB1420A-1999
29. Liu J, He M, Fan L et al (2003) Polyimide resin in advanced electronic packaging. Semicond Technol 28(10):37–41
30. State Administration for Market Regulation (1994) Case outlines for film integrated circuits and hybrid integrated circuits: GB/T 15138-1994. Domestic—National Standard—State Administration for Market Regulation of China (CN-GB)
31. Allan A (2015) International technology roadmap for semiconductors (ITRS). J Appl Phys 86(17):045406
32. Saito A, Nishikawa H (2019) Tin whisker growth mechanism on tin plating of MLCCs mounted with Sn-3.5Ag-8In-0.5Bi solder in 30°C/60%RH. In: 2019 22nd European microelectronics and packaging conference & exhibition (EMPC), pp 1–4
33. Wang X, Zhang F, Chen W et al (2019) Application assessment of China spallation neutron source in the study of atmospheric neutron single-event effects. Acta Physica Sinica 2(5):10
34. JEDEC (2001) Measurement and reporting of alpha particles and terrestrial cosmic ray-induced soft errors in semiconductor devices: JESD89. Foreign-Foreign Academic Associations-(USA) Solid State Technology Association, EIA US-JEDEC
35. Airbus France (2010) Reliability methodology for electronic systems. FIDES Group

Chapter 2
Physical Properties and Reliability Characteristics of Integrated Circuit Packaging

Integrated circuit (IC) packaging protects internal chips and provides mechanical support, electrical connections and heat dissipation. The characteristics of physical properties and reliability of IC packaging materials and structures mainly involve aspects such as electrical performance, thermal properties, mechanical characteristics, hygroscopicity or hermeticity and environmental adaptability.

Chapter introduces 8 typical physical properties of IC packaging from both structures and materials perspectives, which includes routine characteristics such as electrical properties, thermal properties, mechanical properties, hygroscopic properties, sealing characteristics, as well as unconventional features such as radiation resistance, electromagnetic shielding, and growth characteristics of tin whiskers in lead-free plated components. Details regarding the characterization parameters, measurement techniques, and corresponding standards for evaluating physical properties are presented systematacially. Subsequently, the reliability of IC packaging is discussed through 2 major metrics, namely failure rate and lifespan. Where the failure rate quantifies random packaging failures occurring during the effective operational lifetime, whereas, the lifespan characterizes the degradation behaviour of high temperature degradation, thermal fatigue, and vibration fatigue in failure degradation phase. Finally, the environmental adaptability of IC packaging is discussed by considering environmental stresses such as salt atmosphere, thermal shock, extreme temperatures, high humidity and mechanical stress.

B. Zhou et al., *Reliability Technology for Integrated Circuit Packaging*,
https://doi.org/10.1007/978-981-95-3885-0_2

2.1 Characteristics of Physical Properties and Standard Requirements

2.1.1 Conventional Physical Properties

1. Electrical Properties
1) Conductivity and Insulation of Package Materials
(1) Organic Adhesive Materials

Typical organic adhesive materials include conductive adhesives and non-conductive adhesives. Conductive adhesives served as to provide mechanical bonding while simultaneously enable electrical conduction and thermal transfer capabilities. It is noted that typical conductive adhesives used in electronic packaging exhibit conductivities exceeding 10^6 (S/m),with certain high-performance one reaching up to 10^7 S/m. Non-conductive adhesives are generally insulating bonding materials for adhesion purposes.

(2) Substrate Materials

Substrate materials mainly include organic epoxy, ceramics and silicon. Among them, epoxy substrates are conventionally insulating materials such as fiberglass (FR4) or phenolic resin (FR3). The breakdown voltage of epoxy resin used in fiberglass is normally around 30 kV/mm, while some substrates contain inorganic fillers and glass cloth improve the breakdown voltage approaching to 40 kV/mm.

Typical ceramic substrates include aluminum nitride (AlN), alumina (Al_2O_3) and silicon carbide (SiC), where Al_2O_3 exhibits excellent electrical insulating properties but relatively low thermal conductivity. In comparison, AlN preforms an excellent electrical insulation with high heat dissipation properties, making it will-suitable for high-power, multi-leads and large-size packages. One's drawbacks including high sintering temperatures, complex fabrication processes and elevated production costs, which limit quantity production and widespread application of AlN. SiC presents a high thermal conductivity, low thermal expansion coefficients, well dielectric insulation and high mechanical strength. Nevertheless, it's high dielectric constant ($\varepsilon_r = 10$) leads toa narrow applications only for low-frequency packages.

For semiconductor silicon, the conductivity at room temperature is around 10 S/m, which can be enhanced by increased the doping concentration.

(3) Metal Enclosure Materials

Conventional metal enclosure materials are generally conductors including Al, Cu, Mo, W, Cu-W alloys and Cu-Mo alloys with electrical conductivity of over 107S/m. In metal packaging, metal substrates served as large-area ground planes to reduce the capacitance and inductance between signal lines, hence to minimizes crosstalk and electrical noise. Moreover, advantages in using metal enclosure materials are normally concluded as an excellent thermal conductivity for improving the heat dissipation, a superior electrical conductivity for reducing the propagation delay, and a superior shielding capabilities against electromagnetic interference (EMI) and radio frequency interference (RFI).

(4) Molding and Potting Materials

Currently, the majority materials empolyed in molding packaging are silicone and epoxy. By comparison,, epoxy is a more commonly option, with breakdown voltage of around 20 kV/mm. Potting materials mainly include organic silicone potting compounds. The physical properties of different types of organic silicone potting compounds can vary a lot, especially when it comes to insulation performance, waterproofing, temperature resistance, optical properties, adhesion, and hardness. Advancements in materials science have enabled the incorporation of functional fillers into potting compounds, thereby enhancing their thermal conductivity, electrical performance, and magnetic properties.

2) Impact of Interconnect Structure on Electrical Conductivity

The main interconnect structures in package include gold wire bonding, aluminum wire bonding, copper wire bonding, bump bonding, through-silicon vias (TSV) [1] and redistribution layers (RDL) [2]. Gold wire bonding which is the most widely used interconnect structure. It is prone to forming harmful intermetallic compounds at high temperatures. These intermetallic compounds have different lattice constants and distinct thermal and mechanical properties and under thermal stress, atomic migration can lead to the formation of Kirkendall voids at the interface. This increases the interconnect resistance and disrupts the ohmic connection of the integrated circuit.

Aluminum wire bonding is a low-cost bonding technique that has attracted significant attention. However, ordinary aluminum wire is easily oxidized during wedge bonding, forming a hard oxide film. This film degrades both the electrical performance and the mechanical integrity of the bond.

Copper wire bonding has good mechanical properties, as well as excellent electrical and thermal conductivity, making it suitable for chip interconnections. By replacing the expensive gold wire with copper, and the mechanically weaker aluminum wire, the pad pitch can be reduced. However, copper wire's tendency to oxidize may reduce its bond ability and conductivity.

Flip-chip bump interconnects are commonly used in high-density integrated circuit packaging. However, they suffer from issues such as bump cracking, Kirkendall voids in Cu_3Sn microbumps, porous voids, grain boundary embrittlement, intergranular fractures, and Ni/Sn/Ni micro-solder joint voids, all of which negatively affect their conductivity [3].

In TSV interconnect technology, problems like TSV open-circuit faults, axial voids, reduced carrier mobility, and delamination may cause the interconnect circuit's conductivity to degrade [4, 5].

3) Electrical Performance Standards for Package

The electrical performance of integrated circuit packaging is characterized by several key parameters, including inter-lead insulation resistance, interconnect resistance, lead-to-lead capacitance, load capacitance, and inductance. The corresponding technical standards and measurement requirements are summarized in Table 2.1.

Table 2.1 Standards and requirements for electrical performance of integrated circuit packaging

No.	Standard name	Main electrical performance requirements
1	GJB 1420B—2011 *General Specification for Semiconductor Integrated Circuit Packaging enclosure*	Testing insulation resistance between leads and between leads and the base Testing the resistance of leads at external bonding points Measuring the capacitance between leads
2	GB/T 16526—1996 *Test Methods for Capacitance Between Package Leads and Lead Load Capacitance* (Equivalent Standard: SEMI G24–89:1984)	Measuring the lead–to–lead capacitance and lead load capacitance for ceramic, metal and plastic semiconductor packages Lead–to–lead capacitance, 1 MHz Lead load capacitance, 1 MHz
3	SEMI G23–0996:1980 *Test Method for Inductance of Internal Traces of Semiconductor Packages* (Methods for Measuring Inductance of Internal Package Leads)	Techniques for measuring inductance in internal leads for PGA and similar packages Inductance measurement for packages exceeding 0.5nH BGA inductance of the package =measured inductance − residual inductance
4	SEMI G24–89:1984 *Test Method for Measuring the Lead–to–Lead and Loading Capacirrance of Package Leads* (Test Methods for Capacitance Between Package Leads and Lead Load Capacitance)	Designed for testing lead–to–lead capacitance and lead load capacitance in semiconductor packages Suitable for capacitance range from 1 to 100 pF
5	EIA/JEP 123:1995 *Guideline for Measurement of Electronic Package Inductance and Capacitance Model Parameters* (Guidelines for Measuring Inductance and Capacitance Model Parameters in Electronic Packages)	Used to test the inductance and capacitance between external pins of semiconductor device packages
6	SJ/T 11703—2018 *Test Methods for Crosstalk Characteristics in Digital Microelectronic Device Packages*	Outlines methods for measuring the cross–coupling level of broadband digital signals and noise between terminals in digital microelectronic device packages Applicable to various logic series products when the drive and load impedance are already known

2. Thermal Properties
1) Thermal Conductivity of Package Materials
(1) Organic Adhesive Materials

In chip packaging, organic adhesive materials generally use two or more types of polymers as the base to achieve better bonding, conductivity, vibration resistance, and heat transfer properties. Non-conductive adhesives typically use fillers such as silica and polytetrafluoroethylene while conductive adhesives generally use epoxy resin, unsaturated polyester or silicone rubber as the matrix with fillers like carbon (graphite, carbon fiber), metals (Au, Cu, Ag powders) and metal oxides (Al_2O_3). These fillers serve to enhance the electrical and thermal conductivities of the adhesive materials. However, adding high thermal conductivity fillers can potentially decrease the mechanical properties of the material. Currently, the thermal conductivity of commercial silicone adhesives is generally between 0.4 and 0.8 W/(m·K) which is insufficient for the demands of high-power density packaging. The development of adhesive materials with enhanced thermal conductivity has emerged as a key research focus in the packaging industry.

(2) Substrate Materials

The thermal conductivity of substrate materials (such as PCB, ceramics and silicon) directly affects the thermal resistance of the package which in turn influences the internal temperature of the device. For example, ceramics have higher thermal conductivity compared to typical organic materials which means ceramic packaging usually offer better heat dissipation than plastic ones. In PCBs, the thermal conductivity can be improved by increasing the proportion of metal content in the internal layers. Ceramic substrates are typically fabricated via high-temperature direct bonding of copper foil onto the ceramic surface. The main manufacturing processes for ceramic substrates include High Temperature Co-fired Ceramic (HTCC), Low Temperature Co-fired Ceramic (LTCC), Direct Bonded Copper (DBC) and Direct Plate Copper (DPC). When compared to LTCC substrates, HTCC substrates have advantages such as higher wiring density, stronger mechanical strength, better thermal conductivity, and more stable chemical properties [6].

(3) Metal Enclosure Materials

Metal enclosure materials are primarily employed in hermetic packaging devices. The design of hermetic packaging follows specific engineering guidelines concerning the mechanical dimensions, thermal and electrical properties, environmental adaptability, and structural reliability of the metal shell. The material and structural design of the metal enclosure should meet the thermal dissipation requirements of the device. Table 2.2 shows reference performance values for commonly used metal enclosure materials. According to the *Metal Enclosure Design Guidelines* issued by the National Defense Science and Technology Industry Bureau, thermal resistance simulations or experimental verifications should be conducted, where necessary, to ensure compliance with design specifications.

(4) Plastic and Potting Materials

In electronic packaging, the materials commonly used for plastic encapsulation, potting and bottom filling include epoxy resin, silicone resin and polyurethane resin. These materials not only protect the chip from environmental damage but also provide mechanical support and stress buffering. The thermal conductivity of plastic encapsulation materials is relatively low and typically ranging from 0.1 to 2 W/(m·K). The thermal performance of these materials is influenced by the fillers used. For example, adding aluminum nitride filler coated with silica to the plastic material can improve its thermal conductivity. Moreover, the glass transition temperature (T_g) of plastic

Table 2.2 Reference values for the properties of commonly used metal enclosure materials [7]

Base material	Typical code	Density (g/cm^3)	Coefficient of thermal expansion(10–6/°C)	Thermal conductivity [W/(m·K)]
Iron–Nickel–Cobalt Alloy	4J29	8.2	5.3	17
Iron–Nickel Alloy	4J42	8.12	4.8	13
Oxygen–Free Copper	TU1	8.9	17.6	390
Tungsten Copper	WCu85/15	16.4	7.2	180
Molybdenum Copper	MoCu70/30	9.7	8.3	190
Molybdenum	Mol	10.2	5.2	140
Stainless steel	OCr18Ni9	7.93	17.2	16
Aluminum–Silicon Carbide	SiCp/Al–DZ7	<3.1	6.0~8.0	≥160
Nickel	N6	8.89	13.3	82.9

materials is usually lower than 200 °C, meaning that when the operating temperature exceeds this value, the thermal conductivity of the plastic will change.

2) Thermal Resistance
(1) Junction-to-case Thermal Resistance

In most devices, the heat from the junction mainly dissipates downward through the substrate. The equivalent model for junction-to-case thermal resistance is shown in Fig. 2.1.

Here, T_J refers to the junction temperature, T_C is the temperature at the bottom of the package enclosure, θ_{JC} is the junction-to-case thermal resistance, which can be calculated using Eq. (2.1). P_D represents the thermal power dissipation of the device. The junction-to-case thermal resistance includes both the thermal resistance of the chip and the package structure. The thermal resistance of the package can be further separated using a structural function method, which is explained in Chap. 6.

$$\theta_{JC} = \frac{T_J - T_C}{P_D} \tag{2.1}$$

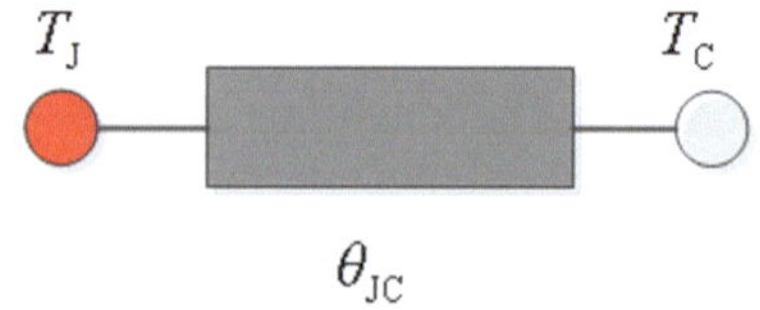

Fig. 2.1 Junction-to-case thermal resistance equivalent model

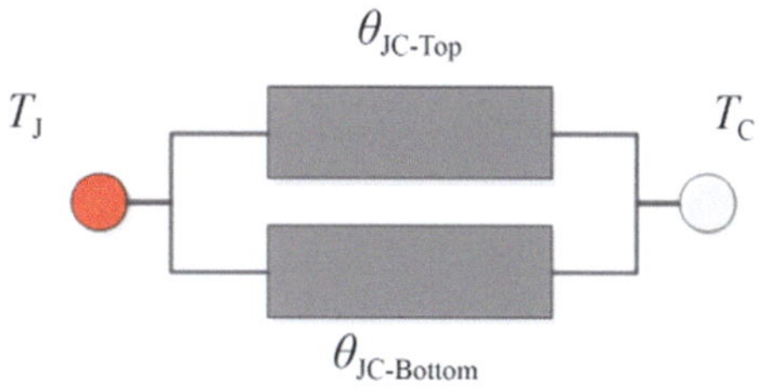

Fig. 2.2 The dual thermal resistance model

(2) Dual Thermal Resistance Model

For some metal and ceramic hermetic packaging, when the chip is in contact with the top of the enclosure via a thermal conductive pad, and the heat flow upwards cannot be ignored, the heat dissipation happens simultaneously through both the top and bottom of the enclosure. In such cases, a dual thermal resistance model can be used to describe the heat dissipation. The dual thermal resistance model is shown in Fig. 2.2.

Here, $T_{C\text{-Top}}$ and $T_{C\text{-Bottom}}$ refer to the temperatures at the top and bottom of the device package enclosure, respectively. $\theta_{JC\text{-Top}}$ and $\theta_{JC\text{-Bottom}}$ represent the thermal resistances from the junction to the top and bottom of the shell, respectively. These can be calculated using Eq. (2.2).

$$\frac{T_J - T_{C\text{ - Top}}}{\theta_{JC\text{ - Top}}} + \frac{T_J - T_{C\text{ - Bottom}}}{\theta_{JC\text{ - Bottom}}} = P_D \tag{2.2}$$

(3) Multi-heat Source Package Thermal Resistance Model

For multi-heat source package, where the heat inside the device primarily dissipates downward through the substrate, it is necessary to consider the coupling effects between the heat sources. Figure 2.3 shows the thermal resistance model for a multi-heat source package with four heat sources arranged in parallel. In this model, T_C is the temperature at the bottom of the package enclosure, T_A is the ambient temperature, T_S is the substrate temperature inside the package, and θ_{SC} is the conductive thermal resistance between the substrate and the bottom of the enclosure [8].

In the equations, $\Delta T_{1,1} \sim \Delta T_{4,4}$ represent the self-heating temperature rise of Heat Sources 1 to 4 respectively. $\Delta T_{i,j}, j$ indicates the thermal coupling temperature rise caused by Heat Source i, $\theta_{1,1} \sim \theta_{4,4}$ are the junction-to-case thermal resistances of Heat Sources 1 to 4 respectively. While $\theta_{i,j}$ represents the coupling thermal resistance between Heat Source j and Heat Source i. ⦵ represents the constant current source (thermal flux) generated by the thermal power dissipation P_D of a heat source and ⦶ is the constant voltage source (temperature rise) created by a heat source acting on a thermal coupling object. The thermal coupling temperature rise occurs at the thermal coupling object due to the coupling thermal resistance ($\theta_{1,2} \neq \theta_{2,1}$). The calculation formula is given by Eq. (2.3).

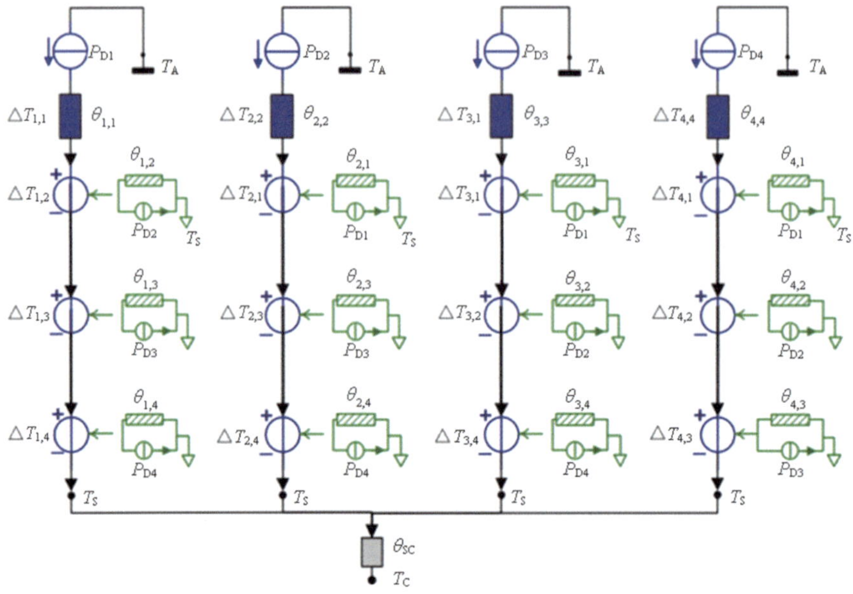

Fig. 2.3 The multi-heat source package thermal resistance model with four parallel heat sources

$$[T_J]_4 = \begin{bmatrix} T_{J1} \\ T_{J2} \\ T_{J3} \\ T_{J4} \end{bmatrix} = \begin{bmatrix} \theta_{1,1} & \theta_{1,2} & \theta_{1,3} & \theta_{1,4} \\ \theta_{2,1} & \theta_{2,2} & \theta_{2,3} & \theta_{2,4} \\ \theta_{3,1} & \theta_{3,2} & \theta_{3,3} & \theta_{3,4} \\ \theta_{4,1} & \theta_{4,2} & \theta_{4,3} & \theta_{4,4} \end{bmatrix} \cdot \begin{bmatrix} P_{D1} \\ P_{D2} \\ P_{D3} \\ P_{D4} \end{bmatrix} + T_c \tag{2.3}$$

3) Thermal Expansion and Interface Thermal Mismatch of Package Materials
(1) Thermal Expansion of Typical Materials

Thermal expansion refers to the change in volume or length of a material as its temperature varies. It changes with both the material type and temperature. Before reaching the glass transition temperature, most materials have a thermal expansion coefficient that increases slowly and linearly with temperature. However, after the glass transition temperature, the rate of increase in thermal expansion can rise by 3–5 times. Table 2.3 shows the thermal expansion coefficients of typical package materials.

(2) Thermal Mismatch at the Package Interface

The thermal mismatch at the package interface is mainly caused by the differences in the thermal expansion coefficients of different materials at the same temperature. At tightly bonded interfaces, a large difference in thermal expansion coefficients between adjacent materials can lead to concentrated thermal stress which then causes delamination or cracking at the material interface. As the power density of the package increases and the structure becomes more complex, it becomes easier to

Table 2.3 The coefficient of thermal expansion for typical package materials [7]

Typical package materials	Coefficient of thermal expansion(10–6/°C)	
	X−axis and Y−axis directions	*Z−axis direction*
SAC305	21	21
Sn63Pb37	18~25	18~25
BT substrate	12.4	57.0
FR4 Substrate ($T<Tg$)	15.8	80~90
FR4 substrate ($T>Tg$)	20	400
PCB	14.5	67.2
SiO_2	0.51~0.59	0.51~0.59
Si	3.2	3.2
Cu_6Sn_5 IMC	16.3	16.3
Cu_3Sn IMC	19.0	19.0

introduce many thermal mismatch interfaces inside the package. Common thermal mismatch interfaces include those between the filling material and the silicon wafer, between the substrate and the bump or pad interface, and the bonding interfaces between chips. The thermal mismatch at these package interfaces can lead to delamination or cracking, which damages the bonding and heat dissipation functions of the interface.

4) Standards and Requirements for Thermal Performance in Package

The thermal performance of integrated circuit packaging is characterized by package thermal resistance and thermal parameters, which include steady-state junction-to-case thermal resistance, transient thermal resistance, steady-state junction-to-ring thermal resistance, transient junction-to-ring thermal resistance, junction top thermal characteristics, and junction board thermal characteristics. The relevant standards and requirements are listed in Table 2.4.

3. Mechanical Properties

In the production, transportation, and usage processes, the integrated circuit (IC) packaging structure faces various types of mechanical problems, which involve many branches of mechanics such as materials mechanics, plasticity mechanics, and fracture mechanics. With the continuous increase in the scale of large-scale integrated circuits and the density of electronic package, the mechanical problems leading to IC packaging failures are becoming more and more severe.

1) Key Mechanical Properties in Integrated Circuit Packaging
(1) Modal Analysis

The modal analysis of integrated circuit packaging is based on vibration theory. We can obtain the natural frequencies, damping ratios and modal shapes of the package structure at different orders by using experimental methods, such as hammer

Table 2.4 Standards and requirements for thermal properties of IC packaging

No.	Standard name	Main thermal properties requirements
1	GB/T 14862—1993 *Test Methods for Thermal Resistance from Semiconductor Integrated Circuit Packaging Junction to Enclosure*	This standard specifies the test methods for the thermal resistance from the integrated circuit packaging junction to the shell It uses thermal test chips to measure the thermal resistance of semiconductor integrated circuit packaging, including ceramic, metal, and plastic types It measures junction-to-case thermal resistance($R\theta$jC), junction-to−case surface thermal resistance($R\theta$jM), and junction-to-reference point thermal resistance($R\theta$jR) Ceramic and metal packages can be tested under a heat sink or in a liquid bath, while plastic packages can only be tested in a liquid bath
2	GJB 548B—2005 *Test Methods and Procedures for Microelectronic Devices*	Determine the thermal performance of microelectronic devices It measures junction temperature, thermal resistance, case temperature, surface temperature, and thermal response time
3	JESD15: 2008 *thermal modeling overview* (Overview of Thermal Models)	The methodology for modeling the thermal behavior of electronic device packages This includes detailed models, compact models and model verification methods
4	JESD15−1 *compact thermal model overview* (overview of compact thermal models)	Compact thermal models (CTM) represent the thermal behavior of electronic packages in the form of a resistive network Two types of compact thermal models: the dual-resistor compact thermal model and the DELPHI compact thermal model, both of which can be used to build thermal models for device packages and calculate junction temperatures
5	JESD15−3: 2008 *Two−Resistor compact thermal model guideline* (dual-resistor compact thermal model guidelines)	The junction-to-case thermal resistance model and junction-to-plate thermal resistance model are used to represent the thermal resistance of single-chip device packages. These models can be developed either through testing methods or by extracting data from validated simulation models

(continued)

Table 2.4 (continued)

No.	Standard name	Main thermal properties requirements
6	JESD15−4: 2006 *DELPHI Compact Thermal Model Guideline* (DELPHI Compact Thermal Model Guidelines)	The DELPHI (Integrated Design Environment Physical Model Development Library) compact thermal model Used for thermal modeling of both single-chip and multi-chip packages. It is limited to steady-state compact models and does not cover dynamic compact models A modeling approach with Boundary Condition Independence (BCI) is employed for compact thermal models. BCI, which is a characteristic of compact thermal models, allows for precise calculations of chip temperatures under various environmental temperatures
7	JESD51−1: 1995 *Integrated Circuits Thermal Measurement Method−Electrical Test Method (Single Semiconductor Device)* Integrated Circuit Thermal Testing Methods—Electrical Testing Methods (Single Semiconductor Device)	This standard outlines an electrical testing method for measuring the thermal characteristic parameters of a single−junction semiconductor device
8	JESD51−2A: 2008 *Integrated Circuits Thermal Test Method Environmental Conditions−Natural Convection (Still Air)* (Integrated Circuit Thermal Testing Methods Environmental Conditions—Natural Air Convection)	Used for measuring the standard junction−to−case thermal resistance of integrated circuits under natural air convection conditions
9	JESD51-3: 1996 *Low Effective Thermal Conductivity Test Board for Leaded Surface Mount Packages* (Low Thermal Conductivity Test Board for Pin-Through Surface Mount)	For testing the junction−to−case thermal resistance characteristics of devices when using PCBs with low thermal conductivity
10	JESD51-4A: 2019 *Thermal Test Chip Guideline (Wire Bond and Flip Chip)* (Thermal Test Chip Guidelines—Wire Bonding and Flip-Chip)	Testing both steady-state and transient thermal resistance Used to compare the thermal performance of packages and to verify the thermal simulations of the package Applied in power topology mapping and transient response studies for complex package structures

(continued)

Table 2.4 (continued)

No.	Standard name	Main thermal properties requirements
11	JESD51-5:1999 *Extension of Thermal Test Board Standards for Packages with Direct Thermal Attachment Mechanisms* (Thermal Test Board Standard Extension for Packages with Direct Thermal Connection)	This specifies the testing requirements for package types where the chips are directly bonded to the PCB
12	JESD51–6: 1999 *Integrated Circuit Thermal Test Method Environmental Conditions–Forced Convection (Moving Air)* (Integrated Circuit Thermal Testing Methods Environmental Conditions—Forced Air Convection)	Used to test the thermal performance of integrated circuits in forced air convection conditions
13	JESD51-7: 1999 *High Effective Thermal Conductivity Test Board for Leaded Surface Mount Packages* (High Thermal Conductivity Test Board for Pin-Through Surface Mount)	Used to test the junction-to-case thermal resistance characteristics of devices under conditions with high thermal conductivity PCBs
14	JESD51–8: 1999 *Integrated Circuits Thermal Test Method Environmental Conditions– Junction–to–Board* (Integrated Circuit Thermal Testing Methods Environmental Conditions—Junction–to–Board)	This involves measuring the junction-to-board thermal resistance(RθJB)and thermal characteristic parameters(ψJB)of integrated circuits
15	JESD51–9: 2000 *Test Boards for Area Array Surface Mount Package Thermal Measurements* (Test Board for Thermal Testing of Surface–Mounted Devices in Array Configuration)	This specifies the requirements for thermal testing PCBs for devices with BGA and LGA chip packages
16	JESD51–10: 2000 *Test Boards for Through–Hole Perimeter Leaded Package Thermal Measurements* (Test Board for Thermal Testing of Through–Hole Pin–Through Insert Package)	This specifies the requirements for thermal testing PCBs for devices with through–hole pin–insert package, such as DIP and SIP

(continued)

Table 2.4 (continued)

No.	Standard name	Main thermal properties requirements
17	JESD51–11: 2001 *Test Boards for Through–Hole Area Array Leaded Package Thermal Measurements* (Test Board for Thermal Testing of Through–Hole Pin–Through Array Package)	This specifies the requirement for thermal testing PCBs for PGA chip package devices
18	JESD51–12: 2005 *Guidelines for Reporting and Using Electronic Package Thermal Information* (Report and Application Guidelines for Thermal Performance Information of Electronic package)	The requirements for suppliers to provide thermal performance information of the package The methods for users to adopt thermal performance data
19	JESD51-13: 2009 *Glossary of Thermal Measurement Terms and Definitions* (Common Terms and Definitions in Thermal Testing)	This provides a unified collection of commonly used terms and definitions in the field of semiconductor thermal testing
20	JESD51–14: 2010 *Transient Dual Interface Test Method for the Measurement of the Thermal Resistance Junction to Case of Semiconductor Devices with Heat Flow Trough a Single Pat* (Junction–to–Case Thermal Resistance Testing Method for Semiconductor Devices in One–Dimensional Heat Conduction Path Conditions: Transient Dual–Interface Method)	This explains the transient dual-interface method for testing junction-to-case thermal resistance of semiconductor devices in a one-dimensional heat conduction path from the heat dissipation junction of the semiconductor to the outer surface of the package
21	JESD51–31: 2008 *Thermal Test Environment Modifications for Multi–Chip Packages* (Thermal Testing Environment Corrections for MCM Package)	Suitable for thermal testing of MCM package Standardizes the method for testing multiple heat sources with independently controlled power

(continued)

Table 2.4 (continued)

No.	Standard name	Main thermal properties requirements
22	JESD51-51: 2012 *Implementation of the Electrical Test Method for the Measurement of Real Thermal Resistance and Impedance of Light−emitting Diodes with Exposed Cooling* (Testing LED Thermal Resistance and Thermal Impedance Using Electrical Methods)	This describes the method for testing the thermal resistance and thermal impedance of LEDs using electrical methods
23	JEP149: 2004 *Application Thermal Derating Methodologies* (Application of Thermal Derating Methods)	Used for the thermal derating design of integrated circuits and package It includes the limit thermal performance requirements Thermal derating requirements are based on reliability standards

excitation and exciter incentive method or simulation analysis method. For example, the natural frequencies and modal shapes of the package lid, bonding wires, and other parts can be tested using a laser vibrometer. This allows us to identify the sensitive resonant frequency range of the package structure and the vibration deformation patterns, thus enabling the design to circumvent resonance frequencies that could compromise device performance. The laser vibrometer is able to measure package structures with a minimum size of 8μm which is very useful for small package structures like bonding wires, where strain gauges cannot be attached.

(2) Vibration and Impact Response Characteristics Analysis

The analysis of vibration and impact response characteristics for integrated circuit packaging is based on mechanical dynamics theory. It employs techniques such as in-situ mechanical monitoring and simulation-based analysis or simulation analysis to obtain the stress, strain and deformation response of the package structure under different accelerations and frequencies. For macroscopic scales, conventional structural mechanics testing methods are typically used including strain gauges, sensors and other testing equipment. However, for package structures at the micron level, like bonding wires, microscopic structural testing methods are used, including laser vibrometer, micro-area residual stress testing and non-contact full-field strain testing. Vibration testing has three main elements: frequency, vibration amplitude or spectrum shape, and duration. Considering these three factors, the vibration resistance design of package structures can be divided into three main areas: avoiding resonance, reducing the response, and fatigue life design.

(3) Interface Mechanical Properties Analysis

There are numerous bonding interfaces in electronic package structures. On a macroscopic scale, different interfaces between materials and structures, such as solder joints, TSV/barrier layers/dielectric layers, thin-film coating interfaces, metal-ceramic bonding interfaces and interlayer interfaces in composite laminates, all have a significant impact on the overall mechanical properties of the package structure and the device performance. On a microscopic level, factors like grain size and grain orientation at the interfaces also play an important role in the overall mechanical properties and structural lifespan of the package.

2) Package Mechanical Testing Techniques

Traditional mechanical testing methods for package primarily focus on four aspects: bending, tensile, impact and vibration fatigue tests.

Bending tests are typically performed using the three-point bending method. In this method, a downward force is applied at the midpoint between two support points. When the sample bends sufficiently and fractures, the test is complete, the measured displacement, compressive force and the location of crack propagation are then used to calculate the fracture strength at the critical area of the sample.

Tensile tests usually follow a uniaxial tensile test approach. The sample is stretched axially at a constant speed using a tensile fixture. When the deformation reaches a certain level and the sample breaks, the test is considered complete. The displacement, stress–strain curve, fracture location and the fracture surface morphology are then used to serve as the basis for deriving the fracture mechanical constitutive equation, fracture strength and the fracture mode of the critical part of the sample.

Impact tests are usually performed using a drop test method, with three common types: free-fall of the packaged product, board-level free-fall and JEDEC board-level free-fall。

Vibration fatigue tests are typically carried out using cyclic mechanical loading, such as sinusoidal vibration, mainly for the solder joints of the interconnects in the package. During the cyclic tensile-compression and shear loading, the solder joints experience repeated deformation which leads to crack initiation. As the cracks propagate, a fracture surface forms while causing an increase in the resistance of the solder joints until the signal can no longer be conducted and ultimately resulting in package failure.

Microscopic structural mechanical testing methods, such as laser vibrometry, micro-area residual stress testing and non-contact full-field strain testing, will be introduced in Chap. 7.

3) Package Mechanical Performance Standards and Requirements

The mechanical performance of integrated circuit packaging is represented by structural strength and reliability, which includes parameters such as the shear strength of the package lid, shear strength/adhesion strength of the chip, bonding strength of internal leads, and the sturdiness of external leads. Relevant standards and requirements can be found in Table 2.5.

4. Moisture Absorption Characteristics

(1) Moisture Absorption of Package Materials

Table 2.5 Standards and requirements for integrated circuit packaging mechanical performance

No.	Standard name	Main mechanical performance requirements
1	GJB 1420A—1999 *General Specification for Semiconductor Integrated Circuit Packaging Enclosure*	Pin strength, pull tests and bending tests Bonding strength, pressure welding wire inspection and coating examination
2	GJB 1420B—2011 *General Specification for Semiconductor Integrated Circuit Packaging Enclosure*	Lead firmness, pull tests and lead fatigue tests Chip shear pull strength Bonding strength
3	GJB 597B—2012 *General Specification for Semiconductor Integrated Circuits*	Bonding strength and chip shear strength Lead firmness and package cover torque, among others
4	GB/T 16525—2015 *Specification for Lead Frame of Plastic Leaded Chip Carrier for Semiconductor Integrated Circuits*	Dimensions and coating thickness of PLCC package lead frames Heat resistance and pin strength, etc
5	GB/T 8750—2014 *Bonding Wires for Semiconductor package*	Diameter and mechanical properties Surface quality, etc
6	GJB 548B—2005 *Microelectronic Device Testing Methods and Procedures* [9] Method 2004.2 - Lead Bonding Firmness, Method 2011.1 - Bonding Strength, Method 2019.2 - Chip Shear Strength, Method 2024 - Glass Sealing Cap Plate Torque Test, Method 2027.1 - Chip Adhesion Strength, Method 2028 - Destructive Lead Pull Test for Pin Grid Array Package, Method 2029 - Soldering Strength for Ceramic Chip Carrier	Material strength: pins and solder balls Interface strength: inner lead bonding, chip adhesion, hermetic cover seal, and solder joint interfaces
7	GJB 7677—2012 *Ball Grid Array Testing Methods*	Solder ball coplanarity Solder ball pull strength Solder ball shear strength Solderability
8	JEDEC JESD22−B112B: 2018 *Package Warpage Measurement of Surface−Mount Integrated Circuits at Elevated Temperature* (Measurement of Package Warpage in Surface-Mount Integrated Circuits at High Temperatures)	Measurement of package warpage during the reflow soldering process BGA packages are more prone to warpage damage due to inherent warpage caused by mismatched thermal expansion coefficients of package materials or moisture absorption

(continued)

Table 2.5 (continued)

No.	Standard name	Main mechanical performance requirements
9	JESD22−B108B: 2010 *Coplanarity Test for Surface−MountSemiconductor Devices* (Coplanarity Testing for Surface-Mount Semiconductor Devices)	Measurement of coplanarity deviations of all leads in surface-mount devices at room temperature
10	JESD22−B110B.01: 2019 *Mechanical Shock−Device and Subassembly* (Mechanical Shock Testing for Devices and Components)	Assessment of resistance to mechanical shock in both free-state and on-board conditions
11	JESD22−B115A.01: 2016 *Solder Ball Pull* (Solder Ball Pull-Off Testing)	Tensile force on device solder balls during manufacturing, transportation, or final use Evaluation of the solder ball's ability to withstand tensile forces
12	JESD22−B117B: 2014 *Solder Ball Shear* (Solder Ball Shear Testing)	Using shear force detection methods to evaluate solder ball integrity Assess the ability of solder balls to withstand mechanical shear forces during device manufacturing, testing, transportation, and final use
13	JESD22−B116B: 2017 *Wire Bond Shear Test Method* (Shear Test Methods for Wire Bonding)	Measure the shear strength of solder balls with diameters between 15 and 76 μm Ensure solder ball height is at least 4 μm
14	JESD22−B119 *Mechanical Compressive Static Stress Test Method* (Static Mechanical Pressure Testing Methods)	Evaluate the pressure tolerance of high-power, large-size devices Simulate mechanical damage caused by heatsink pressure
15	JESD22−B105E: 2018 *Lead Integrity* (Pin Integrity)	Assess the tension and bending stress on package leads Evaluate fatigue at solder joints, lead twisting, and bolt torque resistance

(continued)

Table 2.5 (continued)

No.	Standard name	Main mechanical performance requirements
16	JEP 167A: 2020 *characteristics of Interfacial Adhesion in Semiconductor Packages* (characteristics of Semiconductor Package Interface Adhesion)	Applicable to chip wire bonding package and flip–chip package Includes chip shear, lid shear, lead frame tensile testing, and stud tensile testing Evaluate flip-chip pull strength Tests include peel, wedge, cantilever beam, and three-point bending
17	JESD217.01: 2016 *Test Methods to Characterize Voiding in Pre–SMT Ball Grid Array Packages* (Void Detection Methods for BGA Solder Balls Prior to SMT Assembly)	Applicable to FCBGA, PBGA, CBGA, and CCGA with a ball pitch of no less than 0.5 mm Detect six types of voids in solder balls: microvoids at interfaces, planar microvoids, shrinkage voids, micropore voids, IMC microvoids and pinhole voids Methods: 2D X-ray inspection (simple operation), 3D X-ray inspection (stereoscopic imaging), 5D X-ray inspection (eliminates PCB warpage effects)

Moisture absorption is an important physical property of plastic-encapsulated integrated circuits. Since plastic encapsulation uses polymer materials, there are relatively large gaps between the package materials, which makes it easier for moisture to seep into the micro-cracks (delaminations or cracks) between the chip surface and the encapsulation material. This can result in increased leakage current, unstable charge behavior, or corrosion of the aluminum metallization layer, which ultimately causes electrical leakage on the chip surface and, in severe cases, leads to functional failure of the chip. Additionally, the moisture absorbed by the encapsulating material can cause the package structure to expand, which introduces extra mechanical stress on the chip and potentially causing parameter drift of precision linear circuits. Moisture absorption in package is o a significant reliability concern in plastic-encapsulated integrated circuits.

If the plastic-encapsulated circuit is exposed to long-term conditions of high humidity and temperature, the chip surface will eventually reach a high saturation concentration of moisture and liquid water films after thousands of hours of moisture penetration. When powered, this can accelerate electrochemical corrosion effects.

With the continuous improvement of encapsulation materials and processes, the moisture resistance of devices has been significantly enhanced. Traditional steady-state damp heat reliability evaluations are increasingly ineffective in exposing failure mechanisms. High-Acceleration Stress Testing (HAST) is a widely recognized and effective accelerated moisture and heat testing method. It is an indispensable approach for evaluating the moisture resistance of plastic-encapsulated circuits and modules. HAST can be divided into two types: unbiased HAST (without voltage bias) and biased HAST. The relevant standards for the environmental adaptability of plastic-encapsulated integrated circuits are shown in Table 2.6.

(2) Moisture Sensitivity Level (MSL) Requirements for Plastic Encapsulated Devices

Plastic encapsulated chips, due to their natural moisture-absorbing characteristics, tend to allow moisture to ingress into the package even during storage. As the use of fine-pitch components, ball grid array (BGA) package and 3D stacking technology, the failure mechanisms caused by moisture absorption have attracted significant attention. If a plastic-encapsulated device is exposed to a high-temperature environment, such as during the reflow soldering process, the moisture absorbed inside the package can vaporize rapidly, generating sufficient vapor pressure to induce

Table 2.6 Environmental adaptability standards for plastic encapsulated integrated circuits

No.	Standard name	Moisture absorption applicability scope
1	JESD22–A101D: 2015 *Steady–State Temperature–Humidity Bias Life Test* (Steady-state temperature-humidity-bias life test)	Used to evaluate the reliability of non-hermetic packaging in humid environments, where temperature, humidity, and bias are applied to speed up moisture penetration from the external encapsulating material to the chip surface
2	JESD22-A110E: 2015 *Highly Accelerated Temperature and Humidity Stress Test* (Highly accelerated temperature–humidity stress test)	Used to assess the reliability of non-hermetic packaging in humid environments by applying severe temperature, humidity, and bias conditions to accelerate moisture penetration from the external encapsulation material to the chip surface. The failure mechanism aligns with the 85 °C/85%RH steady-state damp heat test
3	JESD22-A102E: 2015 *Accelerated Moisture Resistance–Unbiased Autoclave* (Unbiased highly accelerated moisture resistance test)	Uses steam condensation or saturated steam environments to evaluate the moisture resistance integrity of non-hermetic packaging
4	JESD22-A118B: 2015 *Accelerated Moisture Resistance–Unbiased HAST* (Accelerated moisture resistance test–unbiased HAST)	Evaluates the reliability of non-hermetic packaged circuits in humid environments without applying bias, to ensure that no failure mechanisms, such as electrochemical corrosion, are accelerated by bias

package failure. Common failure modes include delamination between the encapsulation material and the lead frame, delamination between the encapsulation material and the chip interface, chip damage, and internal cracks. In extreme cases, the "popcorn" effect may occur while causing the chip to swell or even crack.

The moisture absorption characteristics of plastic encapsulated integrated circuits are employed to assess the moisture permeability of both the encapsulation interface and materials. This permeability is expressed using Moisture Sensitivity Levels (MSL) which are classified into eight levels according to the JEDEC J-STD-020E standard [10], ranging from Level 1, Level 2, Level 2a, Level 3, Level 4, Level 5, Level 5a, and Level 6. Moisture sensitivity evaluation typically involves several stages, including sample preparation, initial electrical testing, initial optical microscopy inspection, baking, moisture treatment, reflow soldering, final optical microscopy inspection, final electrical testing and final ultrasonic scanning microscopy inspection. The baking time is related to the thickness of the package and a typical recommended condition is 125 °C/24 h. Different moisture sensitivity levels correspond to different pre-treatment conditions and testing times. The JEDEC J-STD-020E standard specifies the maximum time a plastic encapsulated device can be exposed to certain environmental conditions before it undergoes simulated reflow soldering. Pre-treatment can be performed using accelerated equivalent methods. After pre-treatment, the sample undergoes three cycles of simulated reflow soldering within 15 min to 4 h and followed by final testing. The moisture sensitivity level is determined based on failure modes, such as delamination.

According to the JEDEC J-STD-020E standard, for Level 1 products, there is no usage time limit under conditions of $\leq$ 30 °C/85% RH. For Level 2, 2a, 3, 4, 5 and 5a products, the usage time is limited to 1 year, 4 weeks, 168 h, 72 h, 48 h and 24 h, respectively, when stored in conditions of $\leq$ 30 °C/60% RH. For Level 6 products, the usage time limit under $\leq$ 30 °C/60% RH is determined based on the product's tape and reel label. Products with the corresponding moisture sensitivity level can be reflow soldered without the need for baking pre-treatment, as long as they are within the specified storage conditions and usage time. For instance, Level 2 products can be reflow soldered within 1 year under the $\leq$ 30 °C/60% RH storage condition without requiring baking pre-treatment. The moisture sensitivity level standards for plastic-encapsulated integrated circuits are listed in Table 2.7.

5. Sealing Characteristics
1) Performance Testing of Hermetic Packaging

Hermetic package can be categorized into three types: ceramic, metal and glass. Among the commonly used materials, metal is well-known for its excellent electrical and thermal conductivity, making it highly suitable for hermetic package structures in normal temperature conditions. Ceramic materials such as silicon carbide and silicon nitride, they stand out for their exceptional high-temperature resistance with pressure tolerance and anti-corrosion properties, enabling them to be applied in more challenging environments. On the other hand, glass package offers both optical transparency and hermeticity and making it ideal for optical devices. However, due

Table 2.7 Moisture sensitivity levels for plastic encapsulated integrated circuits

No.	Standard name	Moisture sensitivity level
1	IPC/JEDEC J−STD−020E *Moisture/Reflow Sensitivity Classification for Nonhermetic Surface Mount Devices* (Moisture/Reflow Soldering Sensitivity Classification for Non-Hermetic Surface-Mount Devices)	Identifying the moisture sensitivity levels of non−hermetic SMDs (solder mask defined structures) is important so that proper package, storage, and handling methods can be chosen, which helps to avoid damage during reflow soldering and repairs It is necessary to specify which moisture sensitivity level/pre-treatment level should be applied to qualified SMDs
2	IPC/JEDEC J−STD−033D: 2018 *Handling, Packing, Shipping and Use of Moisture, Reflow, and Process Sensitive Devices* (Operation, package, Transportation, and Usage of Devices Sensitive to Moisture, Reflow Soldering, and Processes)	Standard methods for the operation, package, transportation, and use of moisture-sensitive SMDs should be provided, to prevent reliability degradation after exposure to moisture and reflow soldering After thermal baking, SMDs should be given a package storage life of up to one year The package materials for plastic encapsulated SMD devices include epoxy resin and silicone resin
3	JESD22−A120B: 2014 *Test Method for the Measurement of Moisture Diffusivity and Water Solubility in Organic Materials Used in Electronic Devices* (Test Methods for Moisture Diffusion and Water Solubility of Organic Materials Used in Electronic Device Package)	A quantitative method is required to determine the moisture absorption properties of organic materials used in electronic device package The moisture diffusion rate and water solubility of organic materials should be measured to assess the reliability of surface-mount plastic encapsulated devices under humid conditions and after exposure to high-temperature reflow soldering
4	IPC/JEDEC J−STD−035 *Acoustic Microscopy for Nonhermetic Encapsulated Electronic Components* (Acoustic Microscopy Inspection Method for Non−Hermetic Packaging Components)	The acoustic microscopy inspection method is used to determine interface delamination of non-hermetic packaging components in board-level assembly circuits

to the permeability of gases like hydrogen, helium and neon through glass. it is not as fitting for high-vacuum applications.

(1) Hermeticity Testing

Hermeticity is one of the most essential performance indicators for hermetic package devices with internal cavities. High-reliability electronics set very strict technical requirements on limiting gases like water vapor and hydrogen inside the package. Devices that fail hermeticity standards may pose severe risks to product reliability. For instance, poor sealing may allow water vapor infiltration, leading to internal circuit corrosion, leakage or parameter drifts. To ensure hermeticity, devices must undergo gross leak and fine leak detection tests. The standard GJB 548B—2005 [9]

Method 1014.2 outlines the procedures, methods and failure criteria for hermeticity testing, which are elaborated in Chap. 4.

(2) Bonding Performance Testing

In the lead bonding of metal or ceramic hermetic packagings, the leads are bonded at both ends to the bonding pads on the package enclosure and the bonding area of the chip. Unlike plastic packaging, where the bonding wires are encapsulated within molding materials, hermetic packaging wires are typically suspended. This suspended structure makes them more prone to risks of short-circuit failures, which can happen when wires bridging or collision due to vibration or impact stress during use. Therefore, apart from conducting bonding strength tests, it is advisable to add vibration mode testing for the bonding wires. This allows for an analysis and evaluation of the risks of bridging or collision that could lead to short circuits. GJB 548B—2005 outlines both destructive bond pull test methods (Method 2011.1) and non-destructive bond pull test methods (Method 2023.2).

(3) Excess Detection

Excess particles inside hermetic package are a significant cause of failures. These particles can move freely and potentially damage the internal chip or interconnection structures within the cavity under the influence of vibration or impact. Particle Impact Noise Detection (PIND) is an effective method to identify free particles inside the cavity of hermetic packages. Based on its cavity size, this technique uses a vibrator to apply specified vibration frequencies and amplitudes to the device being tested. The vibrations stimulate the movement of internal free particles and their collisions with the package enclosure generate detectable signals through a transducer. PIND is a non-destructive testing method that is simple, quick and highly effective. GJB 548B—2005 Method 2020.1 provides detailed guidance on the procedure for PIND testing.

2) Stress and Reliability

To ensure the long-term reliability of hermetic package devices, several key conditions must be met: The internal water vapor content needs to comply with product standards or specifications; The hermeticity indicators of the device should meet the requirements after package; the hermeticity must remain intact even after experiencing environmental stresses such as temperature cycling, thermal shocks and mechanical vibrations. Especially during operation, environmental changes can expose the package to various thermal loads, causing thermal mismatches between different materials, these mismatches may lead to failures. At the same time, factors like corrosion, creep, fractures, fatigue cracks, and the delamination of thin films can also result in package failures.

(1) Fatigue Damage and Degradation

Under thermo-mechanical loading, weak spots in hermetic package are likely to suffer fatigue damage which causing the initiation and spread of microcracks. This is a time-dependent process that can lead to parameter drift and eventually cause the device to fail completely due to fractures. Fatigue usually starts in areas where stress concentrates, like the weld interface of the hermetic cover plate or the neck or root of bonding wires. Unlike static damage, fatigue is a progressive process, with its effects accumulating over time. Under a scanning electron microscope, fatigue damage often shows nearly parallel and evenly spaced striation patterns at the damaged interface.

(2) Outgassing of Materials

Within the hermetic package cavity, adhesives used during manufacturing processes are primarily used for fixation. These adhesives, usually composed of epoxy resin mixed with other additives, exhibit excellent properties, such as high bonding strength, good processability, and resistance to various media. However, under high temperatures and prolonged use, some adhesives may oxidize and decompose while releasing trace amounts of volatile substances. This can lead to alterations to the atmospheric composition inside the hermetic cavity. Based on numerous experiments and data analysis, there are three main sources of harmful atmospheres inside the cavity: ambient gases accidentally sealed in during the process; moisture adsorbed on internal materials, chip surfaces, adhesives, or coatings that either decompose or release gases; and gas infiltration through leaks in the seal. By comparison, the release of water vapor and hydrogen is the most detrimental to the reliability of the hermetic package's internal environment.

(3) Brittle Fracture and Degradation Under High-Frequency Vibrational Loads

The hermetic cover plate in package serves to isolate the chip from external environmental interference, but the welding interface of the cover plate is a reliability weak point. When external stresses such as vibration, constant acceleration, impact, or collision are applied, the cover plate may deform. At this point, the solder at the welding joint, which tends to be brittle, is prone to detaching. The detached material becomes free particles, contaminating the sealed cavity. In severe cases, this can reduce the connection strength or cause leakage failures. Such issues are particularly significant under impact loading conditions.

3) Hermeticity Standards for Package

The hermeticity of metal and ceramic-packaged integrated circuits is usually characterized by parameters like leakage rate and internal moisture content. These parameters include standard leakage rate, measured leakage rate, equivalent leakage rate and moisture content percentage. The relevant standards and requirements for the hermeticity of integrated circuits are summarized in Table 2.8.

Table 2.8 Standards and requirements for the hermeticity of integrated circuit

No.	Standard name	Hermeticity and internal moisture content requirements
1	GJB 1420B—2011 *General Specifications for Semiconductor Integrated Circuit Packaging Enclosure*	Hermeticity Testing Leak rate, GJB 548B—2005 Method 1014.2 Sealing, rough leak detection, fine leak detection
2	GJB 597B—2012 *General Specifications for Semiconductor Integrated Circuits*	Apies to semiconductor integrated circuits, including multi-chip integrated circuits Ued to inspect internal moisture content and hermeticity
3	JEP144A: 2020 *Guideline for Internal Gas Analysis for Microelectronic Packages* (Guidelines for Internal Gas Analysis in Microelectronic Packages)	It is suitable for hermetically sealed microelectronic devices, including discrete devices, monolithic, and hybrid microcircuit packages for internal gas analysis The method can measure internal atmospheres in packages with volumes from 0.0002 ~ 200 cc It is necessary to measure moisture content inside the package, as well as other gases such as hydrogen, oxygen, helium and argon This also explains the 10 common moisture-related failure mechanisms in packages
4	JESD22-A109B: 2011 *Hermeticity* (Hermeticity)	This is applicable for leak rate detection in metal and ceramic (hermetic ceramic) packages
5	GJB 548B—2005 *Test Methods and Procedures for Microelectronic Devices*	Method 1018.1 allows a maximum moisture content of 5000 ppm

2.1.2 Special Physical Characteristics

1. Radiation Characteristics

(1) Radiation Sources in Package Materials

The Earth naturally contains significant amounts of radioactive elements such as 235U (0.72%), 238U (99.2%) and 232Th (100%), which are often found in materials used for semiconductor devices, including molding compounds, solder balls and fillers. Additionally, minute quantities of 210Po are always present in the solder joints of semiconductor devices. These heavy radioactive isotopes undergo α decay while constantly releasing α particles with energy levels ranging from 4 and 9 MeV. When these high-energy α particles penetrate the active region of a semiconductor device, they create dense electron–hole pairs along their paths. The electric field inside the device causes the electron–hole pairs to separate and collect at the circuit nodes to produce the interference currents. This can lead to serious problems such as data loss or functional failures in the semiconductor device. The effect of α particles on circuit systems can be very significant. For example, interference from α particles in the CPU's instruction cache could cause soft errors which may stop the CPU from carrying out its intended operations [11].

In a packaged chip, the source of α particle radiation may originate from the package materials, chip materials, lead materials, and the materials used in the semiconductor device manufacturing process. The materials that are closer to the chip tend to have a greater impact from the emitted α particles. For devices manufactured in a highly controlled environment, the primary source of α particle radiation is the package materials rather than those used in the semiconductor fabrication process. Radioactive impurities decay and release α particles, as shown in Fig. 2.4.

(2) The Dangers of α Particle Radiation

The α particle radiation is present in all semiconductor devices, and numerous cases have shown that the soft errors it causes can result in data loss, functional abnormalities, and other forms of damage to electronic devices. Even with substantial efforts to purify the materials and remove radioactive impurities, the international standard can only achieve an ultra-low radiation rate (0.001 cph/cm^2), which still cannot fully prevent soft errors caused by α particles. Moreover, as semiconductor technology advances and the feature sizes of devices continue to shrink, their resistance to α particle radiation rapidly decreases. This is due to the reduction in critical charge, which is caused by the lower operating voltage.

(3) Standard Requirements

The α particle radiation properties of package materials including surface α particle radiation rate measurements at the material level and accelerated radiation tests at the device level. The material-level surface α particle radiation rate measurement is typically conducted according to the JESD221 standard. This measurement aims to obtain data such as the surface radiation rate and energy spectrum of the package materials. On the other hand, the device-level accelerated radiation test is typically performed following the JESD89-2A or IEC60749-38 standards. This test is designed to quickly evaluate the device's response characteristics to α particle radiation and determine the soft error cross-section caused by α particles. The standards for characterizing the α particle radiation properties of package materials are provided in Table 2.9.

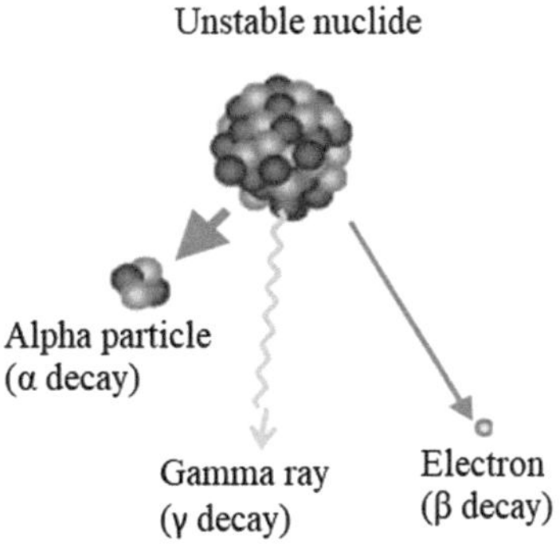

Fig. 2.4 Radioactive impurities decay and release α particle

Table 2.9 Standard requirements for characterizing the α particle radiation properties of package materials

No.	Standard Name	Evaluation requirements
1	JESD221: 2011 *Alpha Radiation Measurement in Electronic Materials* (Measurement method for α particle radiation in electronic materials)	Material-level α particle surface radiation rate measurement It is applicable to gas proportional counters (radiation rate lower than 0.01 cph/cm2) The process of measurement includes preparing the sample to prevent contamination, substrate noise diagnostics and adjustments, setting up test parameters, carrying out the tests, analyzing the data, and reporting the results
2	JESD89−2A: 2007 *Test Method for Alpha Source Accelerated Soft Error Rate* (Accelerated soft error rate testing method based on α particle sources)	Device-level accelerated radiation testing The testing process involves preparing the sample, selecting the source, setting up the experimental layout, processing the data, and generating the final report Usually, artificial α particle radiation sources are selected, with detailed specifications given for the source parameters (type, size, radioactivity, etc.) and the source-device experimental setup The thickness of the α particle source should be at least 0.04 mm The single-event effect testing system is used to monitor soft errors occurring in the device during radiation, including the error count, address, and error data. During the test, proper radiation safety precautions for the radiation source must be observed
3	IEC60749−38: 2008 *Semiconductor Devices—Mechanical and Climatic Test Methods—Part 38: Soft error Test Method for Semiconductor Devices with Memory* (Soft error testing method for semiconductor devices with memory)	

2. Electromagnetic Shielding Properties
(1) Electromagnetic Shielding Package Materials

Electromagnetic shielding materials are protective materials that inhibit the propagation and dispersion of electromagnetic waves. They limit the energy of electromagnetic radiation to a safe range thus reducing its potential harm. These materials are vital for electromagnetic radiation protection. With the growing emphasis on miniaturization, lightweight design and high-speed performance in electronic devices, electromagnetic interference has become more pronounced. To effectively isolate radio-frequency radiation devices and prevent interference from spreading to adjacent components, a conductive thin film is typically applied to the surface of package devices through methods like ion plating, chemical plating or vacuum deposition. Since the thickness of the conductive film is less than $\lambda/4$ (where λ represents the wavelength of the interfering electromagnetic wave), the absorption loss can be considered negligible and the shielding effectiveness mainly relies on reflection loss. The principle of electromagnetic shielding is based on the ability of the shielding

Table 2.10 Standards and requirements for shielding characteristics of organic thin-film coating materials

No.	Standard name	Shielding characteristic requirements
1	GB/T 34,938—2017 *General Technical Requirements for Planar Electromagnetic Shielding Materials*	10 MHz~1 GHz,Shielding Effectiveness≥20dB Thermal Shrinkage≤2% Transparency≥60% Peeling Strength≥280 N/m
2	GB/T 35575—2017: *General Technical Requirements for Electromagnetic Shielding Films*	10 MHz~3 GHz,Level SE−1:>50 dB;Level SE−2:45~50 dB;Level SE−3:40~45 dB;Level SE−4:<40 dB Laminate Contact Resistance≤1.0Ω Peeling Strength≤500 N/m Thermal Shrinkage≤2%

materials to reflect, attenuate and block electromagnetic waves from entering specific areas.

Henkel (Germany), Polymer Science (USA), and Feirongda (FRD) (China) have developed several conductive coating films for package and shielding materials. These materials can be evenly applied to the top and side walls of packaged devices, offering excellent shielding performance (shielding effectiveness refers to the ratio of electric field strength before and after shielding at a specific point), and also exhibit strong adhesion to the encapsulation materials. However, in general, electromagnetic shielding films for package applications are still in the early stages of research and have not been widely implemented yet [12].

(2) Electromagnetic Shielding Requirements for Package Materials

The electromagnetic shielding characteristics of coating materials used in package are mainly regulated in terms of appearance, coating thickness, temperature resistance, peel strength, and shielding effectiveness. For example, the coated material should have a smooth surface, uniform color distribution, and be free from bubbles, cracks, holes, deformation, corrosion, visible impurities, or any processing damage that might affect its usability. The thickness of the coating must be much smaller (less than 1/100) than the conductive wavelength of the sample, and the shielding effectiveness must be $\geq$ 20 dB, among other requirements.

The electromagnetic shielding characteristics of organic thin-film coatings material must comply with the relevant standards specified in Table 2.10.

2.1.3 *Tin Whisker Growth Characteristics*

1. Mechanism of Tin Whisker Growth

Currently, except for some exempted products, lead-free soldering has been fully implemented in the electronics industry. As a result, the issue of tin whisker growth in lead-free components has become one of the major reliability concerns. Tin whiskers are metal filaments that grow spontaneously on the tin-plated layers of component leads, primarily composed of tin. These whiskers pose a risk of short-circuit failure between leads, especially in high-density integrated components.

Although research on tin whisker growth has been conducted for more than 70 years, a comprehensive theory explaining the mechanism of tin whisker growth has not yet been formed. The prevailing view is that internal compressive stress is one of the main driving forces for tin whisker growth. This compressive stress primarily results from the diffusion of copper atoms from the lead substrate to the surface tin plating, and the growth of the intermetallic compound Cu_6Sn_5 during aging (particularly thermal aging). The oxide film on the surface of tin crystals, which differs in orientation from that of the regular tin plating, tends to shear along the grain boundaries. This causes the compressed tin crystals to grow whiskers at the fracture points on the surface. After that, the whiskers continue to grow until the stress is fully released, indicating that there is a growth saturation point for the tin whiskers. Figure 2.5 shows the main mechanism of tin whisker growth. Lee et al. [13, 14] used X-ray diffraction to determine the preferred orientation index of the surface film in the tin plating layer and found that the crystal orientation of tin whiskers differs from that of regular tin crystals.

2. Factors Affecting Tin Whisker Growth

The growth of tin whiskers is not influenced by environmental factors such as air pressure, electric fields, or magnetic fields. It is a spontaneous process that depends on time. They can cause short circuits between adjacent leads or conductors of the component once the whiskers grow long enough. Additionally, under high voltage conditions, tin whiskers might cause arc discharges, or they could bend, deform, or fall off due to airflow, leading to component failure or even product malfunction or accidents.

Tin whiskers are essentially different from dendrites. They are single-crystal structures that form spontaneously on the surface of pure tin due to compressive stress. Factors that may affect the growth of tin whiskers include process stress, external stress, temperature, humidity, grain size, and grain orientation [16].

3. Characteristics and Failure Criteria of Tin Whiskers

According to the JESD22-A121A: 2019 standard and standard industry practices, tin whisker characteristics is usually measured in three aspects: whisker shape, whisker

Fig. 2.5 Major mechanisms of tin whisker growth [15]

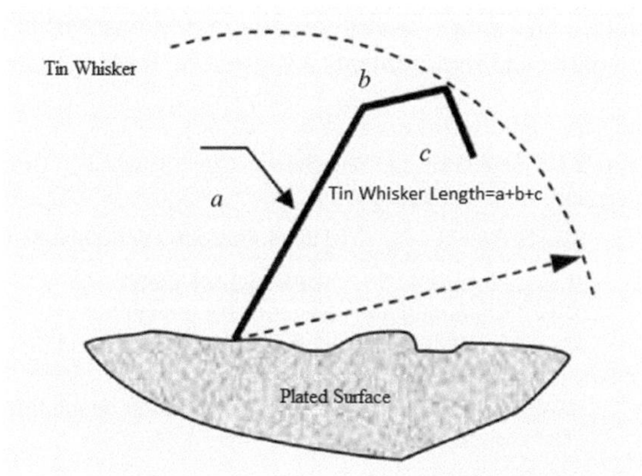

Fig. 2.6 Measurement of tin whisker length

length and whisker surface density. It is generally recommended to use an optical microscope that with a magnification range of 50–300 times or a scanning electron microscope (SEM) with at least 250 × magnification for such measurements. A failure is identified when the whisker length exceeds 50 μm or when the whisker length is greater than or equal to half of the pin pitch of the component being tested. For straight tin whiskers, the actual distance from the growth point on the plated surface to the whisker tip is measured directly. For whiskers with bends in the middle, the straight distances of each segment are measured separately and then summed to get the total length of the whisker. The whisker length measurement is shown in Fig. 2.6.

Since failures caused by tin whiskers can depend on the whisker surface density, measuring the whisker surface density can be a reference for determining whether an integrated circuit will fail. Table 2.11 shows the criteria for low, medium, and high whisker surface density ranges as specified by JESD22-A121A: 2019 [17].

4. Standards for Controlling Tin Whisker Growth on Package Surfaces [18]

Tin whisker growth on lead-free solder surfaces in integrated circuit packaging is characterized by whisker surface density and whisker length under temperature

Table 2.11 Tin whisker surface density evaluation criteria

No.	Tin whisker surface density	Total number of tin whiskers (pieces)	Inspection area of leads, terminals, or samples
1	Low	<10	1 mm^2
2	Medium	10 ~ 45	1 mm^2
3	High	>45	1 mm^2

cycling and temperature-humidity conditions within a specified time. The relevant standards and requirements are listed in Table 2.12.

Table 2.12 Standards and requirements for tin whisker growth on lead-free solder surfaces of integrated circuit packaging

No.	Standard name	Tin whisker growth evaluation requirements
1	JESD22−A121A: 2019 *Test Method for Measuring Whisker Growth on Tin and Tin Alloy Surface Finishes* (The assessment of tin whisker growth on tin and tin−alloy surfaces needs to be performed in accordance with *JEDS201A: 2020*)	Tin Whisker Criteria: Examine the growth tendency of tin whiskers under temperature cycling and humidity conditions by comparing different coatings or processes Measure tin whisker length threshold: 10μm)and surface density(threshold: 20%)
2	JESD201A: 2020 *Environmental Acceptance Requirements for Tin Whisker Susceptibility of Tin and Tin Alloy Surface Finishes*[19] (Test the sensitivity of tin and tin-alloy coatings to whisker growth under specific environmental conditions.)	The acceptance process involves using high-temperature and high-humidity tests to stimulate tin whisker growth, allowing for a proper evaluation of tin whisker sensitivity Test Levels: Category 3: Aerospace–Pure tin and high-tin alloys are strictly not acceptable Category 2: Telecommunications/Servers/Automotive–For products with extended lifespans, like disk drives, the detachment of tin whiskers is a serious issue Category 1: Industrial/Consumer Products–Products with moderate lifespans where tin whiskers are not a significant concern Category 1A: Consumer Products–Products with short lifespans, where tin whiskers receive little attention The minimum allowable tin whisker length needs to be determined
3	JP002: 2006 *Current Tin Whisker Theory and Mitigation Practices Guideline* [18] (Current theories explaining tin whisker formation and practical strategies for mitigation are included)	According to theory of tin whisker formation, the compressive stress responsible for tin whisker growth arises due to irregular growth of Cu6Sn5 on copper-tin plating layers, thermal mismatch at the bimetallic interface, Contaminants in the plating solution and corrosion and oxidation caused by humidity Methods to Suppress Tin Whisker Formation: Non-tin coatings, such as Sn/Pb plating, Sn/Bi plating, or Sn/Ag plating; Barrier Layers: Ni in Sn-Ni-Cu coatings or Ag in Sn–Ag–Cu coatings act as effective barriers; Thermal Treatments: Hot-dipped tin plating, annealing, or matte tin processes; Surface Treatments: Thick tin layers, protective coatings, or other surface modifications

2.2 Reliability Characteristics and Standard Requirements

At present, there are no specific standards for evaluating the reliability of integrated circuit packaging. However, it is possible to the reliability evaluation standards for components and products by using failure rate and lifetime to characterize the reliability of package. The failure rate of integrated circuit package mainly describes the probability of random failure during its operational period while lifetime refers to the wear-out life of the package. Failure caused by overstress is called random failure which corresponds to the random failure period on the bathtub curve. Failures during this period occur randomly and can be simplified as a constant failure rate to represent its reliability. On the other hand, failure caused by prolonged stress is defined o as wear-out failure which corresponds to the wear-out failure period on the bathtub curve. Failures in this stage are primarily due to material aging, fatigue and wear, and the failure rate increases over time. Therefore, wear-out lifetime is commonly used as an indicator to characterize the reliability of the package.

2.2.1 *Package Failure Rate*

Failure rate is an important parameter used to characterize the reliability of package. The failure rate at time t is defined as the probability of failure occurring within a unit time after time t and given that the product has been operating until time t. Figure 2.7 shows a typical failure rate curve, also known as the bathtub curve.

The failure rate of the bathtub curve can be divided into three stages over time: the first stage is the early failure period (Infant Mortality), during which failures are mostly caused by defects in design, raw materials, and the manufacturing process. The second stage is the random failure period (Random Failures) where the main causes of failure are quality defects, material weaknesses, environmental factors and improper usage. During this stage, the failure rate is relatively low and stable and

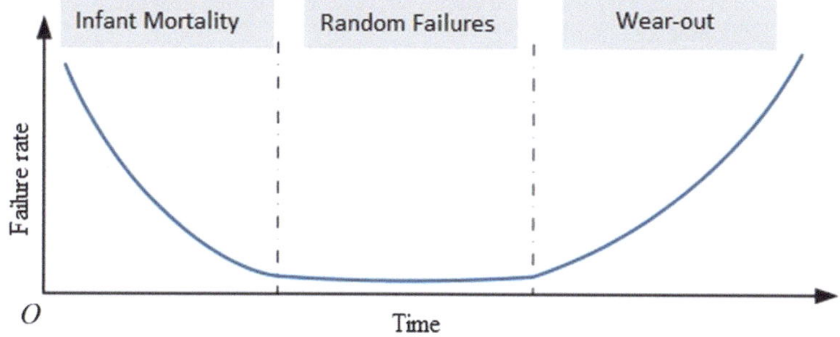

Fig. 2.7 Bathtub curve

often approximated as constant. The third stage is the wear-out failure period (Wear-out), where the main causes of failure are wear, fatigue, aging and other factors while leading to an increasing failure rate over time.

Both domestic and international reliability prediction manuals and standards typically focus on the reliability of the second stage of the bathtub curve and using the failure rate as an indicator to characterize the reliability of related package structures. For example, China's military standard GJB/Z-299C—2006 "*Reliability Prediction Manual for Electronic Equipment*" considers solder joint thermal cycling fatigue failure and uses the surface-mount failure prediction model to evaluate the reliability of the connection between surface-mount components and the circuit board. Equations (2.4) and (2.5) provide failure rates considering temperature stress and the component surface-mount soldering methods.

$$\lambda_{SMT} = A \cdot \lambda_b \tag{2.4}$$

In the formula, A represents the failure adjustment factor, and λ_b is the base failure rate, which can be further expressed as:

$$\lambda_b = \frac{\left[\frac{d}{0.65h} \times |\alpha_s \Delta T - \alpha_{CC}(\Delta T + T_{RISE})| \times 10^{-6}\right]^{2.26}}{3.5} \cdot C_R \cdot \pi_C \tag{2.5}$$

In this formula, d and h are the dimensions related to the solder joints; α_s and α_{CC} are the coefficients of thermal expansion for the circuit board and package materials; ΔT is the difference in the maximum and minimum environmental temperatures; T_{RISE} is the temperature rise caused by power consumption; C_R is the number of power cycles per hour; and π_C is the device lead structure coefficient.

The package type and complexity of integrated circuits are essential factors that determine their package failure rate. The failure rate data for integrated circuit packaging, considering package complexity, from the GJB/Z-299C—2006 standard can be found in Table 2.13.

Table 2.13 Failure rate data for integrated circuit packages considering package complexity from GJB/Z-299C—2006 standard (partial) [20]

Number of pins(piece)	Sealed(10–6/hour)			Non−sealed(10–6/ hour)		
	Dual in−line package(DIP)	Flat package with plug−in Array(FP)	Plug−in array(Includes PGA, BGA)	Dual in−line package(DIP)	Flat package with plug−in Array(FP)	Plug−in array(includes PGA, BGA)
8	0.0259	0.0224	0.0400	0.0337	0.0336	0.0527
24	0.1167	0.1670	0.1495	0.1517	0.1971	0.4146
64	0.4473	1.0040	0.4852	0.5814	0.6396	1.3454
120	–	3.1693	1.0316	–	1.3599	2.8605
280	–	14.9226	2.8517	–	3.7591	7.9070

In 2010, the FIDES reliability prediction standard, issued by the French Defense Association, more thoroughly considered the effects of temperature, humidity, and vibration stress on the failure rate of integrated circuit package. The package failure rate [21] in the FIDES standard is expressed as:

$$\lambda_{packing} = \sum_{i} \lambda_{0-stress\,i} \times \Pi_{stress\,i} \tag{2.6}$$

In this equation, $\lambda_{0\text{-stress}\,I}$ represents the base failure rate for the *i-th* type of stress, and $\Pi_{\text{stress}\,I}$ is the stress coefficient for the *i-th* type. The base failure rates for different stress types can further be represented by the related parameters of the package type as:

$$\lambda_{0-stress} = \mathrm{e}^{-a} \times \left(N_{\mathrm{p}}\right)^{b} \tag{2.7}$$

Here, a and b are constants related to the package enclosure type and the number of pins; N_{p} refers to the number of pins on the package enclosure. The typical package failure rate model parameters for integrated circuits in the FIDES standard are shown in Table 2.14.

2.2.2 Package Wear-Out Life

The IPC 9701A:2006 standard from the American Electronics Circuits Interconnection and Package Association states that the failure rate of package and assembly soldering during the random failure phase of the bathtub curve is very low and is difficult to measure through testing. In practical situations, integrated circuit package failures mostly happen in the third stage, the wear-out failure period. During this stage, the failure rate continuously increases with time, making it impossible to use a constant failure rate model to represent its reliability. As a result, accelerated testing methods are commonly used to assess the wear-out life of the package, and depending on the types of accelerated stress and failure mechanisms of the package, representative life indicators for package reliability are used.

1. High-Temperature Steady-State Life

High-temperature stress is one of the main causes of degradation and failure in integrated circuit package. High temperatures expose defects in the package structure quickly, leading to increased internal mechanical stress, bonding detachment, filling material melting, plastic encapsulation cracking, and solder interface degradation. With the development of high-density package technology, high-density 3D package structures with microbumps have emerged. The small spacing and height make solder joints more sensitive to thermal stress. High temperatures worsen the uneven diffusion rate at the interface, leading to Kendall voids, and the stress confinement effect in small-sized solder joints can cause thermal stress-induced cracks at the solder

Table 2.14 Typical failure rate model parameters for integrated circuit packages in the FIDES standard

Package type	Description	*N*p	λ0−humidity		λTemperature cycling−package case		λ0Temperature cycling−solder joints		λ0−mechanical	
			a	*b*	*a*	*b*	*a*	*b*	*a*	*b*
PDIP	Plastic dual in−line package	8 ~ 68	5.88	0.94	9.85	1.35	8.24	1.35	12.85	1.35
PQFP	Plastic quad flat package	44 ~ 240	11.16	1.76	12.41	1.46	10.80	1.46	14.71	1.46
		240 ~ 304					10.11	1.46	14.02	1.46
PBGA	Plastic ball grid array(BGA) (Ball pitch of 1.27 mm)	119 ~ 352	6.87	0.90	10.36	0.93	7.36	0.93	11.05	0.93
		352 ~ 432					7.14	0.93	10.83	0.93
		432 ~ 729					6.67	0.93	10.36	0.93

interface. High-temperature steady-state life is typically described by the Arrhenius equation.

2. Low-Cycle Thermal Fatigue Life

Thermal variation stress is one of the main causes for integrated circuit (IC) packaging failure. Thermal cycling tests apply a certain number of periodic temperature changes to mainly characterize the failure of package structures due to thermal fatigue. In IC packaging, there is an inevitable mismatch in the thermal expansion coefficients between the chip and the solder layer, and between the chip and the bonding wires. During temperature cycling, materials with different thermal expansion coefficients produce alternating stresses, which cause the materials to bend, deform, and experience creep fatigue. Cracks will form at the junction of two materials with a large difference in thermal expansion coefficients, and these cracks will gradually spread, eventually causing packaging failures such as solder delamination or bond wire detachment. The low-cycle thermal fatigue life is generally described using the Coffin-Manson equation.

3. High-Cycle Vibration Fatigue Life

Vibration stress is an important cause of IC packaging failure, with vibration-induced failures accounting for 20% of the total failure modes in IC devices. The high-cycle vibration fatigue life is characterized by the time or number of cycles it takes for fatigue failure to occur under vibration stress. Under vibration stress, the IC packaging structure experiences irreversible damage. As this damage accumulates, failures such as bond wire detachment, solder joint cracking, airtight cover cracking, and glass insulator cracking can occur. For most IC devices, the type of load they face during actual use is mainly random vibration. Therefore, random vibration tests with specific vibration frequency bandwidths and accelerations are commonly used to assess the high-cycle vibration fatigue life of IC packaging. The high-cycle vibration fatigue life can be described using the Basquin equation, and fatigue life under random vibration is typically represented by the fatigue damage accumulation based on Miner's rule.

4. Electromigration Life

Electromigration (EM) refers to the phenomenon where metal atoms in the microbumps, re-routing layers, and other metallic conductors within the package structure undergo macroscopic movement and diffusion under the influence of electronic wind, when subjected to high current density. In the interconnect structures of integrated circuit (IC) packaging, long-term electromigration can reduce the mechanical properties of these interconnects, leading to the accumulation of anode material to form bumps, while cathode material is depleted, resulting in voids, which ultimately cause package failure. Generally, the current or current density is used as the experimental condition for electromigration life testing, and microscopic changes in the structure are characterized by conduction voltage or contact resistance. In 2003, the International Technology Roadmap for Semiconductors listed electromigration as one of the key factors and core challenges limiting the development of high-density IC packaging technology. Electromigration life is typically described by the Black equation.

5. Packaging Storage Life

IC packaging can protect the chips from external environmental gases, such as moisture and salt atmosphere, which can cause corrosion. In hermetic packaging, if the package leaks and allows environmental gases to enter the IC device, it could cause bond wire corrosion, ion contamination, and ultimately lead to device failure. Ceramic, glass, and metal hermetic packaging are widely used in fields like aerospace, where the hermeticity of the package is crucial. These packages are filled with high-purity nitrogen or other inert gases, and their primary failure mode involves the formation of leaks at the cover seal under the effects of temperature and humidity, allowing external gases to enter the cavity. The plastic encapsulation process represents non-hermetic packaging, where there is no internal cavity and the chip is surrounded by polymer materials. The two main failure modes for plastic-encapsulated devices are the moisture absorption properties of the polymer material itself and the immersion of the encapsulation body at the interface with the lead frame. For packaged devices that are typically in a non-operating storage state for most of their life cycle, the storage life of the IC packaging is usually assessed through accelerated temperature and humidity tests. The accelerated life under humidity is typically described by the Lawson or Peck model.

2.2.3 Failure Rate and Lifetime Standard Requirements

The reliability of integrated circuit packaging can be represented using corresponding failure rate and lifetime parameters such as package failure rate and wear-out lifetime. The related standards and requirements are listed in Table 2.15.

Table 2.15 Standards and requirements for integrated circuit packaging reliability

No.	Standard name	Failure rate and lifetime requirements
1	GJB 7400—2011 *General Specifications for Semiconductor Integrated Circuits for Certified Manufacturing Plant*	Applies to semiconductor integrated circuits, including multi−chip integrated circuits Steady−state lifetime and temperature cycling lifetime
2	GJB/Z−299C—2006 *Electronic Equipment Reliability Prediction Handbook*	Failure rate data for different integrated circuit package structures are provided
3	FIDES Guide: 2009 *Reliability Methodology for Electronic Systems* (Electronic System Reliability Methodology)	The failure rate models and parameters for different integrated circuit package structures are given
4	JESD22−A104E: 2014 *Temperature Cycling* (Temperature Cycling)	Suitable for the thermal fatigue lifetime testing of devices and solder joint interconnects
5	JEP158: 2009 *3D Chip Stack with Through−Silicon Vias(TSVS): Identifying, Evaluating and Understanding Reliability Interactions* (3D Stacked TSV: Identification, Evaluation and Reliability Understanding)	The main failure modes and mechanisms of chip stacking with TSV are introduced Failure modes include thin silicon wafer warping, which leads to open or short circuits in TSV metal interconnects Failure driving factors include soldering temperature, humidity, thermal cycling, and thermal expansion coefficient mismatch, among others Test evaluation methods for 3D chip stacking with TSV test structures are also described
6	JEP156A: 2018 *Chip−Package Interaction Understanding, Identification, and Evaluation* (Chip−Level package Interaction Understanding, Identification and Evaluation)	It provides methods for identifying and assessing the impact of chip−package interaction on product reliability This only involves the interaction between semiconductor package stress and semiconductor chips (L1 package, excluding L2 package)
7	JESD22B113A: 2012 *Board Level Cyclic Bend Test Method for Interconnect Reliability characteristics of SMT ICs for Handheld Electronic Products* (SMT IC Interconnect Reliability Board−Level Cycling Complete Test Method for Handheld Electronic Products)	PCBA components experience various mechanical load conditions during assembly and use, such as repeated bending of the PCB, PCB wiring cracks, solder joint cracking, and SMT IC cracking leading to failure Board−level test methods are used to accelerate the evaluation of SMT IC bending damage characteristics (the maximum size limit for surface−mounted devices is 15 mm × 15 mm)

(continued)

Table 2.15 (continued)

No.	Standard name	Failure rate and lifetime requirements
8	IPC−SM−785 *Guidelines for Accelerated Reliability Testing of Surface Mount Attachments* (Accelerated Reliability Testing Guidelines for Surface−Mount Solder Connections)	It introduces fatigue modes of surface−mount solder connections, reliability prediction, solder joint reliability design, manufacturing/process, and accelerated reliability testing
9	IPC−9701A *Performance Test Methods and Qualification Requirements for Surface Mount Solder Attachments* (Performance Testing Methods and Evaluation Requirements for Surface−Mount Solder Components)	It introduces performance and reliability accelerated testing and evaluation methods for electronic components, including surface−mount solder joints, and categorizes reliability levels and their requirements

2.3 Environmental Adaptability Characteristics and Standard Requirements

2.3.1 High-Temperature Environmental Adaptability

High-temperature environments are one of the primary factors that lead to the failure of integrated circuit packaging. For plastic-encapsulated integrated circuits, high temperatures accelerate the moisture absorption of the encapsulation material. For hermetic seals, high temperatures can cause solder layer degradation, aging of organic materials, formation of holes and cracks, reduction in solder strength and compromise of hermeticity, as well as promote internal material decomposition or outgassing which affects the reliability of the hermetic seal. Additionally, for the interconnections between chips inside the package, high temperatures can accelerate electromigration and stress migration which decreases the overall reliability of integrated circuit products.

For the high-temperature environmental adaptability of integrated circuit package, high-temperature tests are commonly used to assess the impact of time and temperature on thermal failures in the package. For instance, under high-temperature stress, solder bumps experience the growth of intermetallic compounds, leading to degradation of interconnect resistance and an increase in the brittleness of the bumps which in turn affects the reliability of the soldering. The bonding layer of the chip undergoes changes in its microstructure under high-temperature stress, causing a decrease in shear strength and ultimately resulting in a reduced product reliability.

2.3.2 Temperature Cycling Environmental Adaptability

Typical methods used to characterize temperature cycling environmental adaptability include temperature cycle tests, thermal shock tests and power cycle tests. Temperature cycle tests mainly evaluate the ability of the package structure to withstand alternating high and low temperatures, with a relatively low temperature variation rate. Thermal shock tests focus on the package structure's resistance to rapid temperature fluctuations, with a higher temperature variation rate, As a rule moving from extreme low to high temperatures within 10 s. Power cycle tests assess the package structure's ability to endure cycles of high and low power. During power cycling, the heat generated by the chip severely impacts both the chip and package reliability, and these tests are commonly used to simulate the worst-case scenarios in typical applications. Power cycles are usually divided into short and long cycles. Short cycles have a more significant impact on the degradation of bonding interfaces whereas long cycles mainly affect the degradation of solder interfaces.

Because package materials have different coefficients of thermal expansion and temperature variations create stresses between different materials. When these stresses accumulate to a certain extent, they can cause different types of strain. The size of the stress and strain determines the package structure's adaptability to temperature changes. Temperature fluctuations may cause issues like lead fatigue, fractures, and delamination of the chip bonding layer. The temperature cycling environmental adaptability of the package structure is closely related to the physical properties of the package materials, such as the coefficient of thermal expansion, Young's modulus, and Poisson's ratio.

2.3.3 Mechanical Environmental Adaptability

Integrated circuit products encounter various types of mechanical stresses during production, transportation, assembly and usage, including impacts, collisions, vibrations, drops and swings. The mechanical environmental adaptability of IC packaging is evaluated by testing the reliability of the package structure under these mechanical stresses. Common tests for mechanical environmental adaptability include mechanical shock, random vibration, sweep vibration and constant acceleration tests.

Typical mechanical stress parameters include vibration frequency, acceleration and the direction in which the mechanical stress is applied. The vibration frequency selected is determined by the actual application environment. For example, the vibration frequency during automobile transport is generally low, whereas the vibration frequency in rocket launches is more varied, typically ranging from 20 to 2000 Hz. The magnitude of acceleration is used to determine the mechanical strength limits of the package, internal metallization, bonding wires, chip or substrate soldering, and other microelectronic components. Mechanical stress is typically applied along the X, Y and Z axes, with the appropriate direction chosen based on the testing needs. Defects in the structure and mechanics of the package can be identified by conducting mechanical environmental adaptability tests which allowing for reinforcement design of the package structure and the development of vibration isolation and buffering systems.

2.3.4 Environmental Adaptability Standards and Requirements

The environmental adaptability of integrated circuit package is characterized by the results of various environmental tests, including temperature shock, high and low temperature, high-temperature and high-humidity, salt atmosphere, mechanical shock and others. The relevant testing standards can be found in Table 2.16.

Table 2.16 Environmental adaptability test standards for integrated circuit packaging

No.	Standard name	Environmental adaptability requirements
1	GJB 1420B—2011 *General Specifications for Semiconductor Integrated Circuit Package Enclosure*	Thermal shock and temperature cycling tests Humidity and salt atmosphere resistance Solderability Constant acceleration
2	GJB 597B—2012 *General Specifications for Semiconductor Integrated Circuits*	Temperature cycling for inspection Mechanical shock, vibration, and constant acceleration
3	JESD22−A101D: 2015 *Steady−State Temperature−Humidity Bias Life Test* (Steady-state temperature, humidity and bias life tests)	Evaluation of the reliability of non-hermetic sealed integrated circuits in humid environments Applying temperature, humidity, and bias conditions to accelerate moisture penetration into package materials or along package interfaces
4	JESD22-A100E: 2020 *Cycled Temperature−Humidity−Bias with Surface Condensation Life Test* (Temperature-humidity-bias life tests with surface condensation)	Used for hermetic cavity-packaged devices, as an alternative to JESD22−A101or JESD22−A110 test This life test can assess the reliability of non-hermetic or hermetic cavity devices in humid environments after surface condensation, and effectively evaluate the corrosion resistance and ability to resist dendritic growth on the device's surface
5	JESD22-A110E: 2015 *Highly Accelerated Temperature and Humidity Stress Test* (Highly Accelerated Temperature and Humidity Stress Test)	Evaluate the reliability of non-hermetic devices in humid environments Same failure mechanisms as the "Double 85" steady−state humidity life test, but with more severe temperature, humidity, and bias stress, which accelerates moisture penetration through the package material or along the package interface
6	JESD22−A118B: 2015 *Accelerated Moisture Resistance−Unbiased Unbiased HAST* (Accelerated Humidity Test–Without Bias HAST)	Evaluating the Reliability of Non-Hermetically Sealed Devices in Humid Environments Apply temperature and humidity stress under non−condensing conditions to accelerate moisture penetration through package materials or along the interface between materials and metals Unbiased tests can identify failure mechanisms that might be hidden by biased tests, like galvanic corrosion The purpose of the test is to identify failure mechanisms inside the package (T:110~130°C,RH:85%,VP:122 kPa~230kPa), without condensation

(continued)

Table 2.16 (continued)

No.	Standard name	Environmental adaptability requirements
7	JESD22−A102E: 2015 *Accelerated Moisture Resistance −Unbiased Autoclave* (Unbiased Highly Accelerated Stress Test)	Evaluating the Reliability of Non-Hermetically Sealed Solid-State Devices in Humid Environments Perform moisture condensation or saturated vapor testing in a high-pressure vessel. This is not suitable for stacked or carrier-packaged devices, like FR4 materials or polyimide carriers(T:121°C,RH:100%,VP:205kPa)
8	JESD22−A104E: 2014 *Temperature Cycling* (Temperature Cycling)	Testing the Ability of Components and Interconnect Solder Joints to Withstand Thermomechanical Forces from Alternating Iso-thermal High and Low Temperature Cycles $\Delta T \leq 15$°C/min(10~14°C/min),Note: This is an isothermal temperature cycling test, meaning that at any point in the temperature cycle, the sample's temperature remains uniform
9	JESD22−A107C: 2013 *Salt Atmosphere* (Salt Atmosphere, equivalent to MIL750, MIL883)	Salt Atmosphere Corrosion Resistance Equivalent to MIL750、MIL883
10	GJB 2440A—2006 *General Specifications for Hybrid Integrated Circuit Package Enclosure*	Metal and ceramic enclosures applied to hybrid integrated circuits, solid-state relays and fiber optic devices Specifies the quality and reliability assurance requirements for enclosures during production and delivery Identification Test Environment Trials: Temperature cycling, thermal shock, high−temperature baking, humidity resistance, salt atmosphere, mechanical shock and constant acceleration Quality Consistency Test Environment Trials: Temperature cycling or thermal shock, mechanical shock, or constant acceleration
11	JESD22-A122A: 2016 *Power Cycling* (Power Cycling)	The ability of the package structure and solder joints to withstand thermomechanical stress during a non-isothermal high-to-low temperature process. This is a non-isothermal temperature cycling test, where at specific moments in the power cycle, the sample's temperature is non-uniform
12	JESD22−A105D: 2011 *Power and Temperature Cycling* (Power and Temperature Cycling)	Testing and Assessment Under Combined Power Cycling and Temperature Cycling Stress Focuses on the tolerance of flip-chip, ball grid array and chip stacking solder joint interconnects

(continued)

Table 2.16 (continued)

No.	Standard name	Environmental adaptability requirements
13	JESD22−A106B.01: 2016 *Thermal Shock* (Thermal Shock)	Tolerance of Solid−State Device package to Rapid Temperature Changes Failure Criteria: Cracking or damage to the package
14	JESD22−B110B.01: 2019 *Mechanical Shock−Device and Subassembly* (Mechanical Shock Testing of Devices and Components)	Applied to non-hermetic packaged devices, conducted in the X, Y, and Z axes directions Failure Criteria: Cracking of the device package, PCB solder joint interconnection failure, or delamination of the bottom filling material
15	JESD89−2A: 2007 *Test Method for Alpha Source Accelerated Soft Error Rate* (Mechanical Shock Testing of Devices and Components)	Used to determine the αparticle soft error rate in solid-state volatile memory arrays and bistable logic elements
16	JESD221: 2011 *Alpha Radiation Measurement in Electronic Materials* (α Particle Radiation Measurement Method for Electronic Materials)	Applies to materials used in semiconductor manufacturing The radiation rate measured is less than 0.01 cph/cm2

References

1. Murugesan M, Mori K, Nakamura A et al (2020) High aspect ratio TSV formation by using low-cost, Electroless-Ni as barrier and seed layers for 3D-LSI integration and package applications. In: Solid state devices and materials (SSDM)
2. Nimbalkar P, Liu F, Watanabe A et al (2020) Fabrication and reliability demonstration of 5 μm redistribution layer using low-stress dielectric dry film. In: 2020 IEEE 70th electronic components and technology conference (ECTC), pp 62–67
3. Fu Z, Zhou B, Yao R et al (2016) Research on thermal-electric coupling effect of the copper pillar bump in the flip chip package. In: 2016 17th international conference on electronic package technology (ICEPT), pp 1377–1380
4. Mariappan M, Koyanagi M, Fukushima T (2020) Impact of Electroless-Ni seed layer on Cu-Bottom-up electroplating in high aspect ratio (>10) TSVs for 3D-IC package applications. In: 2020 IEEE 70th electronic components and technology conference (ECTC), pp 1736–1741
5. Hwang G, Kalaiselvan R (2017) Development of TSV electroplating process for via-last technology. In: 2017 IEEE 67th electronic components and technology conference (ECTC), pp 67–72
6. Li y, Wang H, Li Y et al (2020) Simulation and thermal fatigue analysis for board level BGA connection of HTCC package. In: 2020 21st international conference on electronic package technology (ICEPT), pp 1–5
7. Xianchun W, Songliang J (1998) Integrated circuit package test manual. In: Electronic package technology series editorial committee electronics industry press
8. Jiancheng W, Xiaoqi H, Bin Z (2014) A method for calculating the junction temperature of MCM devices under convective air-cooling conditions. J Guangdong Univ Technol 31(4):6
9. China electronic technology standardization institute. Microelectronic device test methods and procedures: GJB548B-2005, 2005
10. JEDEC Moisture/reflow sensitivity classification for nonhermetic surface mount devices: JEDEC J-STD-020E-2014[S]. JEDEC solid state technology association, 2014

11. Ebara M, Yamada K, Kojima K et al (2018) Threshold dependence of soft-errors induced by α particles and heavy ions on flip flops in a 65 nm thin BOX FDSOI. In: 2018 IEEE SOI-3D-subthreshold microelectronics technology unified conference (S3S), pp 1–3
12. Saito A, Nishikawa H (2019) Tin whisker growth mechanism on tin plating of MLCCs mounted with Sn-3.5Ag-8In-0.5Bi solder in 30 °C/60%RH. In: 2019 22nd European microelectronics and package conference and exhibition (EMPC), pp 1–4
13. Zhou B, Zhang Z, Li Y et al (2020) Flexible, robust, and multifunctional electromagnetic interference shielding film with alternating cellulose nanofiber and MXene layers. ACS Appl Mater Interfaces 12(4):4895–4905
14. Lee B Z, Lee DN (1998) Spontaneous growth mechanism of tin whiskers. Acta Metall 46(10):3701–3714
15. Xiaoqi H, Yunfei E, Fangfang S (2020) Electronic micro-assembly reliability design (fundamentals). In: Electronics Industry Press
16. Yunfei E, Shaofeng X, Xiaoqi H (2015) Reliability physics. In: Electronics Industry Press
17. JEDEC. Test method for measuring whisker growth on tin and tin alloy surface finishes: JEDEC JESD22-A121A-2008. In: JEDEC soild state technology association
18. JEDEC. Current tin whisker theory and mitigation practices guideline: JP002. In: JEDEC solid state technology association
19. JEDEC. Environmental acceptance requirements for Tin whisker susceptibility of tin and tin alloy surface finishes: JESD201A. In: JEDEC solid state technology association
20. China electronic technology standardization institute. Electronic Equipment reliability prediction handbook: GJB/Z 299C-2006
21. Airbus France. Reliability methodology for electronic systems. FIDES Group, 2010

Chapter 3
Failure Modes, Mechanisms and Reliability of Plastic Packaging

Plastic packaging offers several advantages, including cost-effectiveness, compact design, lightweight properties, and suitability for large-scale industrial production. As a result, plastic packaging technology is widely used in both consumer and industrial electronic systems. The materials and processes used for chip passivation in plastic packaging have reached a high level of maturity. However, the resin materials and structure of plastic packaging still present certain risks. These risks are especially concerning in high-reliability applications. Common issues include corrosion, electrochemical migration, the "popcorn" effect, interface delamination, via holes in system-level and wafer-level packaging, and micro-bumps. This chapter discusses the characteristics, failure modes, failure mechanisms, stresses, and reliability of plastic packaging in detail.

3.1 Overview of the Reliability of Plastic Packaging

Plastic packaging has many advantages. It is small, performs well, offers a wide range of options, and is easy to obtain. It could help advance the miniaturization and lightweight design of equipment in high-reliability fields, such as military and aerospace applications. It can also shorten development cycles. However, plastic packaging has notable limitations in terms of quality and reliability. These differences are clear when it is compared to military-grade hermetic packaging. Plastic packaging is designed for use in environments that are not extreme. It is suitable for devices that are easy to maintain or replace. It was not intended for use in high-reliability fields. Due to the inherent properties of plastic packaging materials, plastic-packaged circuits show weaknesses when compared to hermetically sealed circuits. These weaknesses are apparent in three main aspects [1].

B. Zhou et al., *Reliability Technology for Integrated Circuit Packaging*,
https://doi.org/10.1007/978-981-95-3885-0_3

(1) Narrow Design Temperature Range. The temperature range for plastic-packaged circuits is limited. Commercial-grade circuits operate from 0 to 70 °C, and industrial-grade circuits operate from − 40 to 85 °C. These do not meet the military-grade requirement of − 55 to 125 °C. Additionally, they are not subjected to aging tests.
(2) Poor Moisture Resistance. Plastic packaging, made from polymer-potted plastic packaging is easily affected by moisture. This can cause parameter degradation, material swelling, electrochemical corrosion, and failures like the "popcorn" effect.
(3) Prone to Interface Delamination. Plastic packaging is sensitive to thermal and mechanical stress. Temperature changes, power cycling, or reflow soldering can cause cracks, delamination, or voids.

Due to these characteristics, plastic packaging has limitations in high-reliability applications. It cannot be used without restrictions. Otherwise, it may introduce risks not present in hermetically sealed devices. For example, plastic packaging formed by polymer potting is sensitive to moisture ingress and absorption. If not properly controlled, this can lead to corrosion or parameter drift during storage and assembly. It may also trigger the "popcorn" effect during soldering processes [2].

3.2 Failure Modes and Mechanisms of Plastic Packaging

3.2.1 Failure Modes and Mechanisms Related to Plastic Packaging Materials

The main cause of failure in plastic-encapsulated circuits is moisture ingress. Many other reliability issues also come from this problem. Plastic packaging uses polymer materials with large gaps in the structure. These gaps allow moisture to penetrate into microcracks or delamination between the chip surface and the encapsulating material. This leads to increased leakage current, unstable charges, or corrosion of aluminum metallization. Typically, there are three failure modes when packaging is exposed to humid environments: the "popcorn" effect, corrosion caused by moisture, and electrochemical migration from the combined effects of moisture and bias voltage.

The "popcorn" effect happens because plastic packaging subject to moisture ingress easily. Prolonged exposure to a humid environment reduces the bonding strength between the packaging material and the chip adhesive. Pre-baking is required before soldering to treat the encapsulating material. During the reflow soldering process, the temperature of the device rises quickly to around 240 °C. Moisture trapped in the microvoids inside the packaging vaporizes rapidly, causing a sharp increase in internal pressure. This generates thermal stress. Under the combined effects of thermal stress, moisture stress, vapor pressure, and weakened bonding strength, the packaging material may delaminate or crack. This is known as the "popcorn" effect. It can happen during high-temperature heating or rapid cooling of the device. A typical failure form of the "popcorn" effect is shown in Fig. 3.1. The "popcorn" effect creates delamination on the surface and inside the packaging. This accelerates the corrosion by water and impurity ions. Impurities such as Cl^-, K^+, and Na^+ that enter the Packaging shell can corrode the chip's metallization layer and internal bonding layers. These issues can cause circuit failure. Aluminum metallization corrosion in plastic-packaged chips is shown in Fig. 3.2.

Cracks caused by hygroscopic swelling occur through the following process. Polymer materials in the packaging absorb moisture, which causes expansion and dimensional changes. Hygroscopic swelling creates moisture stress, similar to thermal stress caused by differences in the coefficient of thermal expansion (CTE). Long-term exposure to humid environments greatly reduces interfacial adhesion and accelerates aging. Delamination then occurs at weak interfaces, leading to packaging failure. To reduce the effects of moisture on plastic-encapsulated circuits, several methods can be used. These include optimizing the process and structure of the passivation layer, improving chip packaging quality, and enhancing the water-resistance and moisture-absorption properties of encapsulating materials.

Electrochemical migration (corrosion) caused by bias voltage and moisture is a major failure mode. Its causes are explained below. If plastic packaging remains in

(a) C-SAM image of the "Popcorn" Effect

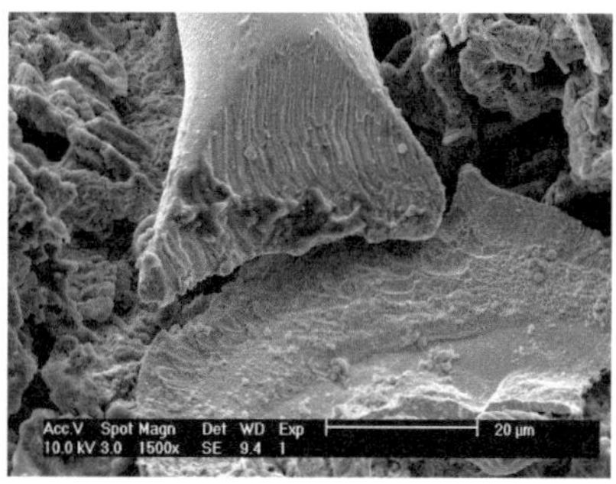

(b) Outer Bonding Point Pull-off

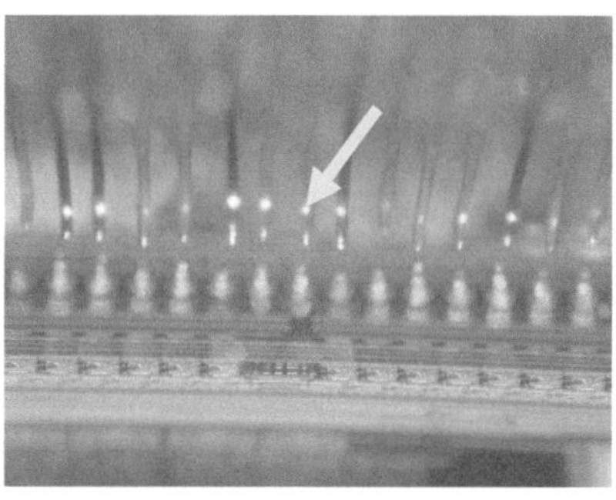

(c) Inner Bonding Point Pull-off

(d) Delamination Between the Chip and Packaging Material

Fig. 3.1 Typical failure modes of the "popcorn" effect in plastic packaging

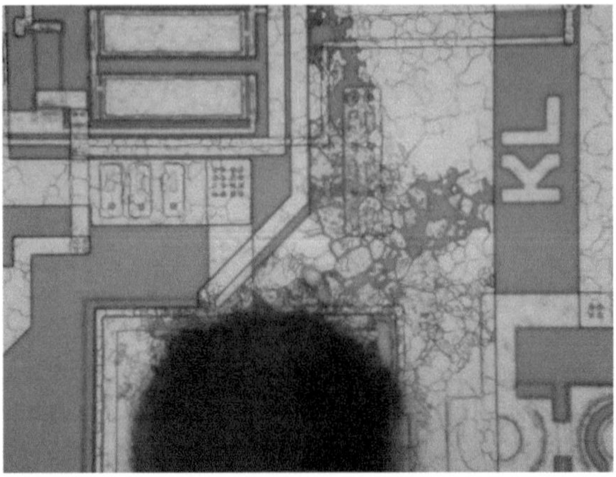

Fig. 3.2 Aluminum metallization corrosion in plastic-packaged chips

harsh environments with high humidity and temperature, moisture can penetrate the chip surface. This results in saturated vapor concentration and a liquid water film on the surface. Such conditions may induce a galvanic effect, causing corrosion. Under electrical bias, this process accelerates electrochemical corrosion.

In packaging, two types of corrosion may occur. The first type happens on the substrate's cathode surface. Here, metal whiskers, like copper whiskers, may grow. In the anode copper's electrolytic solution, metal ions are generated. When moisture creates a migration path, such as through a flame retardant absorbing moisture, these ions migrate to the cathode surface. This process can lead to whisker growth. Reduced solubility of metal ions causes crystallization and whisker formation, eventually leading to an anode–cathode short circuit. The second type of corrosion appears in ball grid array (BGA) packaging. It occurs at the interface between glass fibers and epoxy resin in the substrate or on its surface. This involves the growth of conductive anodic filaments (CAF). When moisture is present, CAF grows from the anode to the cathode along the glass fiber/epoxy resin interface, resulting in failure.

3.2.2 Failure Modes and Mechanisms Related to Packaging Interfaces

Plastic, unlike materials such as glass, ceramics, and metals used in packaging, is a non-hermetic material. Its glass transition temperature ranges from 130 to 160 °C. In civilian applications, plastic-packaged circuits must withstand three temperature ranges: 0–70 °C for commercial use, – 40 to 85 °C for industrial use, and – 40 to 125 °C for automotive use. For plastic packaging, the mismatch in the coefficients of thermal expansion (CTE) between the epoxy molding compound and contacting materials causes thermal stress. This occurs when temperature changes induce compressive or tensile stress at the interface. It is a major source of thermal stress.

Plastic packages contain various materials with different CTEs. These include silicon chips, lead frames, bonding wires, and molding compounds. During packaging, reflow soldering, or in harsh environments with large temperature variations, interface delamination often occurs. Common delamination sites include the molding compound/chip interface, the lead frame/molding compound interface, and the substrate/molding compound interface. Delamination creates safety risks. First, delamination can form cracks in the package. Moisture may penetrate the chip surface through microchannels, causing issues like leakage and corrosion of pads or metallization. Second, delamination reduces mechanical reliability. Cracks may grow under temperature changes or mechanical vibrations. This leads to bond fatigue, intermittent contact failure, or open-circuit failure.

For power devices, interface delamination is more harmful. Voids or delamination at the chip attachment interface increase thermal resistance. When the device operates, heat cannot dissipate effectively. This raises junction temperatures and can burn out the chip. Power devices generate heat during use. Thermal cycling produces periodic shear forces at interfaces due to CTE mismatches. These shear forces cause fatigue in the materials and damage bonding interfaces. This degradation leads to thermal fatigue failure. Bonding wires can break due to thermal fatigue. Pulsed operations induce alternating deformation in wires. Repeated deformation leads to cracks and eventual failure.

Low-temperature delamination and cracking are significant reliability issues in plastic packaging. Mismatched CTEs between molding compounds and lead frames can cause delamination and cracks when the temperature drops from room temperature to low levels during thermal cycling. Thermal stress-induced chip cracks in plastic packaging are shown in Fig. 3.3. The large temperature differences between storage, operating, and packaging conditions make these stresses more evident at low temperatures. Additionally, cracks reduce fracture strength and reliability.

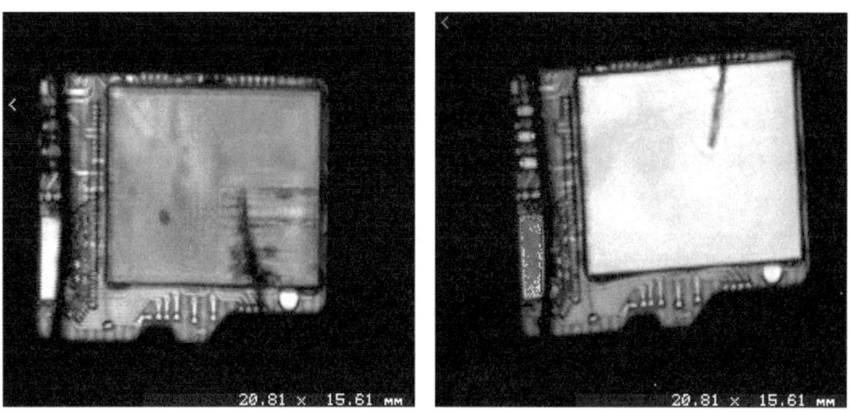

Fig. 3.3 Chip cracks caused by thermal stress in plastic packaging

From the analysis above, plastic packaging has inherent weaknesses due to its material and structural features. It is prone to interface delamination. This creates potential risks in high-reliability applications. These risks have the following characteristics:

(1) Non-convergent system failures: The failure phenomena in the system do not converge. Delamination occurs in random locations. Different locations cause open circuits in different circuit functions. These random delamination sites lead to varying system failures.
(2) Unstable failures: Failures vary with voltage and temperature changes. Their states are often unstable. Delamination can expand over time. Initial delamination may not cause immediate failure. If the device or circuit generates high internal temperatures during operation, temperature fluctuations can extend the delamination. This eventually leads to failure.
(3) Detection of delamination: Interface delamination in plastic packaging can be identified using a Scanning Acoustic Microscope (SAM). SAM effectively locates delamination sites. It also provides standards and criteria, which are suitable for engineering applications.

The main factors affecting delamination include its area, location, and distance. Delamination at chip surfaces or bonding interfaces is highly sensitive to moisture. Higher moisture sensitivity increases delamination. During use, delamination tends to expand further. Devices with higher moisture sensitivity grades show more significant delamination growth. Critical delamination sites include the chip/molding compound interface, the lead frame/molding compound interface, and the substrate/chip bonding interface. For integrated circuits, delamination at the chip or lead frame interfaces can cause parameter drift or bond failures. For power devices, the substrate/chip bonding interface is especially critical.

The layered inspection of the critical areas mainly uses SAM for analysis. SAM works by emitting ultrasonic waves for non-destructive testing. It can penetrate most solid materials and produce detailed images of internal structures. This method helps detect defects like cracks, delaminations, and voids.

3.2.3 Failure Modes and Mechanisms of Flip-Chip Packaging

Flip-chip packaging uses solder bumps to connect chips to substrates. It is suitable for high-density, multifunctional, and high-performance requirements. However, there is a significant difference in the thermal expansion coefficients between chips and substrates. This difference causes thermal stress in the solder bump area. During device operation, this stress gradually accumulates, leading to fatigue damage caused by thermal stress [3]. Over time, cracks form and expand in the packaging structure, eventually causing thermal failure of the chip [4]. The failure modes and mechanisms of flip-chip packaging include the following:

(1) Device failure caused by short circuits between solder balls during reflow soldering.
(2) Device failure caused by solder ball fractures and open circuits under moisture and high-temperature conditions. This occurs because filler materials like epoxy resin between chips and substrates absorb moisture. During reflow soldering, the moisture vaporizes and expands. This creates internal stress, separating bumps from substrates and breaking solder balls, leading to failure.
(3) Device failure caused by voids near solder bumps. This is mainly due to gases released by organic fluxes at high temperatures. These gases become trapped in the solder material, forming voids and eventually causing failure.

3.2.4 Failure Modes and Mechanisms of Bonding Degradation

Wire bonding serves as a bridge between internal chips to external pins or other chips. It ensures smooth input and output signals between chips and external circuits. Wire bonding is widely used due to its simplicity, low cost, and adaptability to various packaging forms.

There are two types of wire bonding: wedge bonding and ball bonding. Ball bonding is more efficient than wedge bonding and is the main form used in plastic packaging. Ball bonding typically uses Au, Ag, or Cu wires, while wedge bonding uses Al or Au wires. In high-power devices, such as automotive electronics, Cu wires are often used for wedge bonding.

Next, I will introduce the failure modes and mechanisms of different types of bonding.

1. Failure Modes and Mechanisms of Au-Al Bonding

(1) Formation of Au-Al Intermetallic Compounds

The failure mechanism of Au-Al bonding involves the formation of Au-Al intermetallic compounds (IMC). These compounds, over time, change into Au4Al or Al5Au2. This transformation creates voids at the bonding interface. These voids form because the diffusion rates of Au and Al atoms are different. The presence of intermetallic compounds increases bonding strength. However, the volume change of these compounds makes the bond more brittle and susceptible to mechanical stress.

The thermal decomposition products of molding plastics can chemically corrode the Au-Al intermetallic compounds. The most significant decomposition products are bromine or bromine-containing molecules from epoxy resin or flame retardants. Bromine molecules (e.g., HBr, CH3Br) corrosion reactions involving intermetallic compounds. This weakens the mechanical strength of the bond and increases the interface resistance. Furthermore, moisture and oxygen accelerate bond degradation. Oxygen causes thermal oxidation, which significantly accelerates lead degradation and failure. Wire bonding degrades more slowly in a vacuum compared to air. Plastic resins absorb moisture in humid environments. This causes the resin to expand and lowers the glass transition temperature. This allows corrosive molecules to reach the bonding area, increasing the failure rate of the plastic packaging. Prolonged exposure to high temperatures causes voids to grow. This makes the bond fragile, increases resistance, and causes circuit failure.

The phase diagram of Au-Al alloys is shown in Fig. 3.4. It shows that Au and Al can form five compounds [2]: Au5A12 (also known as Au8A13), Au4Al, Au2Al, AuA12, and AuAl. These phases are often mixed with the bonding interface, so the observed colors are usually gray, brown, and black. During heating, both Au and Al react until one of them dominates.

The evolution of intermetallic compounds is determined by diffusion. The rate of diffusion of one metal into another is controlled by the number of defects in the lattice, including vacancies, dislocations, and grain boundaries. Metals like Al can diffuse rapidly into Au through the grain boundaries of the Au layer. Figure 3.5 shows the formation and evolution of compounds in an Au-Al thin film system. The formation and evolution are observed in Au-rich, Al-rich, and regions with nearly equal Au and Al content [5]. The final compound's composition is determined by the aging temperature and initial elemental content. Increasing temperature and time will change the composition and proportion of intermediate compounds. Different

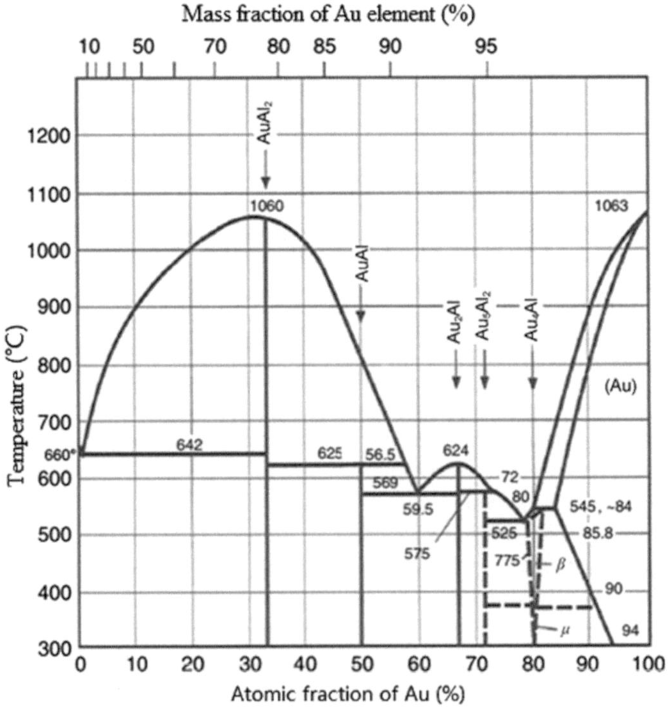

Fig. 3.4 Au-Al alloy phase diagram

compounds have different thermal expansion coefficients. When the compounds change, stress is generated, which causes cracks and eventually leads to bond failure.

(2) Typical Failure of Au-Al Intermetallic Compounds

Kirkendall voids occur when the outward diffusion rate of Au or Al in a region is different from the inward diffusion rate. This difference in diffusion rates creates vacancies. The vacancies accumulate in one area, forming voids. These voids are often found in the Au-rich phase, at the interface between Au_5Al_2 and Au. The diffusion rates vary depending on temperature and the adjacent phases. To form typical Kirkendall voids, the temperature must be raised to over 300 °C for at least 1 h, which results in their formation in the Au-rich region. At temperatures above 400 °C, voids form in the Al-rich region. Alternatively, extending the heating time

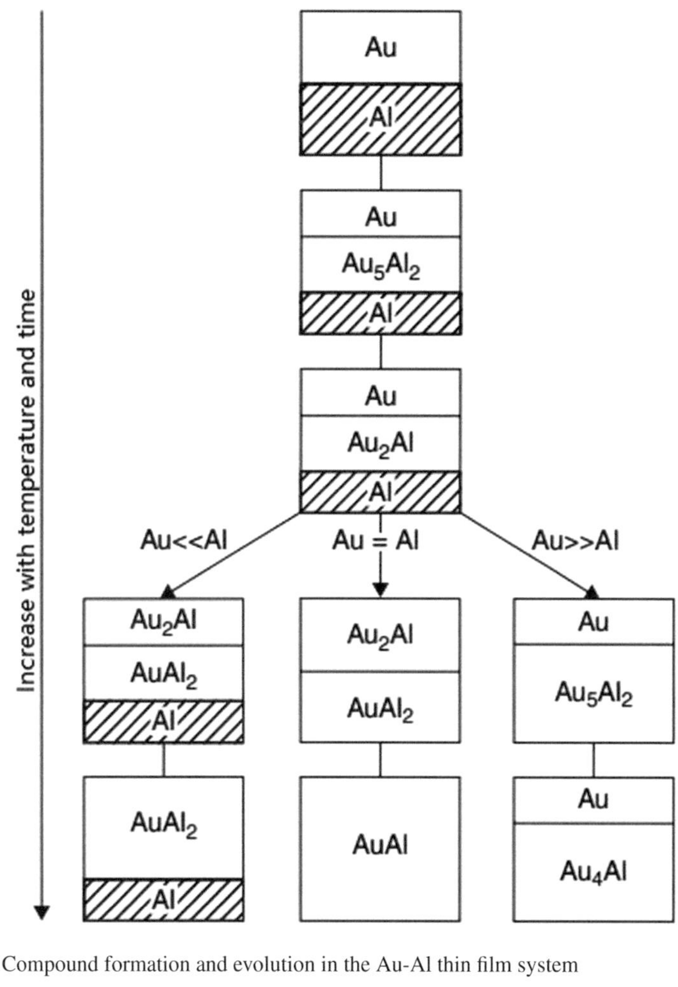

Fig. 3.5 Compound formation and evolution in the Au-Al thin film system

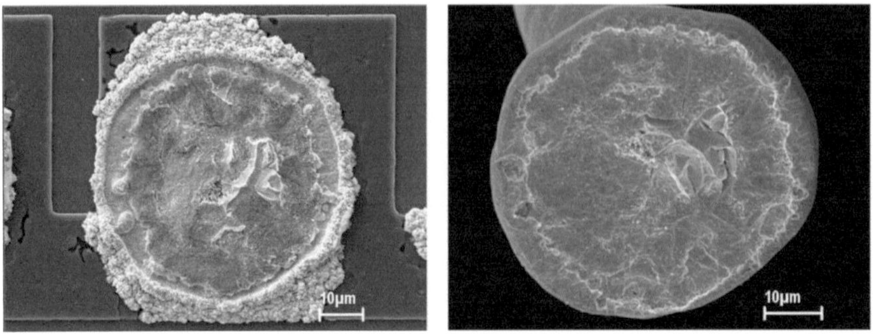

Fig. 3.6 Kirkendall voids in Au-Al bonding after 2000 h of storage at 175 °C

can lead to void formation at lower temperatures [6, 7]. Figure 3.6 shows Kirkendall voids in Au-Al bonding after 2000 h of storage at 175 °C. In addition, under prolonged high-temperature stress (> 125 °C), Kirkendall voids may form due to impurities in the Au layer, cold solder joints, hydrogen atoms, or other defects.

1. Cu–Al Bonding Failure Modes and Mechanisms

(1) Formation of Cu–Al Intermetallic Compounds

Cu–Al intermetallic compounds are much thinner than Au–Al intermetallic compounds. The thickness of Cu-Al intermetallic compounds is about 30 nm [8], while Au–Al intermetallic compounds range from 150 to 300 nm [9]. In high-temperature tests between 150 and 300 °C, the formation rate of Cu-Al intermetallic compounds is about 1/10 of that of Au–Al intermetallic compounds. Cu and Al can form several compounds, such as CuAl2, CuAl, Cu_4Al_3, Cu_3Al_2, and Cu_9Al_4.

The intermetallic compounds formed in Cu-Al wire bonding are very thin and hard to observe using a regular electron microscope. They can be studied with focused ion beam (FIB) technology and high-resolution transmission electron microscopy (TEM) to examine Cu–Al interface reactions. After wire bonding, the Cu–Al interface has two key regions: one is a uniform, dense alumina region, and the other contains Cu–Al intermetallic compounds [10]. The alumina layer acts as a barrier, stopping diffusion between Cu and Al, so intermetallic compounds cannot form in the uniform alumina region. However, intermetallic compounds can form where the alumina film cracks. These discontinuous intermetallic compounds are about 30 nm thick and identified as CuAl2 by electron beam scattering analysis. In regions with alumina film, amorphous alumina connects with crystalline copper oxide.

(2) Evolution of Cu–Al Intermetallic Compounds

The evolution of Cu–Al intermetallic compounds follows the same diffusion mechanism as the Au–Al bonding system. The process of Cu–Al intermetallic compound evolution at 175–250 °C over different times is shown in Fig. 3.7. During aging, Cu_9Al_4 is the second compound to appear after $CuAl_2$ forms. It appears between the copper ball and the intermetallic compound. Cu_9Al_4 and $CuAl_2$ grow together, but Cu_9Al_4 remains the main growing phase before the aluminum disc is fully consumed. Afterward, Cu diffusion becomes dominant, causing $CuAl_2$ to transform into Cu_9Al_4, which is continuously consumed. Therefore, Cu_9Al_4 is the final product. Some researchers have studied the kinetics of Cu-Al compounds. They found that the activation energies for Cu_9Al_4 and $CuAl_2$ are 0.79 eV and 0.63 eV, respectively [11].

(3) Effect of Bonding/Packaging Process on Cu–Al Intermetallic Compounds

The metallurgical characteristics of the Cu–Al interface are influenced by process parameters such as ultrasonic energy, bonding force, bonding time, and substrate temperature [12–14]. Bonding strength is related to the coverage area of the intermetallic compounds. The formation of intermetallic compounds depends on the degree of alumina film fracture on the aluminum disc. The extent of alumina film fracture is influenced by the ultrasonic power during bonding. Ultrasonic vibration causes the oxide films on the Cu and Al surfaces to fracture. This leads to the formation of $CuAl_2$in the areas where Cu and Al are in contact. Xu et al. [15] found that ultrasonic vibration raises the local temperature. The temperature can reach up to 465 °C, which accelerates the formation of compounds.

Bonding time also affects the fracture of the oxide film. Longer bonding times can completely break the oxide film and form continuous layered intermetallic compounds. Additionally, increasing the substrate temperature helps break the oxide film and promotes the formation of intermetallic compounds. Xu et al. [16, 17] tested the effect of pre-applied ultrasonic energy on the formation of intermetallic compounds at the interface. When pre-applied ultrasonic energy is used, intermetallic compounds form around the bonding point and at the center. Without pre-applied ultrasonic energy, intermetallic compounds only form around the bonding point.

3. Failure Modes and Mechanisms of Au–Cu Bonding

The Au-Cu interface can form three tough intermetallic compounds: Cu_3Au, AuCu, and Au_3Cu. The activation energy for these compounds ranges from 0.8 to 1 eV. Some studies have found voids resembling Kirkendall voids in Au–Cu compounds [18]. Other studies used thermo-compression bonding to create Au-Cu bond joints under air and vacuum conditions. They observed that as time and temperature increased, voids appeared at the interface, significantly reducing bond strength. The time required for bond strength to drop to 40% at different temperatures is shown in Fig. 3.8. Figure 3.8 uses a 40% drop in bond strength as the failure criterion. It predicts that a device working at 100 °C would last about 5 years. This shows that Au–Cu bonding meets the lifespan requirements for most electronic products [19]. Other studies found that when Au is thermo-compression onto thick Cu pads, bond strength remains almost unchanged after 3000 h at 150 °C.

For fully bonded Au–Cu interfaces, they can withstand 1000 h of high-temperature and high-humidity tests, 8000 weeks of temperature cycling, and 1000 h of high-temperature storage at 150 °C. These results meet the requirements for modern

Fig. 3.7 Evolution process of Cu–Al intermetallic compounds

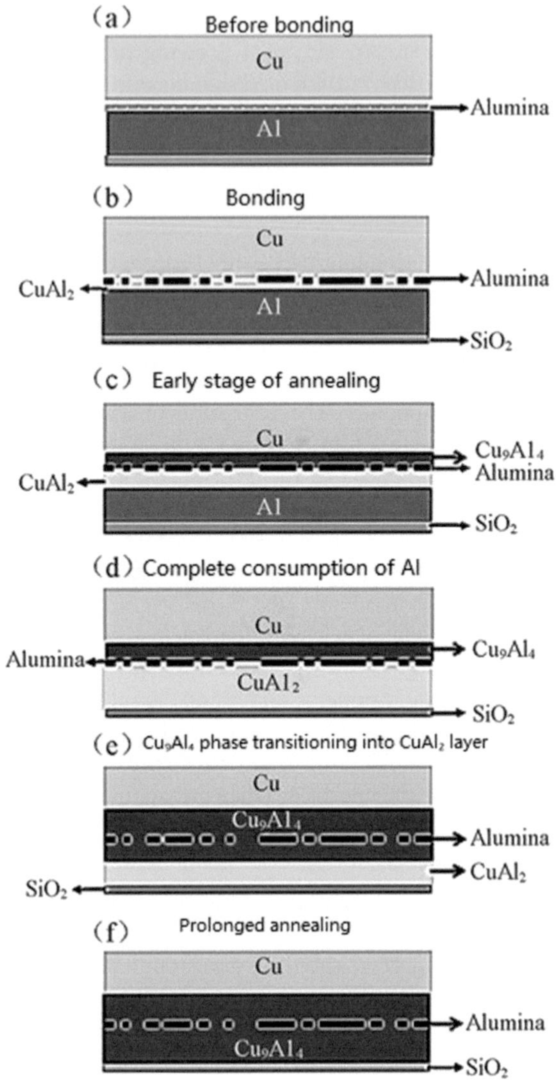

(a) Initial Alumina (Alumina) Film Before Bonding;
(b) CuAl2 Formed During Bonding; (c) Cu9Al4 Appears;
(d) Cu9Al4 and CuAl2 Grow Simultaneously;
(e) Cu9Al4 Continues to Grow by Consuming CuAl2;
(f) Cu9Al4 Becomes the Final Compound Product.

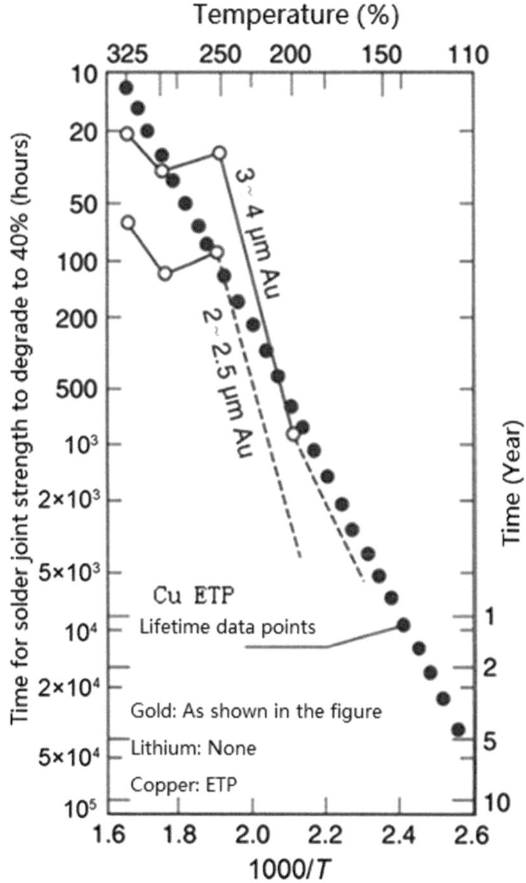

Fig. 3.8 Time for bonding strength to decrease to 40% at different temperatures (the dashed line represents the fitting line)

electronic devices [20]. The bond strength of Au–Cu is significantly affected by microstructure, welding quality, and foreign materials on the Cu pads. When welding the Cu lead frames, it is crucial to remove oils and oxides from the lead frame before bonding.

4. Failure Mechanism of PdAu–Al Bonding Interface

Palladium-plated lead frames are used in the integrated circuit field to replace silver-plated lead frames. This change enhances the adhesion to compounds in the packaging. Palladium-plated lead frames also improve bonding with the solder used in surface-mount packaging. Palladium is plated on copper lead frames, which have a 1.5 μm thick nickel film. The palladium layer is thin. When it melts into the solder, it does not form a large amount of brittle Pd–Sn intermetallic compounds. Thermo-ultrasonic technology is used for bonding on palladium-plated lead frames. This method is similar to bonding on silver-plated lead frames. However, some process parameters, such as power, pressure, and time, need to be optimized. The hardness of the Pd–Ni interface is higher than that of silver-plated lead frames. This causes the splitter's life to shorten. The reliability of this bonding system meets commercial needs. Palladium and gold can fully melt together without forming intermetallic compounds. Both palladium and gold have high electrochemical potential. This makes the bonding interface resistant to corrosion. At 400 °C, palladium slowly forms a green oxide. Therefore, palladium is safe to use as a pad when the temperature is below 300 °C. Palladium is somewhat vulnerable to corrosion from halogens and sulfur. Lead frames should avoid contact with these elements.

5. Failure Mechanism of Ag–Al Bonding Interface

Silver is used as the plating metal for lead frame pads. It is also alloyed with palladium or platinum for metallization layers in commercial applications. In Ag–Al intermetallic compounds, the initial identified phases are Ag_2Al and Ag_3Al. Ag_2Al has a hexagonal close-packed structure. It has stable thermodynamic properties and can form joints with high bonding and shear strength [21–24]. The thickness of the intermetallic compound depends on the diffusion rate of aluminum in Ag_2Al. Ag_3Al is a metastable phase. Research shows that palladium, gold, and zinc can prevent aluminum from diffusing into silver. This prevents the intermetallic compound from becoming too thick [25]. Liao et al. found that when palladium content is below 3.5%, it inhibits the growth of the intermetallic compound. However, when the palladium content exceeds 3.5%, it promotes the growth of the intermetallic compound [26].

The evolution of Ag–Al intermetallic compounds during high-accelerated stress testing (HAST) is observed and analyzed using TEM to examine the compound's morphology and structure. The process of evolution is shown in Fig. 3.9. The initial phase formed is Ag_2Al. In this stage, alumina in the Al (plated) layer is surrounded by the Ag_2Allayer. Over time, silver diffuses further into the aluminum layer, increasing the thickness of the Ag_2Al. At the same time, Ag_3Al forms. The thickness of Ag_2Al and Ag_3Al continues to grow until the aluminum layer is depleted.

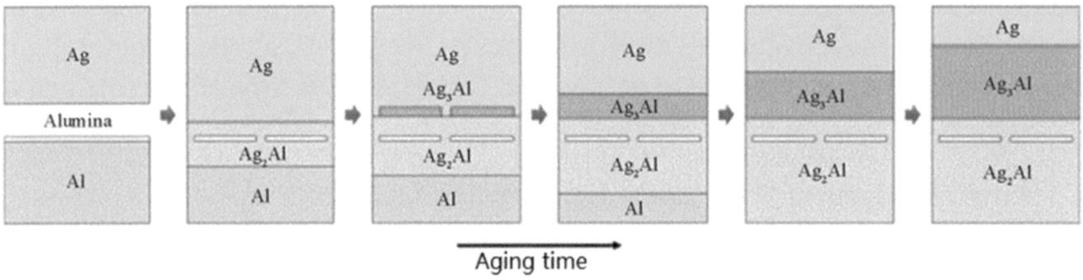

Fig. 3.9 Evolution of Ag–Al intermetallic compounds

3.3 Detection and Analysis of Plastic Packaging

3.3.1 Detection and Analysis of Molding Compounds

Table 3.1 summarizes the material composition of molding compounds (MC) used in plastic-packaged integrated circuits. From Table 3.1, it is clear that there are nine or more chemical substances in the molding compounds. Their composition varies depending on the manufacturer. The most common epoxy resin curing systems in plastic packaging are phenolic, phenol, and phenolic epoxy resin formulas. Epoxy resins are linked together by adding curing agents. More than 100 chemicals can be used to cure epoxy resins. The chemical properties of the molding compound's curing process are closely related to the type of epoxy resin used. Although the reaction between the epoxy resin and the curing agent results in thermosetting polymers, other substances must be added to achieve specific properties for special applications. Accelerators are used to achieve the ideal curing rate during production. Halogenated epoxy resins can also be used as flame retardants.

The addition of fillers in molding compound formulations aims to reduce moisture penetration, increase strength, and improve thermal conductivity and the coefficient of thermal expansion. Coupling agents are added to enhance the bond strength between the filler and the resin, while also reducing moisture penetration. Inorganic flame retardants help lower the concentration of halogenated resins, which are the source of corrosive halogen ions. Release agents are added to improve the separation process of the molding compound from the mold.

With the miniaturization and higher density of packaging, molding compounds must have lower moisture permeability, less stress, and better resistance to the "popcorn" effect [27]. To meet these demands, manufacturers have developed special formulations for different packaging types. For example, to reduce moisture absorption in surface-mount packages, resin materials are modified into mixtures

Table 3.1 Material composition of molding compounds

Component	Content (mass percentage, %)	Material
Epoxy resin	10–20	OCN, Biphenyl
Curing agent (Hardener)	5–15	Amine, Phenol, Anhydride
Accelerator (Catalyst)	< 1.0	Amine, Imidazole, Urea
Filler	60–80	SiO_2, SiN_x, AlN
Flame retardant	1–5	Brominated resin, Antimony oxide
Coupling agent	< 1.0	Silane, Titanate (Esters)
Stress reliever	0–2	Silicone, Butyl Acrylate
Component	Content (mass percentage, %)	Material
Pigment	< 1.0	Carbon black
Release agent	< 1.0	Palm wax, Silicone

of biphenyl epoxy resin and phenolic epoxy resin. To reduce stress in large-size packages, epoxy-silicone resins and polybutadiene are added to the formulation. However, this may increase moisture absorption. Synthetic rubber can be added to improve crack toughness and reduce "popcorn" cracking. Silicon nitride-filled materials coated with silicone resin can be used to increase thermal conductivity.

(1) Material Property Parameters of Molding Compounds.

Manufacturers typically provide analysis certificates for each batch of molding compounds. A typical procurement specification for molding compounds is shown in Table 3.2. Based on these specifications, molding compounds can be selected through chemical and physical property analysis. This analysis is crucial as it affects device performance and reliability. The analysis includes the glass transition temperature (T_g), coefficient of thermal expansion (CTE), filler content, moisture absorption, and ion concentration.

(2) Coefficient of Thermal Expansion.

The coefficient of thermal expansion (CTE) shows the relationship between volume changes in materials and temperature. For molding compounds, the primary factor affecting CTE is the filler content. The T_g of the material may vary based on the curing profile. Measurements indicate that the CTE remains relatively stable both above and below T_g.

The CTE below T_g is defined as CTE1 (α_1), while the CTE above Tg is defined as CTE2 (α_2). The value of α_2 is usually 2–3 times that of α_1. To reduce mechanical stress, the CTE of molding compounds should match other materials in the package. Table 3.3 lists the CTEs of commonly used materials in plastic packaging structures.

Stress in integrated circuit packaging can be estimated by a stress index. The stress index is the product of the CTE and the flexural modulus of the molding

Table 3.2 Typical procurement specifications for molding compounds

Property		Specification limits
Spiral flow		60–90 cm
Gelation time		15–25 s
Thermal hardness		≥ 70 Shore hardness
Gel content		≤ 10 mg/100 g
Coefficient of thermal expansion	CTE1 ($\alpha 1$)	≤ 25 ppm/°C
	CTE2 ($\alpha 2$)	≤ 75 ppm/°C
Glass transition temperature (T_g)		≥ 150 °C
Ash content		70–80% weight percentage
Electrical conductivity of water extract		≤ 150 μΩ/cm
Extractable halogens		≤ 20 ppm
Extractable sodium Ions		≤ 20 ppm
Flammability		UL94-V0

Table 3.3 Coefficients of thermal expansion for common materials in plastic packaging structures

Material	Coefficient of thermal expansion (ppm/°C)	Condition (°C)
Silicon	2.5–3.2	25–180
Copper	17–18	25–180
Aluminum	23–24	25–180
42 alloy	4–6	25–180
Gold	14	25–180
Silicon dioxide (fused)	0.5	25–180
Silicon nitride	1–2	25–180
Die-attach resin	40–60	25–180
Molding compound	10–20, 30–60	25–120, 120–180

compound: Stress Index = CTE (α_1) × Flexural Modulus. If the stress index is below 35, the molding compound is considered low-stress. Low-stress molding compounds often include stress-relieving materials, like silicone oil or butadiene rubber. These materials lower the flexural modulus and reduce the stress index.

Molding compounds can be tailored to achieve specific T_g and CTE values. These values depend on the curing profile and can be adjusted after the molding process. The T_g and CTE of molding compounds are critical factors influencing packaging stress. Monitoring these properties is essential for ensuring the reliability of microcircuits in plastic packaging.

(3) Fillers.

Fillers strengthen molding compounds and improve their hardness. They also reduce costs and enhance thermal conductivity. Fused quartz sand is the most common filler in plastic-encapsulated microcircuits. Fillers lower the thermal expansion coefficient of molding compounds. However, they may increase the flexural modulus, which can cause higher stress. The amount, distribution, particle size, and shape of fillers determine the stress and lifespan of plastic-encapsulated microcircuits.

(4) Moisture Absorption.

Moisture absorption in molding compounds ranges from 0.1 to 0.5% by weight. It depends on temperature, humidity, and the compound's formulation. Reducing moisture absorption is essential to prevent the "popcorn" effect during reflow soldering of surface-mounted devices. Lowering moisture absorption also reduces ion diffusion, preventing corrosion of chip pads and metal layers. Increasing filler content and post-curing can reduce moisture absorption. Biphenyl-based molding compounds absorb less moisture (0.1–0.2% by weight) with high filler levels. This makes them ideal for large-area surface-mounted devices.

(5) Ion Concentration.

Reducing impurity ion levels has improved the reliability of plastic-encapsulated circuits. Advances in resin purification and filler technology have lowered alkaline and halogen ion levels, which cause corrosion.

Molding compounds are not effective moisture barriers. Over time, they absorb moisture and ions. If cracks or delamination occur, moisture can enter through device leads. It can reach the molding compound, chip surface, and pads. Anions like Cl^- and Br^-, and cations like Na^+, K^+, and Sb^+, are impurity ions in molding compounds. These ions can diffuse and dissolve in the chip surface or metal contacts. Moisture with impurity ions accelerates corrosion of chip metals and interconnect structures.

3.3.2 Detection Methods for Interface Delamination in Packaging

Interface delamination in plastic-encapsulated circuits is a major failure mode. Ensuring reliability in these circuits requires careful detection. C-SAM delamination testing for packaging interfaces is essential. Several commonly used methods are described in relevant standards. These include the following seven approaches:

(1) IPC/JEDEC J-STD-035: Acoustic Microscopy for Nonhermetic Encapsulated Electronic Components.
(2) GJB 4027A-2006: Destructive Physical Analysis Methods for Military Electronic Components, Work Item 1103, Ultrasonic Scanning Microscopy for Plastic-Encapsulated Semiconductor Integrated Circuits, Section 2.4.
(3) MIL-STD-1580B: Destructive Physical Analysis for Electronic, Electromagnetic, and Electromechanical Parts, Requirement 16.5.1.3, Acoustic Microscopy.
(4) GJB 548B-2005: Test Methods and Procedures for Microelectronic Devices, Method 2030, Ultrasonic Inspection of Die Attach.
(5) MIL-STD-883G: Test Method Standard for Microcircuits, Method 2030, Ultrasonic Inspection of Die Attach.
(6) PEM-INST-001: Instructions for Plastic Encapsulated Microcircuit (PEM) Selection, Screening, and Qualification, Section 5.3.3, Acoustic Microscopy (C-SAM).
(7) IPC/JEDEC J-STD-020D.1: Moisture/Reflow Sensitivity Classification for Nonhermetic Solid State Surface Mount Devices.

In the above standards, IPC/JEDEC J-STD-035 provides general guidelines for using acoustic microscopy to inspect plastic-encapsulated electronic components. This standard docs not specify defect criteria. Tests following this standard should also refer to defect criteria from other related standards. Work Item 1103 in GJB

4027A-2006, "Ultrasonic Scanning Microscopy for Plastic-Encapsulated Semiconductor Integrated Circuits," corresponds to Requirement 16.5.1.3 in MIL-STD-1580B. These two standards include C-SAM testing as an essential part of destructive physical analysis for plastic-encapsulated circuits. Method 2030 in GJB 548B-2005, "Ultrasonic Inspection of Die Attach," corresponds to Method 2030 in MIL-STD-883G. Both standards provide specific guidelines for ultrasonic inspection of die-attach interfaces. PEM-INST-001 is a regulation from NASA for selecting, screening, and qualifying plastic-encapsulated microcircuits. Method 5.3.3 in this regulation specifies the use of C-SAM to inspect or screen these microcircuits.

C-SAM technology focuses on material inconsistencies, such as voids, cracks, and inclusions. It also examines interface conditions between different materials. In electronic packaging reliability testing, C-SAM detects two types of defects: cracks and voids within the molding compound, and delamination at interfaces. Most related standards, except GJB 548B-2005 and MIL-STD-883G, include criteria for detecting cracks and voids in molding compounds. The defect criteria for these standards are shown in Table 3.4.

C-SAM technology is mainly used to detect seven types of delamination defects, as shown in Table 3.5. The "delamination" at the chip attach interface is typically described as attach voids. The detection method used is the same as for other interfaces, so it is summarized here.

The interface delamination criteria in each standard are listed in Table 3.6. Since the criteria in IPC/JEDEC J-STD-020D.1 differ from those in other standards, they are not categorized by interface. Instead, they are listed separately (see Table 3.7).

Based on the C-SAM inspection method and acceptance criteria for interface delamination in Tables 3.6 and 3.7, Table 3.8 provides typical defects and illustrations of interface delamination.

There are two main preventive control methods for interface delamination in plastic encapsulated circuits. First, strengthen C-SAM screening before assembly

Table 3.4 Standard criteria for cracks and voids in molding compounds

Standard name	Failure criteria
GJB 4027A—2006 MIL-STD-1580B PEM-INST-001	1. Cracks on the plastic Packaging bonding leads 2. Internal cracks extending from the leads to any other internal component (leads, chip, die–attach substrate), with a length exceeding 1/2 of the corresponding pitch 3. Any cracks on the Packaging that cause surface rupture 4. Any voids in the molding compound crossing the bonding leads
IPC/JEDEC J-STD-020D.1	1. External cracks visible under a 40x optical microscope 2. Internal cracks crossing the bonding leads, ball bonding, or wedge bonding 3.. Internal cracks extending from the lead frame to any other internal component (lead frame, chip, die–attach substrate) 4. Internal cracks extending from any internal component toward the exterior of the Packaging, with a length exceeding 2/3 of the corresponding pitch

Table 3.5 C-SAM main detection interfaces and layer descriptions

	Scanning on circuit surface (chip active side facing up)	Scanning on non-circuit surface (chip active side facing down)
Type I Layer: Mold compound and chip surface	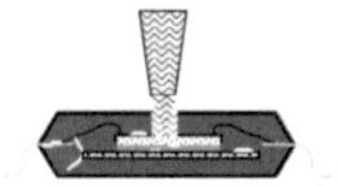	—
Type II Layer: Chip bonding area	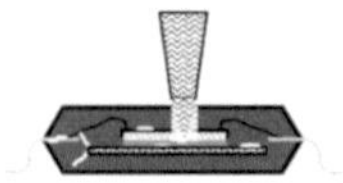	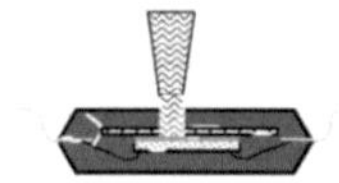
Type III Layer: Mold compound and substrate edge (active side)	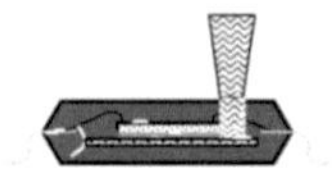	—
Type IV Layer: Mold compound and substrate (non-active side)	—	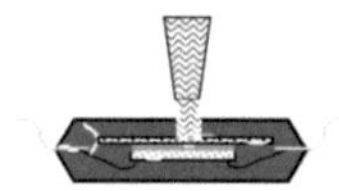
Type V Layer: Mold compound and lead frame	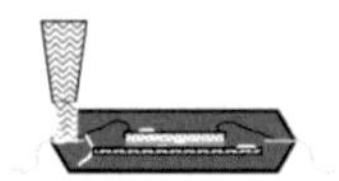	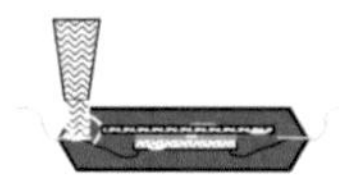
Type VI Layer: Internal multilayer PCB (only for PCB substrate samples)	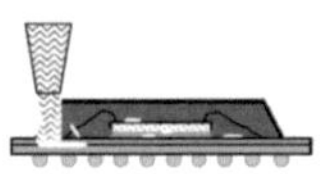	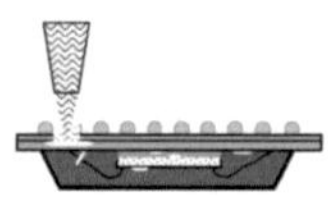
Type VII Layer: Heat sink and substrate	—	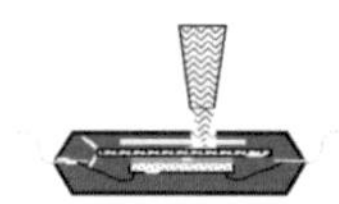

Table 3.6 Criteria for layer delamination in various standards

Interface	GJB 4027A—2006/ MIL-STD-1580B	PEM-INST-001
Mold compound/chip surface	Any measurable delamination between the plastic package and chip	Any measurable delamination between the mold compound and chip surface is a rejectable defect
Mold compound/substrate edge (active side)	—	1. Delamination area on the interface between the substrate and mold compound exceeds half of the rear or upper side edge area. (a reliability−related defect) 2. Any measurable delamination on the interface between the mold compound and wire bonding (on lead frame or substrate) is a rejectable defect
Mold compound/substrate (non-active side)	Delamination area on the interface between the lead terminal pad and plastic package exceeds half of the rear region's area	
Mold compound/lead frame (active side)	1. Delamination of pins in areas with bonding wires1.Delamination of pins in areas with bonding wires 2. Pins completely detached from the plastic package (upper or rear side)	Any measurable delamination on the interface between the mold compound and wire bonding (on lead frame or substrate) is a rejectable defect
Mold compound/lead frame (non-active side)	Pins completely detached from the plastic package (upper or rear side)	—
Interface	GJB 548B—2005/MIL-STD-883G	
Chip bonding area	1. The total void area in the contact zone exceeds 50% of the expected total contact area 2. A single void exceeds 15% of the expected contact area, or a single corner void exceeds 10% of the total expected contact area 3. If the image is divided into four equal quadrants by bisecting two opposite sides, the void area in any quadrant exceeds 70% of the expected contact area in that quadrant	

to prevent the use of products with interface delamination defects. Second, perform baking pre-treatment before assembly. This will help prevent the "popcorn" effect and micro-open bonding circuits during reflow soldering. The main measures are as follows:

(1) Plastic encapsulated circuits should undergo destructive physical analysis (DPA) batch inspection before assembly. Additionally, 100% C-SAM inspection should be conducted.

Table 3.7 Criteria for layer delamination in IPC/JEDEC J-STD-020D.1

Metal lead frame package type	Multilayer board substrate package type (e.g., BGA, LGA)
1. No delamination on the chip active surface 2. No delamination on the wire bonding surface (on lead frame or chip) 3. The change in delamination area of the polymer film layer bridging insulated metal parts must not exceed 10% (validated by TEM) 4. For thermally enhanced packages or devices needing electrical connection on the chip backside, delamination or cracks in the chip bonding interface must not exceed 50% 5. Surface cracks must not extend across the entire length. Surface crack components include the lead frame, tie bars, and heat sink	1. No delamination on the chip active surface 2. No delamination on the wire bonding surface of the multilayer board 3. For encapsulated packages, delamination changes at the encapsulant/multilayer board interface must not exceed 10% 4. Delamination changes at the solder mask/multilayer board resin interface must not exceed 10% 5. Delamination changes inside the multilayer board must not exceed 10% 6. Changes in delamination or cracks in the chip bonding area must not exceed 10% 7. No delamination or cracks at the filler resin/chip interface, filler resin/substrate interface, or filler resin/solder mask interface 8. Delamination caused by surface cracks must not extend across the entire length. Surface crack components include the lead frame, multilayer board, metallized layers of the multilayer board, through–holes, and heat sink

(2) Plastic encapsulated circuits should undergo baking pre-treatment before assembly to remove moisture inside the chip. For circuits with a moisture sensitivity level lower than 2 (according to IPC/JEDEC J-STD-003), a conservative approach should be taken. Treat these circuits as level 2a–5a devices. For devices with a thickness of 2.5 mm or less, bake at 125 °C for 24 h. For devices with a thickness between 2.5 mm and 4.5 mm, bake at 125 °C for 48 h.

3.3.3 Packaging Interface Thermal Resistance and Chip Infrared Thermography Testing Methods

The thermal resistance of the packaging interface and the temperature distribution of the chip are critical for plastic power devices. This section briefly introduces the electrical method for thermal resistance testing of packaging interfaces and the infrared thermography microanalysis technique for chip surface temperature distribution. The detailed principles and methods are described in Chap. 6.

The electrical method is the only non-destructive technique that directly measures packaged devices. However, it cannot provide the peak temperature and temperature

Table 3.8 Typical defects and illustrations of interface delamination

Defects	Illustrations	Defects	Illustrations
Delamination between the plastic package and the chip interface (top view)	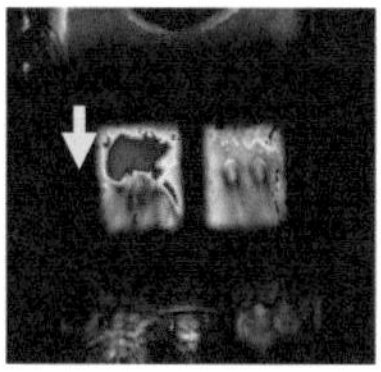	Voids in the chip bonding interface, with any single void area exceeding 15% of the expected contact area (rear view)	
Delamination between the plastic package and the substrate edge (top view)		Delamination between the plastic package and the substrate interface, where the delaminated area exceeds 50% of the substrate area (rear view)	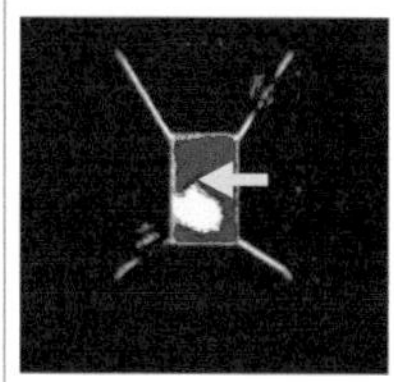
Delamination at the plastic package and lead frame interface, including the wire bonding area (top view)	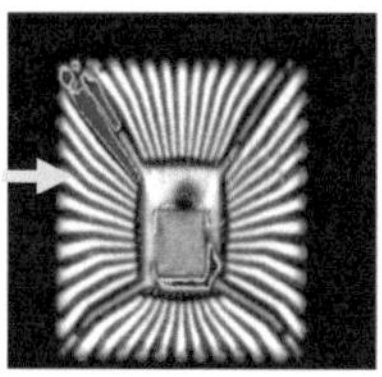	Delamination at the plastic package and lead frame interface, excluding the wire bonding area, where the length exceeds half of the lead frame (top view)	
Delamination at the plastic package and lead frame interface, with the lead frame fully detached from the plastic package (rear view)		Delamination at the plastic package and lead frame interface, where the length is less than half of the lead frame (rear view)	
Delamination at the plastic package and PCB interface, covering about 50% of the interface area (BGA package, top view)	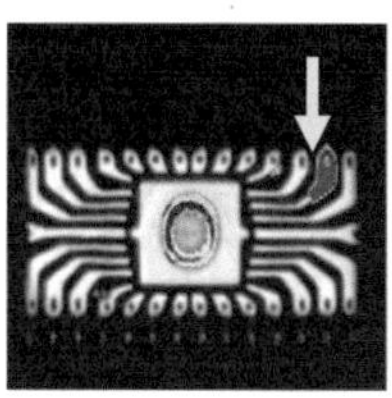	Cracks in the molding compound extending from the substrate to the lead frame (B−scan)	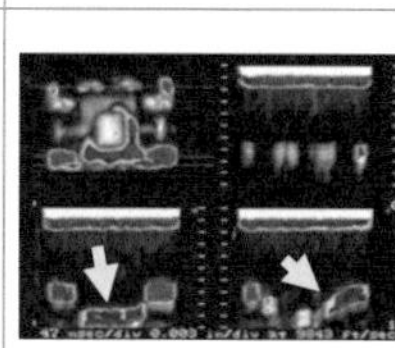

(continued)

Table 3.8 (continued)

Defects	Illustrations	Defects	Illustrations
Adhesive voids at the substrate and heat sink interface, with the total void area less than 10% of the expected contact area		Voids present within the molding compound	

distribution map. It only gives the average temperature of the chip junction. The non-destructive real-time static electrical method can measure the thermal properties of almost all semiconductor devices.

(1) Various semiconductor discrete devices.
(2) Various complex integrated circuits, multi-chip modules (MCM), system-in-package (SiP), and system-on-chip (SoC) structures.
(3) Thermal property testing of complex heat dissipation modules (e.g., heat pipes, fans, etc.).

Infrared thermal imaging camera can detect chip hotspots, verify heat distribution or design, studying heat dissipation, and locate thermal failures. A transient temperature tester can quickly perform infrared thermography testing. When the device is subjected to a pulse or transient signal, the tester can measure the temperature–time variation at a specific point.

3.3.4 Packaging Micro-deformation Detection Technology

For detecting packaging micro-deformation, international standards such as JEDEC (JESD22-B112A) and JEITA (ED-7306) are followed. JEDEC (JESD22-B112A) mentions four methods: shadow moiré, 3D DIC, laser reflection, and reflection moiré. JEITA (ED-7306) mentions two methods: shadow moiré and laser reflection. The shadow moiré method is widely recognized and commonly used in the industry for detecting packaging micro-deformation.

1. Laser Reflection Method

The laser reflection method is a non-contact measurement technique with high accuracy. The principle is as follows: a laser beam shines on each point of the sample's surface. By analyzing the angle between the reflected light and the reference point, the surface morphology of the sample is reconstructed. However, this method has a drawback. It requires large-area point-by-point scanning to get the complete surface

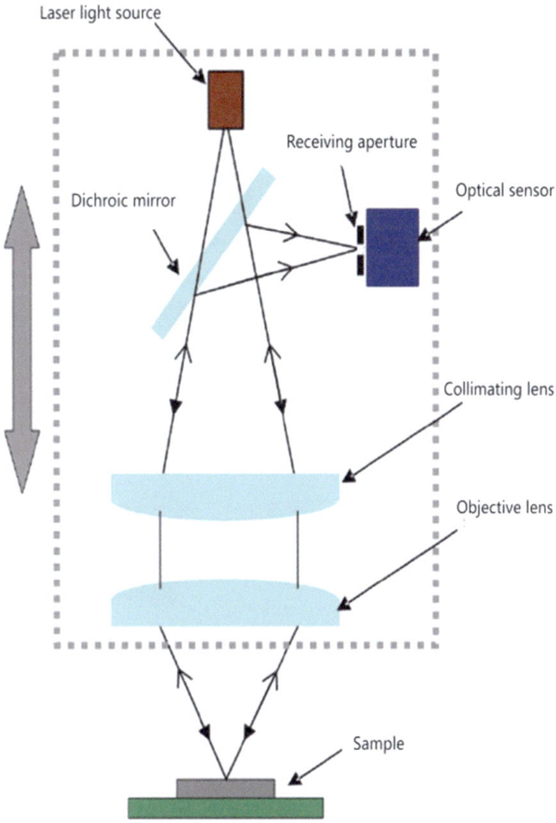

Fig. 3.10 Laser reflection method for micro-deformation detection

image, making real-time measurement difficult. The principle of the laser reflection method for detecting micro-deformation is shown in Fig. 3.10.

2. Shadow Moire Method

The Shadow Moire Method creates moiré fringes through the interference of two gratings with alternating light and dark patterns. The first set of gratings is the sample grating (shadow grating), which can be attached to the sample surface using printing, sticking, or etching methods. The second set of gratings is the reference grating, placed between the light source and the sample. When the sample deforms, the reference grating also deforms. This causes overlapping interference with the reference grating, generating moiré fringes. A computer can analyze the fringe image in real-time, reducing human error and making data processing easier and more accurate.

(1) Principle of the Shadow Moire Method for Detection

Incident light (white light) strikes the sample at a specific angle. The shadow area of the reference grating is projected onto the lower surface of the sample. Moiré fringes are produced by the interference between the reference grating and the shadow grating. The detection principle of the Shadow Moiré Method is shown in Fig. 3.11, and the projection results of the reference grating and shadow grating are shown in Fig. 3.12.

Each moire fringe contains information about different heights. Therefore, the complete moiré pattern provides height information for the entire surface of the sample. This is similar to an elevation map, as shown in Fig. 3.13.

To convert the full-field shadow moire interference pattern into a 3D surface morphology map of the package body, the fringe constant must be accurately calibrated. The fringe constant is determined by the out-of-plane deformation. The calibration formula for the fringe constant is as follows:

$$W = \frac{Np}{\tan\alpha + \tan\beta} \tag{3.1}$$

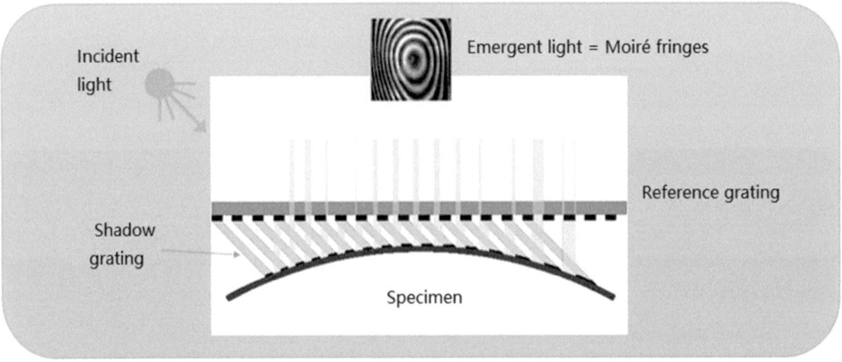

Fig. 3.11 Principle of shadow moiré method for detection

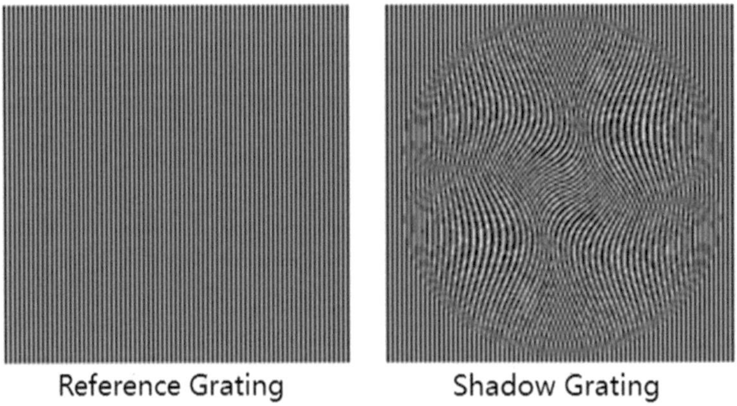

Fig. 3.12 Projection results of reference grating and shadow grating

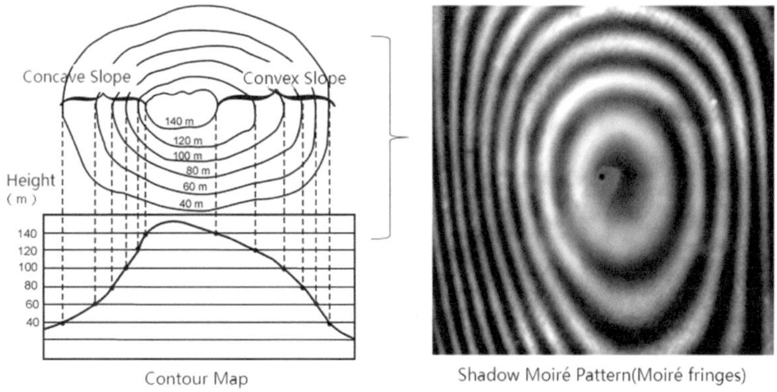

Fig. 3.13 Meaning of moire fringes

In this formula, N represents the fringe order; p represents the grating pitch; α represents the incident angle; β represents the observation angle; W represents the deformation or warping amount. Using phase-shifting techniques, the full-field interference fringes are converted into a 3D surface morphology of the package body, as shown in Fig. 3.14.

In actual measurements, the computer precisely controls the stepper motor to adjust the distance between the sample stage and the grating. This results in a series of discrete phase-shifted interference fringes. Typically, four sets of interference

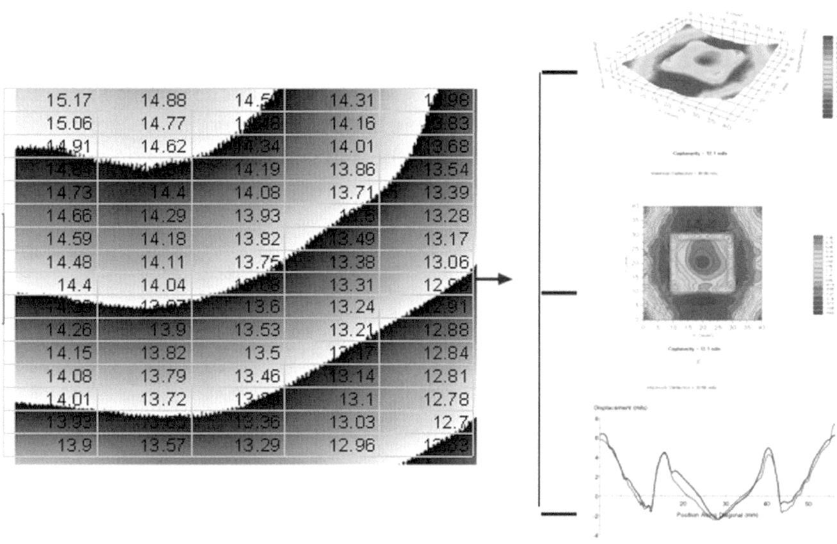

Fig. 3.14 Phase-shifting technique for reconstructing 3D surface morphology of the package body

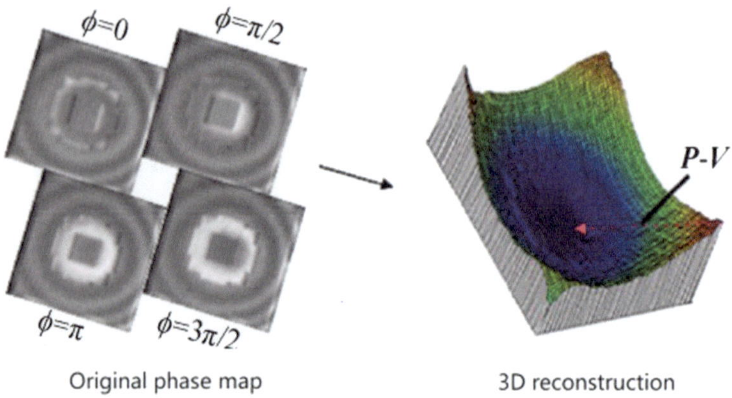

Fig. 3.15 Phase-shifted interference fringes and reconstructed 3D surface morphology

fringes are required to recreate the 3D surface morphology of the sample. The phase-shifted interference fringes and the reconstructed 3D surface morphology are shown in Fig. 3.15.

During testing, adjusting the distance between the grating and the sample generates interference fringes. According to formula (3.1), finer grating pitch enables higher measurement accuracy. However, the physical distance between the grating and the sample must also be correspondingly reduced. It is important to avoid the sample surface contacting the reference grating during testing.

(2) Components of Shadow Moiré Method Detection Equipment

In shadow moiré testing, both the Ronchi grating and the sample are placed in a thermally insulated heating box. A heat source is set beneath the sample to simulate the reflow soldering process. The equipment for shadow moiré micro-deformation detection is shown in Fig. 3.16. It consists of the following main parts (some of which are not shown in Fig. 3.16):

a. Camera for capturing shadow moire fringes.
b. Ronchi grating made of low thermal expansion coefficient glass.
c. Light source for projecting white light.
d. Z-direction stepper sample stage.
e. Computer-controlled imaging system.
f. Sample holder for positioning and aligning the sample.
g. NIST traceable calibration block with stepper height changes.
h. Standard curved glass.
i. Heating unit.
j. Thermocouple.

(3) Shadow Moiré Method Detection Process

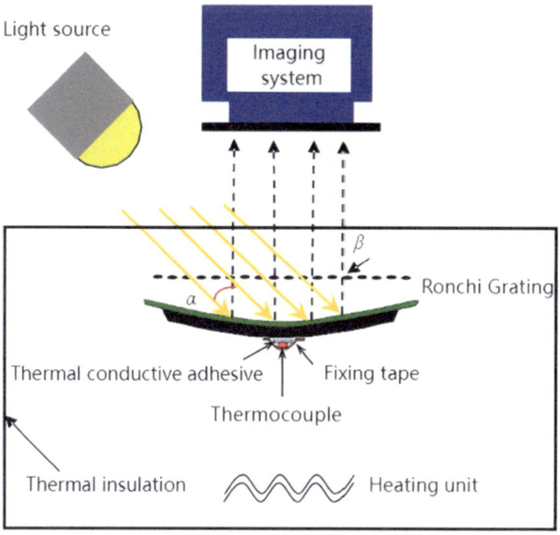

Fig. 3.16 Shadow moiré method micro-deformation detection equipment

The process for BGA package micro-deformation detection using the shadow moiré method is shown in Fig. 3.17. It can be summarized as follows:

a. Pre-processing: This includes removing solder balls from the BGA substrate surface, spray painting, and baking the sample to remove moisture.

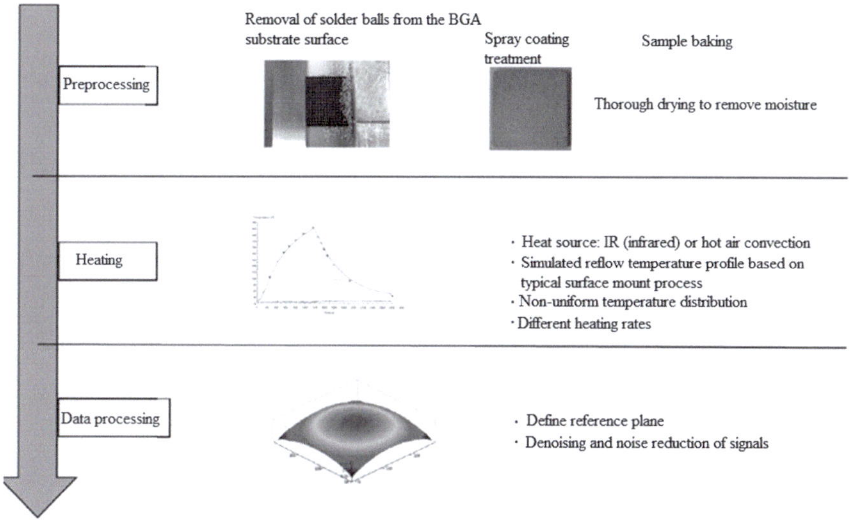

Fig. 3.17 Shadow moiré method BGA package micro-deformation detection process

b. Heating: A typical surface-mount reflow temperature curve is set. The shadow moiré method detection equipment is used to measure the micro-deformation data of the entire substrate surface.
c. Data Processing: Software processes the BGA micro-deformation measurement data. This includes defining the reference plane, removing noise from the signals, and obtaining the 3D reconstructed surface morphology of the BGA substrate.

(4) Shadow Moiré Method Detection Results

The results of shadow moiré method detection should include the following parameters:

a. The warpage of the package at different temperatures. The warpage can be represented in two ways: 3D surface morphology map and diagonal scanning chart. The warpage of the package at a certain temperature is shown in Fig. 3.18.

b. The relationship between warpage and temperature during the heating and cooling process. The vertical axis represents warpage in mils (1 mil = 25.4 μm), while the horizontal axis shows the temperature during the entire heating and cooling process, as shown in Fig. 3.19.

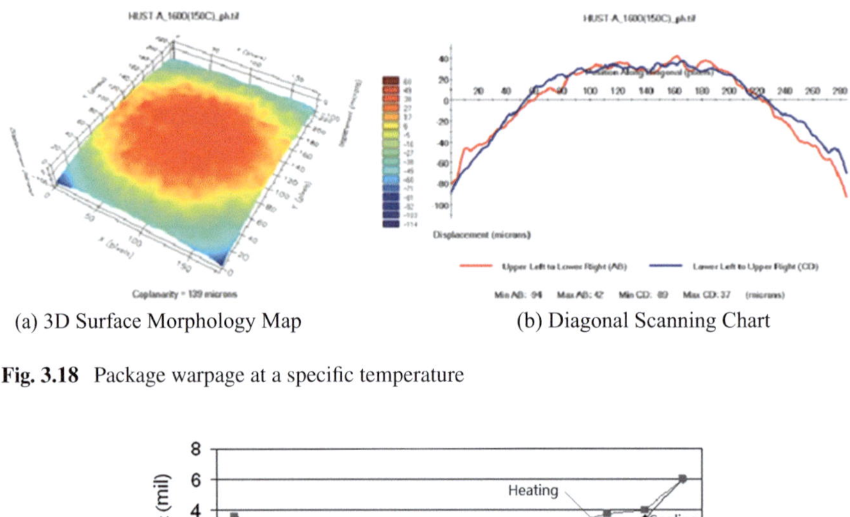

(a) 3D Surface Morphology Map (b) Diagonal Scanning Chart

Fig. 3.18 Package warpage at a specific temperature

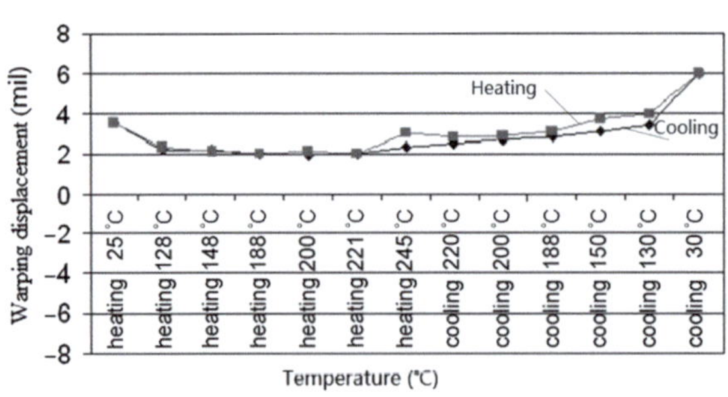

Fig. 3.19 Relationship between warpage and temperature during heating and cooling process

3.4 Stress and Reliability

Plastic-encapsulated circuits face stresses from high temperatures, temperature changes, and electromigration during assembly and operation. This section analyzes the impact of these stresses on the reliability of plastic-encapsulated circuits. It focuses on the effects of high temperature, wet-heat-mechanical stress, and thermal cycling stress on high-density plastic-encapsulated devices.

3.4.1 Wet-Heat-Mechanical Reliability of Plastic Packaging

1. Integrated Stress Analysis of Wet-Heat-Mechanical Reliability

(1) Wet Diffusion Model

Research on plastic packaging reliability has evolved. Initially, studies focused on the effects of heat on the packaging. Later, the focus shifted to the effects of moisture. Research then considered the effects of steam pressure caused by changes in moisture. Finally, the combined effects of moisture, heat, and steam pressure were studied [27–29].

For wet-heat-mechanical reliability stress analysis of plastic packaging, a finite element analysis method integrating moisture, heat, and steam pressure can be used [30]. The strain equation is:

$$\varepsilon_{\mathrm{tot}} = \varepsilon_{\mathrm{sw}} + \varepsilon_{\mathrm{th}} + \varepsilon_{\mathrm{v}} = \beta \times \Delta C + \alpha \times \Delta\theta + P_{\mathrm{v}}(\theta) \times (1 - 2v)/E(\theta) \tag{3.2}$$

In the equation, $\varepsilon_{\mathrm{tot}}$, $\varepsilon_{\mathrm{sw}}$, $\varepsilon_{\mathrm{th}}$ and ε_{v} represent total strain, swelling strain, thermal strain, and vapor pressure strain, respectively. β is the moisture expansion coefficient; C represents the humidity; α is the thermal expansion coefficient, and θ is the temperature; P_{v} is the vapor pressure; E refers to the elastic modulus of the material, while v is the Poisson's ratio of the material.

First, it is assumed that the internal vapor pressure of a multi-microporous material, after absorbing moisture, always reaches the maximum saturated vapor pressure $P_{\mathrm{vg}}(\theta)$ at the current temperature θ (pure saturated vapor pressure). Using its relationship with humidity $C(t)$, the corresponding vapor pressure can be calculated based on Eq. (3.3).

$$P_{\mathrm{v}}(\theta) = \omega\ (t) \times P_{\mathrm{vg}}(\theta) \tag{3.3}$$

In the equation, $\omega(t) = C(t)/C_{\mathrm{sat}}$, denotes the relative saturation humidity, and t is time. After determining the distribution of heat and moisture, a user-defined subroutine extracts humidity C from the moisture model and temperature θ from the thermal model. Using Eqs. (3.2) and (3.3), integrated stress analysis is performed.

For the coupled linear moisture-thermal stress model [31], it is assumed that temperature and moisture are uniformly distributed within the package. In this case, moisture stress can be treated as an additional stress to thermal stress. Thus, the equivalent thermal expansion coefficient α^* is defined as:

$$\alpha * = \alpha + \beta C/\Delta T \tag{3.4}$$

When considering the effect of thermal aging on the viscoelastic behavior of the filler material, the simple superposition of moisture stress and thermal stress is no longer valid.

During the reflow soldering process of plastic packaging, there are five types of integrated stress models: moisture diffusion model, thermal diffusion model, moisture-mechanical model, thermal–mechanical model, and vapor pressure model. The same finite element geometry model is used for all models, but with different boundary conditions, initial conditions, loading conditions, and solution methods.

The transient moisture diffusion equation is similar to the transient thermal diffusion equation and can be expressed using Fick's law.

$$\frac{\partial C}{\partial t} = D\left(\frac{\partial^2 C}{\partial x^2} + \frac{\partial^2 C}{\partial y^2} + \frac{\partial^2 C}{\partial z^2}\right) \tag{3.5}$$

However, unlike temperature, moisture concentration is discontinuous at the interface between two materials. Therefore, w is used as the field variable for humidity, defined as:

$$w = \frac{C}{C_{sat}}, \quad 0 \leq w \leq 1 \tag{3.6}$$

Thus, Eq. (3.5) can be rewritten as

$$\frac{\partial w}{\partial t} = D\left(\frac{\partial^2 w}{\partial x^2} + \frac{\partial^2 w}{\partial y^2} + \frac{\partial^2 w}{\partial z^2}\right) \tag{3.7}$$

For the moisture diffusion model, the initial condition is w=0, and the boundary condition is w=1.

(2) Thermal Diffusion Model.

Similar to moisture distribution, the temperature distribution within the packaging during reflow soldering can be obtained by solving the following equation:

$$\frac{\partial T}{\partial t} = \alpha_T\left(\frac{\partial^2 T}{\partial x^2} + \frac{\partial^2 T}{\partial y^2} + \frac{\partial^2 T}{\partial z^2}\right) \tag{3.8}$$

In the formula, T represents temperature, and α_T is the thermal diffusion coefficient.

(3) Moisture-Mechanical Model.

The mismatch of moisture expansion coefficients (CME) between different materials introduces moisture-mechanical or moisture-expansion stress. This is similar to the thermal–mechanical stress caused by a mismatch of thermal expansion coefficients (CTE). The moisture-mechanical problem can be solved using a method similar to solving the thermal–mechanical problem.

(4) Thermal–Mechanical Model.

A linear elastic thermal–mechanical model is used. The loading temperature ranges from the molding plastic curing temperature of 175 °C to the reflow soldering temperature of 260 °C.

(5) Vapor Pressure Model.

Representative volume elements (RVE) estimate the vapor pressure inside the packaging. A volume element is selected from the moisture-absorbing material. At the microscopic scale, the size of the volume element is sufficient to represent the material's statistical properties. Therefore, a field variable, the void volume fraction f, is defined as:

$$f = \frac{\mathrm{d}V_f}{\mathrm{d}V}, \quad 0 \leq f \leq 1 \tag{3.9}$$

When $f = 1$, delamination occurs at this location. The void volume fraction is a field variable. Its rate of change varies at different locations. If the interface connection is weak, the change is faster at the interface than inside the material. The initial voids are distributed randomly and uniformly, so the initial void volume fraction f_0 is a material characteristic.

The moisture density inside the pores can be defined as:

$$\rho_\mathrm{m} = \frac{\mathrm{d}W_\mathrm{m}}{\mathrm{d}V_f} = \frac{\mathrm{d}W_\mathrm{m}/\mathrm{d}W}{\mathrm{d}V_f/\mathrm{d}V} = \frac{C}{f_0} \tag{3.10}$$

In the formula, dW is the moisture mass in the volume element; dV is the volume element, and C is the local moisture concentration. The transition temperature T_1 is defined as the temperature at which all moisture in the voids turns into vapor:

$$\rho_\mathrm{m}(x_i, T_0) = p_\mathrm{g}(T_1) \tag{3.11}$$

In the equation, $p_\mathrm{g}(T_1)$ refers to the saturated vapor pressure at T_1, and T_0 represents the pretreatment temperature for moisture absorption.

When calculating vapor pressure, there are three distinct cases. In both Case 1 and Case 2, it is assumed that the moisture follows the ideal gas law.

Case 1: The moisture concentration inside the void is low. At the pretreatment temperature T_0, all the moisture converts into vapor.

$$p = \frac{Cp_g(T_0)T}{fp_g(T_0)T_0} \quad T_0 \geq T_1 \tag{3.12}$$

In this equation, p stands for the vapor pressure, and p_g represents the saturated vapor pressure.

Case 2: At the pretreatment temperature T0, the moisture exists in a gas/liquid mixed state. At the reflow soldering peak temperature T, all the moisture transforms into vapor.

$$p = \frac{p_g(T_1)T}{T_1} \quad T_0 \leq T_1 \leq T \tag{3.13}$$

Case 3: Even at the reflow soldering peak temperature T, the moisture does not completely convert into vapor.

$$p = p_g(T) \quad T \leq T_1 \tag{3.14}$$

The initial void volume fraction is needed for calculating vapor pressure. According to Eq. (3.10), when the local moisture concentration C=C_{sat}, the initial void volume fraction can be expressed as:

$$f_0 = \frac{C_{sat}}{\rho_m} \tag{3.15}$$

The stress caused by linear vapor pressure can be estimated as:

$$\varepsilon_p = \frac{1-2v}{E}p \tag{3.16}$$

In the equation, v stands for Poisson's ratio, E is the elastic modulus, and p represents the average vapor pressure. At reflow soldering temperatures, the elastic modulus E drops significantly. It decreases by several orders of magnitude. As a result, vapor pressure stress can become as important as thermal or moisture stress. It is important to note that this expansion is linked to the vapor pressure distribution, not the moisture distribution.

(6) Integrated Stress Model.

Figure 3.20 shows the packaging stress model during the reflow soldering process. The moisture distribution result from the wet diffusion model is used as input for both the steam pressure model and the wet-mechanical model. The temperature distribution from the thermal diffusion model is applied to the steam pressure model and the thermal–mechanical model. The stresses and strains introduced by the steam pressure model, thermal–mechanical model, and wet-mechanical model are combined into one integrated stress model. This model calculates the packaging stress and

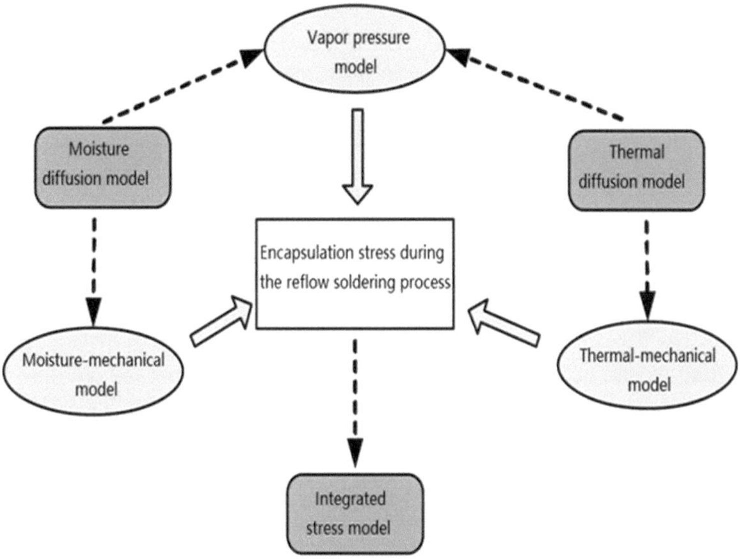

Fig. 3.20 Stress model during the reflow soldering process

strain during reflow soldering. The wet stress and the stress from steam pressure are converted into equivalent thermal stress.

2. Wet-Heat-Mechanical Stress Simulation Analysis of QFN Packaging

(1) Simulation of Moisture Absorption and Desorption Characteristics of QFN Packaging.

QFN (Quad Flat No-lead) packaging is a typical chip-level package, as shown in Fig. 3.21. Compared with other lead-frame packages, QFN packaging has many advantages. These include low cost, small size, better thermal and electrical performance, and high yield. This packaging provides excellent thermal performance because the chip frame and the chip substrate at the bottom are exposed. This design improves heat dissipation. Therefore, QFN is widely used in mobile communication and consumer products. However, despite these advantages, some wet-heat-mechanical reliability issues have been identified during testing and evaluation.

Using the wet-heat-mechanical stress analysis method for plastic packaging, the finite element model of a specific QFN package was meshed. The moisture diffusion of the package under Moisture Sensitivity Level 1 (MSL-1, 85 °C/85%RH/168 h) conditions was simulated. The transient humidity distribution of the QFN package is shown in Fig. 3.22. From the moisture diffusion process in the QFN package, it is clear that: In the initial stage, the relative humidity at the chip bonding layer stays low. This is because both the chip and copper plate are non-hygroscopic, effectively blocking moisture entry. After 168 h of moisture absorption, the humidity of

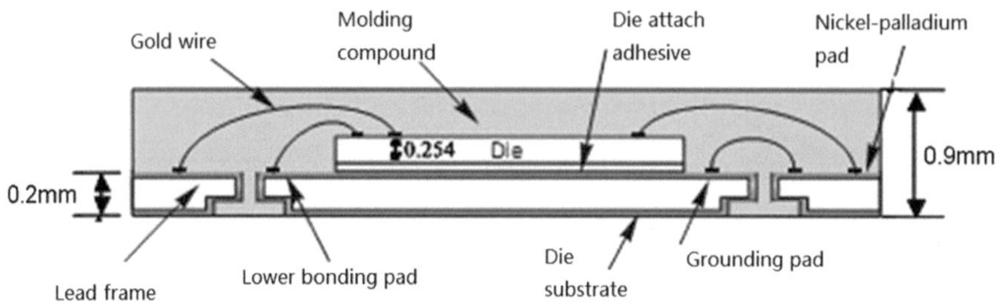

Fig. 3.21 QFN package structure

the molding plastic outside the package reaches saturation, but the bonding layer's humidity remains unsaturated. As shown in Fig. 3.22, there is always a humidity gradient at the package interface until it reaches saturation. Uneven humidity gradients are undesirable. Along with moisture stress, they can lead to stress concentration. When stress reaches a certain level, delamination or chip cracking may occur. Moreover, the non-uniform humidity gradient at the interface can degrade the interface strength. This degradation may occur even when humidity reaches saturation. The combined effect of interface strength degradation and stress increases the likelihood of delamination. Delamination creates a pathway for moisture to enter. When moisture reaches the pins and chip surface, it may cause pin corrosion and damage to the chip passivation layer. In mild cases, this leads to parameter drift; in severe cases, it can cause complete device failure [32].

To analyze moisture diffusion in different parts of the package, the transient humidity distribution at different node positions was plotted. The node selection is shown in Fig. 3.23a, and the transient humidity distribution for each node is shown in Fig. 3.23b. From Fig. 3.23b, it can be seen that the time to reach saturation humidity is strongly related to the position within the package. In the initial stage, the relative humidity of the bonding layer node A is nearly 0, and after 50 h, it begins to absorb moisture slowly. However, after 50 h of moisture absorption, node F reaches near saturation humidity.

Similarly, the moisture desorption analysis for the QFN package was conducted under 125 °C conditions. The results are shown in Fig. 3.24a. It can be observed that the moisture desorption process is much faster than the absorption process. This is because the moisture diffusion coefficient at 125 °C is significantly higher than at 85 °C. The moisture desorption process for each node in the model shown in Fig. 3.23a is illustrated in Fig. 3.24b. The differences in moisture desorption characteristics between nodes are smaller compared to moisture diffusion.

(2) Moisture-Heat-Mechanical Stress Analysis of QFN Package.

Finite element software was used to analyze the stress distribution in the QFN package after an 85 °C/85%RH moisture pre-treatment. The moisture-mechanical stress, thermal–mechanical stress, and moisture-heat-mechanical coupled stress distributions are shown in Figs. 3.25, 3.26 and 3.27 (units are Pa), with the reflow soldering

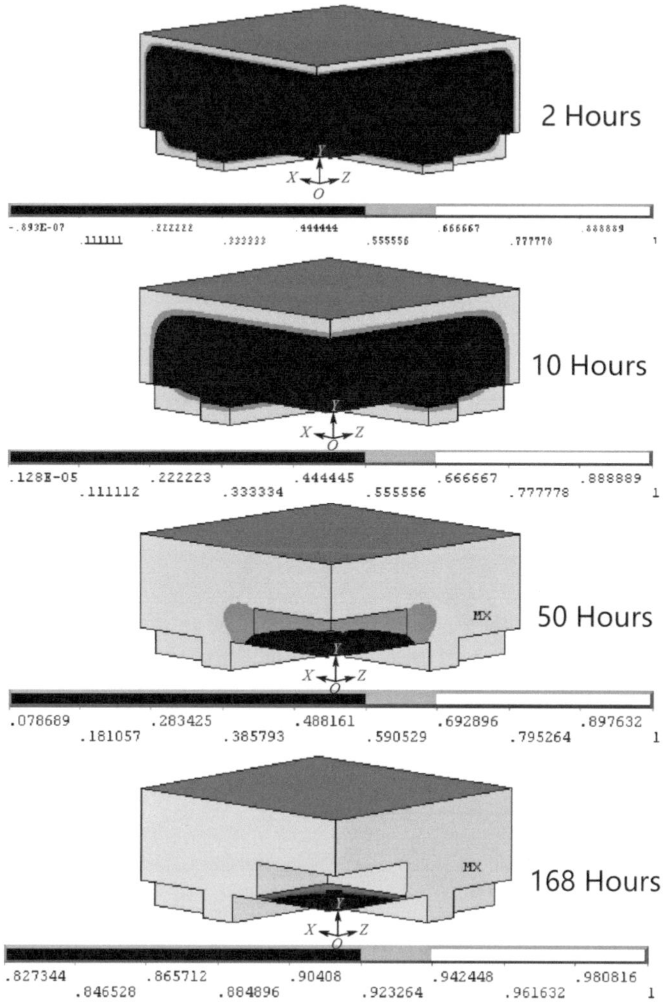

Fig. 3.22 Transient humidity distribution of QFN package

peak temperature at 260 °C. From Fig. 3.25, it can be seen that the maximum equivalent stress occurs at the interface between the chip, adhesive material, and molding plastic. From Fig. 3.26, the maximum thermal–mechanical stress during reflow soldering is 311 MPa. The stress caused by moisture expansion significantly affects the mechanical reliability of the QFN package. If moisture-heat-mechanical stress is ignored during reflow soldering, the impact of reflow soldering stress will be underestimated. From Fig. 3.27, it is evident that the weak point of the QFN package's reliability is at the interface between the chip, adhesive material, and molding plastic. If the stress reaches a certain level and exceeds the adhesive strength of the material interface, cracks will form at this interface.

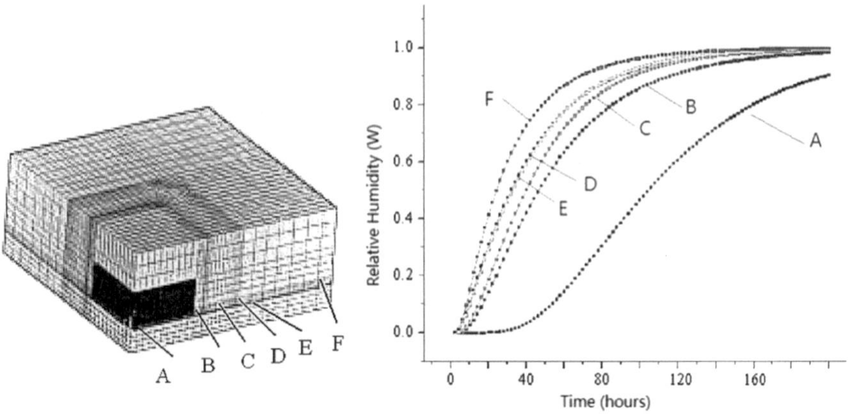

(a) Node Selection for the Model (b) Transient Humidity Distribution at Each Node

Fig. 3.23 Different nodes and their transient humidity distribution

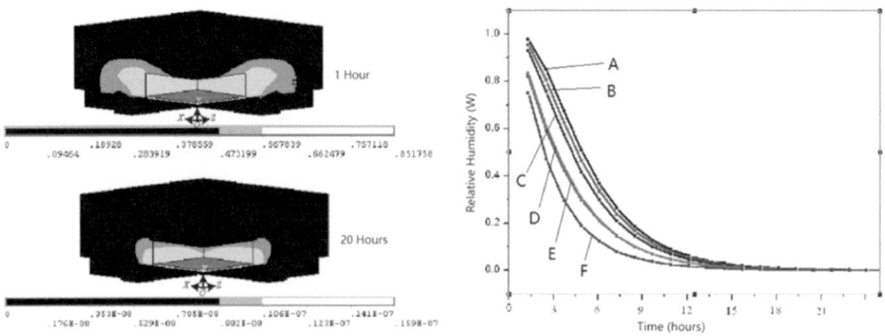

(a) Moisture Desorption Results (b) Moisture Desorption Process for Each Node

Fig. 3.24 Moisture desorption analysis of QFN package

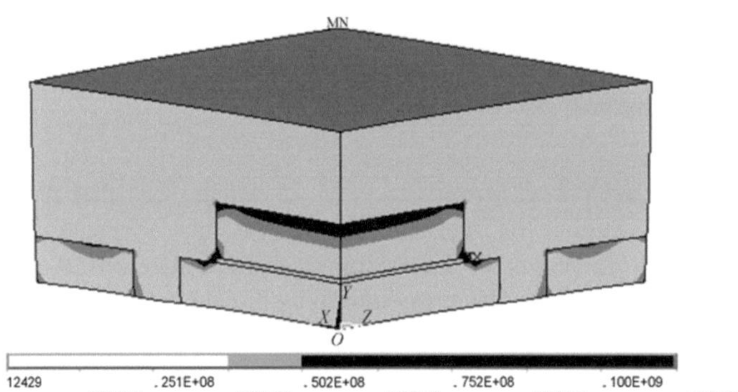

Fig. 3.25 Moisture-mechanical stress distribution in QFN package

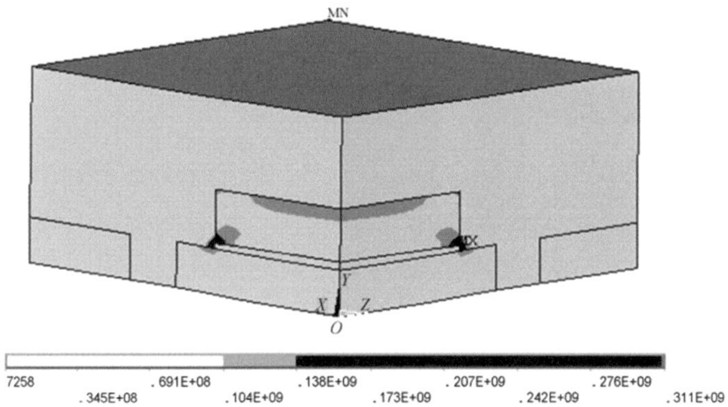

Fig. 3.26 Thermal–mechanical stress distribution in QFN package (peak temperature 260 °C)

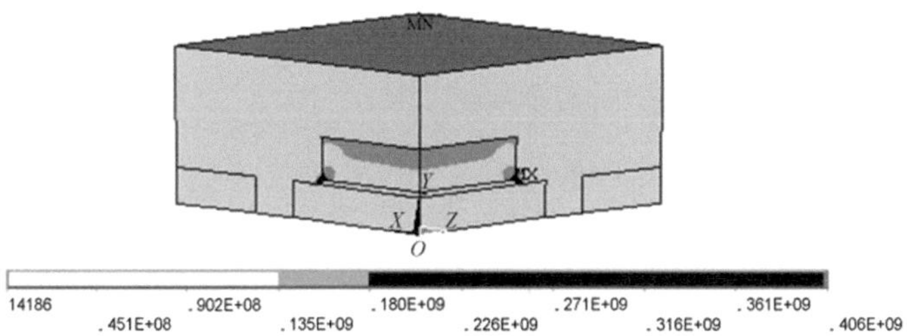

Fig. 3.27 Moisture-heat-mechanical coupled stress distribution in QFN package

(3) Experimental Verification of Moisture-Heat-Mechanical Stress Analysis Results for QFN Package.

A moisture-heat-mechanical stress test was conducted on a QFN package. After 85 °C/85%RH/168 h of moisture pre-treatment, it was inspected using C-SAM and cross-sectioned to ensure no interfacial delamination. The initial interface morphology is shown in Fig. 3.28a. After moisture absorption and three reflow soldering cycles, all devices showed varying degrees of delamination. The C-SAM results are shown in Fig. 3.28b. Delamination occurred at the chip/adhesive material and molding plastic/copper substrate interfaces. Cracks and delamination mainly appeared near the adhesive material close to the chip. Delamination also occurred at the adhesive material and copper substrate interface. This suggests that the adhesive material near the chip and the adhesive material itself experienced considerable stress. Based on the crack damage and propagation trends, the maximum stress occurred at the junction of the chip, adhesive material, and EMC material. This shows that the adhesive material is the weak link in the entire package. This result is consistent with the previous simulation findings.

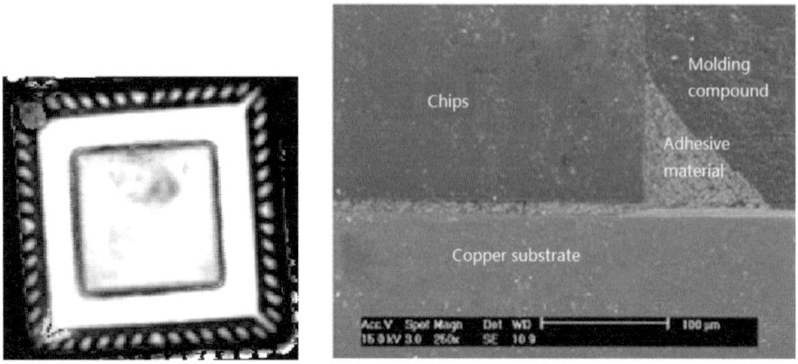

(a) Initial Interface Morphology of QFN (8mm×8mm) Before Testing

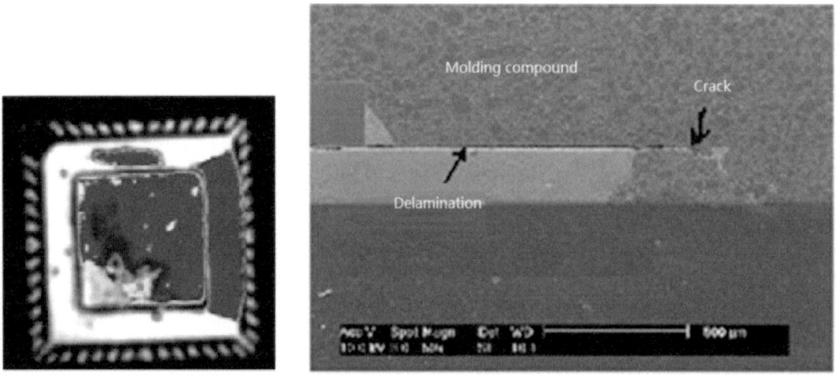

（b）C-SAM Results After Testing QFN (8mm × 8mm)..

Fig. 3.28 Delamination distribution after three reflow soldering cycles for moisture-absorbed QFN package

3. Simulation Analysis of Moisture-Heat-Mechanical Stress in PoP Package

(1) Moisture-Heat-Mechanical Reliability Modeling for PoP Package.

PoP (Package on Package) is a widely used high-density packaging technology. In this section, the moisture-heat-mechanical stress of the PoP package structure is analyzed through simulation. The PoP package structure is shown in Fig. 3.29. It involves stacking two fully tested package blocks together, providing more flexibility for the product.

Non-hygroscopic materials are not included in the analysis of moisture absorption and desorption. Under conditions of 85 °C and 85% RH, the moisture absorption characteristics of molding compound (MC), bonding agent (BA), and substrate material (BT) are listed in Table 3.9. In this table, D represents the moisture diffusion coefficient, C_{sat} is the saturated moisture content, and β is the moisture expansion coefficient.

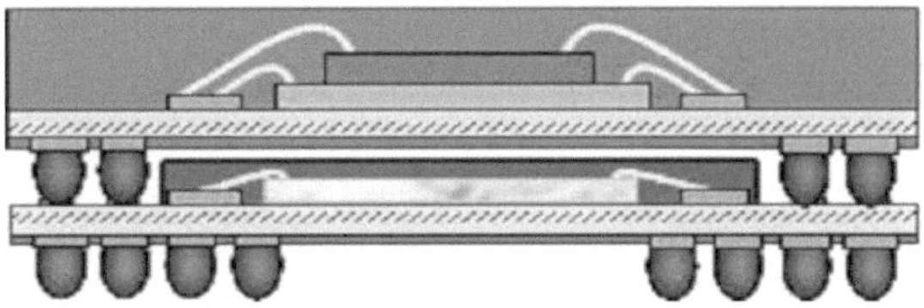

Fig. 3.29 PoP package structure

Table 3.9 Polymer material moisture absorption parameters

Material	D (mm^2/s)	Csat (mg/mm^3)	β (mm^3/mg)
MC	7.55×10^{-7}	2.07×10^{-3}	0.45
BT	1.51×10^{-7}	6.54×10^{-3}	0.69
BA	9.02×10^{-7}	2.32×10^{-3}	0.50

The moisture diffusion coefficient D determines the baking conditions for electronic packaging. The relationship between D and temperature can be expressed by the Arrhenius equation:

$$D = D_0 \exp\left(-\frac{E_\mathrm{d}}{kT}\right) \tag{3.17}$$

Here, D0 is the initial value, Ed is the activation energy, k is the Boltzmann constant, and T is the absolute temperature. In the moisture desorption model, the initial condition is w=1, and the boundary condition is w=0. The moisture desorption characteristics for polymer materials are shown in Table 3.10 [31, 33].

In the package, moisture-induced strain occurs due to differences in the moisture expansion coefficients of different materials. For non-hygroscopic components like chips and solder balls, the moisture expansion coefficient is zero. The relationship between moisture strain and the moisture expansion coefficient is expressed as:

$$\varepsilon = \beta C \tag{3.18}$$

Here, β represents the moisture expansion coefficient, and C represents the moisture content. Equation (3.18) can be rewritten as:

$$\varepsilon = \beta C_\mathrm{sat} \times w \tag{3.19}$$

Table 3.10 Moisture desorption characteristics of polymer materials

Material	E_d (eV)	$D0$ (mm^2/s)
MC	0.31	0.20
BT	0.45	6.0
BA	0.28	1.63

The moisture-thermal–mechanical stress distribution in the package during lead-free reflow soldering is simulated. At temperature T, the thermal strain can be calculated as:

$$\varepsilon = \alpha(T - T_{\text{cure}}) \tag{3.20}$$

In this equation, ε_t is the thermal strain, α is the thermal expansion coefficient, T is the heating temperature, and T_{cure} is the curing temperature.

(2) Moisture-Thermal-Mechanical Reliability Simulation

Based on the JEDEC/JSTD-020D moisture sensitivity level (MSL) classification, the PoP package is simulated under MSL-1 conditions. It absorbs moisture for 168 h. The transient moisture distribution in the PoP package after absorption is shown in Fig. 3.30.

From Fig. 3.30, it can be seen that after 168 h of moisture absorption, the package humidity has almost reached saturation. The only exception is the bonding layer between the two chips of the top-level package. Chips do not absorb moisture. Therefore, moisture enters the bonding layer between the chips at a slower rate. This causes a larger humidity gradient in the bonding layer. The humidity gradient can induce moisture strain, which has been confirmed by simulation results. The results also show that the 168-h absorption time under MSL-1 conditions for PoP packaging is generally sufficient [32].

Moisture desorption at 125 °C was simulated. The moisture desorption status at different times is shown in Fig. 3.31a. From Fig. 3.31a, it is clear that after 20 h of

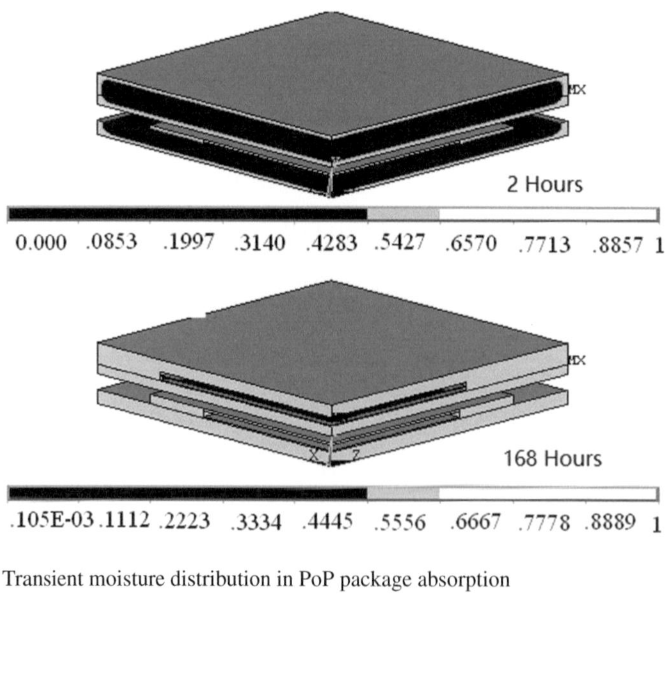

Fig. 3.30 Transient moisture distribution in PoP package absorption

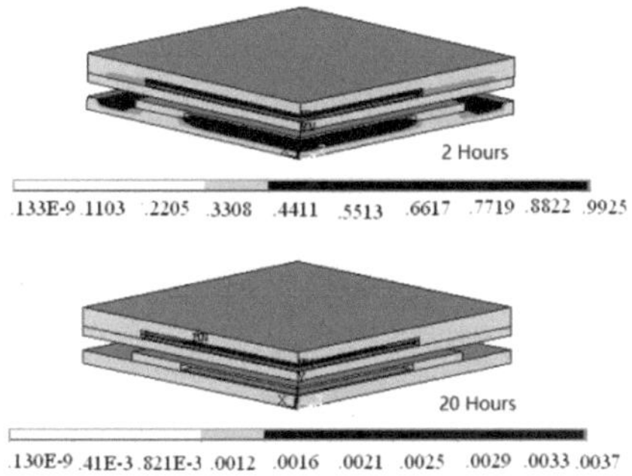

(a) Moisture Desorption in the Package

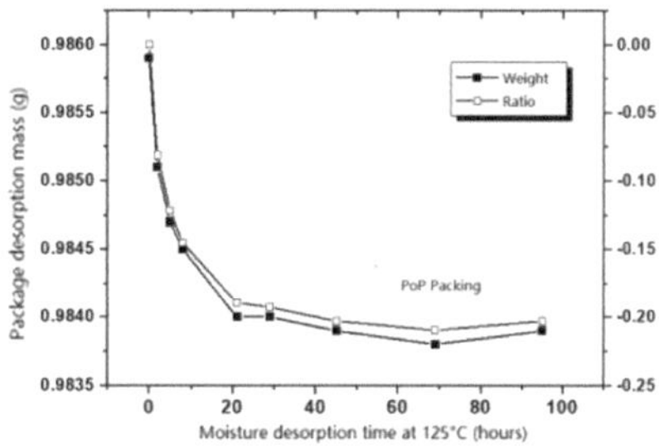

(b) Mass Change Due to Moisture Desorption

Fig. 3.31 Transient humidity distribution and moisture content changes during PoP package moisture desorption

desorption, the moisture in the package is almost entirely removed. The maximum value is only 0.37% of the saturation value. This has a negligible impact on the package. The mass change due to moisture desorption is shown in Fig. 3.31b. Thus, 20 h of desorption time is sufficient to meet the requirements. For surface-mounted devices, the usual pre-treatment baking condition is 125 °C/24 h. However, this is not the optimal condition in all cases.

(3) Moisture Stress Analysis

The equivalent moisture stress distributions for the PoP package and solder balls are shown in Fig. 3.32a, b. Simulation results show that the moisture stress is concentrated on the top-level package chips and solder balls. This is because chips and solder balls do not absorb moisture. However, their moisture expansion coefficients differ from those of the adjacent polymer materials. This difference can cause significant moisture stress at interfaces or contact points during the moisture pre-treatment process.

The chips and solder balls influence the moisture distribution in the polymer materials. They block moisture penetration, creating larger humidity gradients around them. The moisture stress on the top-level package chips is higher than that on the bottom-level chips. This shows that the humidity gradient is a key factor in generating moisture stress.

Figure 3.32 [32] shows that moisture directly affects the reliability of chips and solder balls. Special care is needed during reflow soldering. The vapor pressure in the voids of the packaging materials can cause cracks in chips and solder balls. It may also lead to the "popcorn" effect, which can destroy the entire device.

(4) Reflow Soldering Stress Analysis

During reflow soldering, maximum stress occurs at the peak reflow temperature. Figure 3.33 shows the thermal stress distribution in the PoP package and solder balls at this temperature. Unlike the moisture stress distribution, the peak thermal stress

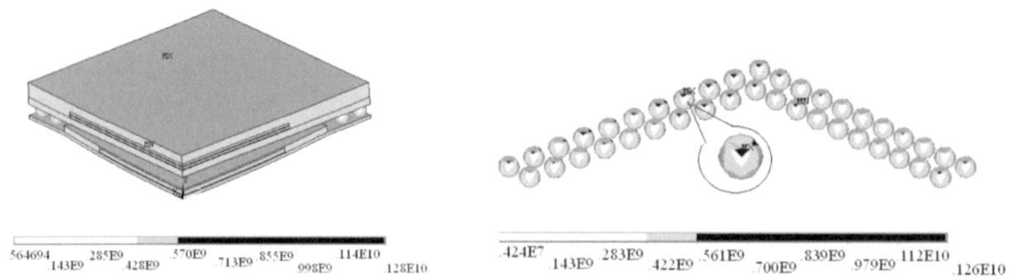

(a) Equivalent Moisture Stress Distribution in PoP Package (b) Equivalent Moisture Stress Distribution in Solder Balls

Fig. 3.32 Equivalent moisture stress in PoP package and solder balls

appears at the junctions of the top-level package chips, bonding layers, and molding compounds. Significant thermal stress is also observed in the bottom-level package. The bonding layers and molding compounds have high thermal expansion coefficients. Chips, in contrast, have low thermal expansion coefficients. This mismatch generates large thermal stress. Initial delamination often starts at the junctions of the top-level package chips, bonding layers, and molding compounds. The peak thermal stress in solder balls is located at their corners, similar to the moisture stress distribution. The thermal stress distribution highlights the solder ball corners as critical reliability points in the PoP package. A comparison of mechanical stress in PoP packaging is shown in Fig. 3.34. Normal stress acts vertically, while shear stress often causes delamination. The equivalent moisture stress in the PoP package is about four times higher than the equivalent reflow soldering stress. This means moisture stress has a significant impact on package reliability and cannot be ignored [32].

The thermal stress distribution in PoP package under lead-free reflow soldering shows:

① The top-level package is the weakest link in reliability. It has high demands for mechanical stability.
② The outer solder balls experience higher moisture and thermal stress than the inner solder balls. The stress is mainly concentrated at the corners.
③ Moisture stress exceeds the thermal stress at peak reflow temperature. The impact of moisture on package reliability cannot be overlooked.

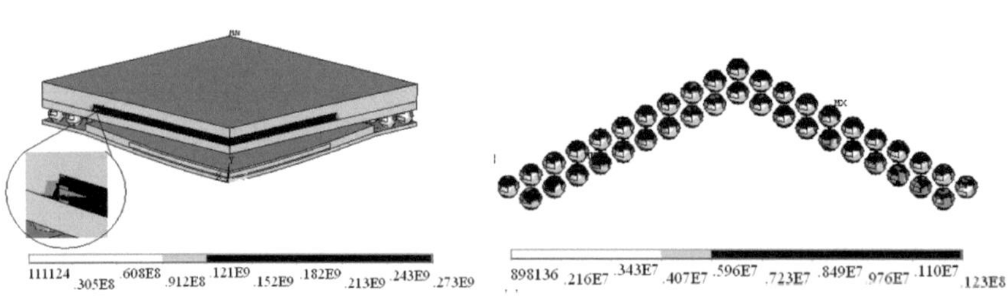

(a) Thermal Stress Distribution in PoP Package (b) Thermal Stress Distribution in Solder Balls

Fig. 3.33 Thermal stress distribution in PoP package and solder balls at peak reflow temperature

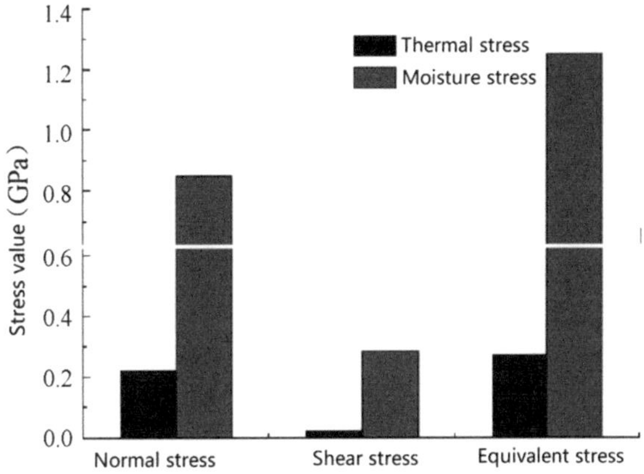

Fig. 3.34 Mechanical stress comparison in PoP package

3.4.2 Stress and Reliability of SiP Packaging

1. Development and Advantages of SiP Technology

System in Package (SiP) refers to assembling multiple active and passive components, along with MEMS, optical devices, and other components, into a single standard package. This forms an independent system or subsystem with unique performance features.

(1) SiP combines previously separate integrated circuits and electronic components on one substrate. This reduces package size and weight, saves materials, lowers I/O pin counts, and improves packaging area efficiency.
(2) Compared to traditional PCBs, SiP technology shortens the distance between metal connections. This reduces parasitic impedance (resistance, capacitance, and inductance) and enhances chip performance.
(3) SiP allows the integration of analog, digital, and RF chips, even with different processing technologies.
(4) SiP simplifies packaging processes and reduces connection points. This improves product reliability and lowers production costs.
(5) The development cycle for SiP products is relatively short.

Reliability is the main challenge in the development of SiP technology. The complex 3D interconnection methods, packaging techniques, and application environments all pose significant threats to the reliability of SiP systems. SiP structures face major reliability issues during production and use, including thermal stress, vibration, drop impact, and moisture stress. Vibration and drop impact are forms of stress caused by external forces. SiP structures are subjected to frequent temperature load shocks during use. These shocks can cause cracks in the micro-bumps, leading to SiP failure. Temperature changes have an ongoing impact on SiP structures, from production to actual use. Therefore, studying the failure modes and mechanisms of thermal stress in SiP structures is crucial for improving their reliability.

2. Stress and Reliability Issues in SiP Packaging

In 3D packaging, solder joint failure is a common cause of failure. The reasons for vertical interconnect solder joint failure include:

(1) Mechanical loads, such as impact, vibration, and inertial loads.
(2) Thermal loads, such as preheating, curing, wave soldering, reflow soldering, and heat generated by the device itself. Thermal shock or thermal cycling loads during testing or use are also factors.
(3) Chemical corrosion, including moisture absorption during packaging, which can cause corrosion or oxidation of solder joints, greatly affecting their reliability.

Therefore, improving the reliability of vertical interconnect solder joints is a key issue in 3D packaging.

3.4.3 Stress and Reliability of WLCSP Packaging

1. Technical Development and Advantages of WLCSP Packaging

CSP (Chip Scale Package) is the next generation of packaging technology, following SiP. According to the Association of Connecting Electronics Industries (IPC), the packaging area of CSP is less than twice the chip area, which greatly improves the integration of the packaging. CSP is typically divided into four types: rigid substrate, flexible substrate, lead frame, and wafer-level assembly. WLCSP (Wafer Level Chip Scale Package) is a type of CSP. The basic process involves first completing the processing on the wafer. Then, the interconnection interface is processed directly on the wafer. After the connections are made using flip-chip technology, testing and aging follow. Finally, the product is separated from the wafer. The volume of the packaged product is almost the same as the bare chip. Unlike traditional packaging, CSP integrates both front-end and back-end processes, optimizing the entire process.

WLCSP introduces re-routing and bump fabrication technologies. It uses standard surface-mount technology to achieve wafer-level interconnection, which increases the interconnection density. Therefore, this packaging technology is a high-density, high-performance stacking technology. Since its invention, it has been widely applied in mobile phones, flash memory, high-speed circuits, drivers, RF devices, logic devices, and analog devices.

Figure 3.35 shows the process flow of WLCSP. It mainly includes two key processes: ① The re-routing process, which changes the soldering zones around the chip into bump zones that are distributed in a planar array on the chip surface. ② The bump fabrication process, which is used to create bumps in the soldering zones, forming a micro-ball array. Methods such as evaporation, chemical plating, electroplating, and solder paste printing can be used. Electroplating is the most widely used method, followed by solder paste printing.

In the re-routing process, a thin dielectric film is deposited on the wafer to passivate the silicon wafer. Benzo-Cyclo-Butene (BCB) or Polyimide (PI) is used as

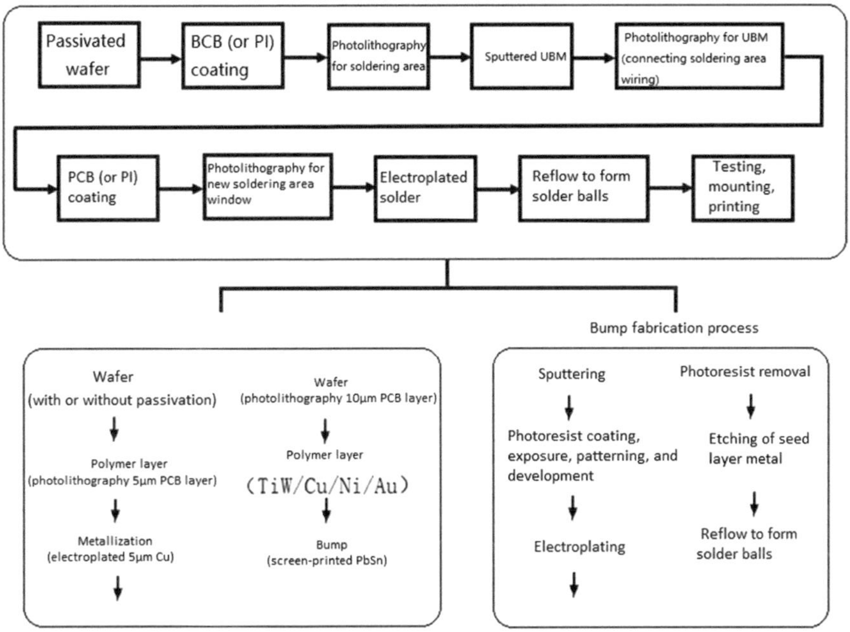

Fig. 3.35 WLCSP process flow

the dielectric material for re-routing, with copper (Cu) as the interconnect metal. The under-bump metallization (UBM) layer is sputtered and deposited beneath the bumps. UBM is a critical interface layer between the metal pad on the chip and the bumps. It provides reliable electrical and mechanical connections and serves as a seed layer for bump deposition. The sputtered Ti layer in UBM acts as an effective diffusion barrier between the metal pad and the bumps. UBM materials include Al/Ni/Cu, Ti/Cu/Ni, or Ti/W/Au.

The entire WLCSP process is compatible with chip manufacturing technology. It only adds re-routing and bump fabrication processes, making it superior to other CSP manufacturing processes in terms of cost and quality control.

2. Stress and Reliability Issues in WLCSP Packaging

In WLCSP packaging, different materials have different coefficients of thermal expansion (CTE). This causes significant thermal stress at the interface and contact points. Such stress can lead to strain energy accumulation, which may result in packaging failure. The reliability of WLCSP mainly focuses on the thermo-mechanical reliability of the solder balls.

Tae-Kyu Lee and colleagues studied the long-term reliability and microstructural evolution of WLCSP Sn–Ag–Cu (Sn–Ag–Cu) micro-solder joints [34]. They investigated the long-term reliability and microstructural evolution of three different solder types (SAC105, SAC305, SAC396) in three different sizes of WLCSP under temperature cycling. In this study, the WLCSP packages with different solders and sizes are shown in Fig. 3.36, and their initial metallographic structure of the solder balls is shown in Fig. 3.37. The lifetime distribution and characteristic lifetime under 0–100 °C temperature cycling stress are shown in Figs. 3.38 and 3.39, respectively.

Finite element simulation analysis of the solder balls shows that stress is concentrated at the interface between the solder ball and the chip. The cumulative crack damage in the WLCSP samples is shown in Fig. 3.40. Cross-sectional analysis of the failed samples reveals that the failure location matches the simulation results. The cross-sectional SEM image of the solder balls in the temperature-cycled failed samples is shown in Fig. 3.41. The crack evolution at the solder ball corners of different sizes of WLCSP packages is shown in Fig. 3.42.

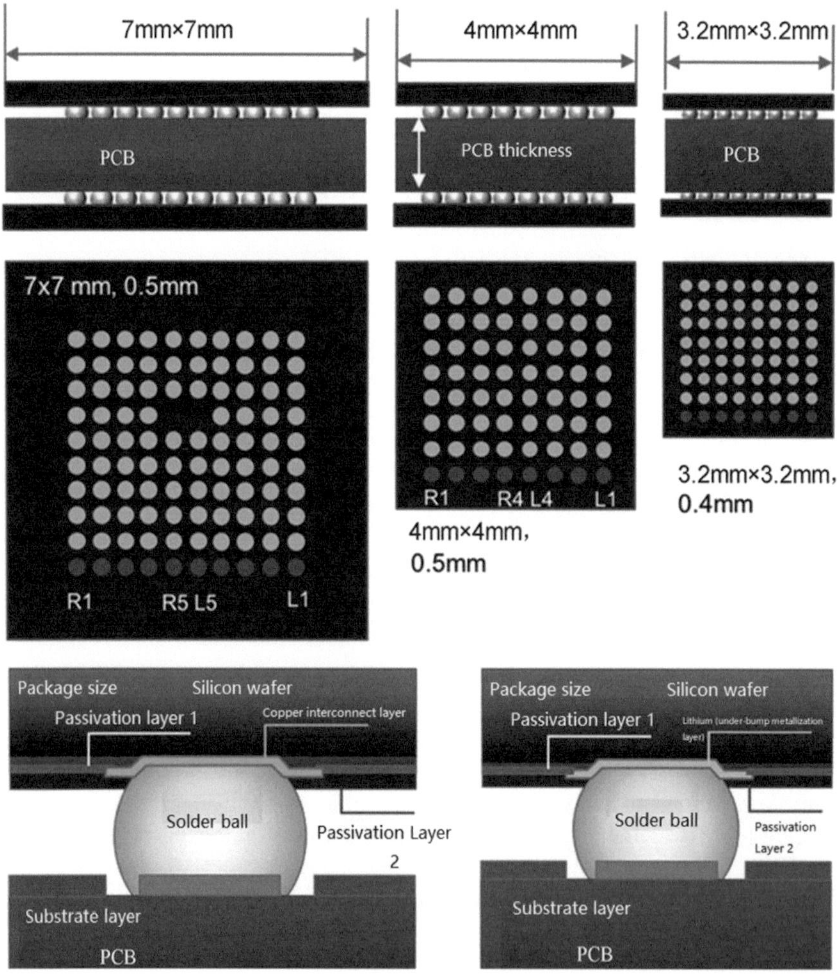

Fig. 3.36 WLCSP packages with different solders and sizes

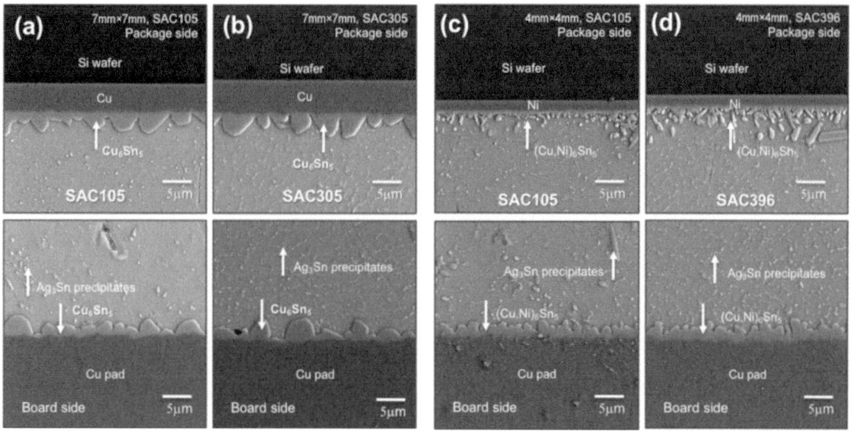

Fig. 3.37 Initial metallographic structure of solder balls

3.5 Typical Failure Cases in Plastic Packaging

Plastic encapsulated circuit products are exposed to stresses such as high temperature, reflow soldering, temperature cycling, humidity, and vibration during use. These stresses can cause failures, such as delamination at interfaces, bond degradation, chip corrosion, and parameter drift. Below are some typical failure cases in plastic-encapsulated circuits.

3.5.1 Corrosion Caused by Moisture Ingress

A power MOSFET showed excessive reverse leakage current during use. Analysis revealed that the lead frame and substrate structures were delaminated from the molding plastic. Scanning Electron Microscopy (SEM) and Energy Dispersive Spectroscopy (EDS) analysis showed significant silver dendrite metal migration on the chip surface. The delamination of the power MOSFET interface led to electrochemical migration on the chip, as shown in Fig. 3.43. The corrosion was caused by the delamination at the interface. Moisture and external contaminants entered the chip surface through the package cracks, leading to electrochemical migration on the chip.

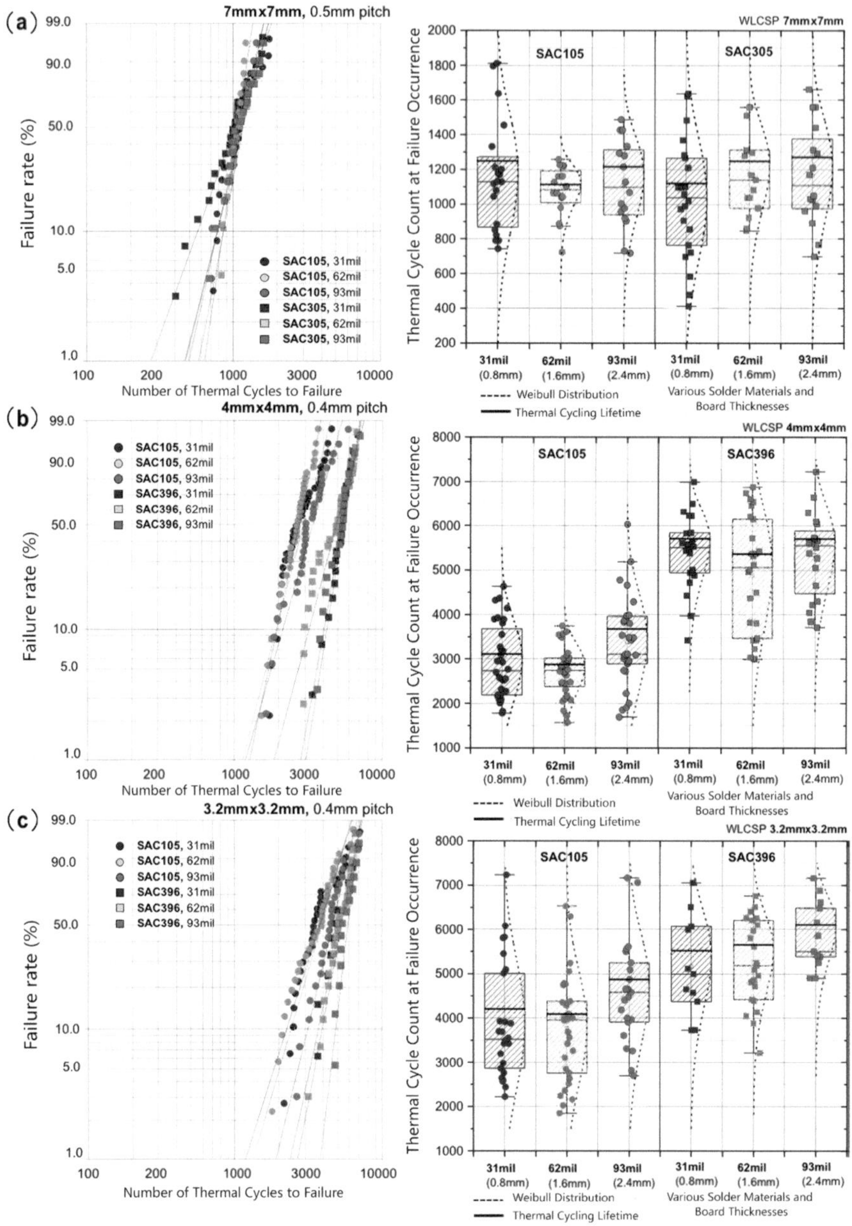

(a) 7mm× 7mm WLCSP (b) 4mm× 4mm WLCSP (c) 3.2mm× 3.2mm WLCSP

Fig. 3.38 Temperature cycling lifetime distribution of WLCSP samples

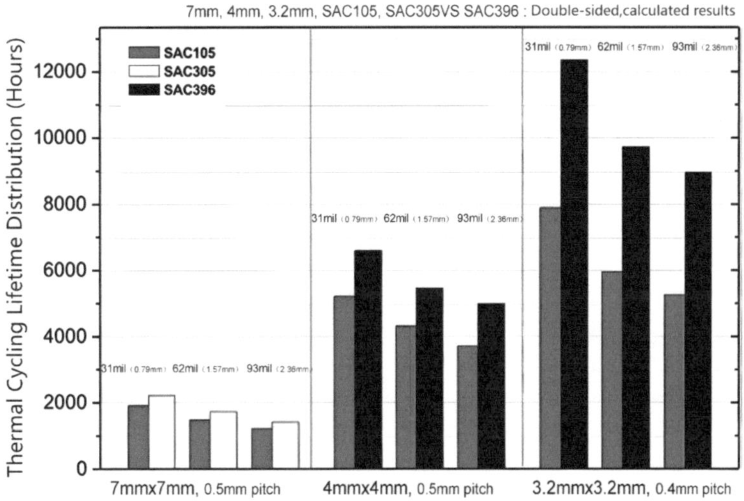

Fig. 3.39 Characteristic lifetime of WLCSP samples under temperature cycling

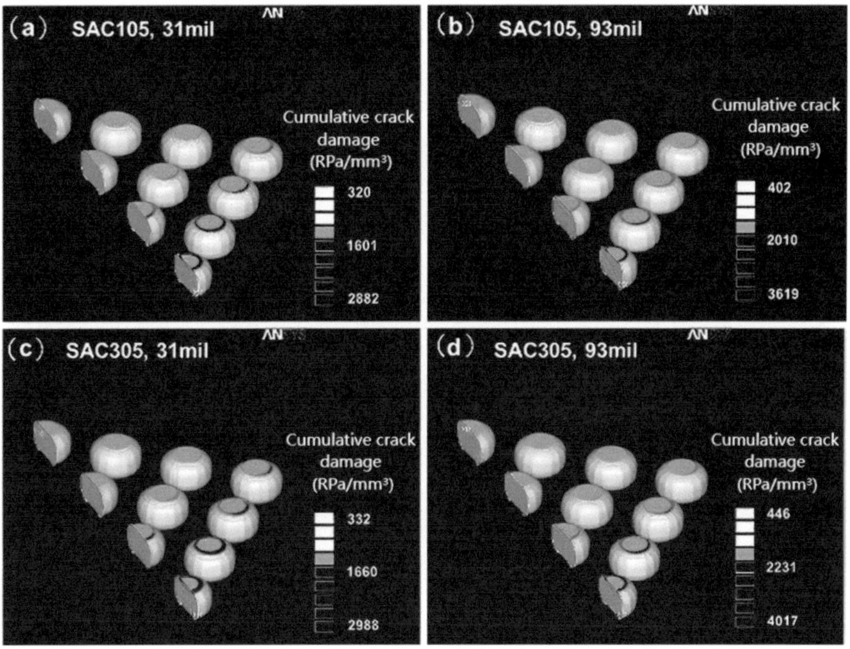

Fig. 3.40 Cumulative crack damage in WLCSP samples

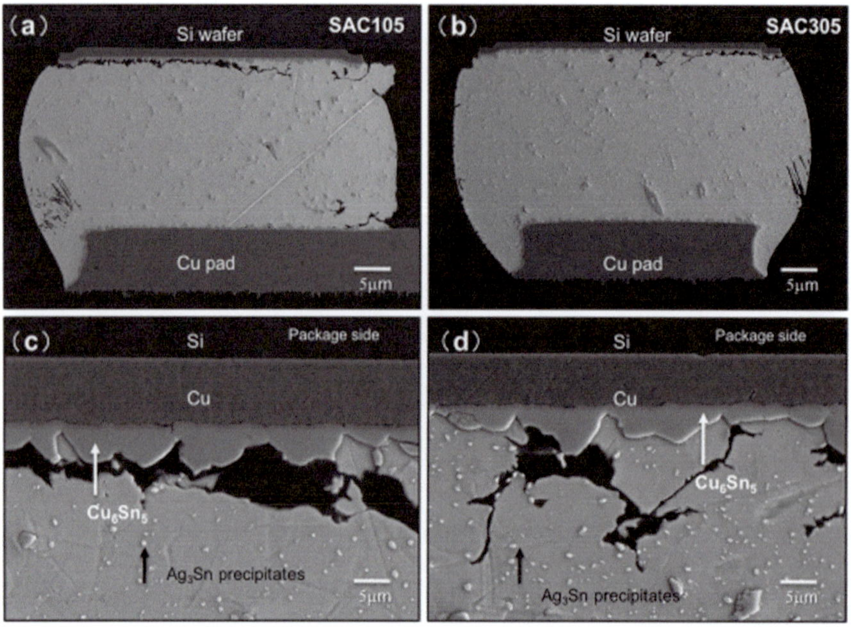

Fig. 3.41 Cross-sectional SEM image of solder balls in temperature-cycled failed samples

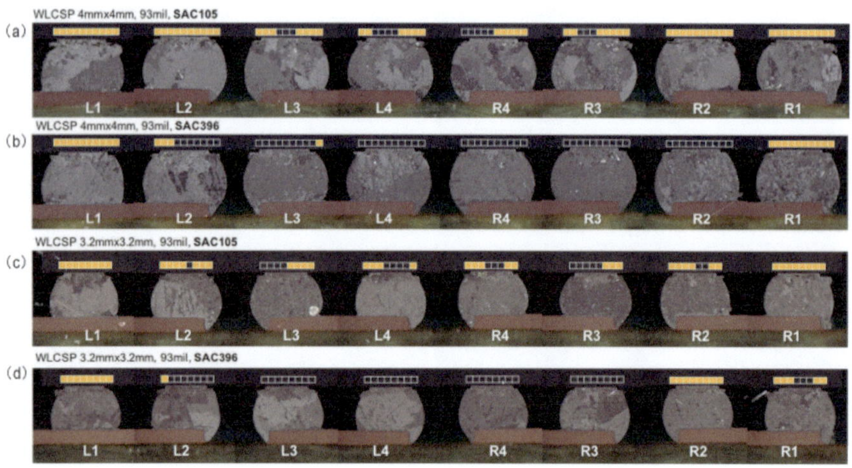

(a) 4mm×4mm SAC105 (b) 4mm×4mm SAC396
(c) 3.2mm×3.2mm SAC105 (d) 3.2mm×3.2mm SAC396

Fig. 3.42 Crack evolution at solder ball corners of different sizes of WLCSP packages

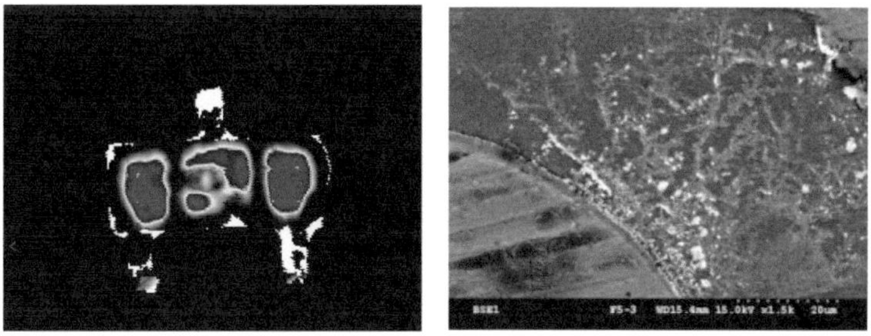

(a) SEM Image of Chip Interface Delamination

(b) Dendritic Metal Migration on the Chip Surface

Fig. 3.43 Electrochemical migration on the chip caused by power MOSFET interface delamination

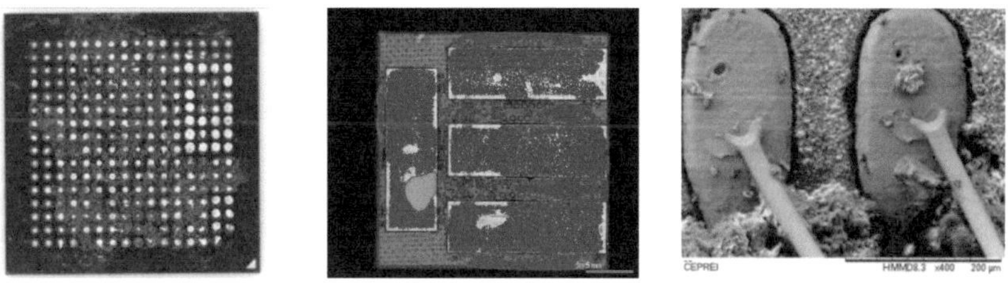

(a) Appearance of the SRAM Chip

(b) Delamination at the Chip-Frame, Molding Plastic, and PCB Interface

(c) Bond Pull-off

Fig. 3.44 Failure morphology of SRAM

A SRAM circuit sample failed after assembly and debugging. The issue was a CPU reset failure, which was traced to the SRAM. Even after reballing and resoldering the SRAM, the fault remained. After replacing the SRAM with a new one, the issue was resolved, and the product worked properly. Analysis showed clear delamination between the chip, frame, molding plastic, and PCB. After opening the package, the bond between the chip and substrate was pulled off. This was caused by the "popcorn" effect. The failure morphology is shown in Fig. 3.44.

3.5.2 Holes and Bond Degradation Caused by High Temperature

A plastic-encapsulated RS-485 chip showed abnormal levels on its enable pin after assembly and powering on. This was an intermittent fault. High-temperature tests at 80 °C extended the fault duration. After returning to normal temperature, the fault persisted for a while before recovering, remaining in an unstable state. Opening the package showed that multiple internal bond balls had separated from their pads. A

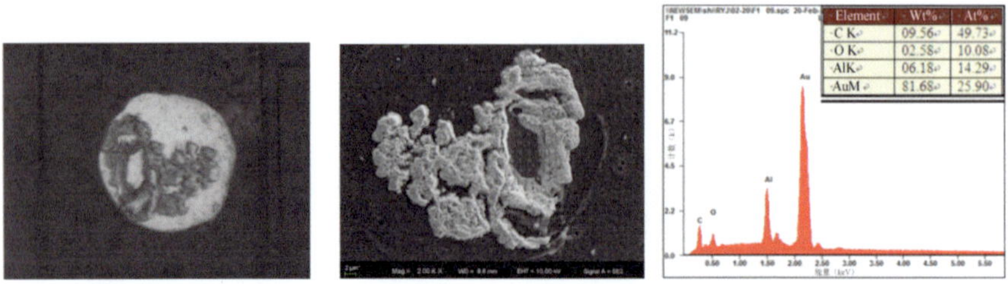

(a) Metallographic Image of Bonding Voids (b) SEM Image of Bonding Voids (c) EDS Spectrum of Au-Al Compounds

Fig. 3.45 SEM and EDS analysis results

brownish substance formed, and one side of the bond balls had a depression. SEM and EDS analysis showed that the brownish substance mainly consisted of Au–Al compounds, with a high concentration of Au. The SEM and EDS results are shown in Fig. 3.45.

This analysis suggests that Au and Al at the bonding interface diffused to form Au-Al compounds. The separation of the bond balls from these compounds showed that the pad had a higher Au content. This means Au had diffused into the pad. Due to organic materials in the solder mask, bubbles formed under high temperature, creating Kirkendall voids [35]. This led to bond cold solder joint failure and intermittent faults.

References

1. Lai P, En Y, Niu F (2006) Risks and countermeasures for plastic-packaged microcircuits in high-reliability applications. Electron Prod Reliab Environ Test (04):53–58
2. Harman G (2010) Wire bonding in microelectronics. McGraw-Hill Education, Wagon Lane, Bingley
3. Lin X, Kong X, En Y et al (2007) Failure analysis and countermeasures for FCBGA-packaged devices. Fail Anal Prev 2(4):4
4. Li X, Yan Y (2005) Electronic packaging solder joint reliability and lifetime prediction methods. Mech Strength 27(4):10
5. Majni G, Ottaviani G, Galli E (1979) AuAl compound formation by thin film interactions. J Cryst Growth 47(4):583–588
6. Liao J, Zhang X, Sureshkumar V et al (2016) Effect of free-air-ball palladium distribution on palladium-coated copper wire corrosion resistance. In: 2016 IEEE 18th electronics packaging technology conference (EPTC). IEEE, pp 152–156
7. Tsai J, Lan A, Jiang DS et al (2014) Ag alloy wire characteristic and benefits. In: 2014 IEEE 64th electronic components and technology conference (ECTC). IEEE, pp 1533–1538
8. Kim SH, Park JW, Hong SJ et al (2010) The interface behavior of the Cu-Al bond system in high humidity conditions. In: 2010 12th electronics packaging technology conference. IEEE, pp 545–549
9. Xu H, Liu C, Silberschmidt VV et al (2009) A re-examination of the mechanism of thermosonic copper ball bonding on aluminium metallization pads. Scripta Mater 61(2):165–168
10. Xu H, Liu C, Silberschmidt VV et al (2010) A micromechanism study of thermosonic gold wire bonding on aluminum pad. J Appl Phys 108(11):19

11. Kim HJ, Lee JY, Paik KW et al (2003) Effects of Cu/Al intermetallic compound (IMC) on copper wire and aluminum pad bondability. IEEE Trans Compon Packag Technol 26(2):367–374
12. Lu C (2014) Review on silver wire bonding. In: 2014 8th international microsystems, packaging, assembly and circuits technology conference (IMPACT). IEEE
13. Ren Z (2016) Silver bonding wire for BSOB (Bond-Stitch-on-Ball)/BBOS(Bonding-Ball-on-Stitch). In: 2016 China semiconductor technology international conference (CSTIC). IEEE
14. Xu H, Liu C, Silberschmidt VV et al (2011) Effect of bonding duration and substrate temperature in copper ball bonding on aluminium pads: a TEM study of interfacial evolution. Microelectron Reliab 51(1):113–118
15. Xu H, Liu C, Silberschmidt VV et al (2011) Effect of ultrasonic energy on nanoscale interfacial structure in copper wire bonding on aluminium pads. J Phys D-Appl Phys 44
16. Xu H, Acuff VL, Liu C et al (2011) Facilitating intermetallic formation in wire bonding by applying a pre-ultrasonic energy. Microelectron Eng 88:3155–3157
17. Xu H, Liu C, Silberschmidt VV et al (2011) Effect of bonding duration and substrate temperature in copper ball bonding on aluminium pads: a TEM study of interfacial evolution. Microelectron Reliab 51:113–118
18. Hai L, Qi C, Zhao Z et al (2010) Reliability of Au-Ag alloy wire bonding. In: Electronic components & technology conference. IEEE, pp 234–239
19. Yoo KA, Uhm C, Kwon TJ et al (2009) Reliability study of low cost alternative Ag bonding wire with various bond pad materials. In: 2009 11th Electronics packaging technology conference. IEEE, pp 851–857
20. Xi J, Mendoza N, Chen K et al (2015) Evaluation of Ag wire reliability on fine pitch wire bonding. In: IEEE electronic components & technology conference. IEEE, pp 1392–1395
21. Tan BW, Niu YH, Wu KS (2014) Silver alloy wire for IC packaging solution. In: 2014 IEEE 36th international electronics manufacturing technology conference (IEMT). IEEE, pp 1–4
22. You CJ, Park SY, Kim HD et al (2014) Study of intermetallic compound growth and failure mechanisms in long term reliability of silver bonding wire. In: 2014 IEEE 16th electronics packaging technology conference (EPTC). IEEE, pp 704–708
23. Cao J, Fan JL, Liu ZQ et al (2015) Effect of silver alloy bonding wire properties on bond strengths and reliability. In: International conference on electronics packaging and iMAPS all Asia conference. IEEE, pp 93–97
24. Mayer M, Xu DE, Ratcliffe K (2016) The electrical reliability of silver wire bonds under high temperature storage. In: Electronic components and technology conference. IEEE, pp 654–659
25. Tanna S, Pisigan JL, Song WH et al (2012) Low cost Pd coated Ag bonding wire for high quality FAB in air. In: 62nd Electronic components and technology conference. IEEE, pp 1103–1109
26. Liao JK. Liang YH, Li WW et al (2012) Silver alloy wire bonding. In: Electronic components and technology conference, San Diego, CA, USA. IEEE, pp 1163–1168
27. Ma XS, Jansen K, Ernst LJ et al (2009) A new method to measure the moisture expansion in plastic packaging materials. In: Electronic components & technology conference. IEEE, pp 1271–1276
28. Jiang Z, Law JS (2008) Effect of non-uniform moisture distribution on the hygroscopic swelling coefficient. IEEE Trans Compon Packag Technol 31(2):269–276
29. Theint EPP, Stephan D, Goh HM et al (2005) High temperature storage (HTS) performance of copper ball bonding wires. In: 2005 7th Electronic packaging technology conference, vol. IEEE, p 6
30. Miao C, Daoguo Y, Lijun Z (2009) The effect of damp heat and steam pressure on the reliability of plastic-packaged devices. Electron Compon Mater 28(1):4
31. Fan X, Jiang Z, Chandra A (2008) Package structural integrity analysis considering moisture. IEEE, pp 1054–1066
32. Liu H, Yang S, Li G (2010) The impact of damp heat on the reliability of PoP packaging. Semicond Technol 35(11):6
33. Lee TK, Xie W, Tsai M et al (2020) Impact of microstructure evolution on the long-term reliability of wafer-level chip-scale package Sn–Ag–Cu solder interconnects. IEEE Trans Compon, Packag, Manuf Technol 99:1594–1603

34. Lee TK, Xie W, Tsai M et al (2020) Impact of microstructure evolution on the long-term reliability of wafer-level chip-scale package Sn–Ag–Cu solder interconnects. IEEE Trans Compon Packag Manuf Technol (99):1
35. Liqiang S (2014) Analysis of voids in thermal conductive solder joints of QFN chips. Electron Technol Softw Eng 11:1

Chapter 4
Failure Modes, Failure Mechanism and Reliability of Hermetic Packaging

Hermetic packaging provides a seal that is impervious to moisture and other contaminants for chips, and is typically used in harsh environments or in fields where high reliability is required, such as military electronic products in the aerospace and aviation industries. For devices with poor hermeticity, moisture can penetrate into the package within a few hours to several days, affecting the performance of the device and even causing failure. Especially for devices operating in special environments, a pressure difference is formed between the device cavity and the outside world, making it easier for the protective gas inside the device to leak out, and for external moisture, harmful gases, harmful ions, and particles to enter the device cavity, leading to leakage, parameter drift, and affecting the performance of the chip, ultimately resulting in frequent device failure issues.

Ideal hermetic packaging is designed to permanently prevent the ingress of contaminants (gases, liquids, and solids). However, such ideal hermetic packaging does not exist in reality. Hermetic packaging is defined as a package in which, after helium is introduced, the leak rate of helium is below a specified rate.

This chapter briefly describes the structural characteristics of hermetic packaging, summarizes the failure modes and mechanisms of hermetic packaging, outlines the performance testing methods for hermetic packaging, summarizes the stress and reliability issues of hermetic packaging, and lists typical cases of hermetic packaging reliability.

4.1 Structural Features of Hermetic Packaging

From the perspectives of the structure and materials of hermetic packaging, the permeability of metals, ceramics, and glass to moisture is very low, about 2 to 3 orders of magnitude lower than that of plastics. The classification of hermetic packaging structures is shown in Table 4.1.

B. Zhou et al., *Reliability Technology for Integrated Circuit Packaging*,
https://doi.org/10.1007/978-981-95-3885-0_4

Table 4.1 Classification of hermetic packaging structures

Sealing type	Schematic diagram	Feature
Metal packaging	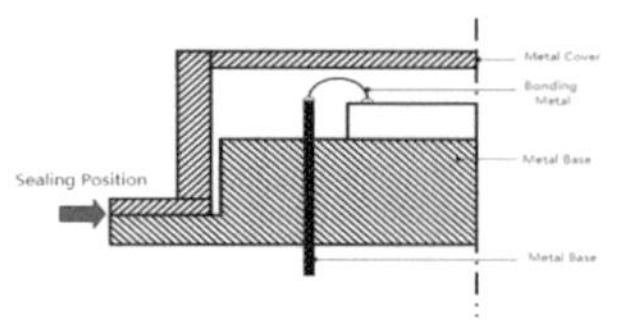	Metal packaging has high hermeticity and good heat dissipation capabilities, but the packaging cost is relatively high
Ceramic packaging	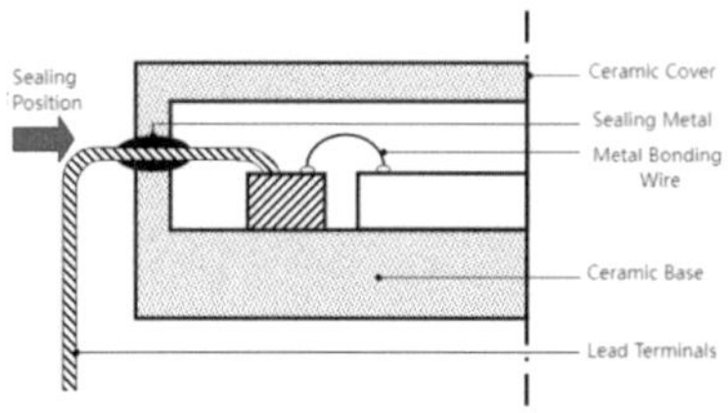	Ceramic packaging has good chemical stability, oxidation resistance, electrical insulation, and density, with a lower cost compared to metal packaging
Glass packaging	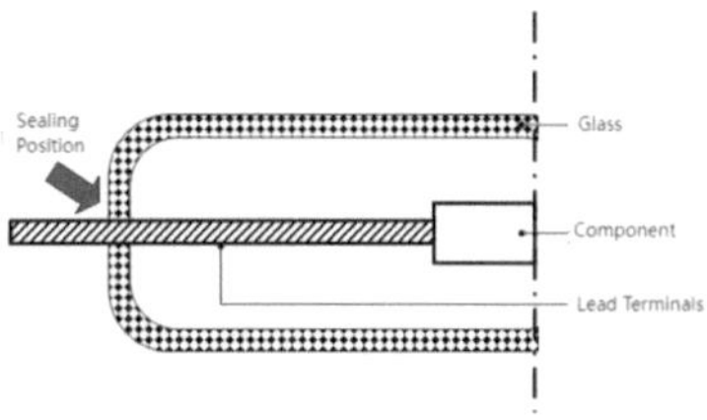	The heat dissipation properties of glass packaging can be altered by adjusting the composition

It can be easily observed from Table 4.1 that among the commonly used hermetic packaging materials, metals possess excellent electrical conductivity and thermal conductivity, making them suitable for the sealing of electronic devices in ambient temperature environments. Ceramics (such as silicon carbide and silicon nitride) used as substrates can be employed in harsh environments, with minimal deformation under high-temperature and high-pressure conditions. Consequently, ceramic packaging exhibits lower thermal stress and a reduced likelihood of cracking. Glass can directly transmit optical signals to devices [1, 2], hence it is applicable for the hermetic packaging of optical devices. However, hydrogen, helium, and neon molecules can permeate through glass, which makes it less suitable for applications in high-vacuum environments. Currently, the industry predominantly utilizes metal and ceramic hermetic packaging.

In metal packaging, three types of covers are used: raised caps, flat covers, and stepped covers. Raised caps are used for flat-style packaging and can be sealed by either fusion welding or soldering [3–8]. Flat covers are used for cavity-style packaging, and the covers are attached to the package body using soldering. Stepped covers are created by etching steps on an ASTM F-15 alloy sheet, resulting in an edge thickness of approximately 0.1 mm.

In ceramic packaging, a sealing structure similar to that of metal packaging is employed, with one difference being that the ceramic packaging for hybrid integrated circuits (Hybrid Integrated Circuit, HIC) typically consists of three layers of alumina. The upper alumina ring requires metallization to facilitate the soldering seal of the cover. Ceramic packaging usually adopts soldering for sealing. During the packaging manufacturing process, a refractory metal or the oxide of several metals, such as W or an alloy of Mo and Mn, is coated around the sealing area on the ceramic surface and then sintered. After sintering, these metallized surfaces are successively plated with nickel and gold. The cover made of ASTM F-15 alloy is electroplated in the same manner and is soldered to the package shell using an Au80/Sn20 alloy under a nitrogen atmosphere.

4.2 Failure Modes and Mechanisms of Hermetic Packaging

The failure modes and causes in hermetic packaging are shown in Fig. 4.1. The failure modes can be summarized into the following six categories: chip attachment, active devices, wire bonding, package sealing, substrates, and contamination. Data collected by the former Rome Air Development Center (now the Air Force Research Laboratory) indicate that defective active devices, interconnect performance, package shell hermeticity, and multilayer substrate quality are the four major factors contributing to the failure of hermetic packaging.

The impact of packaging materials and assembly processes on the actual performance and reliability of hermetic packaging can be categorized based on the root causes of different failure modes. Referring to Fig. 4.2, these root causes lead to three failure mechanisms that result in the loss of hermeticity: Capillary Leak, Permeation Leak, and Outgassing. Capillary Leak occurs when one or a series of microfluidic channels are formed within the package cavity, acting as conduits. Permeation Leak

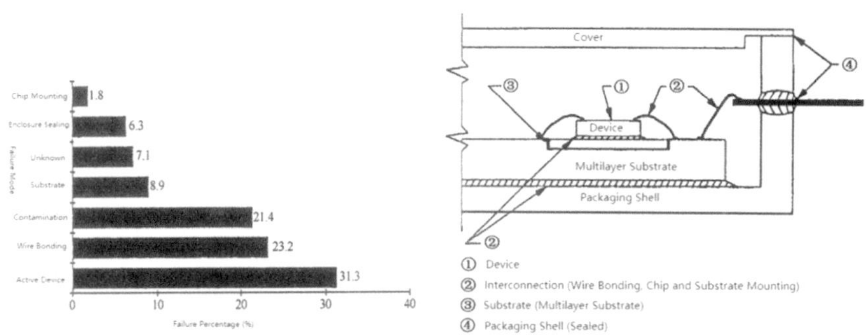

(a) Failure modes of hermetic packaging (b) Key component and causes of hermetic packaging

Fig. 4.1 Failure modes and causes in hermetic packaging

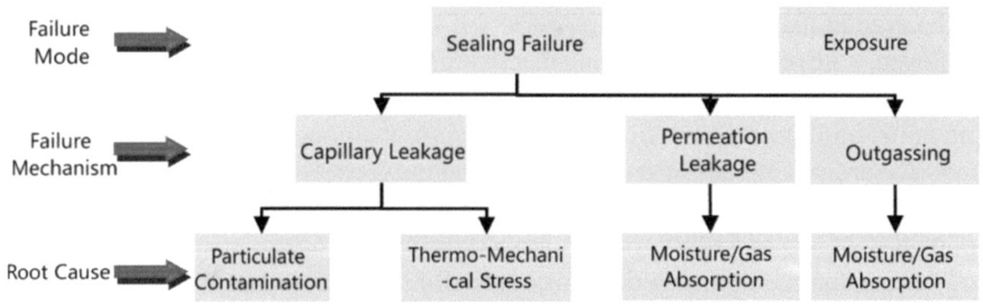

Fig. 4.2 Classification of failure mechanisms related to hermetic packaging failure

is caused by the diffusion of gas or liquid molecules into the material of the package or hermetic cover (or cap), followed by desorption into the cavity. Outgassing is distinct from the other two failure modes; it occurs when gases or liquids present on the surface or within the material are released into the cavity due to external stimuli such as temperature increases. The corresponding failure roots for these three mechanisms are particle contamination, thermo-mechanical stress, and moisture/gas absorption.

Next, the failure mechanisms of hermetic packaging will be introduced one by one, focusing on three key root causes of failure.

4.2.1 Particle Contamination

The failure mechanism of particle contamination primarily arises from friction and wear between internal components of the cavity and fine particles, leading to failure. These failures can occur during both the device assembly and packaging processes, as well as during usage. During assembly and packaging, the gaps between key structures (such as solder joints, leads, and chips) in microelectronic components are typically on the micron scale or even smaller [9–13]. Therefore, even extremely small particles generated during packaging can cause damage to the device. The primary sources of these particles are as follows.

- Particles generated during the chip edge cutting and grinding processes.
- Organic particles and metal particles that may be present during the assembly or clamping process.
- Particles introduced from the assembly and packaging equipment.
- Residual materials from interactions between packaging materials or excess material during the packaging process.

Especially during the chip dicing process, the wafer is cut using a diamond blade while simultaneously applying a strong coolant flow. This process generates a large amount of cutting debris and residues. Contaminants or residues on the chip surface increase surface adhesion, which can hinder the operation of functional components.

Particles generated during the assembly and packaging processes can lead to the following types of failures.

- Blockage of moving structures, preventing their operation.
- Mechanical bridging of structures.
- High-resistance shorts, discharge and electrostatic discharge (ESD), conductive particle failures.
- Changes in the resonance characteristics of ultra-thin structures.
- Obstruction of optical paths, leading to reduced quality of micromirrors, lenses, and optical windows.
- Obstruction of fluid flow in microfluidic devices.
- Impact on surface friction and surface adhesion.

During use, if free particles (movable excess materials) are present inside the cavity of hermetic packaging, such as in integrated circuits or hybrid circuits, these free particles can collide when the device undergoes high-speed directional changes or severe vibrations, leading to failures. The failure mechanism depends on whether the particles are conductive.

① When the particles are conductive metallic materials, they can interfere with the normal operation of the circuit, causing intermittent functioning. In severe cases, they can short-circuit the circuit or prevent it from operating properly.
② When the particles are non-conductive organic materials, if the particles are too large, they may cause deformation of the internal bonding leads of the circuit, potentially affecting signal transmission.

4.2.2 Thermo-Mechanical Stress

During assembly, packaging, and use, thermo-mechanical stress is a significant cause of functional and reliability issues in devices. The primary source of this stress is the thermo-mechanical mismatch between the packaging materials.

In the sealing process, if the packaging leakage rate is sufficiently low, the chip can be bonded in a vacuum, ensuring that each device chip remains in a vacuum permanently. The bonding temperature varies from 200 to 1000 °C, depending on the sealing method used. Of course, each bonding method has its own advantages and disadvantages, and these methods introduce significant residual stress inside the package, which can lead to long-term drift in signal output. Additionally, improper control of bonding process parameters (such as temperature, contact pressure, applied voltage, and time) can result in significant defects at the sealing interface, formation of voids, sealing leakage, and delamination failures.

During use, thermo-mechanical failure caused by the thermo-mechanical mismatch of the materials of internal cavity components or between different materials, along with insufficient interface bonding strength, can result in poor electrical signal performance [14, 15].

The failure mechanisms related to thermo-mechanical loading can be explained as follows.

(1) Fracture of ductile materials

For ductile materials such as metals, they are highly susceptible to the effects of creep. When metal devices are subjected to long-term stress under high-temperature conditions, continuous strain will develop, leading to creep. After a period of load-bearing, creep can easily cause the device to fracture and fail. The temperature affects the creep rate: when the stress level is low and the temperature is below one-third of the material's melting point, no significant creep occurs. However, when the temperature is between 1/3 and 1/2 of the material's melting point, creep accelerates. In this case, even if the stress is below the yield limit, the material is still highly prone to fracture due to creep.

(2) Fracture of brittle materials

For brittle materials such as silicon, in a humid environment and under significant thermo-mechanical cyclic stress, brittle fracture, also known as stress corrosion cracking, often occurs. For example, single-crystal silicon and polycrystalline silicon are commonly used as structural components. Silicon itself does not undergo stress corrosion in a vapor environment, but the surface of silicon is prone to oxidation in air, forming an SiO_2 thin film. This SiO_2 film easily absorbs water molecules from the air, and especially under high electric field conditions, the SiO_2 film undergoes hydrolysis with the water molecule layer. If microcracks form within the SiO_2 film, they will propagate under the combined effects of hydrolysis and tensile stress. The propagation of microcracks accelerates silicon oxidation, promoting the rapid progression of the process, ultimately leading to silicon fracture and failure. This is particularly true for silicon thin films, where the initiation, propagation, and final fracture failure of microcracks all occur within the oxide layer.

(3) Delamination at multilayer material interfaces

The physical property mismatch and process differences between materials in multilayer materials result in high residual stress. The involved processes include curing, molding, and lid sealing, while the service conditions primarily involve long-term storage, high-temperature storage, temperature cycling, and thermal shock. Changes in environmental temperature cause variations in thermo-mechanical stress, and the mismatch in the coefficient of thermal expansion between materials leads to tensile and compressive stresses at the material interfaces. High-low temperature stress cycling causes the interface to develop cracks due to fatigue from the "ratchet effect", which propagate over time, ultimately leading to delamination failure and hermetic packaging failure.

In addition to thermal stress, if the interface is exposed to a high-humidity environment, the moisture will exacerbate delamination caused by thermo-mechanical stress. Moreover, in a humid environment, chemicals can continuously penetrate deeper into cracks through capillary action, intensifying chemical corrosion. This accelerates the propagation of cracks at the interface, leading to delamination. Delamination primarily occurs between the die, die attachment, and the packaging substrate, or between the bonding films in chip-to-chip bonding, providing an easier pathway for moisture ingress.

4.2.3 Moisture/Gas Absorption

For hermetic packaging, the reasons for excessive internal moisture or gas generally fall into four categories. These include excessive moisture or gas in the packaging environment, the absorption of moisture or gas by materials such as packaging shells or die attach, which gradually release the absorbed moisture or gas into the cavity after processes like circuit lid heating, leakage occurring in the circuit before the hermetic test, and, during use, cracks forming at the sealing interface of the hermetic lid or cap due to factors such as creep, brittle fracture, or temperature cycling, random vibrations, and impacts, leading to poor sealing performance. According to GJB 548B—2005 [16] or GJB 128A—1997, domestic military hermetic packaging devices are required to have an internal moisture content of no more than 5000 ppm after being baked at 100 °C for at least 24 h [17–19]. In practical production and usage, the moisture content inside the device needs to be controlled to below 1000 ppm to ensure reliable operation of the device.

The failure modes related to internal moisture in packaging are as follows.

① Corrosion failure: Corrosive substances at the bonding points will cause a significant increase in contact resistance at the bonding interface, reducing the bonding strength, leading to bond separation, open bonding, and ultimately device failure.
② Electromigration and metal migration: Electromigration can result in dendrite formation, metal compound generation, and ion contamination, which can trigger short circuits or device burnouts.
③ Mechanical damage: When microcracks fill with water, surface tension causes the cracks to propagate rapidly, leading to the rapid formation of cracks within the ceramic packaging shell and passivation layer, causing delamination and cracking of the oxide layer, which leads to device failure.
④ Interface delamination: Delamination of the adhesive, organic chip layers, or cracks in the heat sink can cause circuit failure and affect device functionality.
⑤ Current leakage: The higher the moisture content inside the package, the easier it is for moisture to adsorb onto the chip surface, forming leakage paths that increase leakage current. This can even lead to electrochemical reactions of contaminants inside the device, causing parameter drift and degradation.

The absorption of moisture and gases can lead to material diffusion or outgassing effects. On one hand, due to the performance characteristics of barrier coatings and sealing materials, gases can diffuse through the materials, or moisture may propagate along cracks and voids. On the other hand, barrier coatings, adhesives, and etching materials used within the cavity can undergo outgassing during high-temperature cycling or throughout the device's lifetime. The specific failure mechanisms can be explained as follows.

(1) Moisture

The voids present at material interfaces are key locations for moisture condensation. Any minor leakage in hermetic packaging can rapidly alter the moisture content; even for packages with leakage rates just below the detectable limit, moisture penetration can exceed the threshold in just a few weeks. During the preparation, assembly, and packaging process, materials absorb moisture, leading to desorption, water absorption-induced expansion, and delamination failures. In liquid or high-humidity environments, moisture penetration is the most common failure mode, as moisture can migrate along the electrical interconnect structure into the package. Moisture accelerates the aging process of the chip. Additionally, moisture significantly affects the mechanical properties of the chip interconnect structure, adversely impacting reliability. This is because moisture accelerates fatigue mechanisms by forming reactive oxide layers and stress corrosion cracks on the surface of the structure.

The primary hazard of excessive moisture in hermetic packaging is its impact on device lifespan and reliability. The median lifetime of devices under moisture-accelerated life testing is as follows.

$$\tau_{50}(T, \%\mathrm{RH}) = C_1 \exp\left(\frac{N}{\%\mathrm{RH}} + \frac{E_\mathrm{a}}{kT}\right) \tag{4.1}$$

$$\tau_{50}(T, \%\mathrm{RH}) = \frac{C_2}{(\%\mathrm{RH})^n} \exp\left(\frac{E_\mathrm{a}}{kT}\right) \tag{4.2}$$

In the equation, C_1 and C_2 are constants; E_a is the activation energy; N and n are constants ($n > 1$); k is the Boltzmann constant; %RH is the relative humidity; and T is the absolute temperature of the device, whose lifespan decreases as the moisture content increases.

(2) Outgassing

In low-pressure or vacuum environments, devices require hermetic packaging, not only to achieve low loss but also to maintain a suitable internal environment by protecting components from moisture intrusion and contamination. Most micro-resonators, optical systems, and micro-electromechanical systems (MEMS) use high-vacuum hermetic packaging. In these applications, controlling and adjusting the packaging atmosphere is critical to the performance and reliability of the devices.

During various manufacturing steps, liquids, gases, and materials may remain on the surface of the chip. Additionally, excess gas may be retained at the bonding interfaces or inner surfaces. After the sealing and bonding process, different packaging materials within the device (such as chip interconnects, sealing lids, and adhesives) absorb gases, which are subsequently released and contaminate the device cavity. These trapped gases will outgas during the device's operational lifetime and are especially problematic when the device is subjected to high-temperature loads. For example, when using epoxy or cyanoacrylate, the die attachment compounds release gases during curing, producing moisture and organic gases. Hydrocarbons or other reactive substances may also be generated during the hermetic lid sealing process. In vacuum packaging, the typical outgassing mechanism of glass involves the desorption of moisture or gases adsorbed on the glass surface, as well as the outward diffusion of gases that have remained in the glass. These contaminants can damage the internal components of the device or disrupt the required operating.

4.3 Performance Testing of Hermetic Packaging

4.3.1 Hermetic Testing

According to GJB 548B-2005 *Test Methods and Procedures for Microelectronic Devices* trace gas helium fine leak testing failure criteria (see Table 4.2) [16] and MIL-STD-883 *Test Methods and Procedures for Microcircuits* trace gas helium fine leak testing failure criteria (see Table 4.3) [20], systematic sealing tests can be conducted.

Leak testing is generally divided into coarse leak testing and fine leak testing, with fine leak testing conducted first, followed by coarse leak testing. The main methods for testing the hermeticity of microelectronic packaging devices include five types: trace gas helium fine leak testing, radioactive isotope fine leak testing, carbon fluorine compound coarse leak testing, dye penetration coarse leak testing, and weight gain method coarse leak testing.

(1) Trace gas helium fine leak testing

In European and American countries, commonly used testing standards are military standards, such as MIL-STD-883 TM1014 and MIL-STD-750 TM1071. The most

Table 4.2 Failure criteria for trace gas helium fine leak testing (GJB 548B-2005)

Package cavity volume (V)	The equivalent standard leak rate (L) rejection specification value (air)
$V \leq 0.01$ cm^3	5×10^{-3} Pa·cm^3·s^{-1}
0.01 cm$^3 < V \leq 0.4$ cm^3	1×10^{-2} Pa·cm^3·s^{-1}
$V > 0.4$ cm^3	1×10^{-1} Pa·cm^3·s^{-1}

Table 4.3 Failure criteria for trace gas helium fine leak testing (MIL-STD-883)

Package cavity volume (V)	The equivalent standard leak rate (L) rejection specification value (air)
$V \leq 0.01$ cm^3	5×10^{-8} atm·cm^3·s^{-1}
0.01 cm$^3 < V \leq 0.4$ cm^3	1×10^{-7} atm·cm^3·s^{-1}
$V > 0.4$ cm^3	1×10^{-6} atm·cm^3·s^{-1}

commonly used testing methods are helium leak testing and total bubble testing, often used together to assess all potential leak rates. Generally, since total bubble testing is a destructive test, helium leak testing is conducted first. Helium leak testing uses a helium mass spectrometer to measure the leak rate. The helium leak rate is calculated using Eq. (4.3), and this rate can be applied to other gases.

$$L_1 = L_2\sqrt{\frac{M_2}{M_1}} \tag{4.3}$$

In the equation, L_1 and L_2 represent the leak rates of gas 1 and gas 2, respectively; M_1 and M_2 represent the molecular weights of gas 1 and gas 2, respectively.

The helium mass spectrometer requires the package to be pressurized first and filled with helium gas [21]. Most micro devices are vacuum-packaged, so they need to be filled with the trace gas. Based on the cavity volume, the sample should be maintained at a pressure of 45 psi to 75 psi (1 psi = 6.895 kPa) for 2 to 10 h, and then transferred to the helium mass spectrometer for testing within 1 h. If these testing conditions cannot be met, there is a flexible testing condition where all variables can be adjusted to suit the application, which is calculated using the Howl-Mann equation, or Eq. (4.4). The only necessary condition is that the explosion pressure must be at least 29.4 psi.

$$R_1 = \frac{LP_E}{P_O}\left(\frac{M_A}{M}\right)^{1/2}\left\{1 - e^{-[(L_{t1}/VP_O)(M_A/M)]^{1/2}}\right\} \times e^{-[(L_{t2}/VP_O)(M_A/M)]^{1/2}} \tag{4.4}$$

In the equation, R_1 is the leak rate of the trace gas measured in atm·cm^3·s^{-1}; L is the equivalent standard leak rate in atm·cm^3·s^{-1}; P_E is the applied pressure to the sample; P_O is the ambient atmospheric pressure; M_A is the molecular weight of air in the unit of g; M is the molecular weight of the trace gas in the unit of g; t_1 is the exposure time to P_E in the unit of s; t_2 is the residence time in the unit of s; V is the internal volume of the sample cavity in the unit of cm^3.

Military standards provide more flexible failure criteria based on the volume of the package. The minimum cavity volume in MIL-STD-883 TM1014 is 0.01 cm^3, with a failure threshold for leakage in air of 5×10^{-8} atm·cm^3·s^{-1}. In MIL-STD-750 TM1071, for the same minimum cavity volume, the failure threshold for leakage in air is 1×10^{-9} atm·cm^3·s^{-1}. The helium mass spectrometer can measure leak rates between 10^{-9} and 10^{-12} atm·cm^3·s^{-1}.

Before being transferred to the helium mass spectrometer, if helium leaks through a large leakage path from the sample, the package will pass the fine leak test. To avoid misjudgment, a total bubble test is also required for the sample. Prior to immersing the package in a testing liquid (such as FC-84), the package structure is placed in a chamber at a pressure of 5 T or lower for at least 30 min. Then, inert gas is used to pressurize the sample to above 30 psi for several hours. After releasing the pressure, the sample is dried in air and finally immersed in an indicator liquid (such as FC-40) while the temperature is raised to 125 °C. Since the boiling point of the indicator fluid is higher than 125 °C, while the boiling point of the detector fluid is lower than 125 °C, any significant leakage will result in observable bubbles.

Given the known leak rate, the lifespan of a hermetic electronic package can be calculated. Considering the typical micro-cavity volume of 0.1 mm^3 in a hermetic electronic package and assuming the maximum allowable internal pressure is 1 mbar (1 bar = 100 kPa), the required device lifetime is 5 years. According to Eq. (4.5), the leak rate of the package must be less than 6.34×10^{-16} mbar·s^{-1}. This estimate does not account for different leak rates associated with different gases and assumes the package structure is in air at standard atmospheric pressure.

$$L \approx \frac{\Delta P \cdot V}{t} \tag{4.5}$$

In the equation, L is the leak rate in the unit of mbar; ΔP is the change in cavity pressure, in the unit of mbar; V is the cavity volume, in the unit of L; and t is the device lifetime, in the unit of s.

(2) Radioactive isotope fine leak testing

The MIL-STD-883 standard describes the use of Kr-85 radioactive tracer gas for fine leak testing. According to the standard, the sample is pressurized in a Kr-85/air mixture. After the surface gas is flushed away by air, the hermetic package is placed into a scintillation counter. The γ-rays accompanying the β-decay of Kr-85 penetrate the package, and the scintillation counter can measure the amount of Kr-85 that has leaked into the package.

(3) Carbon fluoride compound gross leak testing

The carbon fluoride compound gross leak testing is suitable for testing the leak integrity of packaged devices. The device is first immersed in a low-boiling-point carbon fluoride compound liquid (such as FC-84) and then pressurized. Afterward, the device is immersed in a high-boiling-point carbon fluoride compound liquid at 125 °C. As the low-boiling-point carbon fluoride compound inside the device vaporizes and expands, the gas will escape from the device into the carbon fluoride liquid. Consequently, bubbles will be observed at the leak points of the device.

(4) Dye penetration gross leak testing

Dye penetration gross leak testing is a destructive leak test. The packaged device is first submerged in a dye solution and pressurized for a certain period. Then, ultraviolet light at a specific frequency is applied, and the device is inspected with a magnifying glass to determine whether it has any leaks.

(5) Weight-gaining gross leak testing

Weight-gaining gross leak testing involves immersing the packaged device in a specific liquid, applying pressure for a certain period, and then determining whether the device leaks by measuring the change in mass between its initial and final states.

4.3.2 Bonding Performance Testing

The main methods of interconnecting chips with substrates include wire bonding (WB), tape automated bonding (TAB), flip-chip (FC), redistribution layer (RDL), and through-silicon via (TSV) interposers. Among these, wire bonding is widely used due to its low cost and high flexibility. However, in the civilian sector, the inductance and crosstalk problems of wire bonding at high frequencies are significant, so new technologies like FC, RDL, and TSV interposers have received widespread attention and rapid development [22–24]. For military-grade, high-reliability, aerospace, and avionics applications in hermetically sealed packages, wire bonding remains the mainstream technology for chip-to-substrate interconnection [22, 23].

In hermetically sealed packages, the wire bonds are connected between the chip's solder bumps and the package's pads, creating a suspended structure that does not contact other components within the package. The stronger the wire's resistance to deformation, the better its reliability. Therefore, tests for wire bonding strength under acceleration, vibration, shock, and thermal cycling are critical. Existing bonding strength assessments are mainly based on methods from GJB 128A—1997 *Test Methods for Semiconductor Discrete Devices* (Method 2037) and GJB 548B—2005 *Microelectronics Device Test Methods and Procedures* (Method 2011). In GJB 128A—1997 Method 2037, there are three test methods: test condition A (wire pull force), test condition B (wire pull force), and test condition C (internal wire lead pull force). GJB 548B—2005 Method 2011 includes six test methods: test condition A (bond pull-out), test condition C (wire pull force), test condition D (wire pull force), and others. The following describes the testing methods for straight wire bond strength and non-straight wire bond strength.

(1) Testing method for straight wire bond strength

The schematic diagram for testing the straight bond strength is shown in Fig. 4.3.

In Fig. 4.3b, F_1 represents the tensile load on the bonded wire, while F_2 and F_3 are the corresponding tensile forces on both sides of the bonded wire. The relationship between the three is as follows.

$$F_1 = F_2 \times \sin\theta_2 + F_3 \times \sin\theta_1 \tag{4.6}$$

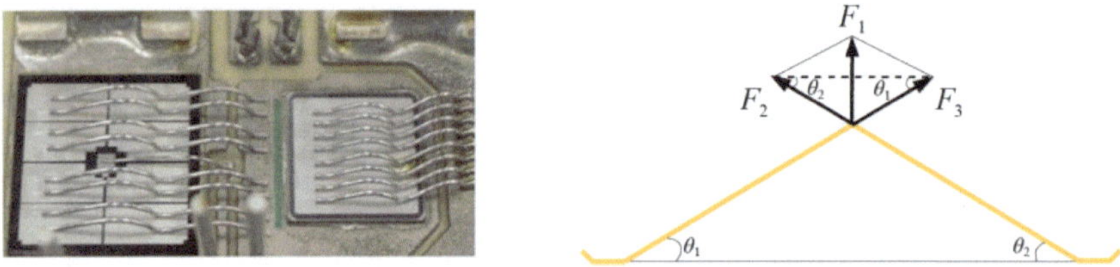

(a) Physical Straight Wire Bond (b) Wire Pull Force Test Schematic Diagram

Fig. 4.3 Schematic of straight wire bond strength testing

(a) Physical Non-Straight Wire Bonding (b) Wire Pull Force Test SchematicDiagram

Fig. 4.4 Schematic diagram of the non-straight wire bond strength testing

When the bonding points on both sides of the wire are in the same plane, $\theta_1 = \theta_2$ and $F_2 = F_3$, then:

$$F_1 = 2 \times F_2 \times \sin \theta_1 \tag{4.7}$$

The bonding wire experiences force in two directions. In the horizontal direction, it bears the tensile force F_{12}, where $F_{12} = F_2 \times \cos\theta_1$. In the vertical direction, it bears the tensile force F_{11}, where $F_{11} = 2 \times F_2 \times \sin\theta_1$. And when the bonding wire fails, the testing instrument will display the force on the vertical bonding wire, which is $F_1 = F_{11} = 2 \times F_2 \times \sin\theta_1$. It can be deduced that if the failure the bonding wire is primarily due to the vertical force, then: $2 \times F_2 \times \sin\theta_1 \geq F_2 \times \cos\theta_1$, which can be simplifies to: $2\sin\theta_1 \geq \cos\theta_1$. On the other hand, if the failure of the bonding wire is primarily due to the horizontal force, then: $2\sin\theta_1 < \cos\theta_1$.

(1) Testing method for non-straight wire bond strength

The testing schematic for non-straight lead bonding strength is shown in Fig. 4.4.

The force applied by the testing equipment to the wire bonding is F_1, and the actual tensile forces experienced by the two sides of the wire bonding are F_2 and F_3. The relationship between these forces is:

$$F_2 \times \cos \theta_1 = F_3 \times \cos \theta_2 \tag{4.8}$$

When θ_1 is much smaller than θ_2, $F_2 \times \sin\theta_1$ is much smaller than $F_3 \times \sin\theta_2$ and can be neglected, so $F_1 \approx F_3 \times \sin\theta_2$. Under the condition where the minimum bonding strength is V, we get:

$$F_1 / \sin \theta_2 \geq V \tag{4.9}$$

That is, when $F_1 \geq V\sin\theta_2$, the product is deemed qualified. Therefore, the failure criterion for non-straight wire bond strength can be modified to:

$$V_{\text{new}} = V \times \sin \theta_2 \tag{4.10}$$

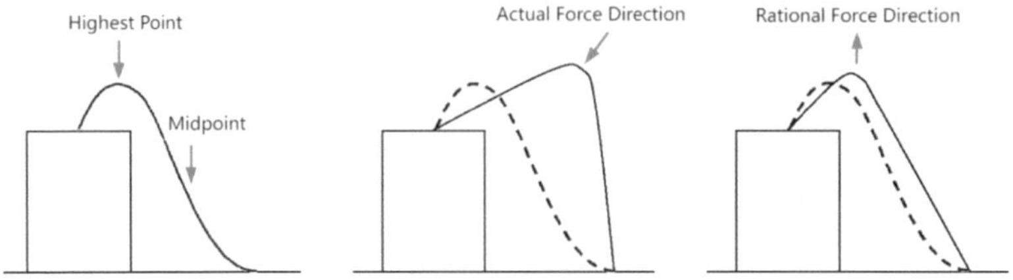

(a) wIRE SHAPE (b) The Actual Force Applied on the Wire (c) The Correct Force Application Position

Fig. 4.5 Schematic diagram of bonding force for non-straight wire bonding point

For non-straight wire bonding, the loading position of the pull hook will have a significant impact on the test results. The bonding pull strength diagram for non-straight wire bonding is shown in Fig. 4.5. Referring to Fig. 4.5, the height difference at both bonding points of the wire leads causes the highest point and the midpoint of the bonding wire to not align. If the bonding wire is loaded at the center according to the methods outlined in GJB 128A—1997 *Semiconductor Discrete Device Testing Methods* Method 2037 and GJB 548B—2005 *Microelectronic Device Testing Methods and Procedures* Method 2011, the bonding wire will deform. During the test, the actual loading force is not perpendicular to the horizontal plane, and horizontal components of force are generated, leading to measurement deviations. Therefore, for non-straight wire bonding pull strength testing, the force application point should be located between the highest point of the wire bend and the midpoint of the wire.

4.3.3 Residues Testing

One of the main causes of failure in hermetic packaging is the presence of particles inside the package. Particle Impact Noise testing (PIND) testing can effectively determine the composition and quantity of particles inside the package. PIND works by using mechanical impacts and vibrations to loosen particles that are constrained or adhered to the hermetic package. After the particles are loosened, a fixed-frequency vibration is applied to induce their movement. Once the particles move within the hermetic package, they will slide and collide randomly with the internal components of the package, generating acoustic and elastic waves. These waves propagate through the package shell, producing a reverberation signal, which can be converted to determine the displacement signal of the particles. Acoustic sensor technology can capture the displacement signal, which is then amplified, collected, processed, and finally displayed through a series of operations.

The randomness of PIND test results is relatively high, and it is often the case that failure cannot be reproduced upon re-testing. Therefore, in PIND testing, a failure is determined when a single failure event occurs.

The particles causing failure in PIND testing mainly include the following types: silicon debris (chips) from the chip edge, ceramic particles from the ceramic package shell, adhesive materials on the chip surface, wire bonding fragments, and excess alloy solder from the package cap.

4.3.4 Other Performance Testing

(1) Optical leak testing method

Optical leak testing is an emerging gas-tightness testing technology, and its principle is illustrated in Fig. 4.6. In an optical leak test, the package device is first placed in a testing chamber, and helium gas is introduced into the chamber to create a certain pressure. Under the pressure of the helium gas in the chamber, the surface of the package deforms inward. If there is no leakage, the deformation of the package surface remains constant. If leakage occurs, helium gas rapidly enters the package interior, reducing the pressure difference between the inside and outside of the package, and the deformation of the package surface begins to recover. In this case, the rate of leak in the package is proportional to the speed of deformation recovery. During the testing process, if a gas-tight package leaks, the built-in digital CCD video camera in the optical leak tester will record the interference fringes of the incident and reflected laser beams on the package surface at different times. By calculating the changes in the interference fringes, the recovery speed of the deformation of the package surface is determined. Based on factors such as the package material, thickness, area, and the internal volume of the gas-tight cavity, the leak rate of the gas-tight package in the helium gas is estimated. Optical leak testing technology can cover both gross leak testing and fine leak testing in a single test, with a testing range from the "open cover state" to 5×10^{-9} atm·cm^3·s^{-1} throughout the process.

Optical leak testing has the following main advantages: ① It can cover both gross leak and fine leak testing in a single test, with a testing range from the "open cover state" to 5×10^{-9} atm·cm^3·s^{-1};

During the optical leak testing process, it is necessary to first obtain the hardness coefficient L_0 of the package shell (measured in the unit of in/psi, where 1 in = 2.54 cm).

$$L_0 = \alpha \times \left(b/ET^3\right) \tag{4.11}$$

In the equation, α is the geometric proportionality coefficient of the package surface; b is the width of the package (in the unit of in); E is the elastic modulus of the package material (in the unit of Pa); and T is the thickness of the package shell (in the unit of in).

The equation for calculating the leak rate L of the package device (in the unit of atm·cm^3·s^{-1}) is as follows:

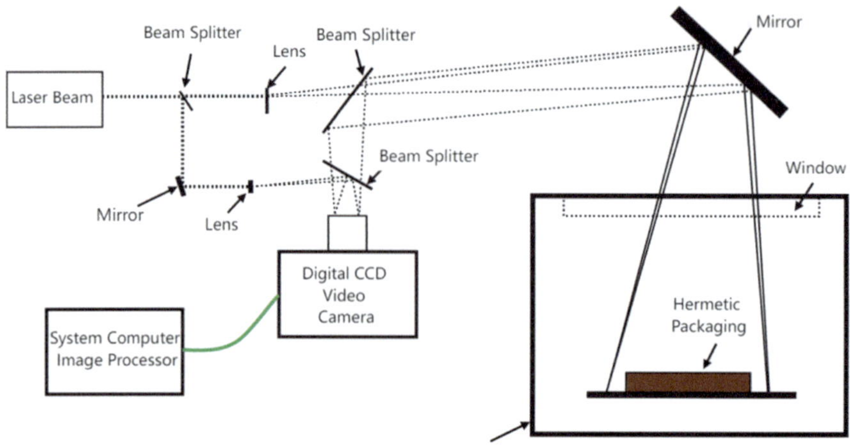

Fig. 4.6 Principle of optical leak testing

$$L = (-V_0/k_2 t) \times \ln\left(1 - d_{yt}/P_0 L_0\right) \tag{4.12}$$

In the equation, V_0 is the volume of the package cavity (in the unit of in^3); k_2 is the leak testing gas constant (1 is for air, 2.67 for helium); t is the test duration (in the unit of s); d_{yt} is the measured deformation of the package shell (in the unit of in); P_0 is the environmental pressure during the test (in the unit of psi); and L_0 is the hardness coefficient of the package shell (in the unit of in/psi).

(2) X-ray testing method

X-ray testing, as a non-destructive testing method, is widely used in the field of welding.

The main principle of X-ray testing is that when X-rays pass through different materials, the differences in density and thickness cause varying degrees of attenuation. In general, areas with higher density and greater thickness absorb more X-rays, while areas with lower density and smaller thickness absorb fewer X-rays. As a result, the X-rays detected by the detector, after attenuation, exhibit different visible light intensities, forming an image that shows varying densities.

4.4 Stress and Reliability

To ensure long-term reliability of hermetic packaging, the following requirements must be met: the environmental moisture content in the operating environment must comply with the required standards; leak testing after hermetic packaging must meet the hermeticity criteria; after experiencing environmental stress loads such as mechanical vibration, temperature cycling, thermal shock, high temperature, and HAST, the hermeticity of the package must meet the evaluation criteria. Particularly during device operation, due to environmental changes, the packaging will experience various temperature loads, leading to the expansion of packaging materials and thermal mismatch, introducing significant stress that may result in device failure. In addition to thermal mismatch, other failure mechanisms include corrosion, creep, fracture, fatigue crack formation and propagation, and delamination of thin films. For example, if any leakage path occurs during the sealing process between two bonded interfaces, moisture may enter the sealed micro-cavity, which can lead to device failure over time. Table 4.4 lists the main influencing factors related to hermetic packaging failure and their corresponding root causes. Here, "electrical" refers to voltage, current, or electric fields; "mechanical" refers to mechanical shock, vibration, acceleration, and tilting, all across the three axes; "thermal" refers to thermal shock, thermal cycling, temperature variation or low-temperature storage, as well as high and low-temperature operation. Some failures require multiple accelerating factors, while some accelerating factors can speed up multiple failure modes.

Table 4.4 Main influencing factors related to hermetic packaging failure and corresponding root causes

Influencing factors	Root causes
Shock; vibration; temperature	Particle contamination
Humidity; heat; pressure	Surface degradation or modification
Maximum applied stress; humidity; high−temperature cycling; number of cycles	Fatigue
High−temperature; maximum strain level; maximum stress level; electrical stress	Creep
Mechanical; electrical; heat; humidity	Microcracks or crack propagation
Mechanical shock; acceleration; vibration	Cracking
Humidity; temperature	Dielectric charging
Humidity; temperature	Dielectric breakdown
Thermal shock; mechanical shock; vibration; acceleration; humidity	Delamination
Temperature; humidity	Outgassing
Humidity; voltage and polarity; temperature	Anodic; corrosion
Temperature; electrical; mechanical; pressure	Elastic or plastic deformation
Temperature; humidity; environmental composition	Oxidation
High-temperature storage; thermal cycling; moisture; pressure	Degradation of hermeticity
Electrical; temperature; humidity; mechanical	Electrical short/ open circuit

4.4.1 Hermetic Packaging Reliability Evaluation Methods

The reliability evaluation of hermetic packaging needs to fully consider the reliability issues arising from moisture absorption, outgassing, and leakage during long-term use or storage. The proposed reliability evaluation process for hermetic packaging is as follows:

(1) Considering actual load conditions such as temperature cycling, thermal shock, high-temperature storage, random vibration, and sinusoidal vibration, the load conditions are refined based on GJB 548B—2005. Using finite element analysis, the influence of different load conditions on the stress distribution of hermetic packaging is initially simulated. The effects of assembly methods, PCB structural parameters, and boundary conditions on the stress at the hermetic sealing interface are analyzed, and sensitive parameters influencing the stress variation at the hermetic sealing interface are identified.
(2) Reliability testing is conducted to study the effects of environmental humidity, temperature, and mechanical loads on the hermetic packaging leakage rate (intermittent leakage degradation at the sealing interface).
(3) Combining a helium mass spectrometer and internal atmosphere analyzer, the leakage rate and internal moisture content of samples at different times during humidity, temperature, and mechanical load tests are compared and analyzed. Prior to testing, a helium mass spectrometer is used for gross leak testing of the initial samples. Gross leak testing is then performed at regular intervals during the test. Finally, moisture content analysis is conducted using an internal atmosphere analyzer on the sample.
(4) Based on the measured leakage rate and moisture content data, the environmental adaptability of the hermetic packaging is assessed. Combining the stress distribution results from simulations, sensitive parameters affecting intermittent leakage degradation at the hermetic sealing interface are identified, and the degradation mechanism is clarified.
(5) Based on basic theories of crack width, length, and gas dynamics, the mechanism of intermittent leakage degradation at the sealing interface is studied. The relationship between leakage rate and crack width and length is quantitatively analyzed using the molecular flow model. According to the molecular flow model, the relationship between leakage rate, temperature, pressure, and other parameters is as follows.

$$R = F\sqrt{\frac{T}{M}}(P_1 - P_2) \tag{4.13}$$

In the equation, R is the leakage rate; F is the flow conductance of the leakage path; T is the thermodynamic temperature; M is the molar mass of the permeating gas; P_1 is the pressure at the high-pressure end of the leakage path; and P_2 is the pressure at the low-pressure end of the leakage path.

For irregular leakage paths in hermetic packaging, it can be approximated as the superposition of multiple small leak holes. Since the pressure difference, temperature, and molecular mass remain almost constant during the testing process, the leakage rate primarily changes with the variations in the length and diameter of the leakage holes. Based on fatigue theory, when the maximum thermal stress at the sealing interface exceeds the material's maximum tensile strength, the diameter of the leakage hole will gradually increase during the test, while the length of the hole will decrease, leading to an increase in the leakage rate as the test duration extends.

(6) By combining the sealing testing method 1014.2 and the internal moisture content testing method 1018.1 from GJB 548B—2005, a degradation life evaluation method for intermittent leakage in hermetic packaging is established.

In fact, three single molecular layers of water (1.2×10^{-7} cm) is sufficient to promote corrosion reactions in the package. Assume the surface area of the hermetic packaging cavity is S, in the unit of cm^2; and the volume is V, in the unit of cm^3.

Then the volume of liquid water $V_{H_2O} = S \times 1.2 \times 10^{-7} cm^3$, and the mass of 1 cm^3 of water is 1 g. Therefore, the mass of liquid water is $S \times 1.2 \times 10^{-7}$ g. The molar mass of water vapor is 18 g, and the gaseous volume is 22.4 L. The volume occupied by these three single molecular layers of water is $\frac{S\times 1.2\times 10^{-7}\times 22.4}{18} \mathrm{L} = S\times 1.49\times 10^{-4} cm^3$. Dividing this by the volume of the packaging yields the water vapor content inside the hermetic packaging

$$\text{Water Vapor Content} = \frac{\textit{Water Vapor Volume} \times 10^6}{\textit{Package Volume}} = \frac{S}{V} \times 1.49 \times 10^2 \mathrm{ppm} \tag{4.14}$$

For molecular flow, $R = F_m\sqrt{\frac{T}{M}}(P_1 - P_2)$. The total amount of gas that leaks into the package from the sealing moment (t_0) to a later moment (t) is the integral of the leak rate over a period of time. The amount of gas leaked into the package Q_{in} can be expressed as follows.

$$Q_{in} = \Delta P_i\sqrt{\frac{T}{M}}\left(1 - e^{-\frac{Rt}{V\Delta P}}\right) \tag{4.15}$$

In the equation, ΔP_i represents the pressure difference at the initial moment, in the unit of atm; t is the time taken for the gas to leak into the package, in the unit of s; and V is the volume of the package, in the unit of cm^3.

Substitute R from (4.14) into Eq. (4.15), we get:

$$Q_{in} = (P_1 - P_2)\sqrt{\frac{T}{M}}\left(1 - e^{-\frac{3.81D^3Tt}{VM(l_0-l_c)}}\right) \tag{4.16}$$

Combining the Eq. (4.14), the degradation time of the hermetic packaging with cavity volume V is $\frac{S}{V} \times 1.49\times10^2 = (P_1 - P_2)\sqrt{\frac{T}{M}}\left(1 - \mathrm{e}^{-\frac{3.81D^3Tt}{VM(l_0-l_c)}}\right)$. Solving for time t, we get:

$$t = -\frac{VM(l_0 - l_c)}{3.81D^3T}\ln\left(1 - \mathrm{e}^{-\frac{149S}{V(P_1-P_2)}\sqrt{\frac{M}{T}}}\right) \tag{4.17}$$

The Eq. (4.17) represents the intermittent leakage degradation life of hermetic packaging under a constant leak rate, in which $(l_0 - l_c)$ is related to the molecular flow leak holes in the packaging structure.

4.4.2 Degradation Features of Hermetic Packaging Under Combined Humidity and Temperature Loads

Under pure thermo-mechanical loading, weak points in hermetic packaging are highly prone to fatigue damage, leading to the initiation and propagation of microcracks. This phenomenon is time-dependent and can cause device parameter drift, eventually resulting in breakage and complete device failure. Fatigue typically starts at locations with localized stress concentration, such as the welding interface of the hermetic cover. The fatigue damage at the welding interface is essentially a cumulative process of material degradation, characterized by the following features: ① cracking may occur even when cyclic stress is far lower than the welding material's strength limit, due to the accumulation of time-dependent damage; ② before cracking occurs, the plastic material does not exhibit significant residual deformation. In fact, such damage becomes more pronounced under high humidity conditions or even saturated humidity. Therefore, it is necessary to investigate the thermal-variability adaptability of the hermetic packaging's sealing interface under saturated humidity conditions. Figure 4.7 shows the schematic of a recommended constant-humidity and temperature-change testing platform designed to evaluate the hermetic packaging's performance under constrained stress conditions. By monitoring changes in the moisture content within the cavity, the intermittent leakage degradation of the hermetic packaging can be assessed.

Under pure thermo-mechanical loads, the weak points of hermetic packaging are highly susceptible to fatigue damage, leading to the initiation and propagation of microcracks. This is a time-dependent phenomenon that can cause device parameter drift and ultimately fracture, resulting in complete device failure. Fatigue typically begins at locations with local stress concentration in the material, such as the welded interface of the hermetic lid. Fatigue damage at the welding interface of hermetic packaging is essentially a process of cumulative material damage, mainly characterized by the following features: ① Cracking can still occur due to time-dependent damage accumulation, even when the cyclic stress is much smaller than the strength limit of the welding material; ② Before cracking occurs, the plastic material does not

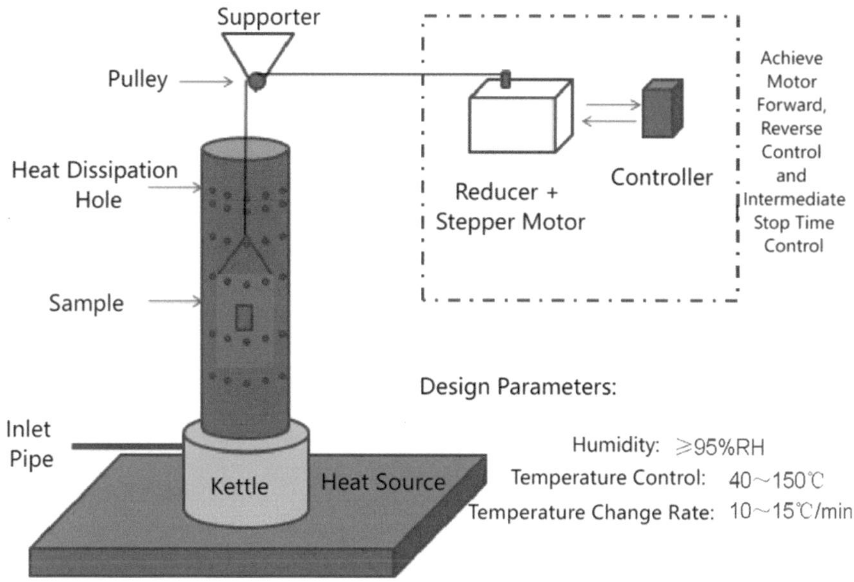

Fig. 4.7 Schematic diagram of the constant-humidity and temperature-change testing platform

cause significant residual deformation. In fact, this type of damage becomes more pronounced under high humidity or even saturated humidity conditions. Therefore, it is necessary to examine the thermal variation adaptability of the sealing interface of hermetic packaging under saturated humidity. Figure 4.7 shows the recommended constant humidity temperature variation test platform for assessing the hermetic packaging's constraint stress conditions. The degradation of intermittent leakage in hermetic packaging is evaluated by monitoring the change in water vapor content inside the cavity [25] (Figs. 4.8, 4.9 and 4.10).

4.4.3 Outgassing Behavior of Adhesive Under High-Temperature Load

In hermetic packaging, the adhesive material is sometimes made from an epoxy resin mixed with additives. The composition of this adhesive material is unstable, and at high temperatures, it can lead to the oxidation and decomposition of organic compounds, releasing volatile substances that change the internal atmosphere of the hermetic package. Based on numerous experimental results and data analysis, it has been found that there are three main sources of outgassing within hermetic packaging: Surrounding gases entering the package cavity during the sealing process; Gases or moisture adsorbed or dissolved on the surface of various materials within the package (such as the lid, inner walls of the packaging shell, chip, adhesive, bond wires, etc.)

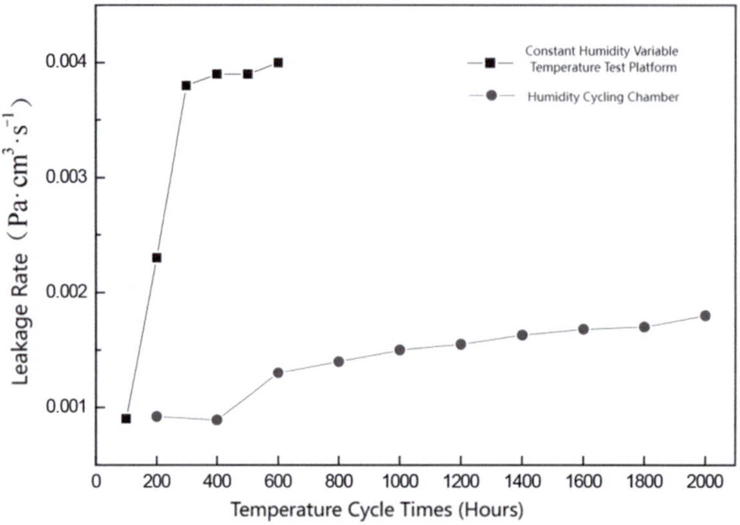

Fig. 4.8 The trend of leakage rate variation with the number of thermal cycles

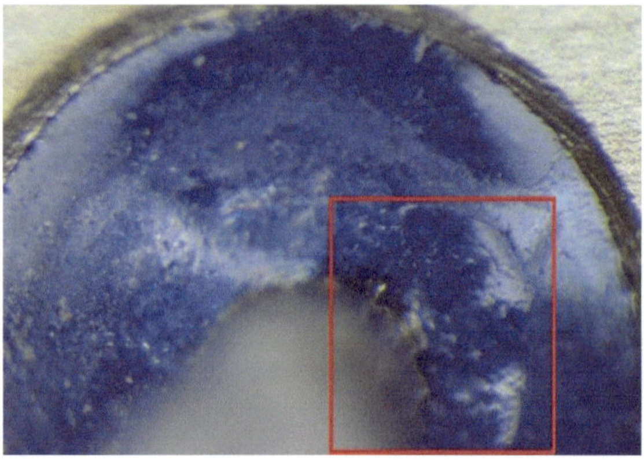

Fig. 4.9 Lead terminal—glass insulator interface crack

that precipitate during the high-temperature sealing process; Gas leakage from the hermetic package, causing external gases to enter. Among these, moisture-induced corrosion and hydrogen-induced hydrogen poisoning have the most significant impact on the reliability of internal components of hermetic packages.

The impact of moisture on components can be summarized in two aspects. On one hand, increased moisture can cause corrosion of the chip metallization. The water molecules on the chip's surface, under the influence of an electric field, undergo electrolysis and corrode the chip surface, affecting the device's reliability. On the other hand, increased moisture affects the electrical performance of the chip. This is because the residual water film on the chip's surface contains positive and negative ions, resulting in conductivity. Under voltage load, ionic conductivity can lead to a

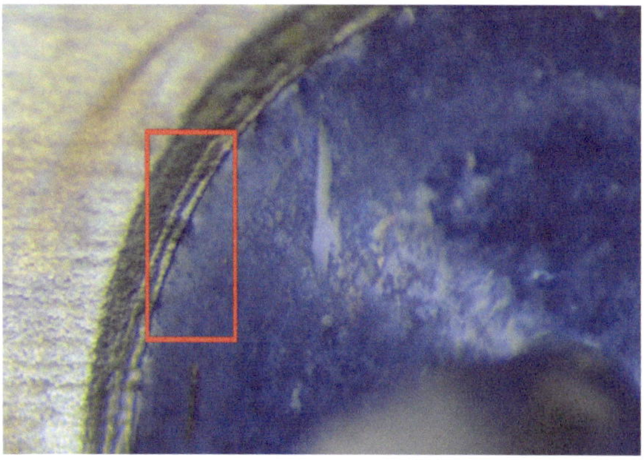

Fig. 4.10 Glass insulator—casing interface crack

significant increase in leakage current, causing drift in device electrical parameters and degradation in electrical performance. The impact of moisture varies for different types of devices. For bipolar devices, moisture leads to a sharp increase in reverse leakage current of the PN junction, a decrease in reverse breakdown voltage, and a reduction in the current amplification factor. For MOSFET devices, the migration of positive and negative ions induces opposite charges on the chip's surface, leading to the formation of an inversion conductive channel in the depletion region of the chip's surface, causing a degradation in transconductance and threshold voltage shift.

The impact of hydrogen on devices mainly manifests as hydrogen poisoning, which is particularly notable in microwave devices. During the manufacturing process of microwave devices using iron-based or nickel-based alloys and electroplating layers, hydrogen gas is often used, and iron-containing metals and nickel coatings can adsorb hydrogen on their surfaces. If the device is placed in hermetic packaging, the hydrogen content can reach several percentage points of partial pressure after being released at high temperatures. Research shows that even a hydrogen partial pressure as low as a few hundred millipascals can cause degradation in GaAs devices. The fitting curve showing the variation of hydrogen content with time and temperature is shown in Fig. 4.11, reflecting the release behavior of hydrogen during high-temperature storage and its impact on GaAs samples. From Fig. 4.11, it can be seen that at storage temperatures of 125 and 150 °C, the hydrogen content increases with the length of storage time. As the temperature rises to 175 °C, the hydrogen content shows an initial increase, followed by a decrease, and then a slow increase. After 1008 h of storage, the hydrogen content sharply decreases. This is speculated to be due to the device's hermeticity deteriorating after 1008 h of storage at 175 °C, leading to the release of hydrogen from the device and a rapid decrease in its content. Hydrogen has a significant impact on the lifespan of GaAs devices. The failure process of most GaAs devices can be well predicted using Eq. (4.18).

$$t = AP^n \exp\left(\frac{E_a}{K_B T}\right) \tag{4.18}$$

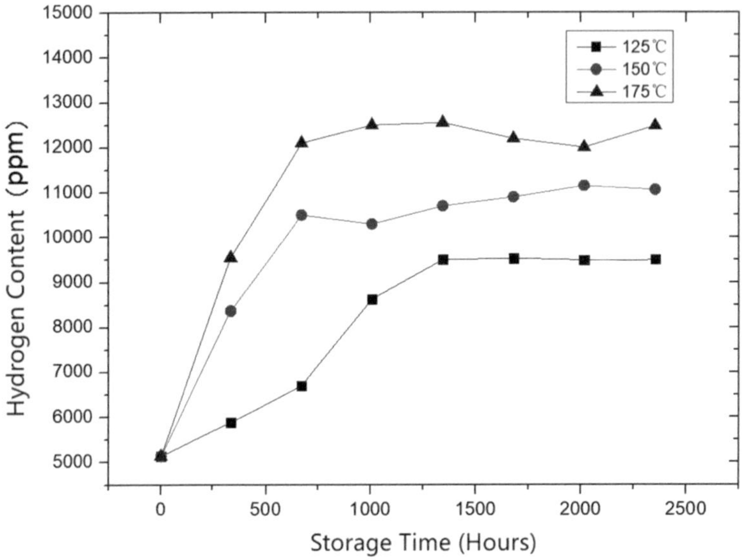

Fig. 4.11 Fitting curve of hydrogen content variation with time and temperature

In the equation, t represents the mean lifetime (in the unit of hours) when the source-drain saturation current (I_{DSS}) decreases by 10%; A is a constant (5.46 × 10^{-6}); P is the hydrogen partial pressure; $n = -0.7935$; E_a is the activation energy (0.73 eV for hydrogen-induced degradation); K_B is the Boltzmann constant (8.615 × 10^{-5}); and T is the environmental thermodynamic temperature (in the unit of K).

If the condition for the storage or service life of the GaAs device is n years, then according to Eq. (4.19), the threshold value of the hydrogen partial pressure at a certain temperature T can be obtained.

$$P = \exp\left[\frac{\left(\ln\frac{t}{A} - \frac{E_a}{K_B T}\right)}{n}\right] \tag{4.19}$$

In the equation, the meaning of the parameters are the same as those in the Eq. (4.18).

4.4.4 Brittle Fracture Under High-Frequency Vibration Loads

In hermetic packaging, the hermetic lid can block the influence of the external atmosphere on the internal structure of the package, significantly increasing the reliability of the devices inside. However, weak points at the connection interface of the hermetic lid remain a critical factor affecting hermeticity. For example, the alloy solder at the weld seam can overflow and extend to the chip end [24, 26]. Under external stresses (such as high-temperature loads, vibration loads, constant acceleration loads, shock loads, collision loads, etc.), the thin lid may undergo significant deformation, causing the brittle solder near the lid to detach. These detached solder particles can become free particles and contaminate the hermetic cavity. In more severe cases, a decline in the connection quality at the hermetic lid interface can result in leakage, leading to the contamination of the internal structure by the external environment, which can cause rapid device failure. To gain a clearer understanding of the failure of the hermetic lid, the finite element method was used to simulate the crack propagation process at the flat lid and T-shaped lid connection interfaces (see Figs. 4.12 and 4.13). Clearly, the hermetic lid is prone to failure at the edges of the board. The flat lid tends to develop cracks at the four corners, with cracks extending from the edges toward the center. The T-shaped lid is prone to cracking at the edge of the lid, with cracks also extending from the edges toward the center.

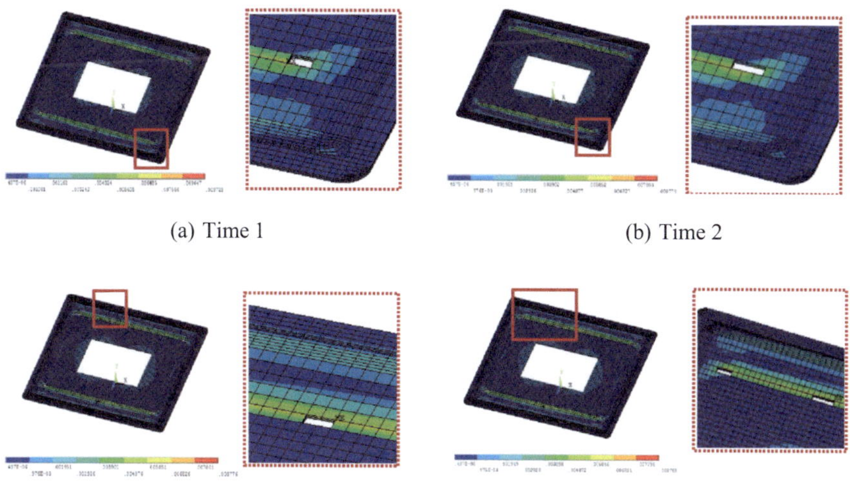

(a) Time 1 (b) Time 2

(c) Time 3 (d) Time 4

Fig. 4.12 Failure and crack propagation of the flat hermetic cover

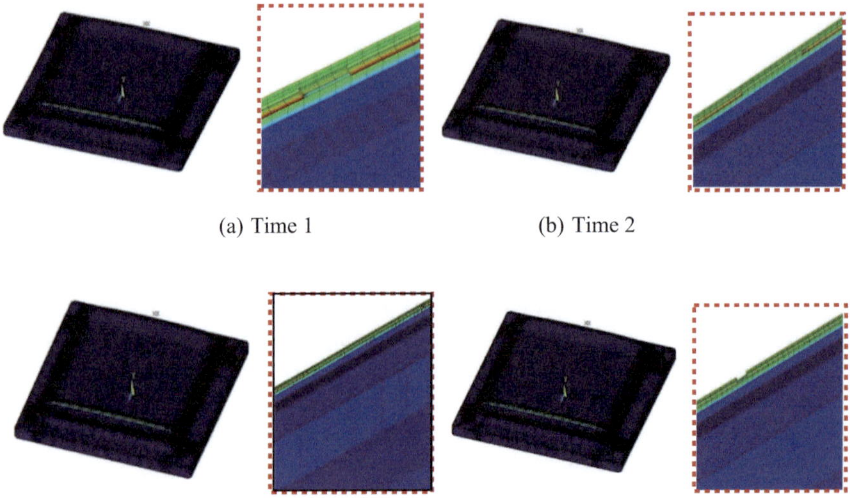

(a) Time 1 (b) Time 2

(c) Time 3 (d) Time 4

Fig. 4.13 Failure and crack propagation of the T-shape cover

4.5 Typical Failure Cases of Hermetic Packaging

4.5.1 Analysis of the Intermittent Leakage Degradation Mechanism at the HIC Metal-Glass Packaging Interface

This case study combines virtual simulation and experiments to analyze the intermittent leakage degradation mechanism of the metal-glass packaging interface of a hermetic HIC under temperature variation ranging from − 55 to 125 °C and 100% RH. Figure 4.14a shows a cross-sectional slice of the key lead-out terminal of the hermetic HIC. It is evident that there are significant cracks at the metal-glass packaging interface, which corresponds closely to the Von-Mises stress distribution in the finite element simulation shown in Fig. 4.14b. Preliminary analysis suggests that the damage mechanism of the metal-glass packaging interface is crack propagation caused by cyclic stress fatigue degradation.

To further clarify the relationship between leakage rate and cracks at the metal-glass packaging interface, the kinetic theory of gas flow under existing crack width and length conditions needs to be applied. Based on gas dynamics, the type of gas flow can be classified according to the mean free path (mfp) and leak path diameter (D) as follows: (a) when mfp/D is less than 0.01, the flow is viscous; (b) when mfp/D is greater than 1.0, the flow is molecular; (c) when mfp/D is between the above two values, the flow is transitional. The mfp of the gas is:

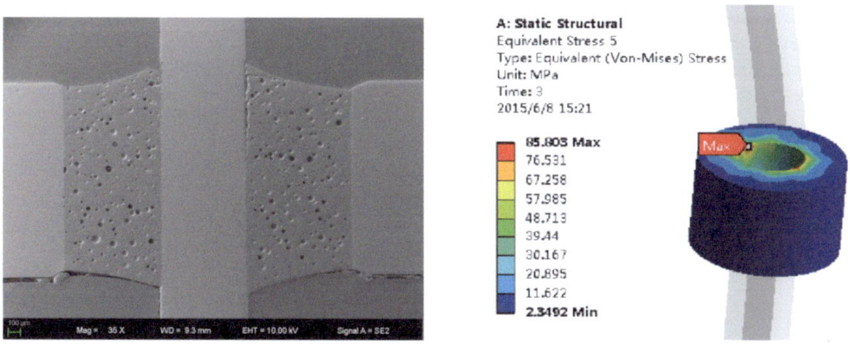

(a) Cross-section Slice Image (b) Von-Mises Stress Distribution (125°C)

Fig. 4.14 Cross-sectional slice image of the key lead terminal of the hermetic HIC and Von-Mises stress distribution map of the metal-glass packaging interface

$$\text{mfp} = \frac{kT}{\sqrt{2}\pi P\sigma^2} \tag{4.20}$$

In the equation, k is the Boltzmann constant; T is the thermodynamic temperature; P is the pressure; and σ is the molecular diameter. The method used in this experiment is the helium mass spectrometry leak testing. During the leak testing process, helium is the main gas flowing into the package, and the specific parameters are as follows: $T = 298$ K, $k = 1.38 \times 10^{-16}$, $P = 9.11 \times 10^5 Pa$, $\sigma = 2.2 \times 10^{-8}$m. Substituting these into Eq. (4.20), we get mfp $\approx 2.1 \times 10^{-5}$m. Based on this, it is determined that the diameter of the interface cracks observed in this experiment is less than 21 μm, and the leakage mode is molecular flow. For irregular leak paths, they can be approximated as the superposition of multiple fine leak holes. When the length of the fine leak holes is much larger than their diameter, the flow conductance is as follows.

$$F = \frac{3.81D^3\sqrt{\frac{T}{M}}}{l} \tag{4.21}$$

In the equation, D is the leak hole diameter; l is the leak hole length. The length of the molecular flow leak hole, l, is equal to the original length of the glass insulator minus the crack length, specifically:

$$l = l_0 - l_c \tag{4.22}$$

In the equation, l_0 is the original length of the glass insulator ($l_0 = 1.5$ mm); l_c is the crack length obtained from SEM in the experiment (see Fig. 4.15).

Substituting Eqs. (4.21) and (4.22) into the Eq. (4.13), we get:

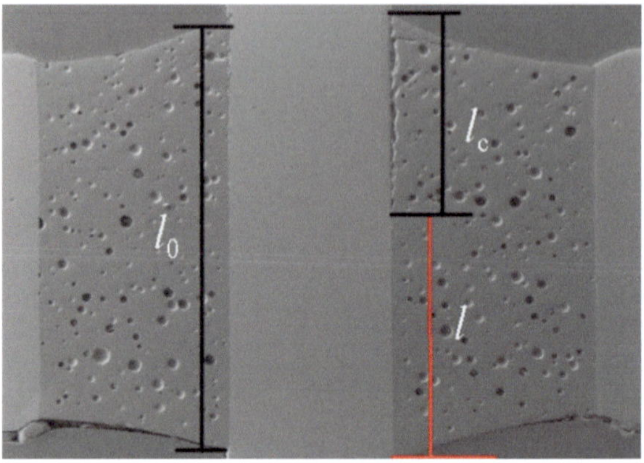

Fig. 4.15 Marking of the crack length at the metal-glass packaging interface of the glass insulator

$$R = \frac{3.81D^3T}{(l_0 - l_c)M}(P_1 - P_2) \tag{4.23}$$

Since the pressure difference, temperature, and molecular mass remain essentially constant during the testing process, the leakage rate primarily changes with the variations in the leak hole length and diameter. The leakage rate is proportional to the cube of the bottom leak hole diameter and inversely proportional to the bottom leak hole diameter. Based on fatigue theory, when the maximum thermal stress of the glass insulator exceeds the maximum tensile strength of the glass, the diameter of the glass insulator's leak hole continuously increases as the test progresses, and the leak hole length gradually decreases. This leads to an increase in the leakage rate as the test time extends.

4.5.2 Non-contact Online Monitoring of Random Vibration of Hermetic Cover

The hybrid microcircuit metal hermetic packaging on the PCB will experience random loads, i.e., disturbance stresses, in a random vibration environment. Under prolonged disturbance stress, the hybrid microcircuit metal hermetic packaging may experience fatigue failure, leading to cracking of the cover of the hybrid microcircuit metal hermetic packaging. The random vibration test of the hermetic packaging component is shown in Fig. 4.16.

The 3D precision laser vibration measurement system was used in the experiment to analyze the resonance behavior of the hermetic packaging module under random vibration conditions. The test bandwidth was 5 kHz, with 3200 spectral lines and a frequency resolution of 1.5625 Hz. The layout of the sample measurement points and the test site are shown in Fig. 4.17.

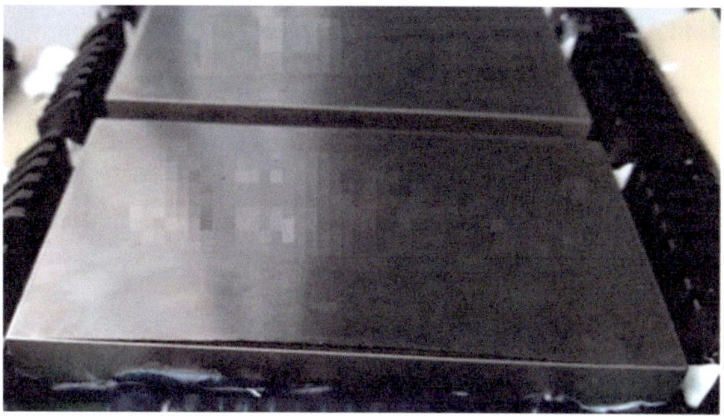

Fig. 4.16 Random vibration test of the hermetic packaging component

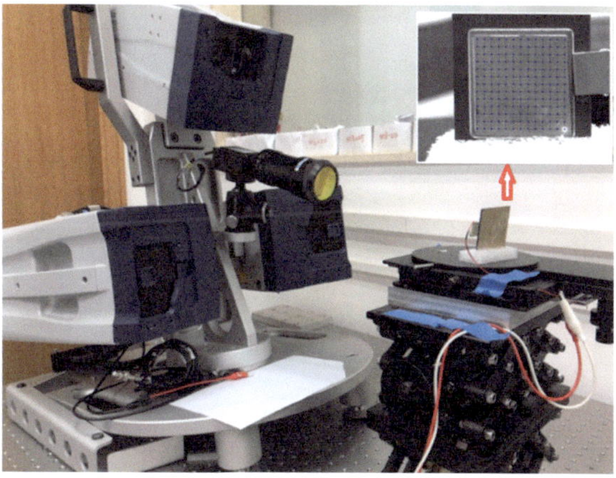

Fig. 4.17 Layout of the sample measurement points and the test site

Table 4.5 shows the comparison between the experimental and simulation results of the module shell's natural frequencies. Within the 0–2000 Hz frequency range, there are eight resonance frequencies, which are 737.33, 754.49, 1035.04, 1050.77, 1417.97, 1546.40, 1805.80, and 1833.97 Hz. The error between the experimental results and finite element simulation results is less than 5%. The maximum stress occurs at the center of the edge of the cover, and the maximum equivalent stress is 15.587 MPa.

Clearly, during the 0–2000 Hz vibration test with the lead terminals fixed, resonance damage may occur at the cover plate welding layer. In the production or disassembly or repair processes of this model's shell, resonance of the cover plate and base may occur. If ultrasonic intensity and time are not properly controlled, resonance damage may result in the cover plate welding layer, internal chips, and base insulator, potentially leading to residual stress accumulation. The finite element method (FEM) was used to analyze the residual stress distribution caused by the welding process in the hermetic welding structure. The distribution of welding residual stress along the surface of the coval alloy at L1 and L2 is shown in Fig. 4.18. Through calculation and analysis, it was found that uneven heating conditions and thermal

Table 4.5 Comparison between the experimental and simulation results of the module shell's natural frequencies

Natural frequency	Modal test results (Hz)	Finite element simulation results (Hz)	Mode shape test structure	Simulation mode shape results	Error (%)
1st level	737.33	734.96		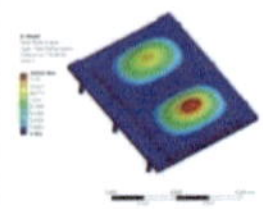	0.33
2nd level	754.49	735.72		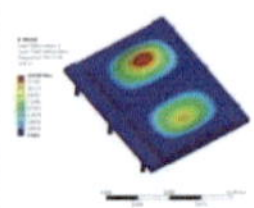	2.5
3rd level	1035.04	1074.30		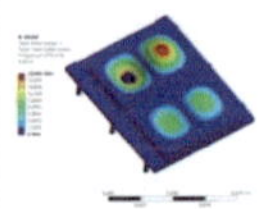	3.8
4th level	1050.77	1074.50		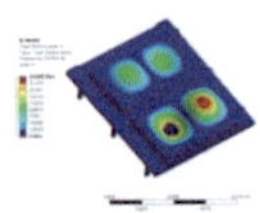	2.3
5th level	1417.97	1476.40		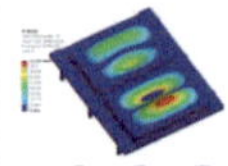	4.1
6th level	1546.40	1479.90		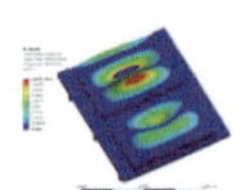	4.3
7th level	1805.80	1811.20		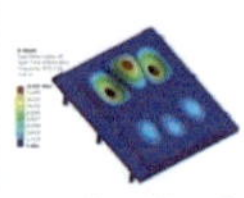	0.3
8th level	1833.97	1815.40		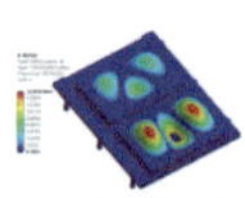	1.0

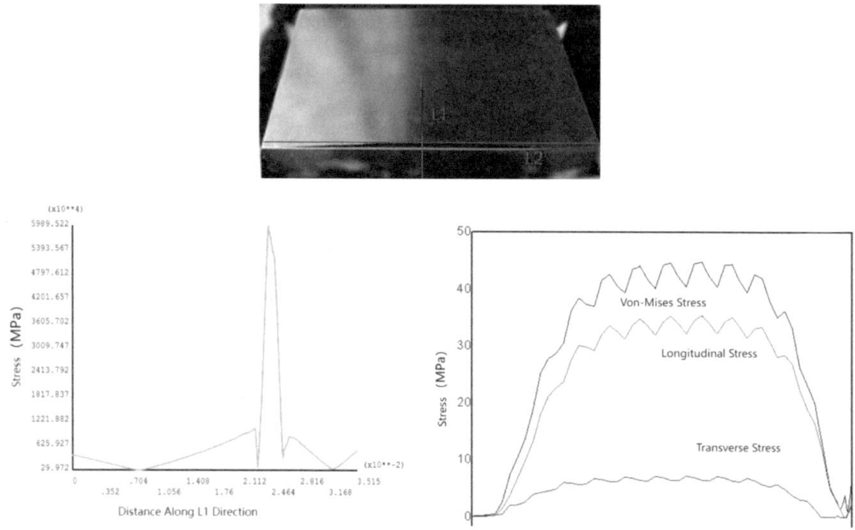

Fig. 4.18 Distribution of welding residual stress along the surface of coval alloy at L1 and L2

conductivity differences lead to temperature imbalances in the weld. The highest residual stress level reaches 64.9 MPa, located at the center of the weld structure. The residual stress distribution along the surface of the coval alloy follows the same pattern as the residual stress at the weld center, exhibiting a serrated variation with the maximum value at the center. The stress distribution analysis results indicate that the longitudinal welding residual tensile stress is concentrated at the center and ends of the weld, while compressive stress predominates at the other locations.

References

1. Yakabe S, Matsui H, Saito Y et al (2020) Multi-channel single-mode polymer waveguide fabricated using the mosquito method. J Lightwave Technol 39(2):547–556
2. Ali M, Watanabe A, Kakutani T et al (2020) Heterogeneous integration of 5G and millimeter-wave diplexers with 3D glass substrates. In: 2020 IEEE 70th electronic components and technology conference (ECTC), pp 1376–1382
3. Tao Z, Bob T, Ode M et al (2005) Tin-lead solder and its applications in the field of electronic device packaging. Electron Packag 5(8):4
4. Jueqi Y (1987) Binary alloy phase diagrams. Shanghai Scientific and Technical Publishers, Shanghai
5. Jun X, Qiang H, Gang L et al (2009) Sn-Bi series low-temperature lead-free solders and their development trends. Electron Process Technol 30(1):1–4
6. Tao W (2005) Low-temperature soldering process control of gold-tin solder. Integr Circuit Commun 23(3):4
7. Lake J, Wild R (1983) Some factors affecting leadless chip carrier solder joint fatigue life. J Mater Process Contin Innov SAMPE, 1406–1424

8. Baelmans M, Meyers J, Nevelsteen K (2003) Flow modeling in air-cooled electronic enclosures. In: 2003 IEEE 19th semi-therm symposium, pp 27–34
9. Yamamoto K, Koga S, Seino S et al (2020) Low loss BT resin for substrates in 5G communication module. In: 2020 IEEE 70th electronic components and technology conference (ECTC), pp 1795–1800
10. Kim J-H, Yoon K-S, Oh H-D et al (2020) Study on advanced substrate for double-side package to reduce module size. In: 2020 IEEE 70th electronic components and technology conference (ECTC), pp 1904–1909
11. Hoque MA, Haq MA, Chowdhury MM et al (2020) Evolution of the properties of SAC-Bi-Ni-Sb lead free solder during mechanical cycling. In: 2020 IEEE 70th electronic components and technology conference (ECTC), pp 2048–2057
12. Chang N, Chung CK, Wang Y-P et al (2020) 3D micro bump interface enabling top die interconnect to true circuit through silicon via wafer. In: 2020 IEEE 70th electronic components and technology conference (ECTC), pp 1888–1893
13. Brandtner T, Pressel K, Floman N et al (2020) Chip/package/board co-design methodology applied to full-custom heterogeneous integration. In: 2020 IEEE 70th electronic components and technology conference (ECTC), pp 1718–1727
14. Nayini M, Horn T, Patel J et al (2020) Coupled thermo-mechanical simulation methodology to estimate BGA reliability of 2.5 D packages. In: 2020 IEEE 70th electronic components and technology conference (ECTC), pp 1653–1658
15. Jacqueline S, Bunel C, Lengignon L (2020) Outstanding reliability performances of silicon capacitors for 200 °C automotive applications. In: 2020 IEEE 70th electronic components and technology conference (ECTC), pp 2133–2138
16. China Electronics Standardization Institute (2005) Test methods and procedures for microelectronic devices: GJB548B-2005
17. Aihua R, Zixue L, Weipeng N (2005) The impact of hermeticity on the water vapor content inside circuits. Electron Compon Appl 7(2):3
18. Songliang J (2002) The impact and control of water vapor content in packaging. Electron Packag 2(6):12–14
19. Zhenqiu G, Fa L (2001) Control of water vapor content inside sealed semiconductor devices. Semicond Technol 7:3
20. Microcircuit Test Methods and Procedures (2017) MIL-STD-883
21. Park H, Park M, Seo HK et al (2020) Development of CMOS-compatible low temperature Cu bonding optimized by the response surface methodology. In: 2020 IEEE 70th electronic components and technology conference (ECTC), pp 1474–1479
22. Nishad VK, Sharma R (2020) First principle analysis of Li-doped armchair graphene nanoribbons for nanoscale metal interconnect applications. In: 2020 IEEE 70th electronic components and technology conference (ECTC), pp 2278–2283
23. Kumar R, Pathania S, Guglani S et al (2020) Role of grain size on the effective resistivity of cu-graphene hybrid interconnects. In: 2020 IEEE 70th electronic components and technology conference (ECTC), pp 1620–1625
24. Zhongwei H, Shousheng L (2006) MCM-C metal hermetic packaging technology. Electron Packag 6(009):1–6
25. Wude H (2000) Prevention of glass insulator fracture in sealed relays. Mech Electr Compon 04:19–21
26. Qinong Z, Jusheng M (1995) Ways to improve the reliability of metal-glass package integrated circuit enclosures. Electron Prod Reliab Environ Test 5:19–24

Chapter 5
Failure Modes, Failure Mechanisms, and Reliability of 3D Packaging

In the past decades, with the guidance of Moore's law, the size of transistors has continuously shrunk, while chip performance and computational capabilities have steadily improved. The relentless revolution has driven the sustained development of 2D planar integrated circuits. However, as the integrated circuit (IC) industry advances, the electronics manufacturing sector is approaching the physical limit of traditional CMOS process, making it increasingly challenging to enhance performance through transistor scaling. According to the *International Technology Roadmap for Semiconductors (ITRS)*, 3D packaging based on vertical interconnect stacking has emerged as a critical research direction in the "Post-Moore era." 3D packaging stacks multiple chips or systems (e.g., image sensors, MEMS, RF modules, memory) vertically, enabling the creation of more compact, multifunctional, and intelligent systems. It offers effective solutions for emerging fields such as 5G, IoT, and AI.

Primarily, 3D packaging methodologies include CoC, PoP/PiP, and 3D Through-Silicon Via (TSV)/Through-Glass Via (TGV) integration. Up to now, 3D packaging technology has been extensively researched by universities, research institutes, and packaging-testing companies. Nevertheless, several reliability challenges must be addressed before widespread adoption, such as heat accumulation due to high power density, coefficient of thermal expansion (CTE) mismatch in heterogeneous integration, and structural delamination. This chapter first outlines the current development status and key structural features of 3D packaging. It then introduces the failure modes and failure mechanisms of 3D packaging structures, covering reliability issues in 3D chip stacking, 3D package stacking, and TSV-based 3D integration. Finally, classic 3D packaging failure cases are introduced.

B. Zhou et al., *Reliability Technology for Integrated Circuit Packaging*,
https://doi.org/10.1007/978-981-95-3885-0_5

5.1 Development and Mainstream Technology of 3D Packaging

In 1965, Gordon Moore, one of the founders of Intel, proposed that the density and performance of transistors on integrated circuits with fixed prices would approximately double every 18 months (known as Moore's Law) [1]. Moore's Law emphasizes the continuous shrinking of lithographic scales and the integration of functions within a single chip in a 2D plane. However, 3D packaging technology gradually emerged in the 1970s, initially stacking chips with different functions vertically along the Z-axis and then integrating them through vertical interconnect methods. Based on their implementation methods, 3D packaging structures are divided into 3D chip stacking, 3D package stacking, and 3D stacking based on TSV/TGV interposers [2]. Figure 5.1 illustrates the development of typical 3D packaging structures. After 2010, the development of advanced 3D packaging technologies such as 2.5D interposers, 3D integrated circuits (3D IC), and fan-out (FO) packaging has further enhanced the level of advanced 3D packaging. Advanced 3D packaging structures such as Chiplet, Chip on Wafer on Substrate (CoWoS), and Integrated Fan-out, (InFO) have been proposed and applied successively. Over the past 50 years, the interconnect line width has improved from 1000 to 1 μm, and even to sub-micron levels, with a performance improvement of over 1000 times [3].

The development trend of advanced packaging technologies is shown in Fig. 5.2, which mainly includes seven important technologies: Flip-Chip Ball Grid Array (FCBGA) packaging, Wafer Level Chip Scale Package (WLCSP), Fan-Out Wafer Level Package (FOWLP), Embedded IC, 3D Wafer Level Package (3D WLP), 3D Integrated Circuit (3D IC), and 2.5D Interposer. Most of these packaging technologies are closely related to Wafer Level Package (WLP), which allows the realization of high-performance and multifunctional packaging devices while fully utilizing the advantages of WLP, such as mass production, low cost, and small size. From

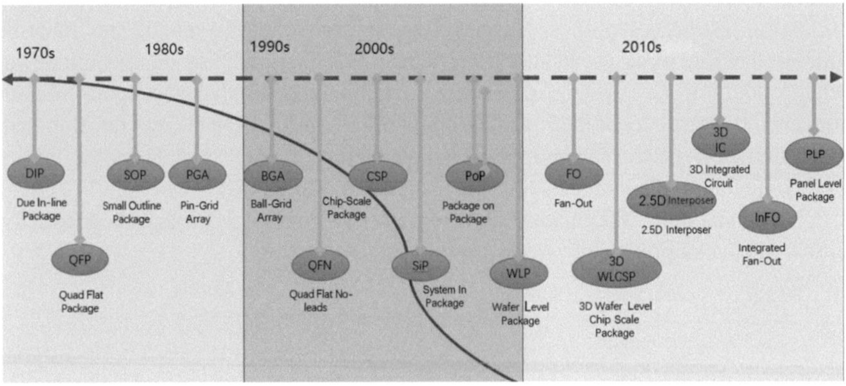

Fig. 5.1 Development of typical 3D packaging structures [3]

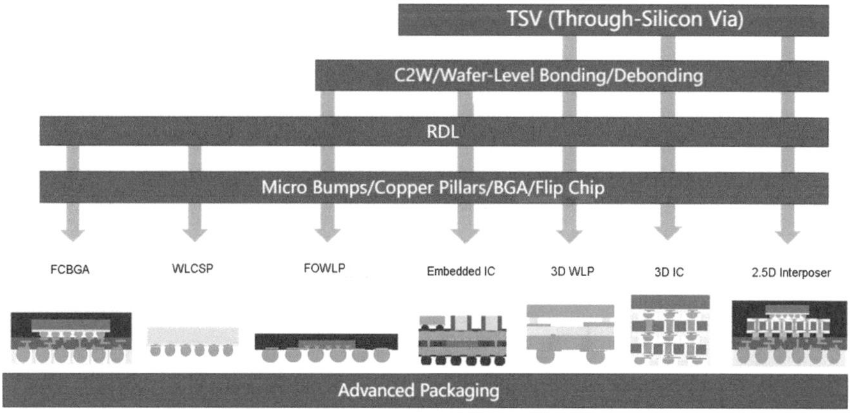

Fig. 5.2 Development trend of advanced packaging technologies

Fig. 5.2, it can be seen that the main processes supporting the above advanced packaging technologies include TSV, redistribution, bonding/debonding, etc. To meet the increasingly complex 3D integration requirements, advanced packaging technologies themselves are continuously innovating and developing [3, 4].

5.2 Main Structural Characteristics of 3D Packaging

5.2.1 3D Chip Stacking Structure

1. Wire Bonding Structure

Wire Bonding (WB) chip stacking utilizes the method of adhesion to first attach two or more bare chips to a substrate and then interconnect the chips with the substrate through wire bonding. Subsequently, redistribution is performed on the substrate, and electrodes are drawn out from the bottom ball grid array microbumps. Finally, the packaging is completed through resin molding, hermetic sealing, or non-resin molding [5]. A six-layer bare chip stacking diagram is shown in Fig. 5.3.

According to the chip stacking method, wire bonding chip stacking can be divided into four types of structures: pyramid structure, cantilever structure, side-by-side structure, and filled structure [6–8].

(1) Pyramid structure

In the pyramid structure, bare chips are stacked layer by layer from bottom to top in descending order of size, forming a shape similar to a pyramid. This packaging is called pyramid structure stacking packaging. The packaging is shown in Fig. 5.4. There is no explicit limit on the number of chip layers in this stacking packaging.

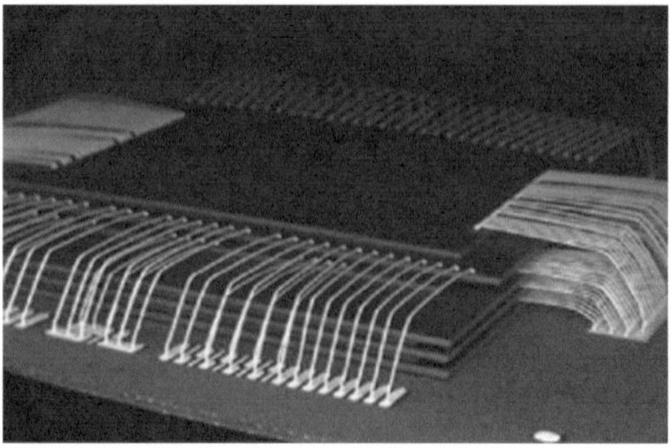

Fig. 5.3 Six-layer bare chip stacking diagram

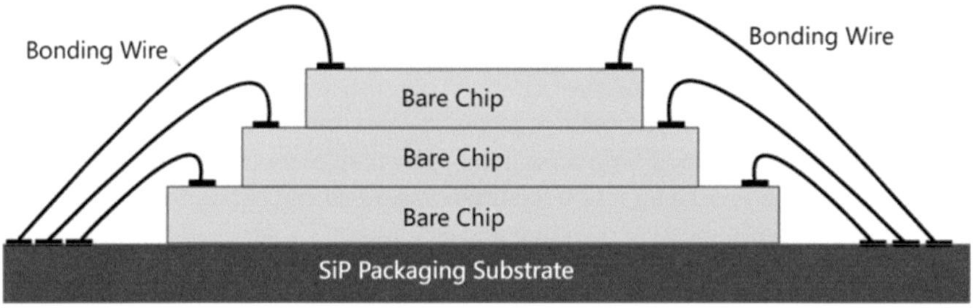

Fig. 5.4 Pyramid structure stacking packaging

It mainly considers the impact of the number of chip layers on the thickness of the packaging body and the heat dissipation between stacked chips.

(1) Cantilever structure

Cantilever stacking packaging is a method for bare chips of equal size or where the upper chip in the stack is larger. Generally, inserting dielectric between chips to elevate the upper chip is helpful for the wire bonding of the lower chip. There is no explicit limitation of the number of layer for this type of stacking packaging, but the thickness of the packaging body will limit its stacking height. At the same time, heat dissipation of the stacked chip structure needs to be considered. Figure 5.5 shows cantilever stacking packaging.

(2) Side-by-side structure

Side-by-side stacking packaging refers to stacking multiple small bare chips on a large bare chip. As shown in Fig. 5.6, since the small bare chips on the upper layer cannot be directly bonded to the bottom SiP stacking substrate, a silicon interposer is inserted between the large and small bare chips to achieve interconnection between the small bare chips and the SiP packaging substrate. The small bare chips are firstly stacked side by side on the silicon interposer, and then connected to the silicon

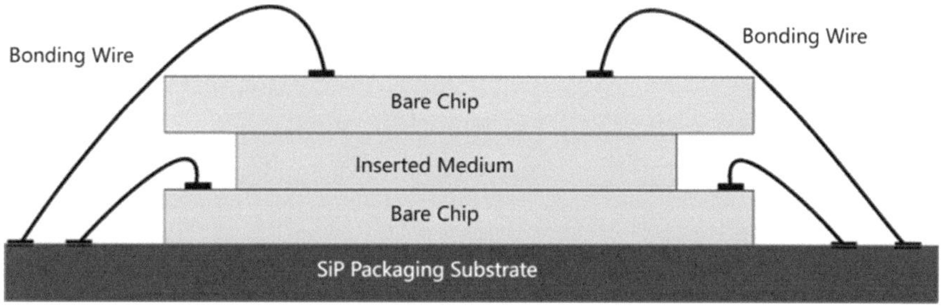

Fig. 5.5 Cantilever stacking packaging

interposer layer with a redistribution layer through bonding wires. Finally, the large bare chip is connected to the SiP packaging substrate via bonding wires, forming the side-by-side stacking packaging structure.

(2) Filled structure

Filled stacking packaging stacks two bare chips of the same size together and separates them with a polymer ball-type adhesive, as shown in Fig. 5.7. This packaging method is commonly used for memory chips, such as SRAM and flash memory. The filled stacking packaging process is simple, but it is prone to reliability issues such as de-bonding, low wire arc, and thermal mismatch between stacked layers.

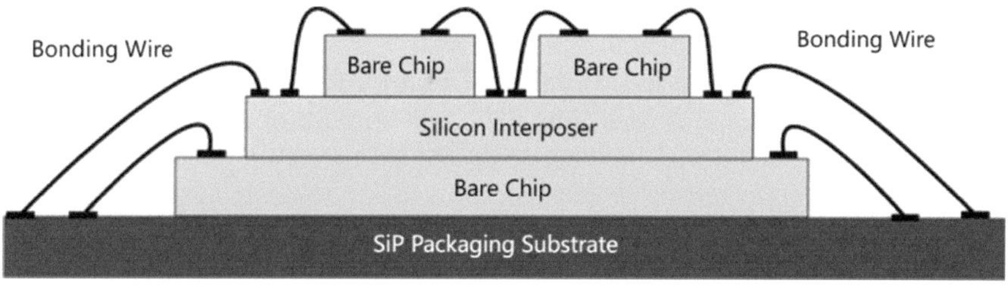

Fig. 5.6 Side-by-side stacking packaging

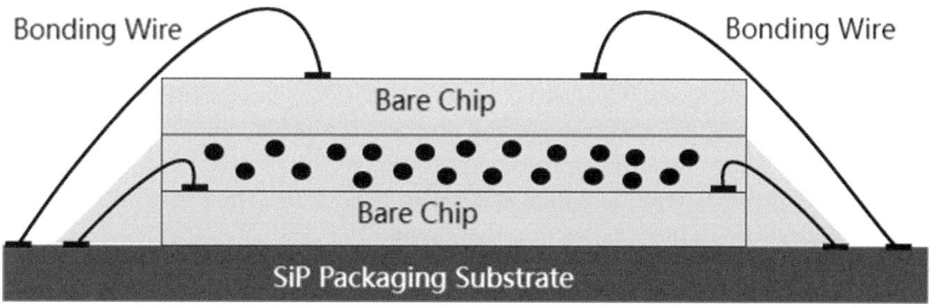

Fig. 5.7 Filled stacking packaging

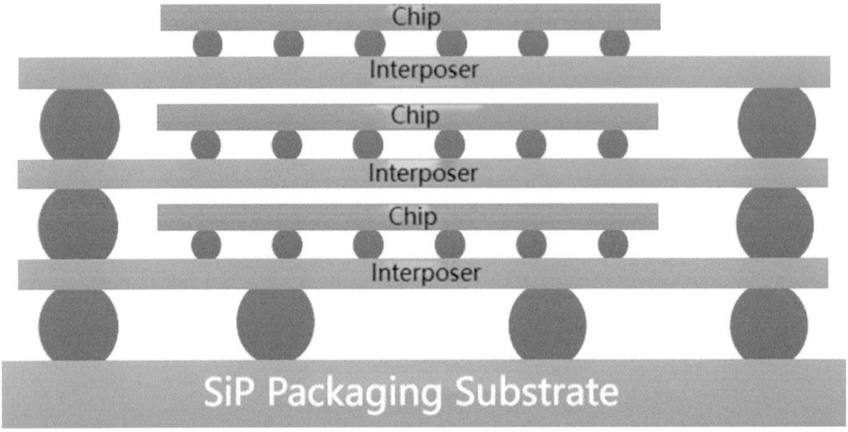

Fig. 5.8 Flip-chip stacking packaging

2. Flip-Chip Structure

Flip-Chip (FC) stacking technology is a novel leadless microelectronics packaging technique. Compared to traditional wire bonding packaging structures, the Flip-Chip stacking packaging first deposits solder balls on the I/O PADs and then flips the chip and heats it to form a connection with the substrate, as shown in Fig. 5.8. In the traditional structure, the electrical side of the chip connected to the substrate through wire bonding faces upwards, while the electrical side of the Flip-Chip faces downwards, essentially flipping the former over.

In 1960, IBM developed the world's first Flip-Chip, effectively improving the reliability of components. At that time, the main purpose of IBM's design of the Flip-Chip was to reduce wire bonding costs and improve chip reliability. Initially, Flip-Chips used copper bumps, which later evolved into solder bumps made through the bumping process on chips and then inverted and attached to the substrate, with the solder joints assembled through reflow soldering. In recent years, with the development of packaging technology and the demand for chip miniaturization, Flip-Chip has evolved into a high-performance packaging interconnection method. Currently, System in Package (SiP), Multi-Chip Module (MCM), image sensors, microprocessors, hard disk drives, medical sensors, and other devices have widely adopted Flip-Chip technology. The main Flip-Chip interconnection technologies are Controlled Collapse Chip Connection (C4), Direct Chip Attach (DCA), and Flip Chip Adhesive Attachment (FCAA). C4 technology first prepares C4 solder joints on the front or side of the substrate. After the chip and substrate are stacked, vertical interconnection between layers is completed through reflow soldering. The Flip-Chip stacking structure based on C4 interconnection is shown in Fig. 5.9.

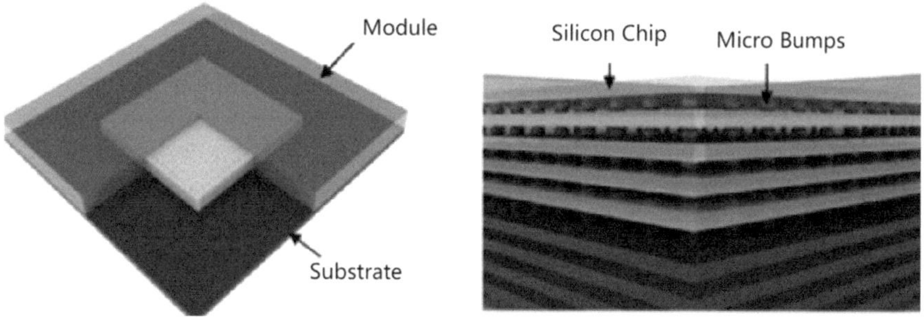

Fig. 5.9 Flip-chip stacking structure based on C4 interconnection

5.2.2 3D Package Stacking Structure

Stacking two or more packaging bodies together to form a multi-chip packaging structure is called 3D package stacking. The commonly used structure is the stacking of two packaging bodies. Depending on the packaging structure, 3D package stacking can be divided into Package in Package (PiP) and Package on Package (PoP). The two forms of 3D package stacking are shown in Fig. 5.10 [9, 10]. The mainstream implementation method in current industrial production is to directly stack the two packaging bodies together and then connect each chip to the substrate through wire bonding.

PiP is also known as device-in-device [11]. The chips inside the package are first stacked on the substrate through wire bonding, and then the stacked chips are bonded between the substrates with gold wires to form a device-in-device. PiP has the advantage of low profile and can be used with standard SMT processes. However, PiP has some drawbacks, such as low packaging yield. If any chip in the package fails, the entire packaging device will fail.

PoP is a 3D package stacking technology developed for integrated circuits in mobile devices. PoP consists of two stacked packaging bodies, with the lower packaging body and the upper packaging body interconnected through a solder ball array. Compared with traditional 3D chip stacking technology, PoP is slightly larger in size, but system companies have more component suppliers, and since the components of the lower and upper packaging bodies of PoP have already passed packaging

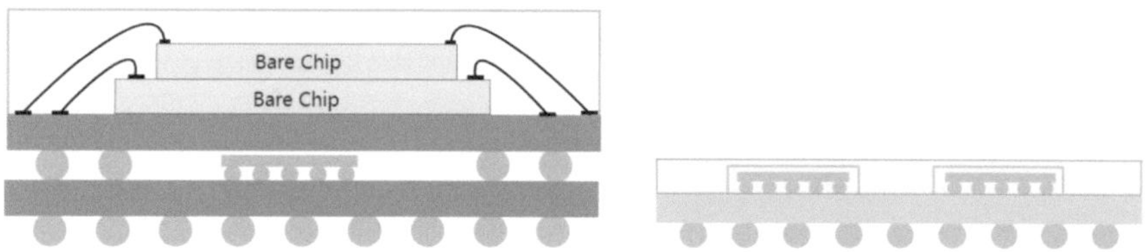

Fig. 5.10 Two types of 3D package stacking

tests, the yield is guaranteed. Therefore, PoP's system integration has both supply chain flexibility and cost control advantages. In recent years, with the further thinning of packaging bodies, warpage has become a key factor affecting PoP assembly yield. The trend of thinning makes the warpage problem more prominent, which has become a bottleneck hindering the thinning development of PoP. Therefore, various new technologies and materials are constantly emerging to reduce the warpage of packaging bodies [12–17].

5.2.3 3D TSV Packaging Structure

3D TSV packaging technology achieves electrical connections between different chip layers by creating vertical conductive channels in silicon wafers. This technology significantly reduces the length of interconnects, thereby reducing delay, capacitance, and inductance, and enabling high-speed, low-loss signal transmission between chips. It is considered the main technology for implementing 3D TSV packaging [18–20]. Figure 5.11 shows 3D TSV and stacked devices based on 3D TSV technology.

3D TSV technology can stack chips with different functions (such as analog circuits, logic circuits, RF circuits, memory, and MEMS) to achieve 3D heterogeneous integration of different chips [21–24]. Compared with other packaging structures, the heterogeneous integration structure has significant advantages in cost and performance. Today, 3D TSV technology is recognized worldwide as the fourth-generation interconnect technology. In the actual device design and production process, the trend towards achieving small-sized 3D packaging integration through 3D TSV technology is inevitable. Compared with traditional 2D wire interconnects, 3D TSV packaging has many advantages.

(1) Small size and low weight: Compared to traditional single-chip packaging, the size and weight of 3D TSV packaging are reduced to 1/50–1/40 of the original; compared to MCM technology, the size of 3D TSV packaging is reduced to 1/6–1/5 of the original, and the weight is reduced to 1/13–1/2 of the original.

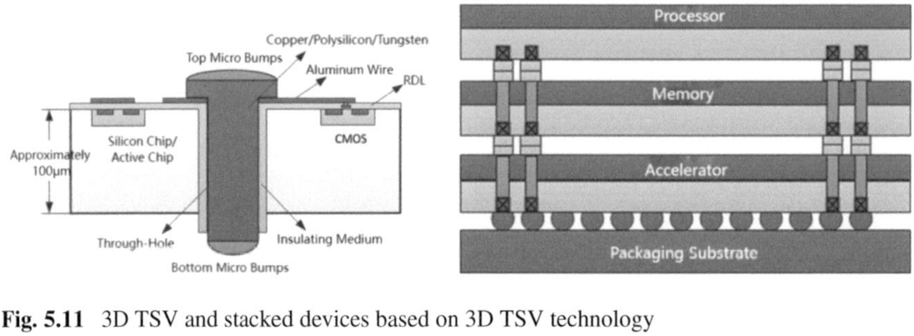

Fig. 5.11 3D TSV and stacked devices based on 3D TSV technology

(2) High integration density: Compared to 2D packaging, multiple chips can be stacked within the same package using 3D TSV technology. With the development of 3D TSV technology, higher aspect ratio TSVs will further increase packaging density, and the number of chips stacked vertically will continue to increase. Complex multi-system integration can even be achieved within a single package.
(3) Lower power consumption and signal delay: Through 3D TSV technology, vertical interconnections between chips can be realized, significantly reducing the length of interconnects and effectively reducing signal delay caused by long interconnects. It has been reported that signal delay can be reduced to the picosecond level using 3D TSV interconnections. In the microwave RF field, the reduction of interconnect length can effectively reduce device power consumption and interconnect delay, improving system operating speed.

Although 3D TSV packaging has many advantages, it also has some shortcomings, mainly including: ① Devices packaged with 3D TSV interposers have high interconnect density, small size, and high power density, so thermal reliability is an inevitable issue for 3D TSV packaging structures; ② The TSV manufacturing process is complex (including through-via, SiO_2 dielectric layer, Ti/Cu seed layer, and electroplated Cu layer), resulting in high packaging costs.

5.3 Failure Modes and Failure Mechanisms of 3D Packaging

5.3.1 Common Failure Modes of 3D Packaging

1. Failures Caused by Chip Stacking Processes

Wire bonding and flip-chip bonding are two commonly used methods in chip stacking packaging. Compared to single-chip packaging, chip stacking packaging has lower reliability, with failure modes mainly including chip cracking, delamination, bonding failure, wire shorting and breakage, thinning process defects, and solder joint failure [25–31].

(1) Chip cracking. Causes for chip cracking are as follows: one is excessive mechanical stress and the other is CTE mismatch between the chip and packaging materials. During processes with significant temperature changes, such as reflow soldering, shear and tensile stresses at the heterogeneous interfaces can lead to chip cracking. A schematic diagram of chip cracking is shown in Fig. 5.12.
(2) Delamination. Delamination can occur at interfaces such as between the chip and lead frame, between the lead frame and molding compound, and between the solder joint and substrate. Schematic diagram of chip delamination is shown in

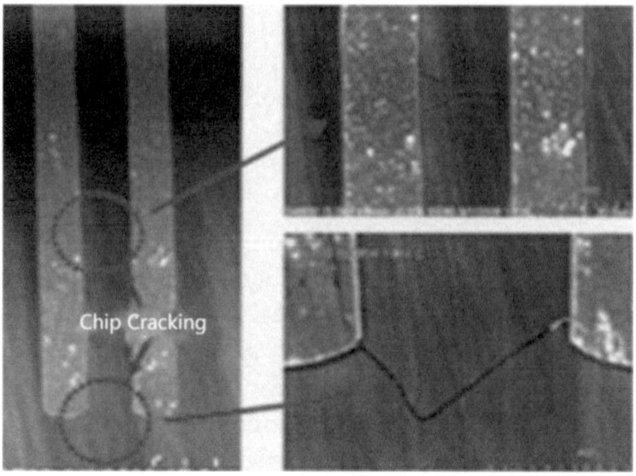

Fig. 5.12 Schematic diagram of chip cracking

Fig. 5.13. The causes of delamination mainly include thermal mismatch, interfacial reactions (such as oxidation, moisture, contamination, etc.), mechanical loading, internal pressure, and volume expansion or contraction.

(3) Bonding failure, wire shorting and breakage. Bonding failure in packaging structures is mainly manifested as open circuits at bonding points or broken bonding wires. The causes of open circuits at bonding points include Au-Al compound failure, poor bonding quality, thermal fatigue, and corrosion. The causes of broken bonding wires include large current melting and mechanical stress-induced breakage. In packaging structures with multiple stacked chips, the number of bonding wires increases with the number of chip layers, thereby increasing the risk of wire shorting. Moreover, in a plastic package of a specific thickness, the space between bonding wires becomes smaller as the number of stacked chip layers increases, especially for low wire arc, which significantly increases the risk of wire shorting. Currently, three types of low wire arc bonding

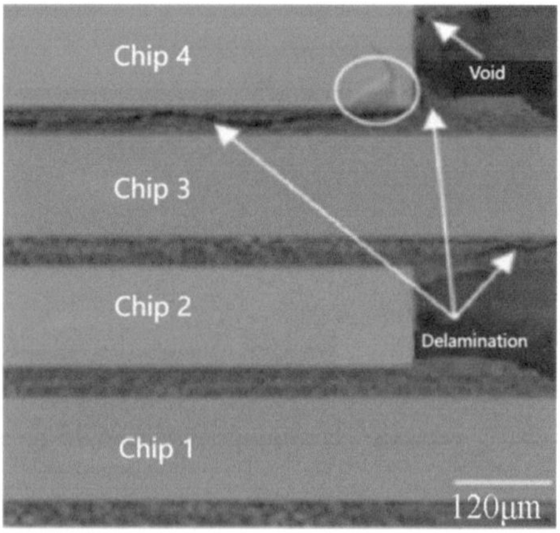

Fig. 5.13 Schematic diagram of chip delamination

processes are used in the industry: standard forward bonding process, stacked forward bonding process, and stacked reverse bonding process. Stacked reverse bonding is the main low wire arc bonding technology used for internal chip interconnection in 3D packaging, but excessive wire retraction during the stacked reverse bonding process can lead to wire neck cracks, which may cause failure.

(4) Thinning process defects. Compared to traditional 2D packaging processes, 3D packaging processes require thinning of stacked chips. The main defects of the thinning process are surface roughness and warpage. First, thinned wafers are prone to breakage due to their small thickness. Second, backside grinding of chips often results in uneven surfaces, creating localized stress that can reduce product reliability and shorten service life. When dicing, wafers are susceptible to chipping (thin and brittle wafers may extend chipping from the backside to the front side, leading to wafer chipping). Even if chipping is minor and not detected initially, it can affect the reliability of the device during service.

(5) Solder joint failure. Solder joints in chip interconnections may experience solder joint breakage, Kirkendall voids in Cu_3Sn microbumps, grain boundary embrittlement and intergranular fracture in Ni/Sn/Ni microbumps, and voids in solder joints.

2. Failure Caused by Package Stacking Processes

The 3D interconnect processes in package stacking mainly include: ① solder paste printing; ② placing bottom package body; ③ placing top package body; ④ reflow soldering, underfilling, and inspection [31–36]. Failure modes caused by these processes mainly includes warpage and solder joint failure.

(1) Warpage

To reduce the overall thickness of the stacked packaging structure, the substrate needs to be thinned as much as possible. During the reflow soldering process, the substrate expands due to temperature increase, causing significant warpage of the bottom package body. At the same time, due to the mismatch of CTE between packaging materials, warpage can occur during the reflow soldering process or during device service. The warpage during the assembly reflow soldering process is shown in Fig. 5.14. The upper figure is a schematic diagram of "frowning" warpage, where the molten solder is squeezed together, forming a solder bridge (short circuit); the lower figure is a schematic diagram of "smiling" warpage, where the external solder joints are pulled apart, which may lead to a gap (open circuit). Warpage of chips or wafers can pose significant challenges to assembly processes, assembly yield, solder joint reliability, and applications.

(2) Solder joint failure

① Solder joint breakage: The interconnect solder joints related to package stacking include first-level solder joints between individual package structures and second-level solder joints connecting the package to the PCB. The material for microbumps is typically lead-free solder based on Sn, such as SnAgCu, SnAg,

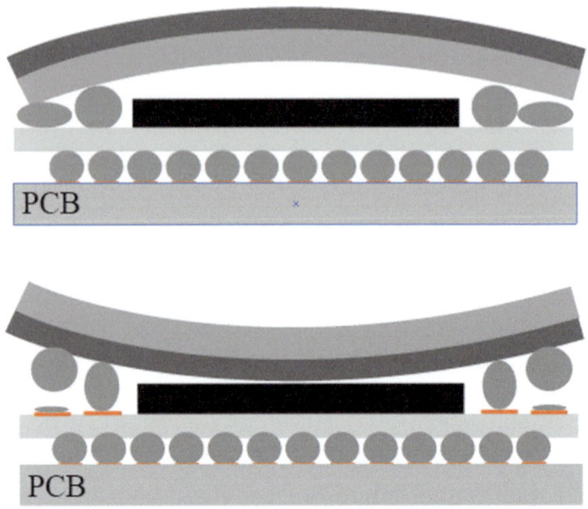

Fig. 5.14 Warpage during the assembly reflow soldering process

SnCu, and SnAgCu-X, where X represents the fourth element. During solder joint interconnection, an intermetallic compound (IMC) forms in the middle, which has poor mechanical properties and is prone to breakage under mechanical impact and vibration.

② Kirkendall void and porous voids in Cu_3Sn microbumps: During high-temperature storage tests or current stress tests, Kirkendall voids are commonly observed in Cu/Sn microbumps, where excessive consumption of Cu leads to the formation of microvoids in the Cu_3Sn layer. Kirkendall voids are a serious reliability issue; if not controlled, they can coarsen along the Cu_3Sn/Cu interface and induce crack formation, potentially leading to device failure. Under corrosive heat treatment or current stress, porous structure may form within Cu/Cu_3Sn/Cu microbumps.

③ Grain Boundary Embrittlement and Intergranular Fracture: Ni/Sn/Ni microbumps, under the influence of temperature and current, can form microvoids that migrate along the Ni_3Sn_4 grain boundaries, leading to grain boundary embrittlement. From a mechanical reliability perspective, intergranular fracture may occur along grain boundaries containing a large number of microvoids and impurities.

④ Void Formation in Ni/Sn/Ni Microbumps: The impact of reverse IMC growth often leads to an uneven interface, which in turn causes Sn whiskers during heat annealing. The diffusion of Sn atoms is relatively fast, and the asymmetric atomic flux leads to vacancy aggregation at the Sn whisker site. When the remaining Sn atoms are completely consumed, the supersaturation of vacancies leads to void formation. Voids within Ni_3Sn_4 microbumps can severely reduce mechanical reliability.

2. Failures Caused by TSV Wafer Fabrication

The 3D packaging interconnect technology based on TSV interposers mainly consists of five process steps: TSV wafer fabrication, wafer thinning, ultra-thin wafer dicing, wafer bonding and debonding, and 3D chip stacking packaging. Many reliability-related issues arise in these five process steps.

(1) Failures caused by TSV wafer fabrication

① TSV open or short circuit faults: When forming TSVs by plasma etching of the silicon substrate, the inherent non-uniformity of plasma density distribution can lead to TSV open and short circuit faults. TSV etching process defects can cause residual silicon at the wafer edge, leading to open circuit faults in edge TSVs; excessive etching in the wafer center can cause leakage or short circuit faults. After dielectric deposition, if the dielectric material is not fully etched by plasma etching to remove the material at the bottom of the TSV, residual dielectric at the TSV bottom can cause an open circuit. If the dielectric material is over-etched, local interconnects within the Back End of Line (BEOL) may be damaged, leading to TSV leakage.

② Void lines along the TSV axis: When electrodepositing metal copper, improper current density and additive usage can generate void lines along the TSV axis. During the post-deposition annealing process, these void lines may grow and degrade the mechanical and electrical properties of the TSV filler.

③ Short circuit between copper and silicon substrate: The morphology of the TSV sidewalls can affect subsequent packaging processes. Rough sidewalls of the TSV can lead to short circuits between the metal copper in the TSV and the silicon substrate, causing leakage. Additionally, microcracks within the TSV dielectric layer may be another cause of TSV leakage.

④ Wafer warpage: Due to the mismatch of CTE between metal, dielectric layer, and silicon substrate in the TSV structure, stress concentration can occur at the points of the dielectric layer and barrier layer at the bottom of the TSV. Compared to traditional wafers, TSV wafers have interconnect structures on both sides. If the residual stress on both sides is unbalanced, wafer warpage may occur. In actual process operations, wafer warpage can be reduced by minimizing the residual stress difference between the two sides of the wafer or by increasing the wafer thickness.

⑤ Dielectric delamination: After the TSV is filled with metal, chemical mechanical polishing (CMP) is used to remove the copper overlayer on the TSV surface. Similar to the indentation effect, the Low-k dielectric in the TSV may crack or delaminate due to mechanical stress during the chemical mechanical polishing process.

⑥ Reduced carrier mobility: Stress generated between copper and silicon surfaces can adversely affect the electrical performance of devices near the TSV, reducing carrier mobility due to the piezoresistive effect.

⑦ Copper protrusion: After heating and cooling, copper near the interface at the end of the TSV generates maximum Von-Mises stress, leading to plastic deformation, which is the fundamental cause of copper protrusion. After thermal cycling, Cu–SiO_2–Si interface dielectric cracking and roughening of the top metal line of the TSV may occur.

⑧ Deformation of RDL/BEOL structures or delamination of TSV end-cap layers: Because the ends of TSVs are typically connected to RDL/BEOL structures, protrusions or intrusions of copper in the through-vias pose a significant risk to the integrity of these structures. Stress induced by protrusions at the via ends can cause deformation of the RDL/BEOL structures or delamination of the end-cap layers of TSVs, posing a substantial challenge to device reliability.

(3) Failures caused by wafer thinning

In 3D packaging, to accommodate more chips within a single package, wafers need to be thinned, sometimes requiring a thickness less than 200 μm, or even less than 50 μm.

① Wafer warpage: During the grinding process, various defects may occur, such as scratches, cracks, debris, and amorphous or polycrystalline surface damage layers. These defects can reduce the fracture toughness of silicon and increase wafer warpage. Warpage can lead to chip fractures, solder joint detachment, and other reliability issues, as shown in Fig. 5.15. Additionally, wafer warpage can complicate subsequent packaging processes, such as the patterning of TSVs.

② Residual stress accumulation: During wafer thinning, residual stresses are present on the wafer. These residual stresses include compressive and tensile stresses, which depend on the conditions of dielectric deposition, metal plating, and wafer thinning.

Wafer-level ripple: In addition to wafer-level bending, wafer-level ripple is a phenomenon that often occurs after wafer thinning and high-temperature processing. This poses a significant challenge for subsequent backside processing steps.

(3) Failures caused by ultra-thin wafer dicing

① Chip cracking and microcracks: Dicing ultra-thin wafers can easily cause chips to crack. Cracks can be divided into two types: one is through-cracks, which

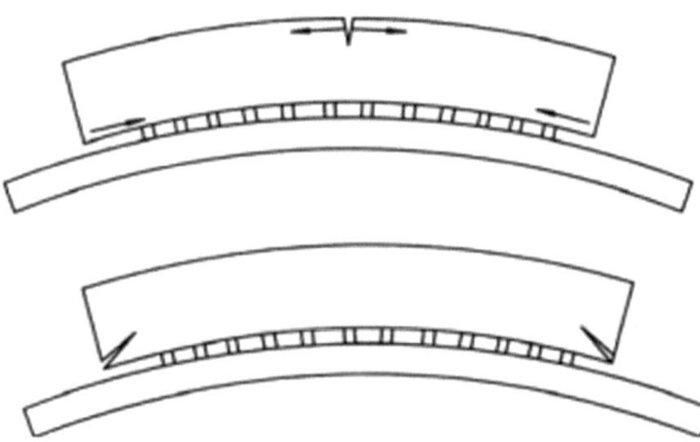

Fig. 5.15 Fracture caused by wafer warpage

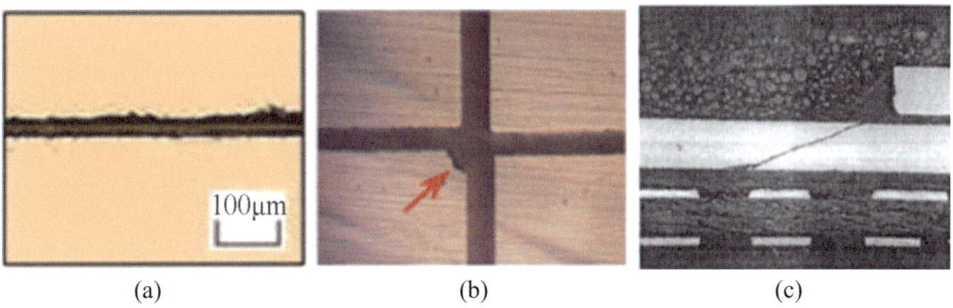

Fig. 5.16 Chip cracking and microcracks

can directly lead to chip failure if they affect the active circuit area, as shown in Figs. 5.16a, b; the other is microcracks, which, although they may not affect the active circuit area and are difficult to detect through optical inspection, may lead to severe reliability issues, as shown in Fig. 5.16c

② Chipping and microcracks on silicon sidewalls: High metal content in the scribe lines can increase the load on the blade, leading to more debris formation, which may cause microcracks on the silicon sidewalls. The size and frequency of debris, the thickness of the wafer, and the residual stress of the mold all determine the reliability risk. A typical reliability failure is the cracking of the wafer caused by the interaction between assembly and testing. In special cases, the fillet of the molding compound or epoxy resin may crack.

③ Dielectric cracks and delamination: Due to poor mechanical properties, some low-k dielectrics are at risk of cracking and delamination during mechanical dicing.

④ Chip open circuit: During laser dicing, silicon and low-k dielectrics may produce material debris that redeposits on the wafer surface, leading to open circuits during chip interconnection, as well as epoxy resin delamination and solder joint cracking.

(4) Failures caused by wafer bonding and debonding

The fabrication of 3D packaging devices often requires temporarily bonding the wafer to a carrier wafer, followed by debonding.

① Bonding quality degradation: Surface contamination or condensation water in nanoscale or sub-nanoscale trenches on the wafer surface can negatively affect the wetting of the fluid on the bonding surface, leading to degraded bonding quality.

② Surface stripes: During spin coating, large particles that hinder normal fluid flow can form stripes in the coating.

③ Wafer warpage, increased total thickness variation, delamination, and cracking: During the bonding and debonding processes, the wafer undergoes multiple thermal cycles. The difference in CTE between structural layers can cause stress in the wafer stack during cooling, leading to wafer warpage, increased total thickness variation, delamination, and even cracking.

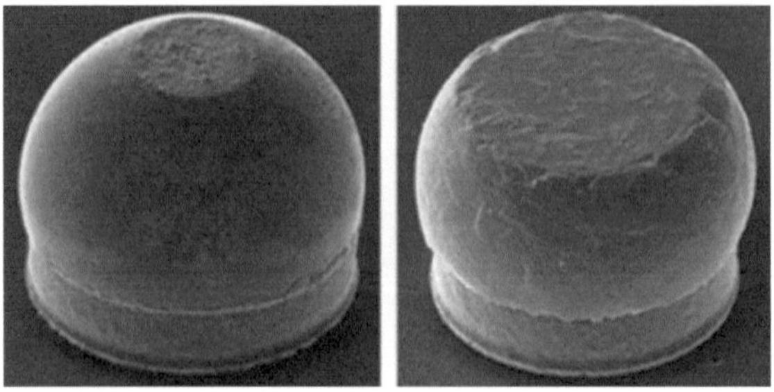

Fig. 5.17 Slight flattening (left) and severe flattening (right) during the reflow soldering process

Non-contact open circuit: During bonding and debonding, as the solder volume decreases, the problem of solder diffusion to the bottom under-bump metallization (UBM) becomes more severe. At temperatures close to the melting point of the solder, the circular microbump is flattened by the hard chuck material after reflow soldering, which can lead to non-contact open circuits during chip interconnection. The microbumps show slight flattening (left) and severe flattening (right) during the reflow soldering process, as shown in Fig. 5.17.

(3) Failures caused by 3D chip stacking packaging.

① Bending, delamination, and fracture failures: 3D chip stacking packaging is achieved by stacking two or more chips vertically. After thinning, the strength and stress resistance of the chips are reduced. Therefore, the difference in CTE between the lead frame and the chips, as well as excessive placement force, can lead to bending, delamination, and even fracture failures of the chips. In ultra-thin stacking packaging, die attach films are used instead of die attach adhesive, but the length of the die attach film is generally shorter than that of the chips. During molding, holes may appear at the edges of the die attach film. These voids in ultra-thin stacking packaging, as shown in Fig. 5.18, are the initial points for cracks or delamination, leading to device failure

② Void formation in chip gaps, solder joint shorts, electrochemical migration or corrosion: Generally, the chip interconnection process includes flux spraying (or dipping), chip pick-and-place, reflow soldering, and cleaning (or flux removal).

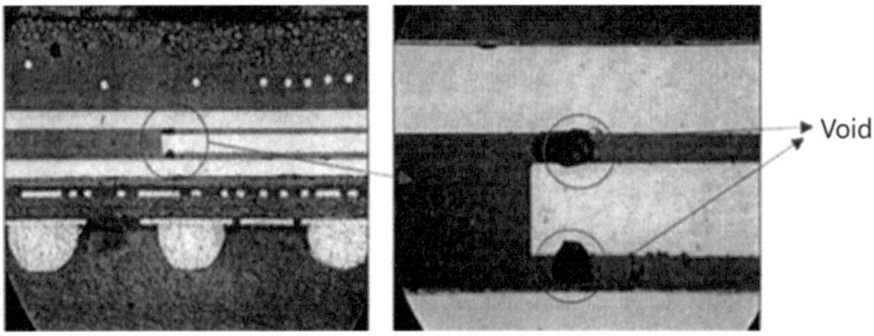

Fig. 5.18 Voids in ultra-thin stacking packaging

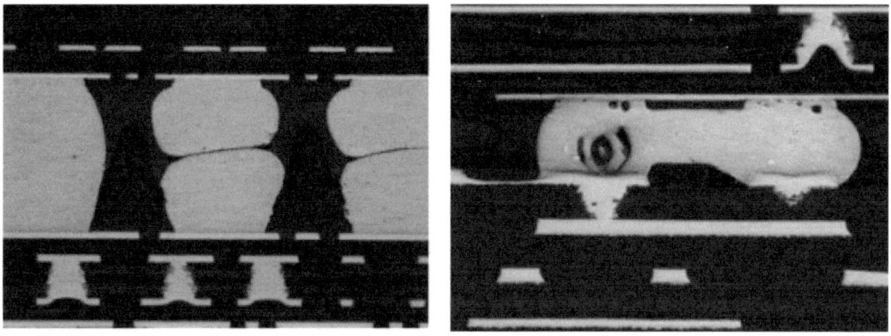

Fig. 5.19 The "pillowing effect" of solder joints and bridging morphology

During this process, flux may remain in the chip gaps, causing void formation in the chip gaps, solder joint shorts, electrochemical migration, or corrosion, which pose reliability risks.

③ Solder joint cracking, soldering defects, voids, bridging, and poor wetting: The packaging structure of 3D packaging devices is relatively complex, such as multi-layer bare chip stacking and package stacking. These new packaging structures increase the number of small and highly dense solder joints. Compared with traditional packaging, the requirements for solder joints in new packaging are stricter. Common solder joint failure modes include solder joint cracking, soldering defects, voids, bridging, and poor wetting. The "Pillowing Effect" of solder joints and bridging morphology are shown in Fig. 5.19.

④ Chip misalignment, chip tilt, and solder joint open circuit: In 3D packaging, the large number and multiple layers of chips require the chips to be thinner. Therefore, the large-scale reflow soldering process used in traditional 2D packaging cannot be used in 3D packaging. This is because reflow soldering generates high temperatures, and the warpage extending from the chips and substrates due to high temperatures can overcome the surface tension of the solder, causing chip misalignment and leading to chip tilt and solder joint open circuits.

5.3.2 Failure Mechanisms of 3D Packaging

Based on the rate of damage accumulation, the failure mechanisms of 3D packaging can be divided into two categories: overstress and wear, as shown in Fig. 5.20. Overstress failures are often instantaneous and catastrophic; long-term accumulation of damage leads to wear failures, which first manifest as performance degradation before device failure. Furthermore, based on the type of load that triggers the failure, the failure mechanisms of 3D packaging can be classified into mechanical, thermal, electrical, radiation, and chemical types. In the reliability analysis of 3D packaging structures, failures are usually classified according to the type of load, with failure time being a key parameter [37].

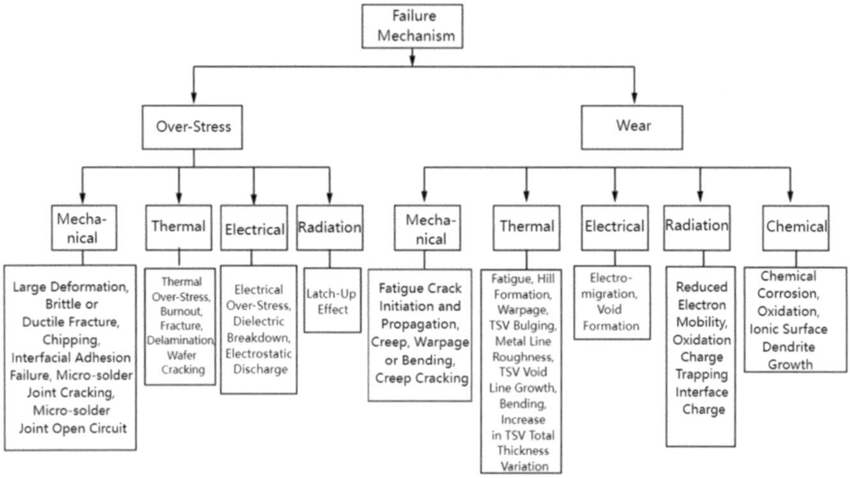

Fig. 5.20 Classifications of failure mechanisms

Mechanical loads include physical impacts, stresses, and inertial forces applied to the 3D packaging body. The responses of 3D packaging structures and materials to mechanical loads mainly include large deformations (elastic and plastic deformation), brittle or ductile fracture, chipping, delamination (delamination between chips, between dielectric and chips, or other interfaces, delamination of TSV filler, delamination between TSV and RDL/BEOL structures, delamination of packaging, etc.), poor interface adhesion, micro-solder joint cracking, micro-solder joint open circuit, warpage or bending, fatigue crack initiation and propagation, creep, and creep cracking.

Thermal loads include high-temperature heating in 3D interconnection processes (such as thermal compression bonding of microbumps, reflow soldering of package stacking, pre-heating for chemical mechanical polishing in TSVs, dielectric growth, etc.), heating processes of individual chips or packaging bodies themselves, and thermal load effects from application environments. Dimensional changes in materials due to thermal expansion are all caused by external thermal loads, and physical properties such as creep rates are also altered by thermal loads. The majority of packaging structure failures are due to local stress concentration caused by mismatch of CTE. Additionally, combustion of flammable materials within devices can cause excessive thermal load and lead to failure.

The main causes of electrical loads are electrical shocks, voltage instability or current fluctuations caused by sudden oscillations during current transmission, electrostatic discharge, electrical overload, excessive input voltage, and high current. Reliability issues caused by external electrical loads include dielectric breakdown, surface breakdown of voltage, thermal dissipation of electrical energy, and electromigration. Moreover, electrical loads can increase electrochemical corrosion, induce dendritic growth, and subsequently lead to leakage and thermal degradation.

Reliability issues related to chemical loads are primarily caused by environmental factors such as chemical corrosion, oxidation, and ion surface dendritic growth. The performance degradation of plastic-encapsulated devices due to moisture permeation through the encapsulant is a major concern.

When moisture from the environment permeates into the package, it can extract residual catalysts from the package, forming new products that enter various interfaces of the chip, such as metal pads and semiconductor structures, thereby causing device performance degradation. For example, flux residues remaining on the device after assembly can migrate to the chip surface through the package, bringing reliability issues. In addition, long-term exposure of encapsulation materials to high-temperature and high-humidity environments can lead to degradation of materials such as epoxy polyimide, an effect also known as reversion. Since the degradation of molding compounds may take months or years, accelerated testing is generally used to determine whether molding compounds are prone to degradation failure [37].

5.4 Reliability of 3D Packaging Technology

5.4.1 Reliability of 3D Chip Stacking Technology

3D chip stacking packaging is a multi-chip stacking package that vertically stacks multiple chips in the same package structure, interconnected through wire bonding or flip-chip methods. The 3D chip stacking structure can effectively reduce the size of chip packaging and improve device performance. However, chip stacking and small size can lead to increased power density, which in turn causes thermal reliability issues. This section will discuss the reliability issues of 3D chip stacking technology [38, 39].

1. Wafer Warpage Issues

Currently, mechanical grinding is commonly used for wafer thinning. Ultra-thin wafer grinding requires coarse grinding first, followed by fine grinding with smaller diamond particles (generally less than 20 μm). The schematic diagram of wafer grinding thinning is shown in Fig. 5.21. Since the silicon used for chips is a single-crystal wafer, with silicon atoms arranged periodically in a diamond lattice structure, when the backside of the wafer is thinned by mechanical grinding, a certain thickness of damage layer will be formed on the backside of the wafer, and the thickness of the damage layer is proportional to the radius of the diamond particles. The damage layer on the backside of the wafer will disrupt the lattice arrangement of the single-crystal silicon inside the wafer, creating internal stress. The ability of the wafer to resist internal stress decreases as the thickness of the wafer decreases. When the thickness is too small, residual stress can cause wafer warping.

In the packaging process of single chips, the thickness of the chips is generally greater than 300 μm, so the residual stress and surface wear introduced by the

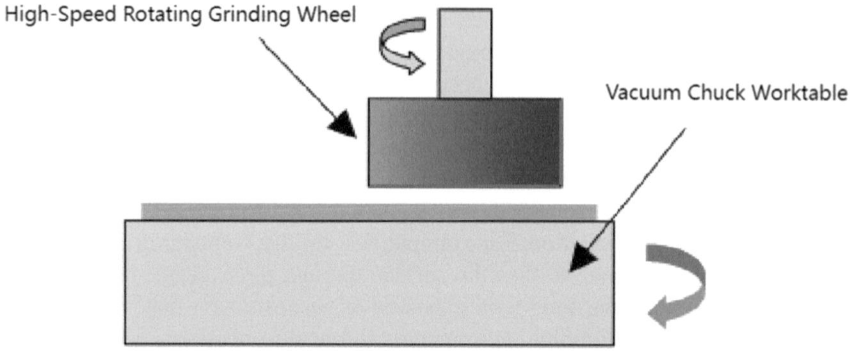

Fig. 5.21 Schematic diagram of wafer grinding thinning

grinding process can be neglected in terms of their impact on chip performance. The entire thinning process is divided into two stages: coarse grinding and fine grinding. Coarse grinding generally uses larger diamond particles (> 40 μm) to quickly reduce the thickness of the wafer, but it forms a damage layer of about 1.5 μm on the backside of the wafer. Fine grinding usually selects smaller diamond particles (< 20 μm), and the fine grinding process is slower, resulting in a thinner damage layer on the wafer surface (about 0.5 μm). For thicker wafer thinning, the coarse grinding process can generate significant residual stress inside the silicon wafer, causing warpage after thinning. For example, a 6-inch wafer can warp up to 200 μm after thinning. However, for chips with a thickness greater than 300 μm after thinning, the circuits of the chips are not affected by subsequent packaging processes.

For 3D stacked packaging chips, the thickness must be controlled within 200 μm. If the same thinning process parameters used for single-chip packaging are adopted, 3D stacked packaging chips will experience greater warpage deformation. For example, an 8-inch wafer thinned to 200 μm can warp up to 1500 μm or even more.

In actual packaging processes, to reduce the impact of wafer thinning on 3D stacked packaging chips, smaller diamond particles are generally used for fine grinding to lower the roughness of the damage layer on the backside of the thinned chips to below 0.2 μm. Fine grinding can effectively remove the damage layer caused by coarse grinding. However, fine grinding with small particles can lead to high grinding pressure, which generates a large amount of heat, causing warpage and damage to the chips. For example, for an 8-inch wafer, the warpage after fine grinding with small diamond particles is about 180 μm [24, 25]. Figure 5.22 shows the backside of wafers ground with diamond particles of different sizes. It can be seen that smaller diamond particles can effectively reduce the damage on the backside of the wafer. When the grinding particle size is reduced to below 6 μm, the roughness of the wafer backside is significantly decreased, and a mirror-like effect can be achieved.

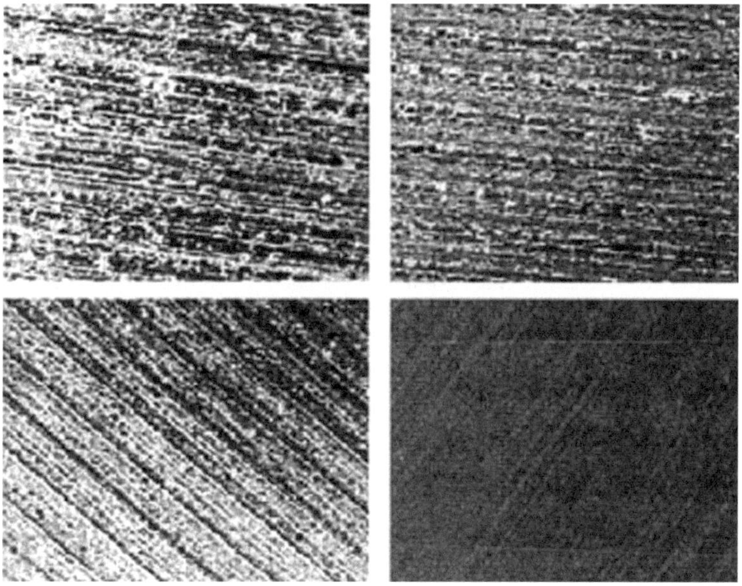

Fig. 5.22 The backside of wafers ground with diamond particles of different sizes

2. Impact of Wire Bonding Between Chips

Wire bonding in stacked chips includes forward bonding and reverse bonding. Bonding with the first-level solder joint placed on the chip metal bonding pad is called forward bonding, while reverse bonding refers to placing the second-level solder joint on the chip metal bonding pad, with the lead as the first-level solder joint.

The distance from the top wire to the surface of the plastic package limits the height of wire bonding and the wire arc between different layers in stacked chips. Excessive wire arc can lead to wire exposure or short circuits between wires, especially in multiple stacked chip packages where the number of wires increases with the number of chip layers. In a plastic package of a specific thickness, the requirement for small arc wires is particularly prominent, and the space for wires is increasingly limited. Additionally, the thickness of the chip is often greater than the maximum arc of the wire.

Compared to reverse bonding, forward bonding is more efficient due to the elimination of the ball planting process and easier control of the spacing between wires. However, the wire arc cannot be effectively controlled. Reverse bonding can effectively control the wire arc and forms a larger gap at the second-level bond pad of the

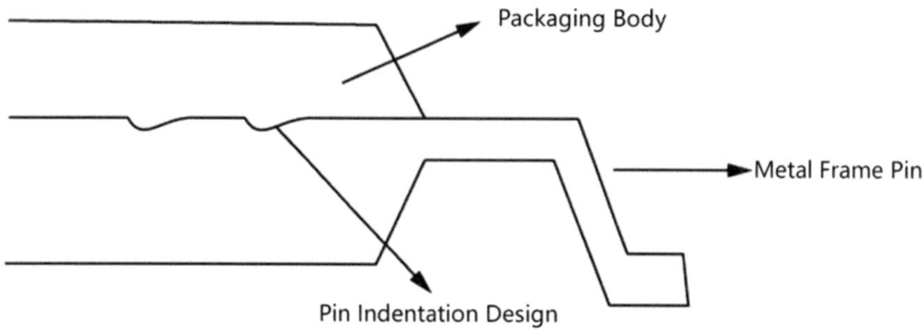

Fig. 5.23 Schematic diagram of the lead frame slot design

lead. However, excessive wire retraction during the bonding process can lead to wire neck cracks, which can cause reliability issues.

3. Impact of Moisture Ingress on Chip Stacking Packaging

For 3D chip stacking packaging structures, the increase in the number of internal chip layers and lead frames, and the reduced distance from the surface of the top chip to the package surface, make the surface chips more susceptible to moisture ingress. This affects the moisture sensitivity level and reliability of the product. To address these issues, molding compound with low moisture absorption rate are often chosen for 3D packaging to reduce the probability of chip delamination [29].

In addition to the moisture resistance of the molding compound mentioned above, the interface between the metal frame lead and the plastic package is also a significant concern. The schematic diagram of the lead frame slot design is shown in Fig. 5.23. By modifying the structure of the lead frame, the adhesion between the molding compound at the edge of the package and the metal frame can be increased to prevent moisture from entering the package through the interface and causing device performance degradation or failure.

5.4.2 Reliability of 3D Package Stacking Technology

With the widespread application of 3D package stacking in mobile devices, research on the reliability of package stacking has become increasingly important. This section will discuss the reliability issues of 3D package stacking.

1. Thermal Warpage Issues in PoP Package Stacking

PoP package stacking is typically assembled through reflow soldering to achieve interconnection between solder joints and to connect the device to the PCB. Packaging materials mainly include epoxy molding compounds, die attach adhesive, chips, lead-free solder, substrates, and PCBs. Due to the mismatch of CTE between packaging materials, warpage can occur during reflow soldering or in use. The biggest challenge for PoP package stacking technology is warpage, which poses a significant threat to assembly processes, solder joint reliability, and applications. Between assembly yield and reliability, warpage has a greater impact on the former, mainly because solder joints in a molten state do not have load-bearing capacity. Therefore, the shape formed after solder joint solidification is controlled by the surface tension of the molten solder joint, the warpage geometry of the individual packaging body and the PCB. The warpage of the top packaging body determines the yield of the first-level solder joints, while the warpage of the bottom packaging body and the underlying PCB determines the yield of the second-level solder joints. To reduce the height of the entire PoP package stacking, the substrate is thinned as much as possible. During reflow soldering, the substrate expands due to temperature increase, leading to significant warpage of the bottom packaging body.

Warpage is dynamic and changes with the temperature profile of the package stacking reflow soldering process, with relative displacement of solder balls reaching 100–150 μm. Currently, domestic and international researchers have conducted detailed studies on various factors affecting warpage, which can be optimized from three aspects: structure, material, and process. For chips, epoxy molding compounds, and substrates, designing rational packaging structures and processes and selecting appropriate materials can reduce the thermal mismatch between chips and substrates. At present, some researchers are considering alternative solutions to warpage, such as using thinner chips and substrates, using films to attach chips, designing materials with good performance, using thinner epoxy molding compounds, and flip-chip interconnection structures to reduce the net height of interconnection between chips and substrates.

The choice of packaging materials has a significant impact on the warpage of PoP package stacking. Compared to the Young's modulus of materials, the CTE of epoxy molding compounds and substrate materials has a greater effect on warpage, with the CTE of epoxy molding compounds being the most critical factor. The warpage direction of the top and bottom packaging bodies can be controlled by the choice of materials. To reduce warpage in PoP devices, more and more packaging materials are emerging, and the thermal properties of materials are continuously being optimized, specifically shown as lower CTE for substrate materials and higher CTE for epoxy molding compounds. The lower the CTE value of the substrate material, the more effectively PoP package warpage can be reduced; the higher the CTE value of the epoxy molding compound, the greater the improvement in PoP package warpage. Currently, the CTE value of substrate materials at room temperature has decreased from the original 15–17 ppm to 5–7 ppm; the CTE value of epoxy molding compounds, which was originally around 8–10 ppm, has now reached 12–15 ppm for ultra-high CTE epoxy molding compounds, and the new generation has reached 22–25 ppm. Various high CTE epoxy molding compounds are also continuously being researched. The warpage issue caused by thinning chips can be significantly improved by using new packaging materials. However, as package stacking technology develops and packages become thinner, controlling warpage by optimizing the performance of packaging materials becomes less effective.

At present, a combination of experience and modeling methods is generally used to predict the warpage behavior of package stacking technology during the reflow soldering process. Chiavone et al. used a 3D contour tool to analyze the separation issues between warped packages, as changes in gaps can affect the geometry of solder joints. Lall et al. proposed a comprehensive modeling method to predict the warpage behavior of PoP packages, compiling a series of related material properties, geometric shapes, and reflow soldering parameters into a finite element model to predict the warpage behavior of PoP packages. In addition, they designed a statistical model to predict the probability distribution of warpage as a function of input parameters.

2. Reliability of interconnection solder joints in package stacking

Interconnection solder joints related to package stacking include the first-level solder joints on individual packaging bodies and the second-level solder joints connecting the bottom packaging body to the PCB. The solder material for microbumps is typically lead-free solder based on Sn, such as SnAgCu, SnAg, SnCu, or SnAgCu-X, where X represents the fourth element. The solder composition plays a crucial role in the quality and reliability of the package. For a given UBM, adjusting the solder composition and soldering temperature distribution can significantly affect performance. Here, we will discuss the reliability issues of microbumps used in package stacking, taking Cu/Sn/Cu microbumps and Ni/Sn/Ni microbumps as examples.

(1) Cu/Sn/Cu microbumps

Li et al. prepared Cu/Sn/Cu microbump samples by controlling the compression of two polished copper surfaces with a 25 μm thick tin foil in between, under different reflow soldering temperatures. They monitored the microstructural evolution of Cu_6Sn_5 during the continuous liquid-phase reaction of Sn in the molten state. The FIB images of Cu/Sn/Cu microbumps are shown in Fig. 5.24, with Cu_6Sn_5 forming a scallop-like morphology at the Cu/Sn interface. Cu_6Sn_5 grows simultaneously on two opposite interfaces, and when the two opposite Cu_6Sn_5 grains come into contact, the parallel grain boundaries unexpectedly disappear. In Fig. 5.24c, no lateral grain boundaries are observed, only longitudinal columnar grain boundaries. The columnar Cu_6Sn_5 is further consumed by Cu to form a more stable Cu_3Sn phase, with finer columnar grain structures. When the residual Cu_6Sn_5 is completely consumed, unlike Cu_6Sn_5 grains, the two opposite Cu_3Sn grains have both lateral and longitudinal grain boundaries. The fundamental reason for the coarsening of opposite Cu_6Sn_5 grains into single grains with the disappearance of lateral grain boundaries is still unclear. However, the disappearance of lateral grain boundaries is undoubtedly beneficial to the mechanical reliability of microbumps, as lateral grain boundaries passing through the microbumps can induce crack formation and propagation, reducing the microbump's resistance to fracture.

During high-temperature storage or current stress tests, Kirkendall voids are commonly observed in Cu/Sn solder joints, where excessive consumption of Cu leads to the formation of microvoids in the Cu_3Sn layer. The formation of Kirkendall voids is caused by the unbalanced mutual diffusion flux of Cu and Sn atoms, with

(a) 5min reflow soldering at 260°C (b) 10min reflow soldering at 300°C (c) 480min reflow soldering at 300°C (d) 5min reflow soldering at 480°C

Fig. 5.24 FIB images of Cu/Sn/Cu microbumps

Cu atoms diffusing much faster than Sn atoms. Adding Co or Ni to lead-free solder can effectively reduce Kirkendall voids. Moreover, controlling the plating solution or using nanotwinned Cu as UBM material can significantly reduce the nucleation of microvoids. Kirkendall voids are a serious reliability issue; if not controlled, they tend to coarsen along the Cu_3Sn/Cu interface and induce crack formation. The cross-section of SAC solder/Cu pad after Ar^+ sputter etching is shown in Fig. 5.25. Undoubtedly, Kirkendall voids are undesirable and harmful to the mechanical and electrical properties of microbumps, which needs to be strictly avoided.

Wang et al. conducted a thermal annealing test at 170 °C on Cu–TSV/Sn–Cu solder/Cu microbumps and found that after 1000 h of annealing, Kirkendall voids at the interface between Sn–Cu solder and Cu–TSV in the Si interlayer could induce crack formation. Kirkendall voids preferentially nucleate and grow at the TSV side of the Cu–TSV interface. This is attributed to the different potentials between the Cu–UBM on the chip side and the Cu–TSV in the interposer. In the latter, Cu is electroplated into the through-via, and organic impurities (such as S) may be trapped in the via and segregated at the end of the Cu–TSV. Therefore, Kirkendall voids, due to heterogeneity, preferentially nucleate at the Cu/Cu_3Sn interface, which may be why Kirkendall voids are only observed at the Cu–TSV interface and not on the chip side. The backscattered SEM images of Cu–TSV/Sn–Cu solder/Cu microbumps at different stages of thermal aging at 170 °C are shown in Fig. 5.26.

(a) Before reflow soldering

(b) After 500 thermal cycles, Kirkendall voids form and occupy Approximately 10% of the interface

(c) After 1000 thermal cycles, Kirkendall voids grow and induce crack formation at the interface

Fig. 5.25 Cross-section of SAC solder/Cu Pad after Ar^+ sputter etching

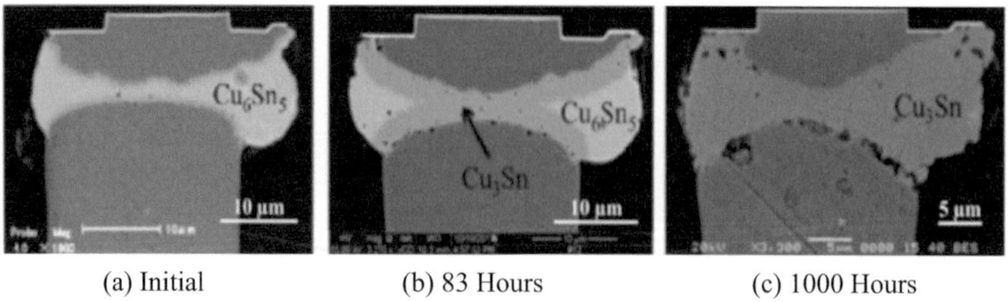

Fig. 5.26 Backscattered SEM images of Cu–TSV/Sn–Cu solder/Cu microbumps at different stages of thermal aging at 170 °C

In addition to Kirkendall voids, Chen et al. reported that porous voids could form inside Cu/Cu_3Sn/Cu microbumps under corrosive thermal annealing or current stress conditions. In porous microbumps, Cu_3Sn can exist in two forms: a pore-free Cu_3Sn layer forming Kirkendall voids at the Cu/Cu_3Sn interface and a porous Cu_3Sn layer sandwiched between the pore-free counterparts. The two forms of Cu_3Sn are shown in Fig. 5.27. Wang et al. conducted a corrosive high-temperature storage test on Cu wire/Sn/Cu wire micro-joints with infinite Cu content at 210 °C. The study found that after one week of annealing, the mechanical strength of the Cu wire dropped sharply from 162 to 60 MPa.

(2) Ni/Sn/Ni microbumps

For typical Ni/Sn/Ni microbumps, Ni is used as a barrier layer to prevent the formation of IMC, especially under reflow soldering or thermal compression bonding conditions. Therefore, a sufficient amount of unreacted solder can be left as a mechanical buffer layer to maintain the ductility of the microbumps. The SEM image of Ni/Sn/Ni microbumps after isothermal annealing at 170 °C is shown in Fig. 5.28. For the initial Ni/Sn/Ni microbumps, after reflow soldering, the solder in the center of the Ni/Sn/Ni microbumps more or less transforms into Ni_3Sn_4_IMC, with the remaining solder being pushed to the periphery of the microbumps. After further thermal annealing, the Ni/Sn/Ni microbumps are eventually completely transformed into Ni_3Sn_4-IMC, with a series of microvoids forming along the Ni_3Sn_4 grain boundaries, which can easily

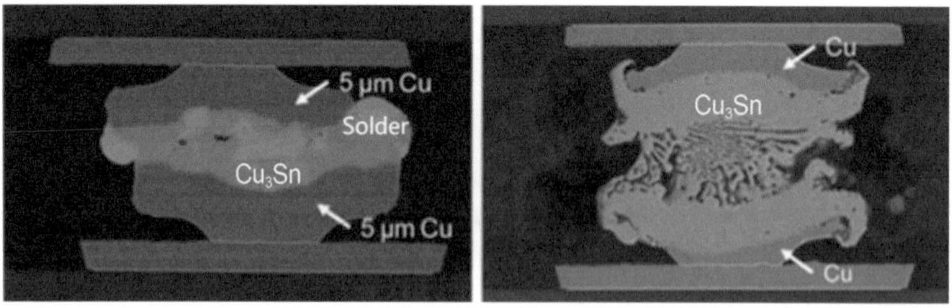

Fig. 5.27 Two forms of Cu_3Sn

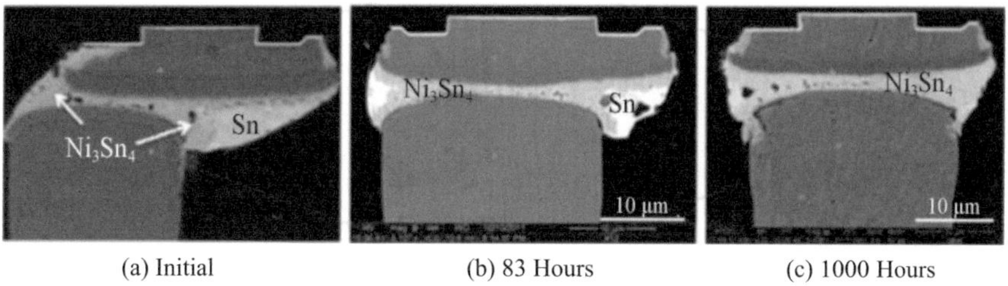

Fig. 5.28 SEM image of Ni/Sn/Ni microbumps after isothermal annealing at 170 °C

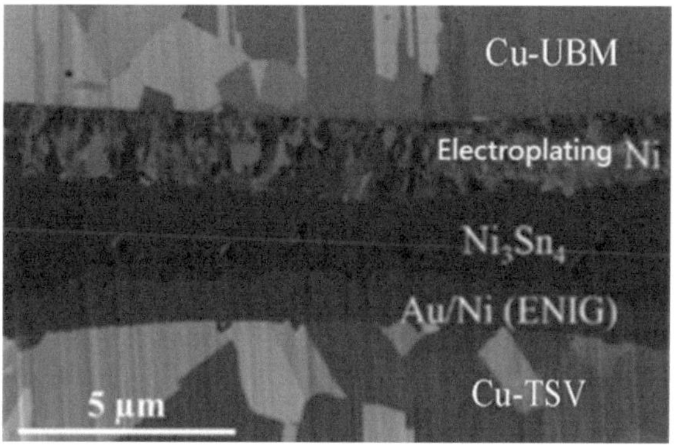

Fig. 5.29 FIB cross-section image of the Ni/Sn/Ni microbump-TSV joint

induce grain boundary embrittlement due to trapped impurities. From a mechanical reliability perspective, intergranular fracture may occur along grain boundaries containing a large number of microvoids and impurities. Chuang et al. provided a detailed description of the impurity migration phenomenon at the Ni_3Sn_4 grain boundaries, and the FIB cross-section image of the Ni/Sn/Ni microbump-TSV joint is shown in Fig. 5.29.

The chemical reaction between Ni and Sn in Ni/Sn/Ni microbumps is a reliability issue. When the solder is completely transformed into Ni_3Sn_4, volume shrinkage can no longer be adjusted by reducing the thickness of the solder due to the brittle nature of the IMC. Instead, void formation occurs as the solder thickness decreases. In Ni/Sn (10 μm)/Ni microbumps after isothermal annealing at 180 °C, shown in Fig. 5.30 as cross-section SEM image of Ni_3Sn_{4m}icrostructure evolution, microvoids were found in the Ni_3Sn_4 microbumps after 240 h of annealing. To avoid the microcracking or smearing effects that may be caused by mechanical polishing, ion beam milling is generally used to polish the cross-section. Figure 5.30 clearly shows that in the formation of Ni/Sn-IMC, voids originate from Sn vacancies. For Ni/Sn/Ni microbumps, Sn atoms diffuse faster than Ni atoms. During thermal annealing, the growth of IMC is a flux-driven maturation process, where larger grains grow at the expense of smaller grains. The reverse IMC impact-induced growth effect often forms an uneven interface, leading to Sn whisker formation during thermal annealing. The faster diffusion

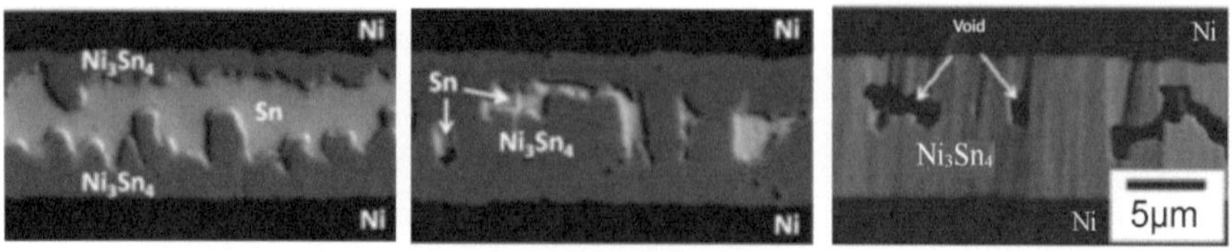

(a) 72 Hours (b) 192 Hours (c) 240 Hours

Fig. 5.30 Cross-section SEM image of Ni_3Sn_4 microstructure evolution

rate of Sn atoms and the asymmetric atomic flux lead to vacancy aggregation at the Sn whisker site. When the remaining Sn atoms are completely consumed, the supersaturation of vacancies leads to the formation of voids. Undoubtedly, voids inside Ni_3Sn_4 microbumps severely reduce mechanical reliability. In the UBM of Ni_3Sn_4 microbumps, the mechanical compliance of Ni (elastic modulus $E = 186$ GPa) is lower than that of Cu (elastic modulus $E = 120$ GPa). Based on the above discussion, Ni_3Sn_4 microbumps seem to have more severe mechanical reliability issues than Cu microbumps: trapped impurities lead to grain boundary embrittlement and volume shrinkage, causing voids or microcracks. To address mechanical reliability issues, the main approach is to introduce an Ni barrier layer to inhibit IMC formation. However, for fine-pitch interconnection applications in 3D IC technology that require solder joint thicknesses less than 5 μm, completely IMC-formed joints are inevitable. Therefore, using Ni as UBM material does not necessarily alleviate complete IMC formation and may actually lead to serious mechanical reliability issues.

(3) Impact of mechanical shock and vibration on solder joint reliability

To date, PoP packaging devices have been widely used in mobile communication equipment. During use, mechanical stress constantly affects the operation of PoP packaging devices, and can cause solder joint cracking. Compared with other surface-mount devices, the interconnection solder joints in package stacking are more susceptible to damage in mechanical shock and vibration environments. Package stacking has a larger mass and higher center of gravity than surface mounting, which can impose more load on the second-level (bottom) solder joints of the PCB. Jingen Luan studied the impact of drops on the performance of two-layer PoP packaging devices, and the results showed that the most failure-prone solder joints are located at the outermost periphery of the bottom packaging body, with fractures occurring at the interface between the PCB pad and the IMC being the main mode of failure. J. Y. Lee investigated the effect of adding underfill on the drop performance of two-layer PoP packaging devices and found that underfill can significantly improve the drop performance of the devices, with underfill having a lower Young's modulus that can mitigate the deformation caused by drops. Vicky Wang and Dan Maslyk et al. conducted drop tests on two-layer PoP packaging devices and found that no failures occurred when both the top and bottom packaging bodies were filled with underfill; failures occurred when only the bottom packaging body was filled with underfill, with the failure location being in the top packaging body; SAC125 solder joints had better drop performance than SAC305 solder joints. New low-Ag lead-free solders have been developed to replace SAC305, SAC396 and other alloys to improve the drop impact performance of face-array packaging, including package stacking. These new solders are low-Ag alloys, such as SAC105 (98.5Sn–1.0Ag–0.5Cu) and SAC125 (98.3Sn–1.2Ag–0.5Cu). McCormick et al. compared the TMF performance of high-Ag and low-SAC alloys over a temperature range of − 40 to 125 °C. Although high-Ag solders like SAC305 and SAC405 performed better in temperature cycling than low-Ag (SAC105 and SAC125) solders, the low-Ag solders (SAC105 and SAC125) had better drop impact performance. Moreover, surface treatment techniques can affect the drop impact performance of package stacking, as the high strain rate characteristics of deformation "force" the failure path to leave the bulk solder and reach the solder/pad interface. Fractures typically occur in the IMC, and their development directly depends on surface treatment techniques and solder alloys. Y. Ejiri et al. used a rapid shear test to simulate the load applied to solder joints during high strain rate events (such as drop tests) and found that the combined effect of IMC thickness (depending on solder composition, surface treatment techniques, and aging conditions) and overall solder strength plays a role in suppressing the overall failure behavior of solder joints. These test results are crucial for explaining the relationship between solder microstructure and strength and the extent to which interface physical metallurgy controls interconnection fracture behavior in complex package stacking structures.

5.4.3 Reliability of 3D TSV Packaging Technology

3D TSV technology enables vertical interconnection between chips, which helps reduce signal delay and increase packaging density, making it a promising direction for future large-scale integrated circuit packaging. However, the manufacturing process of TSVs introduces many new types of defects, which in turn cause reliability issues for TSVs. This section will discuss the reliability issues and impacts of TSVs.

1. Stress in TSVs

(1) Sources and impact of stress.

The stress in TSVs and adjacent Si mainly comes from growth stress and thermal–mechanical stress, the latter of which is caused by the mismatch of CTE between Cu in the through-via and the surrounding Si. When electroplated Cu radially grows on the Cu seed layer from the via sidewall, growth stress is generated, which can create fine seams along the TSV axis. These fine seams may leave a thin void line along the TSV axis, which can grow during the annealing process after deposition (see Fig. 5.31), leading to the loss of mechanical and electrical properties of the TSV filler. This is because, at room temperature, the electroplated Cu grains grow after annealing, eliminating grain boundaries and causing volume shrinkage of Cu in the TSV [40]. This may cause any existing voids to grow during the annealing

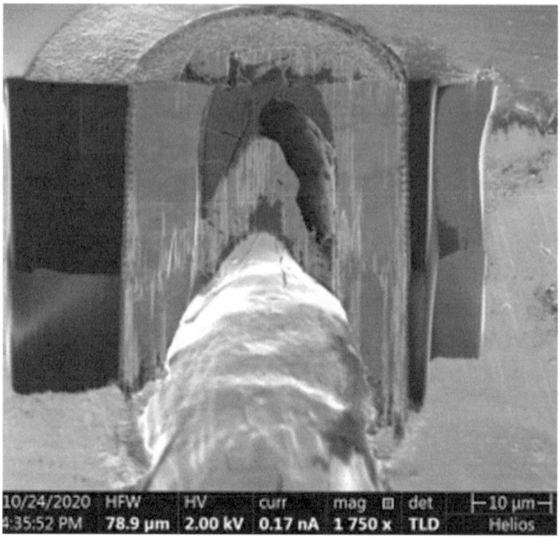

Fig. 5.31 Void line in TSV [41]

process and the pre-annealing process for chemical mechanical polishing. During annealing, lattice vacancies diffuse towards the voids near the TSV axis, causing the voids to grow and eliminating the radial compressive stress gradient around the original defects in Cu [41].

Due to the large difference in CTE between Cu and Si ($\alpha_{Si} = 2.8 \times 10^{-6}$/K, $\alpha_{Cu} = 17 \times 10^{-6}$/K), the difference in thermal expansion or contraction of Cu and Si during heating or cooling processes leads to the generation of thermal–mechanical stress. In the use of electronic devices, thermal cycling caused by fluctuations in joule heat continuously occurs, and 3D packaging devices are no exception. Electroplated Cu deposited on Si is often under residual tension (at ambient temperature). During heating, the residual tension is first elastically released, followed by an increase in compressive stress. As the temperature rises, the yield strength and creep resistance of Cu decrease, thereby eliminating stress through plastic yielding and creep. Therefore, at the peak temperature, there is almost no residual tension. During the subsequent cooling process, tensile stress gradually increases (radial stress in TSV). The temperature range and thermal cycling rate determine the details of this behavior.

At room temperature, Si surrounded by the TSV array is under compressive stress, which shows a significant gradient along the TSV axis. When close to the TSV end, the compressive stress in Si decreases. Assuming that the radial and axial stresses in the Si surrounding the TSV do not depend on the axial position relative to the TSV, it can be inferred that the axial stress σ_{zz} in the middle of the Si at ambient temperature is negative (σ_{zz} is positive in Cu). It should be noted that smaller TSV diameters (larger TSV aspect ratios) show more significant axial stress gradients, Moreover, the stress in the Si surrounded by the TSV array is usually greater than that outside the TSV array, and these stresses change from tensile (caused by growth and annealing) to gradually compressive during annealing at elevated temperatures after

Cu electroplating. After annealing, After annealing, Cu's volume expansion increases the radial compressive stress in Si at the interface but turns it into tensile stress away from the interface. Conversely, in addition to increasing grain size, annealing also reduces the compressive stress in Cu [42]. The radial compressive stress in Cu–TSV or filler can also accumulate due to thermal cycling. Moreover, TSVs with high impurity content seem to cause high residual stress. It has been reported that there is a large radial tensile stress at the Si–Cu interface, which may make these parts prone to failure in TSV components. For a given diameter, the radial stress at the Cu/Si interface usually increases with the depth of the TSV. Although the dependence of radial stress on TSV diameter is not monotonic, the depth of the TSV will affect the stress state of the TSV component.

The stress generated on the surfaces of Cu and Si has an adverse effect on the electrical performance of devices near the TSV, reducing carrier mobility due to the piezoresistive effect [43, 44]. This necessitates an isolation zone (KOZ) in Si, where active devices cannot be placed within a few micrometers of each TSV. The KOZ is proportional to the square of the TSV diameter, and larger aspect ratio TSVs also have larger KOZs [45, 46]. A lower yield strength of Cu microstructure reduces the KOZ size, which first increases with increasing yield strength until the TSV plastically yields, then remains stable [47].

After heating and cooling, Cu near the interface at the end of the TSV generates maximum Von-Mises stress, leading to plastic deformation, which is the root cause of Cu protrusion. The maximum principal stress (tensile) is radially distributed at the interface near the middle of the TSV and in the surrounding axial dielectric layer. Signs of dielectric cracking may appear at the axial Cu/SiO_2/Si interface after thermal cycling, and roughening of the top metal line of the TSV may occur. TSV dielectric cracking and roughening of the top metal line are shown in Fig. 5.32 [48]. However, when the package is cooled from the stress-free temperature (150 °C), each chip undergoes warpage, fundamentally changing the stress and displacement relative to the individual chip. The radial interface stress becomes compressive due to the convex curvature, reducing the tendency for interface and dielectric cracking and shifting the critical failure location to the Cu column near the microbump interface in the 3D package. In fact, the failure probability near the Si-Cu microbump and the required KOZ diameter increase with the diameter of the TSV and Cu column in the 3D package. In addition to defects in the TSV (including the complexity of electrical performance and potential interface and dielectric cracking), induced stress also causes plastic-related phenomena of Cu protrusion, severely affecting reliability.

(2) Microstructure and stress

Generally, at room temperature, Cu–TSV has a large initial tensile stress, which becomes a smaller tensile stress after annealing at 200 °C and cooling to ambient temperature. The larger initial tensile stress is caused by grain boundary elimination during annealing and device manufacturing processes, as previously mentioned, which is undesirable from a reliability perspective because it induces significant stress in Si [49–51]. Subsequent annealing treatment reduces the compressive stress

(a) Before Cycling (b) After 100 Cycles (c) After 200 Cycles

Fig. 5.32 TSV dielectric cracking and roughening of the top metal line

in Cu (initially larger grain areas expand during annealing, possibly due to relaxation associated with high-temperature plasticity and creep). It has been found that Cu–TSV has a random structure both before and after annealing. However, after annealing, voids and cracks form at different locations in the TSV, as shown in Fig. 5.33, thereby reducing stress. This may be due to the diffusion of vacancies towards existing defects under the radial compressive stress gradient within the TSV during the annealing process. The stress distribution within the TSV filler is quite uneven, and stress concentration may occur at grain boundaries depending on the structure, morphology, and distribution of Cu grains.

(3) Metal protrusion

In actual service, chips are affected by self-heating and ambient temperature. At different temperatures, the difference in CTE between Cu and Si in the TSV can generate shear stress at the end of the TSV. If the shear stress is large enough, it can cause plastic deformation of Cu within the TSV. In most cases, plastic deformation of the metal at the end of the TSV causes Cu extrusion, and the degree of extrusion is proportional to the temperature. In actual manufacturing processes, Cu extrusion is prone to occur during chemical mechanical polishing before or after annealing, as well as during heating for dielectric deposition processes related to RDL/BEOL structures. After heating to high-temperature annealing/manufacturing temperatures and cooling, the stress within the TSV exceeds the yield strength of the entire top of the TSV, causing yielding at the top of the TSV [42]. The accumulation of plastic deformation and creep strain at the top of the TSV can lead to Cu protrusion (or Cu protruding from the top), as shown in Fig. 5.34.

During the through-via process, the interfacial shear stress generated after cooling of the manufacturing temperature is symmetric along the length of the through-via and concentrated at both ends. Therefore, for through-vias, interfacial sliding and Cu extrusion are also symmetric at both ends. For blind vias, the interfacial shear stress increases from the blind via end to the open end, resulting in greater protrusion at the open end. It should be noted that interfacial diffusion-induced interfacial sliding under external shear stress is a mechanism that adapts to the differential deformation of Cu and Si at the interface, requiring sufficiently high temperature and long duration

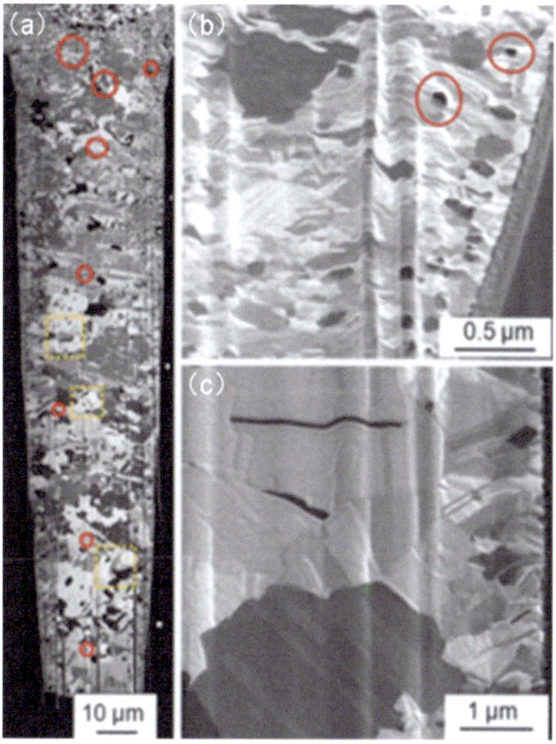

Fig. 5.33 Voids and cracks forming at different locations in TSV

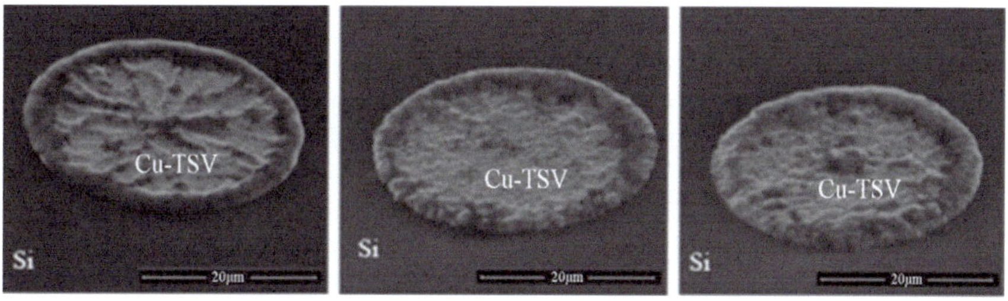

Fig. 5.34 Cu protrusion in TSV

to activate the diffusion process. This is why, under rapid cycling conditions, uniform Cu protrusion occurs due to the non-elastic deformation of Cu, without significant interfacial sliding; whereas, when cycling slowly over a larger temperature range, steps form at the interface due to sliding. The CTE of Cu is much larger than that of Si, and when cooling from high temperatures, Cu undergoes significant relative contraction, which can be adjusted by sliding at the interface. In this case, the end of the Cu through-via may contract rather than protrude. Contraction is due to Cu deformation, but the interface step is due to diffusion-regulated interfacial sliding.

However, it should be noted that as the number of thermal cycles increases, the stress generated in the metal filler reaches saturation due to progressive strain hardening, and the protrusion or contraction rate of the Cu through-via end decreases during thermal cycling.

The end of the TSV is mostly connected to the RDL/BEOL structure, so the protrusion or intrusion of the Cu through-via poses a significant threat to the integrity of these structures. The stress caused by the protrusion of the via end can lead to deformation of the RDL/BEOL structure or delamination of the end-cap layer of the TSV, as shown in Fig. 5.35. Due to the potential serious reliability issues, the impact of Cu protrusion has been widely studied, and the influence of various process parameters has been noted, such as TSV pitch/diameter, Cu coverage after electroplating, and annealing conditions. Under normal circumstances, most Cu protrusion phenomena are uniform or overall (10–30 nm protrusion), with only a small portion of TSVs showing single-grain extrusion after high-temperature annealing. The average protrusion of larger diameter TSVs is greater, but pitch seems to have little effect. However, although the average protrusion of the TSV array is larger for larger TSVs, the maximum Cu protrusion seems to be independent of the TSV diameter, indicating that the diameter of the TSV has little impact on the reliability of the BEOL structure. Cu coverage usually has no effect, but a higher pre-annealing temperature for chemical mechanical polishing reduces Cu protrusion. An additional annealing process after pre-annealing for chemical mechanical polishing significantly reduces Cu protrusion. Recent studies have shown that the surface roughness frequently observed at the end face of TSVs after thermal cycling is related to the sliding of grain boundaries along the Cu non-coherent boundary. However, non-coherent boundaries are prone to sliding due to their higher interface energy, with a misorientation of about 59°. The interface energy of coherent boundaries is much lower and does not slide. This means that if the electroplating process of the TSV can be controlled to produce only coherent $\Sigma 3$ grain boundaries near the upper surface, Cu protrusion caused by grain boundary sliding can be largely eliminated.

(4) Heating rate dependence of Cu protrusion

During thermal cycling, the extent and mechanism of Cu protrusion depend on the heating/cooling rate and the temperature range of the packaging thermal cycle. There have been studies on the effect of heating rate on Cu protrusion during the heating process, in which samples were first rapidly heated to 300 °C at a heating rate of 0.1 °C/s, and then further heated to 425 °C at various heating rates, with a TSV pitch of 200 μm in the samples. No changes occur on the TSV surface below 300 °C, but after holding at 425 °C for 90 min, existing surface features begin to protrude. As shown in Figs. 5.36 and 5.37, when heated at a rate of 0.02 °C/s, there is almost no change during the 425 °C holding period [52]; however, at a higher heating rate (0.05 °C/s), due to grain boundary sliding, grains protrude along the edges and center of the TSV. Over time, these features continue to grow. It should be noted that multiple mechanisms (including grain boundary sliding, interfacial sliding, and creep) are at work at all heating rates, although the dominant mechanisms vary depending on the rate and temperature range. Grain boundary sliding, interfacial sliding, and creep

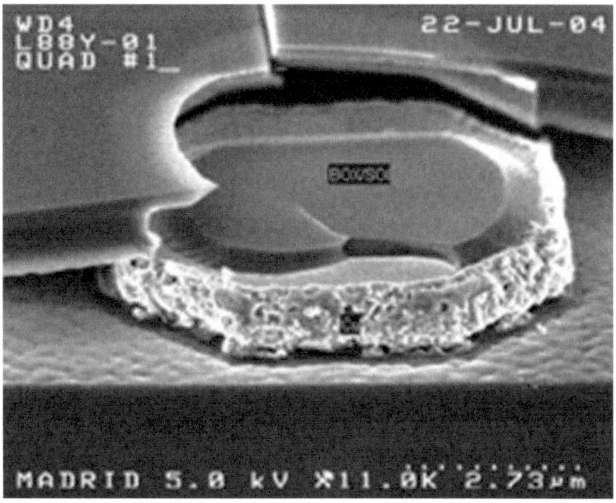

Fig. 5.35 Cu protrusion in TSV causing chip delamination

are all heating rate-dependent (diffusion-dependent) processes. Due to this diffusion dependence, as well as the dependence on the stress state of the TSV and interface at any given moment (the driving force for these processes), the relative kinetics of the above mechanisms depend on both temperature and heating rate.

Figure 5.38 shows the topography of TSV after 5 thermal cycles from room temperature to 300 °C and from room temperature to 425 °C at a heating rate of 0.01 °C/s. Even at the same heating rate, the temperature range of the thermal cycle has a significant impact on the main mechanism of Cu protrusion (grain boundary

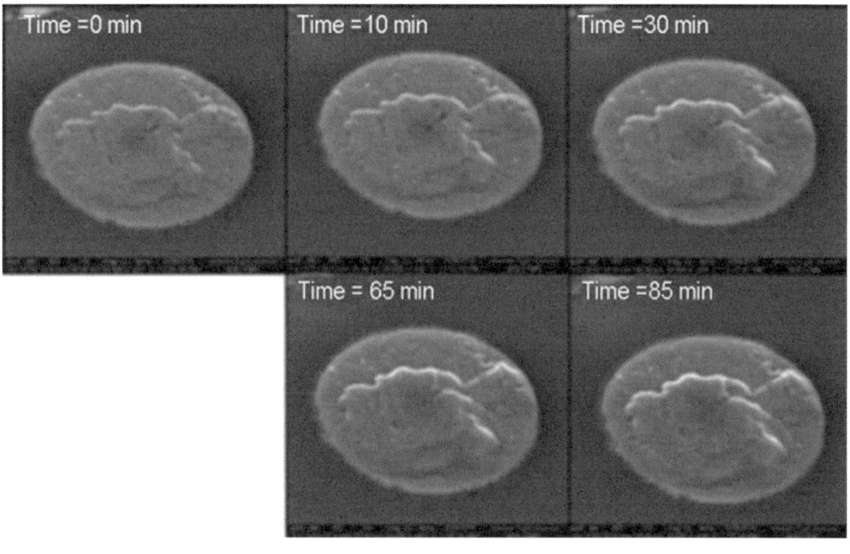

Fig. 5.36 In-situ observation of single Cu–TSV surface during 425 °C holding period at a heating rate of 0.02 °C/s

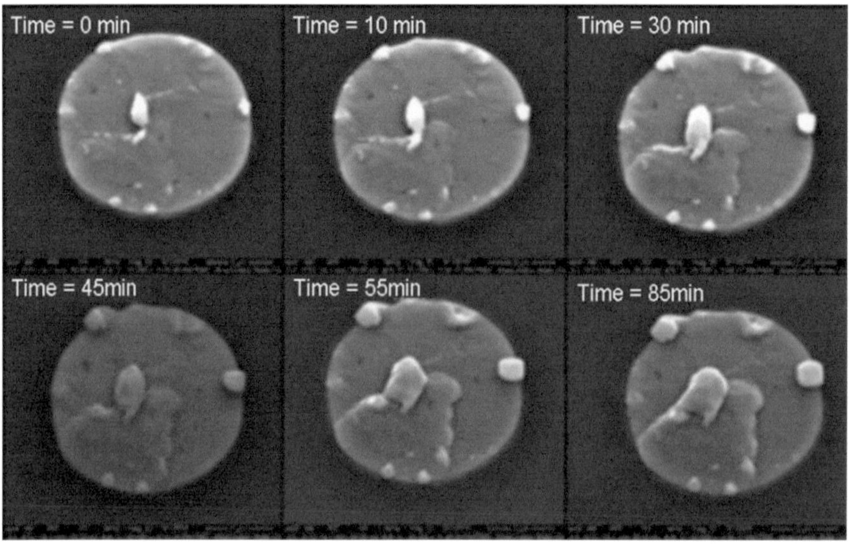

Fig. 5.37 In-situ observation of single Cu–TSV surface during 425 °C holding period at a heating rate of 0.05 °C/s

sliding at 300 °C and interfacial sliding at 425 °C). At this lower heating rate, the creep mechanism (grain boundary sliding and interfacial sliding) dominates, with grain boundary sliding dominant at low temperatures and interfacial sliding dominant at high temperatures.

In summary, when the heating rate is high, the temperature range is large, especially with a lower peak temperature, non-plastic deformation dominates. At medium heating rates and temperature ranges, there is sufficient thermal activation and diffusion time, and the stress in the TSV is high enough for grain boundary sliding to be

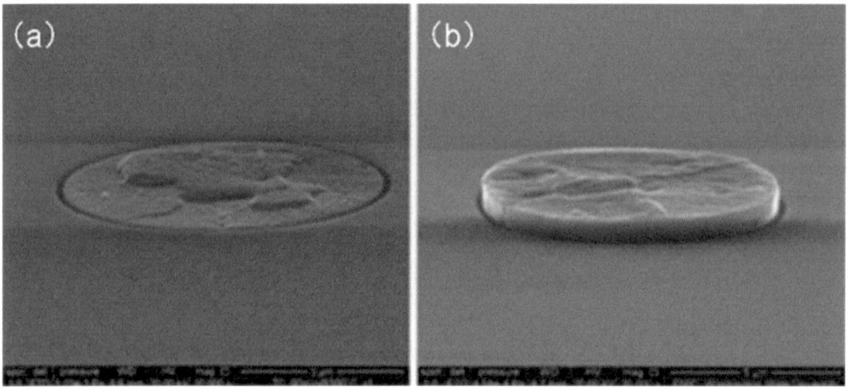

Fig. 5.38 Topography of TSV

the dominant mechanism, leading to significant surface undulations at the top of the TSV. At higher peak temperatures and slower heating rates, the stress in the TSV is significantly reduced (due to stress relaxation from dislocations or diffusion creep), and interfacial sliding dominates. The interaction between stress state, heating rate, and peak temperature usually results in a superposition of the above three effects, although one or two of them may be dominant.

2. Electromigration-related effects

Although electromigration is an important reliability issue for metal interconnects in electronics, especially in BEOL structures, TSVs are generally less susceptible to electromigration-induced failures. This is primarily due to the relatively large cross-section of TSVs, which reduces current density. However, at the junction of TSVs with RDL/BEOL structures, the combination of current and the complex stress state above and below the TSV can not only lead to substantial diffusion effects but also to void growth related to electromigration. The impact of stress gradient, electric potential gradient, and temperature gradient on atomic flux divergence (AFD) is analyzed using finite element methods. At the top and bottom UBM contacts with TSVs, AFD related to diffusion flow and electromigration is usually significant [53]. In most cases, the stress gradient caused by joule heat leads to AFD phenomena, while a smaller portion of AFD is caused by the electric potential gradient. TSVs are susceptible to electromigration and void growth. Generally, reducing the stress gradient will decrease void growth.

The results of electromigration tests conducted on devices with thin and thick UBM at the top and bottom of TSVs (see Fig. 5.39) indicate that no voids form inside the TSV due to electromigration, regardless of the current direction. However, significant void nucleation is observed at the intersection of the UBM in the RDL structure and the end of the TSV in the downstream direction of electron flow. It is speculated that these voids form because the TiN barrier layer prevents Cu atoms from migrating from the TSV to the voids, but migration from the void region may occur in the Cu line at the end of the TSV. Therefore, the TiN layer, which acts as an effective diffusion barrier, is actually the root cause of electromigration void nucleation. In thin Cu lines, voids occupy the entire thickness of the thin Cu line. For thicker Cu lines, the current crowding effect at the point where electrons leave the TSV is significant, leading to voids in the thicker Cu lines as well. In thin Cu lines, voids are located at the interface between the TiN and Cu line. Through electromigration testing, it has been found that voids exist in the back Cu line downstream of the electron flow direction, but not at the SiN-Cu interface, and the opposite is true at the TiN-Cu interface. However, the small voids present on the TSV side remain unchanged before and after the electromigration test [54, 55]. Therefore, although electromigration does not easily damage the TSV itself, it can severely affect the RDL/BEOL structure reliability in the device.

Under electromigration conditions, interfacial sliding can cause different types of "complications." External current can enhance or weaken interfacial sliding caused by interfacial shear stress, depending on the direction of the applied electric field.

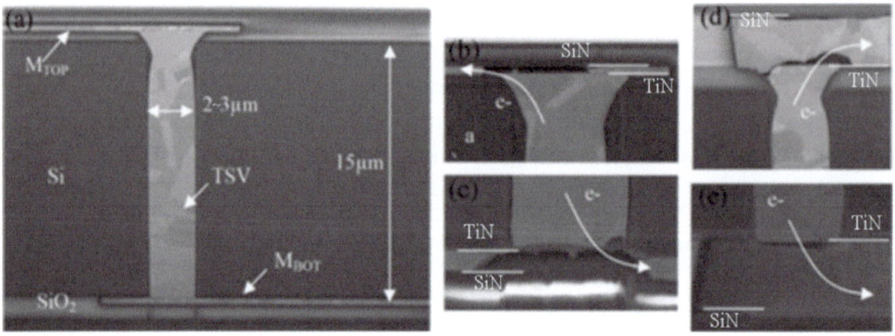

Fig. 5.39 Results of electromigration testing

Even if there is no shear stress or very little shear stress at the interface, electromigration can drive interfacial sliding along the TSV-Si interface. As shown in Fig. 5.40, under electromigration and constant current with thermal cycling conditions, the end of the TSV protrudes in the direction of electron flow and indents in the opposite direction. Under the influence of the applied current, Cu diffuses along the interface in the direction of electron flow, causing the position of the Cu filler relative to Si to change over time. Electromigration-induced interfacial sliding is asymmetric and continuously accumulative. This can pose potential serious reliability challenges, especially when the current density through the through-via increases as the TSV diameter decreases [49, 50]. It should be noted that although this type of through-via migration occurs in samples without RDL vias under electromigration conditions, the presence of RDL/BEOL dielectric can limit this migration, thereby mitigating it. Since the RDL/BEOL dielectric has a low elastic modulus, the effect of any external constraints is limited, and the impact of this phenomenon on the stability of the RDL/BEOL structure requires further study.

Studies on Si interposers with several TSVs and UBMs have shown that there are two main mechanisms of electromigration-induced damage. Typically, because there is a BEOL structure at the top and RDL, Cu pads, and solder joints at the bottom, the

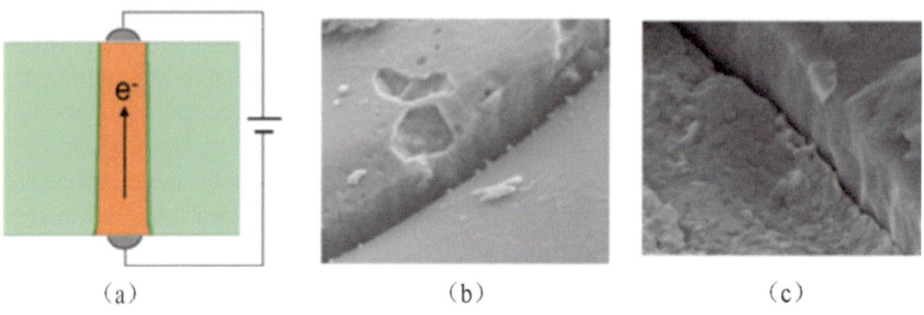

Fig. 5.40 **a** Schematic diagram of electromigration and thermal cycling (25–425 °C) test arrangement. Thermal cycling + electromigration with current density of 5.2×10^4 A/cm^2, TSV end protrusion. Reverse current density of -5.2×10^4 A/cm^2, Cu intrusion into Si

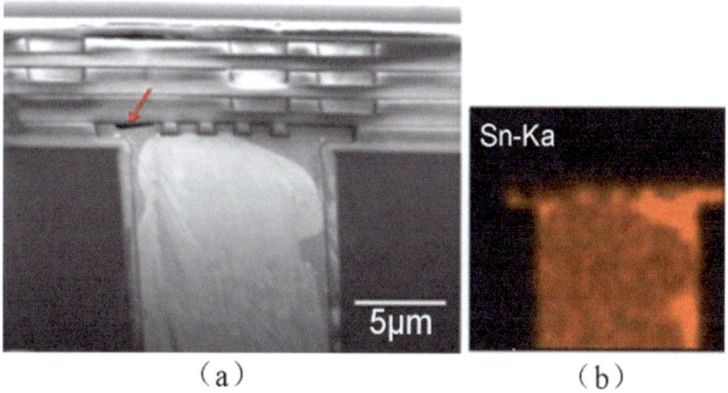

Fig. 5.41 Damage caused by electromigration

protrusion and contraction of TSVs are inhibited. However, under extreme electromigration conditions (200 °C, TSV current density of 1.5 × 105 A/cm^2, for 20 days), voids were observed in the metal 1 (M1) layer, where electron flow converges after leaving the larger cross-sectional TSV and diverges again before leaving. These voids are indicated by arrows in Fig. 5.41a [56]. These electron flow crowding-induced electromigration voids are consistent and usually form under high-density interconnect structures and high current densities. The damage observed under extreme electromigration conditions is caused by Sn diffusion into the TSV, as shown in Fig. 5.41b. It is clear that Sn has electromigrated along the electron flow direction from the Sn-based solder joint, through the UBM below the TSV, along the TSV-Si interface, and into the TSV [57]. Under extremely high current densities, elements in the solder alloy can cause significant degradation of device electrical performance due to alloying of Cu, and the volume change caused by alloying can lead to mechanical instability. As mentioned above, the test conditions in these experiments are more stringent than those experienced by packages containing TSVs, but they indicate the types of damage that packages may experience as current density and temperature increase.

5.5 Case Studies of 3D Packaging Failures

5.5.1 CoWoS 3D Packaging Structure Failure Case

CoWoS first connects the semiconductor chip to the wafer through the Chip on Wafer process, and then connects the C2W chip to the substrate to form CoWoS. The typical CoWoS packaging structure is shown in Fig. 5.42 [19, 58, 59].

1. Reliability of Cu Interconnects in TSV Interposers

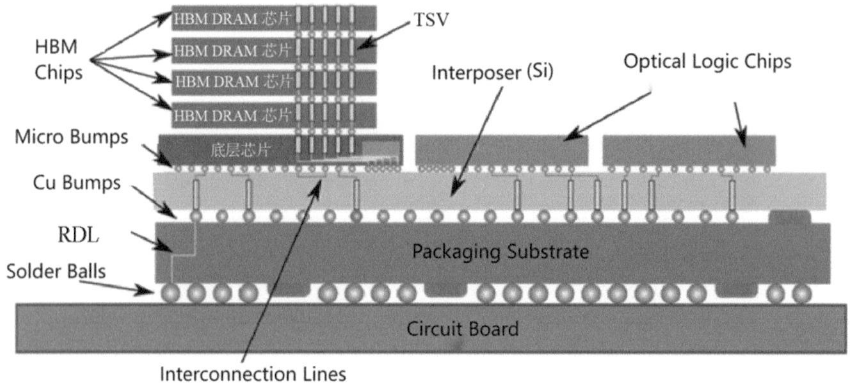

Fig. 5.42 Typical CocWoS packaging structure [12]

Since the interposer uses high-density Cu interconnects, inserting a large number of Cu-TSVs below the Cu interconnects poses many reliability issues for the product. The volume of Cu in the TSV is 200,000 times that of the Cu in the interconnect. After typical TSV electroplating with Cu fill, subsequent chemical mechanical polishing and packaging processes may cause cracks or short circuits in the Cu interconnects. However, these can be improved by controlling the electroplating process and annealing the microstructure of Cu.

To ensure the reliability of the Cu interconnects below the TSV, this case proposes a complete set of reliability test structures and defines a new design rule for reliable Cu interconnects to TSV. Reliability tests include Cu interconnect electromigration, stress migration, intermetallic dielectric TDDB, and V_{bd}/TDDB of MiM decoupling capacitors. After optimization of the process, all test items have met the reliability criteria with good margin [19, 20].

2. Microbump Reliability

Microbumps are one of the key supporting components for CoWoS 3D IC integration, and the microbump structure, interface engineering, and underfill material all have a significant impact on the reliability of microbumps. The reliability issues of microbumps are discussed in two parts: the thermal–mechanical reliability and the elevtromigration of them.

(1) Thermal–mechanical reliability of microbumps

Taking advantage of the CoW process, chips are connected to the TSV microbumps on the entire wafer. However, due to the small size of the microbumps, the peel stress at the interface between the Al pad and the microbump is the main issue for structural reliability. During the design process, the structure of the microbump and the choice of underfill material can be optimized through finite element modeling. The thermal resistance between the microbump and the substrate is used as an indicator to optimize the manufacturing process. The interfacial resistance of the microbump

Fig. 5.43 Interfacial resistance of microbump [51]

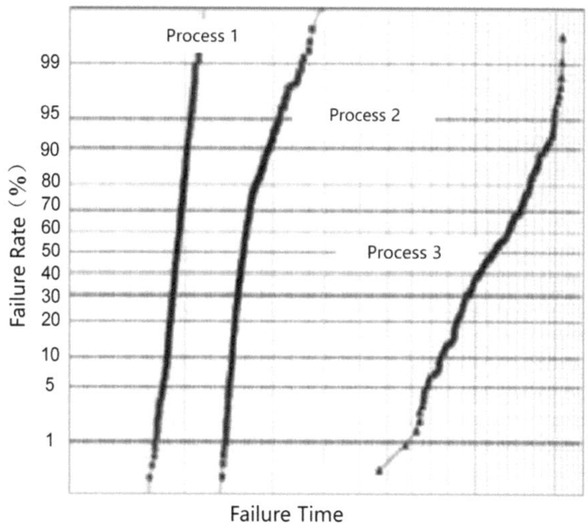

is shown in Fig. 5.43. Interface optimization can improve the interfacial bonding strength between the microbump and the Al pad.

Table 5.1 shows the results of the temperature cycling test for CoW microbump connections. It can be seen from Table 5.1 that the processing technology of Process 1 effectively improves the interfacial bonding strength between the microbump and the Al pad. The typical failure mode of microbumps under temperature cycling conditions is shown in Fig. 5.44.

2. Electromigration of microbumps

To test the electromigration behavior of microbumps, two types of microbump test structures were designed: one is a single microbump Kelvin structure; the other is a daisy chain structure with 22 microbumps. The microbump resistance changes with electromigration stress in the single microbump Kelvin structure can be monitored with high precision, which can better understand the electromigration behavior of microbumps; while the daisy chain structure is more statistically significant and can understand the average behavior of multiple microbump structures.

Figure 5.45 shows the daisy chain structure for electromigration test.

3. Reliability of TSVs

Table 5.1 Temperature cycling test results for CoW microbump connections

Test condition	Process 1 (failures/total samples)	Process 2 (failures/total samples)	Process 3 (failures/total samples)
200 Cycles	0/20	11/20	20/20
500 Cycles	0/20	15/20	20/20
1000 Cycles	0/20	20/20	20/20

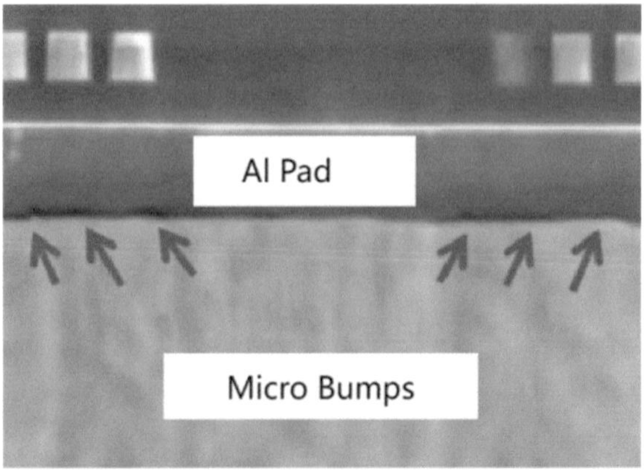

Fig. 5.44 Typical failure mode of microbumps under temperature cycling conditions [52]

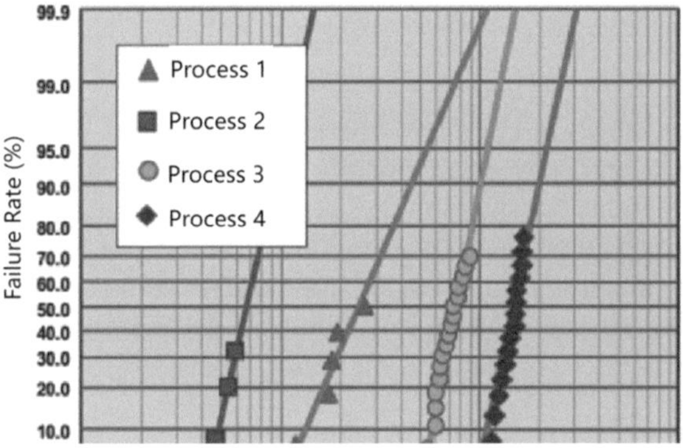

Fig. 5.45 Daisy chain structure for electromigration test [51]

For the TSV interposer in CoWoS, one end is connected to the first layer of Cu interconnects on the front side, and the other end is connected to the C4 bumps on the back side. The TSV interposer plays a key role in signal transmission by connecting the C4 bumps and RDL. At the same time, the TSV interposer provides the power and ground connections between the C4 bumps and the chip.

Figure 5.46 shows the TSV and C4 bump electromigration test structure. The structure includes two key interfaces of the TSV interposer: the chemical mechanical polishing surface of the TSV interposer and the interface with the first layer of Cu interconnects on the front side, and the connection interface with the C4 bumps on the back side. By optimizing these two interfaces, the electromigration performance of the structure can be improved. Two TSVs and two C4 bumps are connected to form a test structure. A current of 500 mA is applied to this structure at 160 °C to measure electromigration. The cumulative failure distribution of TSV electromigration is shown in Fig. 5.47. Process 1 and Process 2 represent different electroplating parameters and different backside grinding processes, respectively. It can be seen that the device lifetime of Process 2 has obvious improvement. The microstructure of the TSV and C4 bump interface is shown in Fig. 5.48. Voids are formed between the TSV and C4 bump in the samples of Process 1, but no damage is observed in the samples of Process 2. The failure mode of Process 2 samples is typical electromigration failure of C4 bumps (see Fig. 5.49), where the Cu pad on the substrate is completely consumed on the cathode side. After applying the same current stress to the TSV and Cu interconnects, no electromigration degradation is found at the interface.

4. Reliability of components

After the above process reliability analysis and optimization, the reliability of the integrated CoWoS component is tested at the component level, including temperature cycling condition-B (TCB, − 55 to 125 °C), high-temperature storage (HTS, 150 °C), and unbiased high-acceleration stress test condition-A (uHAST, 130°C/85% RH) and 3-level pre-treatment (30°C/60% RH/192 h + 260 °C reflow soldering × 3).

The results show that CoWoS has a good reliability margin. It has passed the industrial standard TCB 1000 cycles, HTS 1000 h, and uHAST 96 h tests, and extended to TCB 2500 cycles, HTS 2000 h, and uHAST 1000 h without failure.

Moreover, the thermal–mechanical stress risk of the entire microbump IMC is assessed: after high-temperature storage for 1000 h, followed by 250 cycles of TCB and 250 h of high-temperature storage. Additionally, after increasing the TCB test to 500 cycles and the HTS test to 500 h, no failures occurred. The IMC of the entire microbump has good thermal–mechanical reliability under the protection of the underfill.

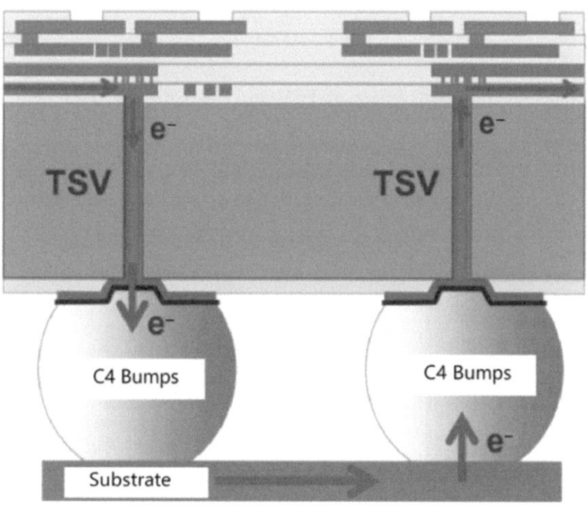

Fig. 5.46 TSV and C4 bump electromigration test structure [51]

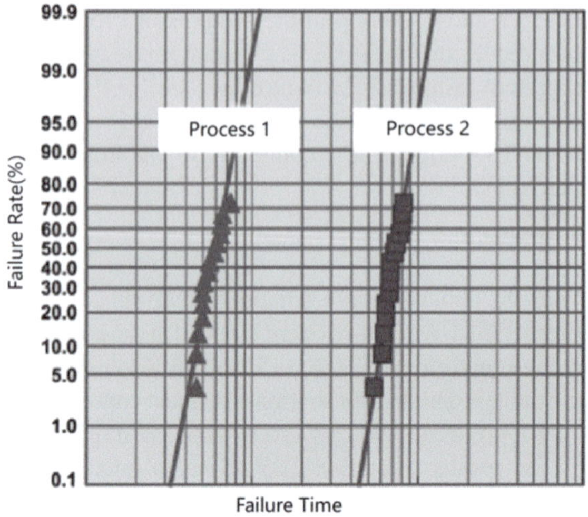

Fig. 5.47 Cumulative failure distribution of TSV electromigration [51]

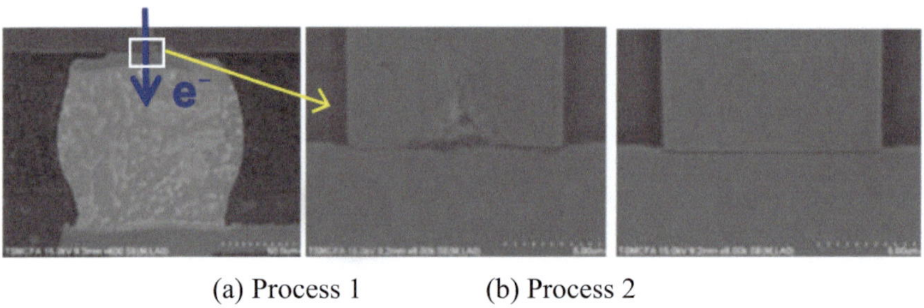

Fig. 5.48 Microstructure of TSV and C4 bump interface [51]

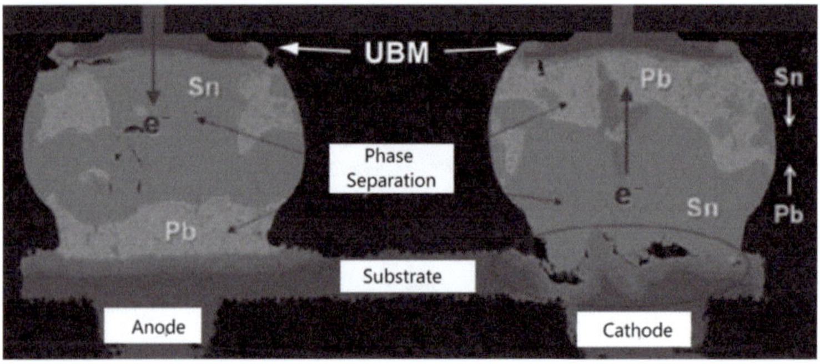

Fig. 5.49 Typical failure mode of C4 bump electromigration [51]

5.5.2 Fan-Out Packaging Failure Case

Fan-out packaging adopts the traditional embedded wafer-level ball grid array packaging scheme, and its packaging stack structure is shown in Fig. 5.50. Fan-out packaging has advantages such as high routing density, small pin pitch, thin packaging thickness, and low high-frequency transmission loss, and has been recognized as a mainstream packaging technology in recent years. However, due to its novelty, fan-out packaging still faces many reliability issues that need to be solved [60].

Currently, fan-out packaging technology mainly faces two key process challenges: warpage and chip offset. In fan-out packaging, if any problem occurs in the molding, chip bonding, RDL, and microbump processes, it will lead to the failure of the entire chip package [61]. Warpage is mainly caused by the mismatch of CTE between different materials. The epoxy resin material used in chip packaging will expand and contract due to temperature changes, and when the CTE of other materials are mismatched, delamination or fracture may occur at the contact interface. Chip offset refers to the deviation of the chip from its original designed position during the placement and molding processes due to material properties, equipment precision, and process parameters.

National Center for Advanced Packaging Co., Ltd. has carried out the complete manufacturing of fan-out packaging structures and back-end assembly processes for daisy chain chips, and conducted electrical characterization, reliability testing, and failure sample analysis on packaging samples from different batches and under different process parameter conditions [62]. The reliability testing includes dividing the chips that have undergone MSL-3 pre-treatment into three categories, followed by temperature cycling (TC) tests (500 cycles), and finally high-temperature storage (HTS) tests for 1008 h and high-acceleration stress tests (HAST) for 144 h. The reliability test results of fan-out packaged chips are shown in Table 5.2.

It can be seen from Table 5.2 that the test pass rate of chips is directly related to the magnitude of the stress applied in different test items. The greater the stress, the higher the proportion of failed chips, and the lower the test pass rate.

The fracture morphology of failed chips after TC testing is shown in Fig. 5.51.

From Fig. 5.51, it can be seen that although there is delamination between the medium and the conductor, it does not lead to the fracture of the conductor itself. This

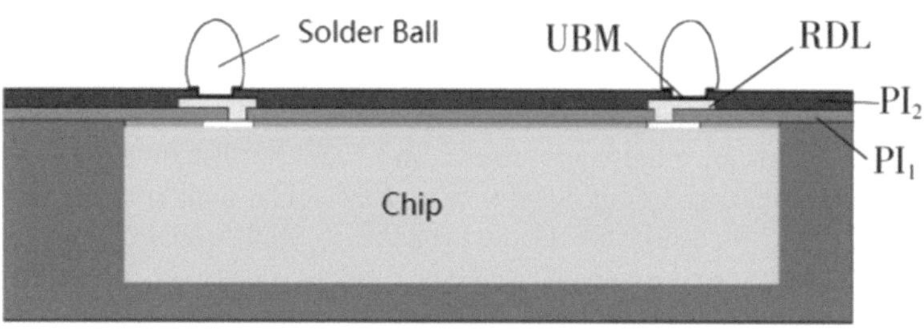

Fig. 5.50 Fan-out packaging stacking structure

Table 5.2 Reliability test results of fan-out packaged chips

Test type	Number of tested chips	Number of failed chips	Test pass rate (%)
Pre-treatment	150	60	60.0
TC test (500 cycles)	30	4	86.7
HTS test (1008 h)	30	3	90.0
HAST test (144 h)	30	2	93.3

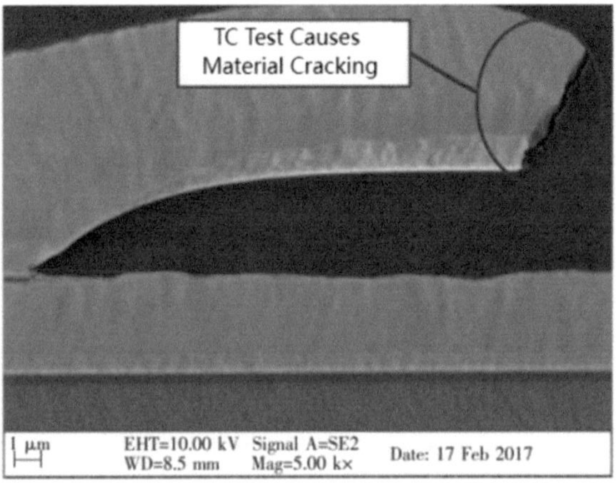

Fig. 5.51 Fracture morphology of failed chips after TC testing

is because the materials experience expand and contract under high/low temperature impact during TC test, thereby causing interfacial delamination. Therefore, to resist the stress that causes interfacial delamination, it is necessary to improve the toughness and adhesion strength of the interfacial material.

The comparison of the average resistance of each chip after HAST testing is shown in Fig. 5.52.

To compare the resistance changes before and after the test, the cross-section of the solder joints in the normally resistant chips was analyzed using SEM equipment before pre-treatment. The cross-section SEM image of the solder joint is shown in Fig. 5.53. Partial voids exist in the IMC combined with the solder joint and UBM, but the solder joint has not completely collapsed. The location of the voids is at the grain boundary of the IMC. Observing the morphology of the solder joint, it is considered that the flux coverage is uneven, which prevents sufficient wetting of the solder joint during reflow soldering, thereby generating voids.

After pre-treatment, more voids appear, and even the part in contact with the pad is broken. The cross-section SEM image of the solder joint after pre-treatment is shown in Fig. 5.54. According to the principle of void formation, after experiencing pre-treatment (pre-treatment includes 3 reflow soldering), the grains in the solder grow and coarsen. The plastic deformation of the solder causes microvoids at the grain boundary between the solder and the pad. Subsequently, as the thermal

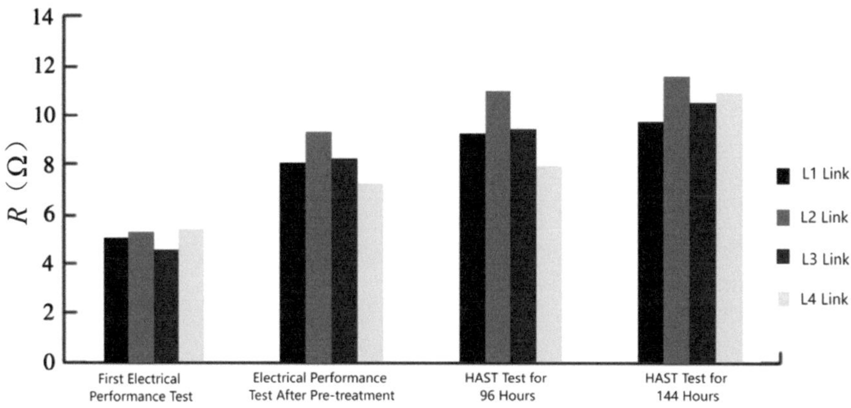

Fig. 5.52 Comparison of average resistance of each chip after HAST testing

Fig. 5.53 Cross-section SEM image of solder joint

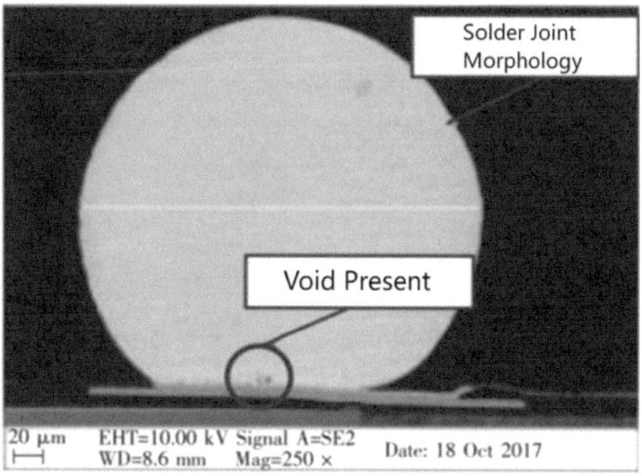

cycle progresses, the voids enlarge and increase, forming clusters of voids, eventually leading to the formation of microcracks, and with the increase of microcracks, macrocracks are formed, thereby enlarging the voids at the interface and increasing their impact on resistance.

After 144 h of HAST testing, the average resistance of the 4 chains increased by about 5 Ω. After 500 cycles of TC testing, the average resistance of the 4 chains also increased by about 5 Ω. After 1008 h of HTS testing, the average resistance of the 4 chains also increased by about 5 Ω. To analyze the reasons for the increase in resistance, the solder joint structure and delamination conditions after TC and HTS testing were observed. The cross-section SEM image of the solder joint after TC testing is shown in Fig. 5.55.

It can be seen from Fig. 5.55 that after TC testing, the IMC between the pad and the solder joint has become significantly thicker, thereby increasing the resistance. After ultrasonic scanning, it was found that several chips had delamination. The observed

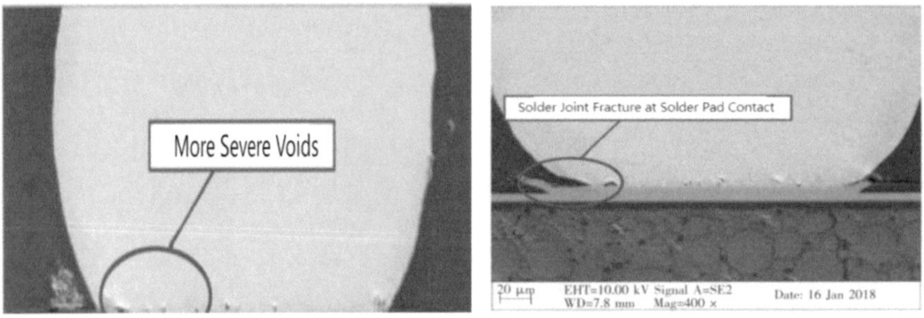

(a) Voids (b) Solder joint and pad contact fracture

Fig. 5.54 Cross-section SEM image of solder joint after pre-treatment

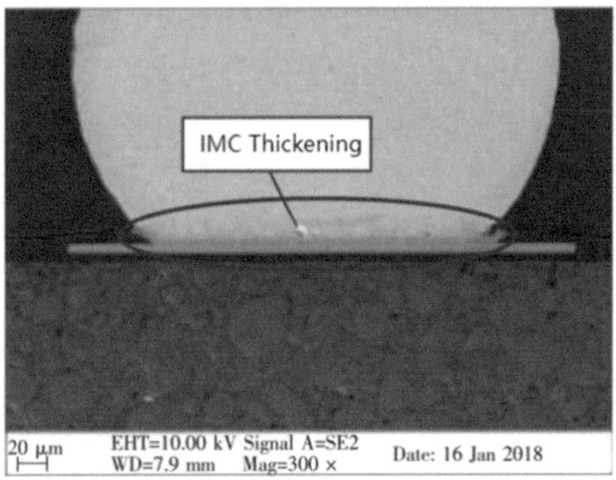

Fig. 5.55 Cross-section SEM image of solder joint after TC testing

results of the ultrasonic scanning microscopy (SAM) are shown in Fig. 5.56. It can be seen that after TC testing, looking from the front of the chip, delamination occurred in the second layer, with the delamination occurring between PI and epoxy molding compounds. Although after 500 cycles, the delamination between PI and epoxy molding compounds did not affect the continuity of the link, if TC testing continues, it may cause more serious delamination, thereby causing RDL fracture.

5.5.3 TSV Structure Failure Case

TSV technology is one of the key interconnection technologies in 3D integration. It creates vertical electrical connection channels that penetrate chips or wafers to achieve vertical interconnection between chips and wafers, and serves the functions of signal conduction, heat dissipation, and mechanical support. TSV has advantages such as good electrical performance, low power consumption, high interconnection

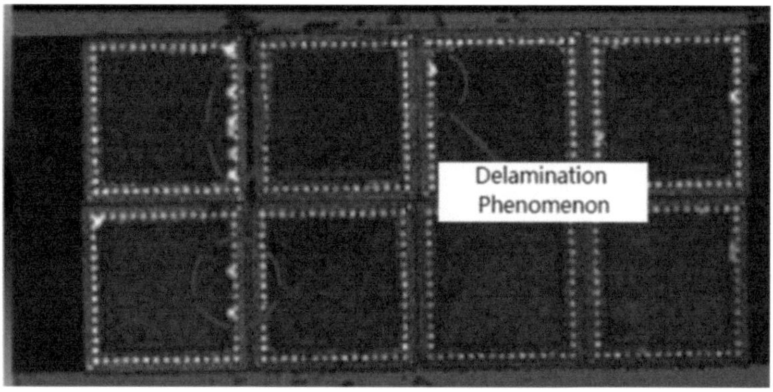

Fig. 5.56 SAM results

density, and small size, and has been widely used in various fields. However, there are still many thermo-mechanical reliability issues to be solved. This case combines ion-beam material removal to measure the residual stress gradient at the TSV interface at nanometer resolution in the as-received state, after annealing, and after temperature cycling. The microstructural changes of the Cu seed layer at the TSV interface were measured using SEM and Electron Backscatter Diffraction (EBSD), clarifying the stress relaxation mechanism related to microstructural evolution. To verify the credibility of the experiment, finite element and SRIM program simulations were used to discuss the impact of Focused Ion Beam (FIB) processing on the experimental results.

1. Ion-beam material removal method

The structural parameters of TSV are shown in Fig. 5.57, with a diameter of 30 μm, depth of 100 μm, and pitch of 200 μm. The materials at the TSV interface are Cu-TSV, TiW barrier layer, SiO_2 insulating layer, and Si substrate.

Fig. 5.57 Structural parameters of TSV

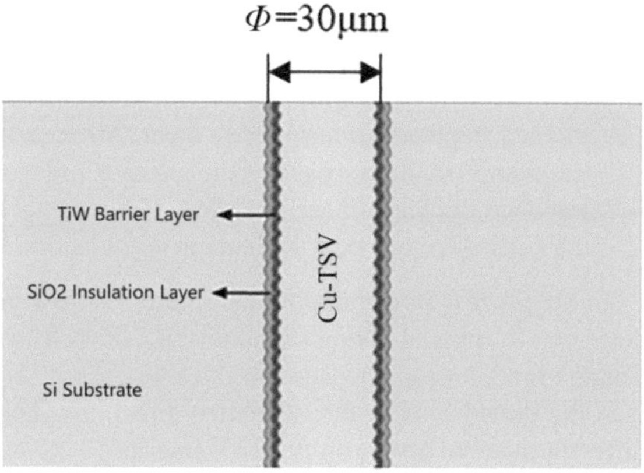

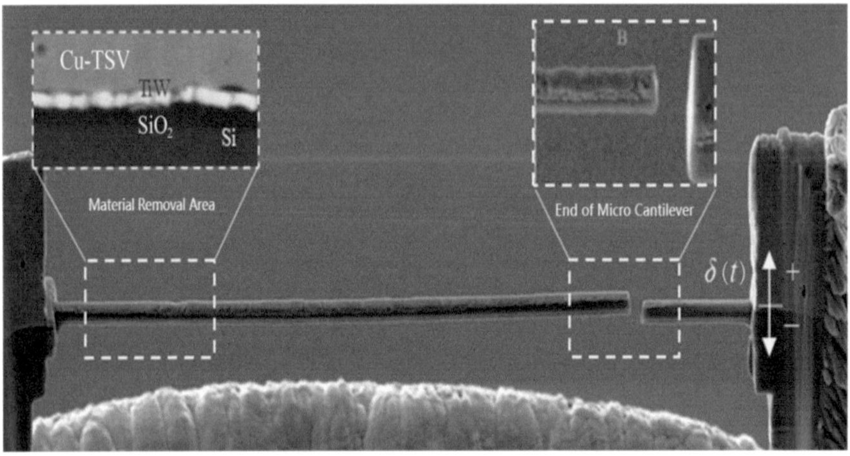

Fig. 5.58 TSV interface micro-cantilever beam structure

In this case, three groups of samples were prepared. The first group is the TSV samples that have just been electroplated (untreated); the second group of TSV samples underwent 10 temperature cycles (0–125 °C, with both heating and cooling rates at 10 °C/min, and dwell times at both high and low temperatures for 2 min each); the third group of TSV samples underwent annealing at 250 °C, with annealing conditions of heating at a rate of 10 °C/min to 250 °C, holding for 30 min, and then cooling to room temperature with the furnace.

FIB was used to fabricate the TSV interface micro-cantilever beam structures for the three groups of samples, as shown in Fig. 5.58.

The main process for measuring the residual stress inside the TSV interface micro-cantilever beam structure using the ion-beam material removal method is as follows.

(1) One end of the initial micro-cantilever beam structure is cut open. Due to the existence of internal residual stress, the micro-cantilever beam will warp to a certain extent. The initial warpage δ_{origin} at the end of the micro-cantilever beam is measured using SEM.
(2) Using FIB, the interface material of the TSV is gradually removed in the material removal area (ILR Area) of the micro-cantilever beam. The Cu-TSV layer is thinned step by step until the Si layer. After each layer of material is removed, the change in the warpage of the micro-cantilever beam $\delta(t)$ and the remaining interface thickness d are recorded. The residual stress inside each layer of the TSV interface can be calculated in combination with relevant theories.

In the process of calculating the interface stress gradient, it is necessary to first determine the biaxial Young's modulus $E_{\text{b\_f}}$ of each layer of material and the substrate Young's modulus $E_{\text{b\_s}}$, because the thickness of each layer of material is much smaller than the dimensions in the other two directions. The biaxial Young's modulus of different material layers on the TSV interface is calculated as shown in Tables 5.3,

5.4 and 5.5, with $E_{Cu} = 155.47$ GPa, $E_{TiW} = 129$ GPa, E_{SiO_2}=72.1 GPa, and E_{Si} = 131 GPa. According to $E_b = E/(1 - \upsilon)$, where υ represents the Poisson's ratio, the biaxial Young's modulus of different material layers on the TSV interface is calculated.

After obtaining the biaxial Young's modulus of each layer of material, the distribution of residual stress at the interface of the three test samples can be calculated based on the warpage measured by SEM, as shown in Fig. 5.59.

As shown in Fig. 5.59a, for the untreated samples, the compressive stress generated at the Cu-TSV/TiW interface is mainly caused by the magnetron sputtering process used to prepare the Cu seed layer at the TSV interface. In the sputtering process, to maintain visible glow discharge, argon gas with a pressure of about 13.3 Pa is

Table 5.3 Biaxial Young's modulus of different material layers on the untreated TSV interface

Number of Layers	0–3	4	5–6	7	8–9	10	11–12
Material Composition	Si Layer	Si/SiO2 Layer	SiO2 Layer	SiO2/TiW Layer	TiW Layer	TiW/Cu Layer	Cu Layer
*E*b (GPa)	201.54	119.42	86.87	138.36	215.00	216.63	222.10

Table 5.4 Biaxial Young's modulus of different material layers on the TSV interface after temperature recycling

Number of Layers	0–2	3	4–6	7	8–9	10	11–12
Material composition	Si Layer	Si/SiO2 Layer	SiO2 Layer	TiW/SiO2 Layer	TiW Layer	TiW/Cu Layer	Cu Layer
*E*b (GPa)	201.54	109.63	86.87	183.71	215.00	219.94	222.10

Table 5.5 Biaxial Young's modulus of different material layers on the TSV interface after annealing

Number of layers	0	1	2
Material composition	Si/SiO2 Layer	TiW/SiO2 Layer	TiW/Cu Layer
*E*b (GPa)	189.68	159.18	223.74

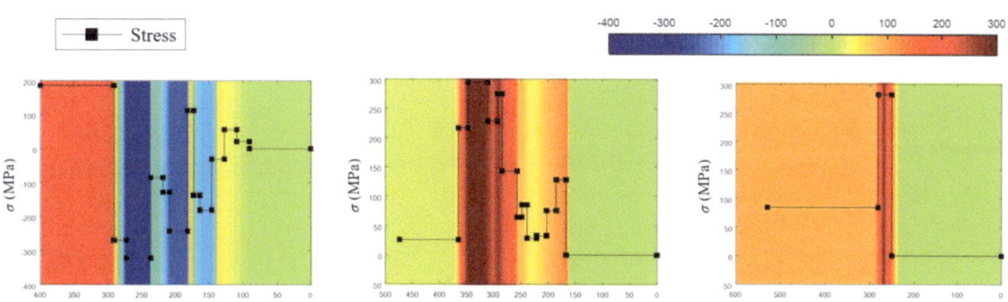

(a) (b) (c)

Fig. 5.59 Distribution of residual stress in three groups of TSV test samples introduced. Under the argon gas pressure during the sputtering process, compressive stress is generated in the Cu-TSV, which is consistent with the experimental measurement results.

Figure 5.59b shows the stress distribution of the TSV interface after temperature cycling. It can be seen that after 10 temperature cycles, the residual stress at the interface has changed from the compressive and tensile residual stresses in the untreated state to tensile stress. The residual stress in the Si far from the Cu-TSV position has decreased from 128.03 to – 0.34 MPa, approaching a stress-free state. The residual stress in the Cu-TSV has decreased by about 86.07%, from 184.93 MPa in the untreated state to 25.76 MPa.

Figure 5.59c shows the stress distribution of the TSV interface after annealing. During the annealing process, the residual compressive and tensile stresses inside the TSV interface all change to tensile stress. In the Si, the residual stress is – 0.29 MPa, approaching a stress-free state. The residual stress in the Cu–TSV has decreased by about 54.83%, from 184.93 MPa in the untreated state to 83.54 MPa.

It can be concluded from Figs. 5.59a–c that both 250 °C annealing and temperature cycling with a 125 °C temperature difference can reduce the residual stress inside the Cu-TSV and change the residual stress gradient curve of the entire TSV interface.

2. Microstructure of TSV Interface

To further explain the results of the ion-beam material removal method measuring the stress distribution at the TSV interface, the microstructure of the Cu seed layer at the TSV interface was characterized using SEM and EBSD, attempting to use the changes in microstructure to explain the changes in stress distribution.

(1) Microstructure of the TSV interface

The failure modes of the Cu–TSV/TiW/SiO_2/Si interfaces of the three groups of TSV test samples were observed using SEM, as shown in Fig. 5.60.

It can be seen from Fig. 5.60 that after temperature cycling, voids have appeared in the Cu–TSV near the interface. The possible reason for the formation of these voids is that under thermal load, the creep deformation within the crystal causes the grains to slip, and some unstable three-grain boundary points releasing stress and generate voids. When a certain number of voids are formed in the Cu–TSV, the voids will join together to form cracks, as shown in Fig. 5.60c.

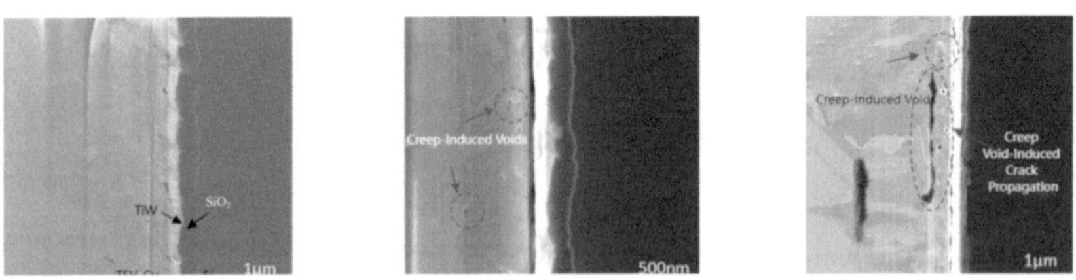

(a) Untreated Sample (b) Sample after 10 Temperature Cycles (c) Sample after 250°C Annealing

Fig. 5.60 SEM image of three groups of TSV test sample

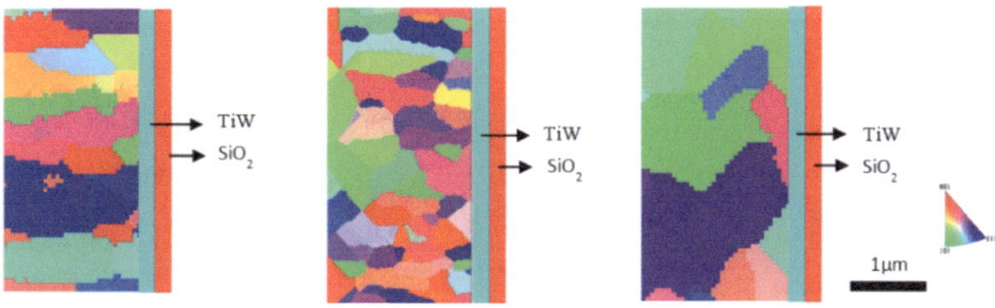

(a) Untreated Sample (b) Sample after 10 Temperature Cycles (c) Sample after 250°C Annealing

Fig. 5.61 EBSD images of three groups of TSV test samples

(2) Crystal structure near the TSV interface

The crystal structure near the TSV interface was captured using an EBSD probe. Since TiW and SiO_2 are amorphous, no EBSD images can be obtained. The EBSD images of the three groups of TSV test samples are shown in Fig. 5.61.

It can be seen from Fig. 5.61 that both temperature-cycled and annealed samples have undergone recrystallization. For the temperature-cycled samples, the grain size of Cu near the interface has decreased, while for the annealed samples, the grain size has increased. This may be because when the Cu is subjected to lower stress or shorter duration, recrystallization occurs within the crystal first, reducing the grain size. When the stress is higher or the duration is longer, the grains will grow further after becoming smaller. The average grain size of Cu–TSV under different test conditions is shown in Table 5.6.

The recrystallization caused by thermal load may lead to deformation within the crystal, thereby changing the distribution of average misorientation between grains and altering the local stress state. To further study this phenomenon, the data obtained from EBSD were processed using Channel 5 software to generate the misorientation distribution map at the TSV interface, which can be used to determine the degree of crystal deformation in the TSV interface region. The misorientation distribution maps of the three groups of TSV test samples are shown in Fig. 5.62.

For Cu–TSV, the average misorientation near the interface is larger, due to the high stress at the interface induces crystal deformation, dislocation accumulation, and strain concentration. Under thermal load conditions, high temperature and thermal stress will further promote crystal deformation, where creep can cause some grains to slip. Once the slip reaches a certain degree, unstable grain boundary points (such as three-grain boundary points) will generate voids, releasing stress and alleviating crystal deformation, resulting in a decrease in dislocation density and stress. When the thermal load conditions intensify or the duration increases, more voids will be

Table 5.6 Changes in average grain size of Cu–TSV under different test conditions

Treatment method	Untreated	10 temperature cycles	250 °C annealing
Average grain size (nm)	226.851	128.773	495.11

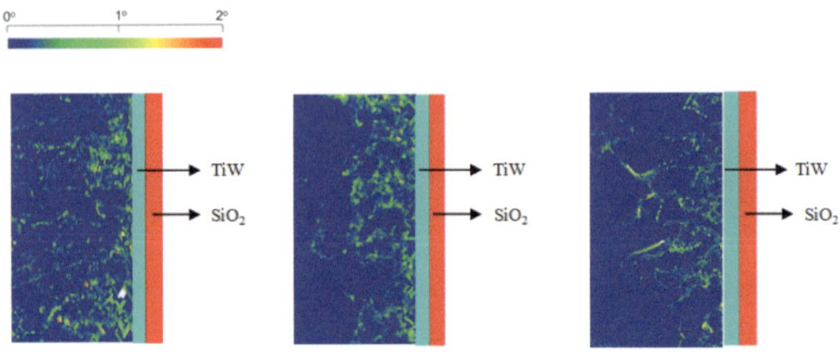

(a) Untreated sample (b) Sample after 10 temperature cycles (c) Sample after 250°C annealing

Fig. 5.62 Misorientation distribution diagrams of three groups of TSV test sample

formed in the Cu–TSV at the interface, and these voids will join together to form cracks.

References

1. Moore GE (1998) Cramming more components onto integrated circuits. Proc IEEE 86(1):82–85
2. Arden WM (2002) The international technology roadmap for semiconductors—perspectives and challenges for the next 15 years. Curr Opin Solid State Mater Sci 6(5):371–377
3. Lau JH (2011) Overview and outlook of through-silicon via (TSV) and 3D integrations. Microelectron Int 28(2):8–22
4. Karnezos M (2004) 3D packaging: where all technologies come together. In: IEEE/CPMT/SEMI 29th international electronics manufacturing technology symposium (IEEE Cat. No. 04CH37585), pp 64–67
5. Miettinen J, Mantysalo M, Kaija K et al (2004) System design issues for 3D system-in-package (SiP). In: 2004 Proceedings of 54th electronic components and technology conference (IEEE Cat. No. 04CH37546), pp 610–615
6. Song B (2006) Reliability evaluation of stacked die BGA assemblies under mechanical bending loads. University of Maryland, College Park
7. Brunner J, Qin IW, Chylak B (2004) Advanced wire bond looping technology for emerging packages. In: IEEE/CPMT/SEMI 29th international electronics manufacturing technology symposium (IEEE Cat. No. 04CH37585), pp 85–90
8. Gerber M, Dreiza M (2006) Stacked-chip-scale-package-design guidelines-design optimization helps to avoid manufacturing problems, to maximize product performance, and to achieve lowest packaging cost. EDN 51(12):79–86
9. Jing L, Kailin P, Weitao Z et al (2011) Stacked packaging technology. Semicond Technol 36(02):161–164
10. Li Y, Niu TF, Coderre J (2007) The application of 3G-driven component stacking assembly technology. China Electron Mark (Basic Electron) (07):70–76
11. Schmidt K, Rauchensteiner D, Voigt C et al (2021) An automated optical inspection system for PIP solder joint classification using convolutional neural networks. In: 2021 IEEE 71st electronic components and technology conference (ECTC), pp 2205–2210
12. Gagnon P, Fortin C, Weiss T (2019) Package-on-package micro-BGA microstructure interaction with bond and assembly parameter. In: 2019 IEEE 69th electronic components and technology conference (ECTC), pp 306–313

13. Chong SC, Ching EWL, Siang SLP et al (2020) Demonstration of vertically integrated POP using FOWLP approach. In: 2020 IEEE 70th electronic components and technology conference (ECTC), pp 873–878
14. Guo K, Ran H, Wang J (2021) Experimental tests and stress analysis of SnPb solder joints in a ceramic PoP device. In: 2021 22nd international conference on electronic packaging technology (ICEPT), pp 1–5
15. Kim J, Kim K, Lee E et al (2021) Chip-last HDFO (high-density fan-out) interposer-PoP. In: 2021 IEEE 71st electronic components and technology conference (ECTC), pp 56–61
16. Xia D, Wang B, Liao C et al (2021) Warpage behavior study and optimization for ultra-thin POP memory with multi-stacked chips. In: 2021 22nd international conference on electronic packaging technology (ICEPT), pp 1–5
17. Yao Y, Cui Z, Ran H et al (2021) Fatigue life predictions of SnPb solder ball in A ceramic PoP device. In: 2021 22nd international conference on electronic packaging technology (ICEPT), pp 1–5
18. Lee K, Koyanagi M (2015) TSV characteristics and reliability: Impact of 3D integration processes on device reliability, three-dimensional integration of semiconductors. Springer, pp 201–233
19. Garrou P (2009) Researchers strive for copper TSV reliability. Semicond Int
20. Farooq M, Graves-Abe T, Landers W et al (2011) 3D copper TSV integration, testing and reliability. In: 2011 International electron devices meeting, pp 7.1.1–7.1.4
21. Kawano M, Wang X-Y, Ren Q et al (2021) One-step TSV process development for 4-layer wafer stacked DRAM. In: 2021 IEEE 71st electronic components and technology conference (ECTC), pp 673–679
22. Jourdain A, Schleicher F, De Vos J et al (2020) Extreme wafer thinning and nano-TSV processing for 3D heterogeneous integration. In: 2020 IEEE 70th electronic components and technology conference (ECTC), pp 42–48
23. Zhou J-Y, Wang Z, Wei C et al (2020) Three-dimensional simulation of effects of electro-thermo-mechanical multi-physical fields on Cu protrusion and performance of micro-bump joints in TSVs based high bandwidth memory (HBM) structures. In: 2020 IEEE 70th electronic components and technology conference (ECTC), pp 1659–1664
24. Seo S-K, Jo C, Choi M et al (2021) CoW package solution for improving thermal characteristic of TSV-SiP for AI-Inference. In: 2021 IEEE 71st electronic components and technology conference (ECTC), pp 1115–1118
25. Liao XP, Gao L (2016) The application of stacked chip wire bonding technology in ceramic packaging. Electron Packag 16(2):5–8
26. Li X, Zhou Y, Sun CS (2008) Research progress on failure mechanisms of plastic encapsulated microelectronic devices. Semicond Technol 33(2):98–101
27. Fu GR, Zhang Z (2005) Failure analysis technology. Phys Test: Phys 41(4):212–216
28. Liu B, Li YH, Zhang XC et al (2006) Application of lock-in infrared thermography in non-destructive testing. Non-destr Test 30(3):12–15
29. Feng LC, Tao N, Xu C (2010) Lock-in thermography and its application in non-destructive testing. Infrared Laser Eng 39(6):1120–1123
30. Xu C, Huo Y, Li YH et al (2007) Basic experimental study on lock-in thermography non-destructive testing method. Non-destr Test 29(12):728–730
31. Lin XL, Yun YF, Yao RH (2016) A defect location method for 3D stacked ICs. J South China Univ Technol (Nat Sci Ed) 44(5):36
32. Lan YQ (2014) Reliability of three-dimensional stacked chip packaging. Xi'an University of electronic Science and Technology, Shaanxi
33. Zhu WM (2012) Thermal and mechanical reliability of stacked chip packaging. Guangdong University of Technology, Guangdong
34. Lin NA (2013) Reliability and failure analysis technology research of system-level packaging (SiP). South China University of Technology, Guangdong
35. Wai AX (2011) Research on advanced 3D stacked chip packaging technology and reliability. Fudan University, Shanghai

36. Ni JF, Wang JJ (2004) Study on the failure mechanism of thin chip fragmentation in IC cards. Semicond Technol (04):40–44
37. Dasgupta A, Pecht M (1991) Material failure mechanisms and damage models. IEEE Trans Reliab 40(5):531–536
38. Gao ST, Yang KW (2004) New microelectronic packaging technology. Electron Packag 4(1):10–15
39. Zhou TAI (2018) Research on the development trend of microelectronic packaging technology. Modern Inf Technol 2(8):2
40. Harper J, Cabral C Jr, Andricacos P et al (1999) Mechanisms for microstructure evolution in electroplated copper thin films near room temperature. J Appl Phys 86(5):2516–2525
41. Kong L, Lloyd J, Yeap K et al (2011) Applying X-ray microscopy and finite element modeling to identify the mechanism of stress-assisted void growth in through-silicon vias. J Appl Phys 110(5):053502
42. De Wolf I, Croes K, Pedreira OV et al (2011) Cu pumping in TSVs: effect of pre-CMP thermal budget. Microelectron Reliab 51(9–11):1856–1859
43. Athikulwongse K, Chakraborty A, Yang J-S et al (2010) Stress-driven 3D-IC placement with TSV keep-out zone and regularity study. In: 2010 IEEE/ACM international conference on computer-aided design (ICCAD), pp 669–674
44. Ho PS, Ryu SK, Lu KH et al (2011) Reliability challenges for 3D interconnects: a material and design perspective. Presentation at the 3D Sematech Workshop, Burlingame, p 17
45. Trigg AD, Yu LH, Cheng CK et al (2010) Three dimensional stress mapping of silicon surrounded by copper filled through silicon vias using polychromator-based multi-wavelength micro Raman spectroscopy. Appl Phys Express (2.265) 3(8):086601
46. De Wolf I, Simons V, Cherman V et al (2012) In-depth Raman spectroscopy analysis of various parameters affecting the mechanical stress near the surface and bulk of Cu-TSVs. In: 2012 IEEE 62nd electronic components and technology conference, pp 331–337
47. Jiang T, Ryu S-K, Im J et al (2013) Impact of material and microstructure on thermal stresses and reliability of through-silicon via (TSV) structures. In: 2013 IEEE international interconnect technology conference-IITC, pp 1–3
48. Okoro C, Lau JW, Golshany F et al (2013) A detailed failure analysis examination of the effect of thermal cycling on Cu TSV reliability. IEEE Trans Electron Devices 61(1):15–22
49. Dutta I, Kumar P, Bakir M (2011) Interface-related reliability challenges in 3-D interconnect systems with through-silicon vias. JOM 63(10):70–77
50. Kumar P, Dutta I, Bakir M (2012) Interfacial effects during thermal cycling of Cu-filled through-silicon vias (TSV). J Electron Mater 41(2):322–335
51. Meinshausen L, Liu M, Lee TK et al (2015) Reliability implications of thermo-mechanically and electrically induced interfacial sliding of through-silicon vias in 3D packages. In: International electronic packaging technical conference and exhibition, vol 56895. American Society of Mechanical Engineers, p V002T02A002
52. Yang H, Lee T-K, Meinshausen L et al (2019) Heating rate dependence of the mechanisms of copper pumping in through-silicon vias. J Electron Mater 48(1):159–169
53. Tan Y, Tan CM, Zhang X et al (2010) Electromigration performance of Through Silicon Via (TSV)-A modeling approach. Microelectron Reliab 50(9–11):1336–1340
54. Frank T, Moreau S, Chappaz C et al (2012) Electromigration behavior of 3D-IC TSV interconnects. In: 2012 IEEE 62nd electronic components and technology conference, pp 326–330
55. Frank T, Moreau S, Chappaz C et al (2013) Reliability of TSV interconnects: electromigration, thermal cycling, and impact on above metal level dielectric. Microelectron Reliab 53(1):17–29
56. De Messemaeker J, Pedreira O, Vandevelde B et al (2013) Impact of post-plating anneal and through-silicon via dimensions on Cu pumping. In: 2013 IEEE 63rd electronic components and technology conference, pp 586–591
57. Harren S, Deve H, Asaro R (1988) Shear band formation in plane strain compression. Acta Metallurgica 36(9):2435–2480

58. Tu K-N (2011) Reliability challenges in 3D IC packaging technology. Microelectron Reliab 51(3):517–523
59. Lin L, Yeh T-C, Wu J-L et al (2013) Reliability characterization of chip-on-wafer-on-substrate (CoWoS) 3D IC integration technology. In: 2013 IEEE 63rd electronic components and technology conference, pp 366–371
60. Watanabe Y (1999) Electrical transport through Pb (Zr, Ti) O3 pn and pp heterostructures modulated by bound charges at a ferroelectric surface: ferroelectric pn diode. Phys Rev B 59(17):11257–11266
61. Hou F, Lin T, Cao L et al (2017) Experimental verification and optimization analysis of warpage for panel-level fan-out package. IEEE Trans Compon, Packag Manuf Technol 7(10):1721–1728
62. Xu J, Sun Y, Sun P et al (2018) Reliability testing method and verification of fan-out packaging structure. Semicond Technol 43(10):787–794

Chapter 6
Thermal Performance and Analysis Techniques of Integrated Circuit Packaging

Heat generation in integrated circuits is inevitable during operation. To ensure performance and reliability, it is crucial to minimize heat generation and dissipate it efficiently. Although advances in chip manufacturing have reduced the power consumption of individual transistors, power density has increased. Smaller feature sizes and higher packaging densities make heat dissipation a growing challenge. Thermal analysis helps assess the heat characteristics of microelectronic devices. Cooling technologies and structural optimization control the temperature of heat-generating components. Keeping internal temperatures within safe limits ensures stable and reliable operation. As a result, thermal management has become a key research focus in the industry.

According to GJB 548B—2005, the thermal performance of microelectronic devices includes junction temperature, thermal resistance, shell temperature, and thermal response time. Packaging thermal performance mainly refers to thermal resistance, which measures a package's heat dissipation capability under different assembly processes. A thorough thermal performance analysis requires an understanding of heat transfer principles and thermal analysis methods. It is also important to consider internal heat generation, heat coupling, and application requirements. This chapter introduces heat transfer theories, common thermal failure modes in packaging, key thermal performance parameters, mainstream analysis methods, and factors affecting thermal performance.

B. Zhou et al., *Reliability Technology for Integrated Circuit Packaging*,
https://doi.org/10.1007/978-981-95-3885-0_6

6.1 Thermal Effects of Integrated Circuits

6.1.1 Thermal Issues in Integrated Circuits

The development of microelectronics follows Moore's Law and its extensions. Process nodes continue to shrink, and integration density keeps increasing. Devices now offer better performance and more advanced structures. Failure mechanisms are also evolving, along with analysis techniques. High-density, high-frequency, and low-loss packaging has become widely used. During operation, internal interconnects, polysilicon layers, and transistors generate resistance to electric current. This resistance produces Joule heating. Without proper cooling, the device's thermal balance will be disrupted. The junction temperature will rise until a new balance is reached. If it exceeds the device's limit, the circuit may stop working or overheat and fail. Effective heat dissipation is essential for reliable packaging.

As circuits become smaller and more complex, thermal management becomes more challenging. The average heat flux density of high-density circuits and power devices now exceeds 500 W/cm^2. In some areas, it surpasses 1000 W/cm^2. Accurately measuring thermal performance and identifying hot spots are critical for effective heat management. Proper material selection and heat transfer mechanisms help maintain safe operating temperatures [1].

Thermal design and reliability are now key concerns in high-density packaging. Smaller chips and higher power increase power density, leading to higher junction temperatures. This affects performance and may cause failure. 3D heterogeneous packaging further complicates heat management. It increases thermal mismatch between materials, causing degradation and creep over time. These changes eventually lead to cracks and interface delamination. Compared to traditional planar structures, 3D packaging is harder to analyze. Standard surface thermal analysis methods are no longer effective.

6.1.2 Classification of Thermal Effects in Integrated Circuits

With the rapid development of integrated circuits, especially very large-scale integration (VLSI), chip size continues to shrink. The power dissipation of individual transistors decreases. However, as the number of transistors increases and operating frequency rises, the total power dissipation of packaged devices grows significantly. As high-density packaging technology advances, interconnect linewidths and bump diameters become smaller. This leads to a higher current density, making Joule heating more significant. Therefore, power dissipation and Joule heating are the two main factors that cause severe thermal effects in integrated circuit packaging. Typically, power dissipation results from electrothermal effects in active structures, while Joule heating occurs in passive structures due to electrical resistance converting part of the energy into heat.

1. Power Dissipation

Integrated circuits are mainly composed of complementary metal-oxide-semiconductor (CMOS) technology. Power dissipation is directly related to the capacitance of logic units, the square of the operating voltage, and the operating frequency, as shown in Eq. (6.1) [2].

$$P = NCV^2 f \tag{6.1}$$

where, P is the power dissipation of CMOS (W); N is the number of transistors per chip; C is the capacitance of each logic unit (F); C is the operating voltage (V); f is the operating frequency (Hz). Although capacitance decreases as feature sizes shrink and operating voltage drops, the rapid increase in transistor count and frequency continues to raise power dissipation. Reducing input capacitance and voltage swing can help lower power dissipation. However, due to the growing number of transistors and higher operating frequencies, controlling the rise in power dissipation has become increasingly difficult.

2. Joule Heating

Interconnect structures in integrated circuits have electrical resistance. When current flows through these conductors, Joule heating occurs. James Prescott Joule, a British physicist, discovered this phenomenon in 1840. Its quantitative relationship follows Joule's Law:

$$Q = I^2 Rt \tag{6.2}$$

where, Q is the heat generated (J), I is the current (A), R is the resistance (Ω), t is the time (s).

As current increases, Joule heating effect in power transmission lines becomes more significant. Due to differences in material properties, heat generation and dissipation vary, leading to large temperature gradients inside the package. To study the temperature rise caused by Joule heating in micro-interconnect bumps, Fu [3] conducted an experiment. The micro-interconnect bump samples were prepared using grinding and polishing to obtain a metallographic cross-section. A current stress of 1.5×10^4 A/cm^2 was applied under a temperature-controlled platform set at 70 °C. The hotspot temperature on the flip-chip surface and copper pillar interconnect cross-section was measured and analyzed. The copper pillar bump cross-section and infrared thermal imaging test results are shown in Fig. 6.1.

The temperature distribution of micro-bump packaging under 70 °C and 1.5×10^4 A/cm^2 is shown in Fig. 6.2. Figure 6.2a presents the hotspot temperature at the cross-section of two interconnected copper pillar bumps. Ignoring the boundary effect during measurement, the highest Joule heating temperature rise recorded was 0.3 °C. The hotspot appeared at the solder cap joint of the copper pillar bump. The electrical resistivity of copper pillars, Sn1.8Ag solder, and Cu_6Sn_5 intermetallic compound (IMC) is 1.7 μΩ·cm, 13 μΩ·cm, and 17.5 μΩ·cm, respectively. The solder

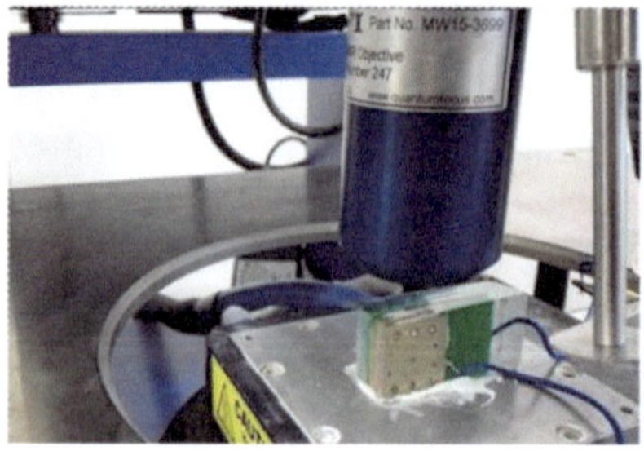

Fig. 6.1 Cross-section sample of copper pillar bumps and infrared thermal imaging test

resistivity is 7.6 times higher than that of the copper pillar bump [4, 5]. According to Joule heating calculations, the temperature rise is proportional to the resistance value. Figure 6.2b shows the surface temperature of 224 interconnected copper pillar bumps. The surface temperature rise caused by Joule heating reached 31 °C. This indicates a significant temperature accumulation effect. The temperature increase due to Joule heating grows rapidly as the number of bumps increases.

Under the effect of current stress, the Joule heating effect causes the temperature of the copper pillar bumps to be higher than the ambient temperature set by the heating platform. Due to differences in thermal conductivity and electrical resistivity of the materials, the Joule heating generated by current conduction causes a noticeable thermal gradient in the copper pillar bumps. Figure 6.3 shows the relationship between temperature and current density. As the current density increases, the hotspot temperature of the copper pillar bumps and the maximum internal temperature difference increase in a parabolic manner. This happens because the temperature rise due to Joule heating is directly proportional to the square of the current density. The chip temperature also rises quickly with the increase in current density. When the current

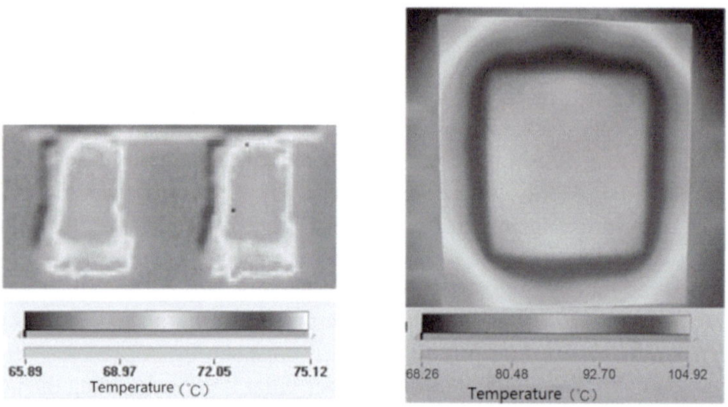

(a) Hotspot temperature at the cross-section of two interconnected copper pillar bumps

(b) Surface temperature of 224 interconnected copper pillar bumps

Fig. 6.2 Temperature distribution of micro-bump packaging (70 °C, 1.5×10^4 A/cm^2)

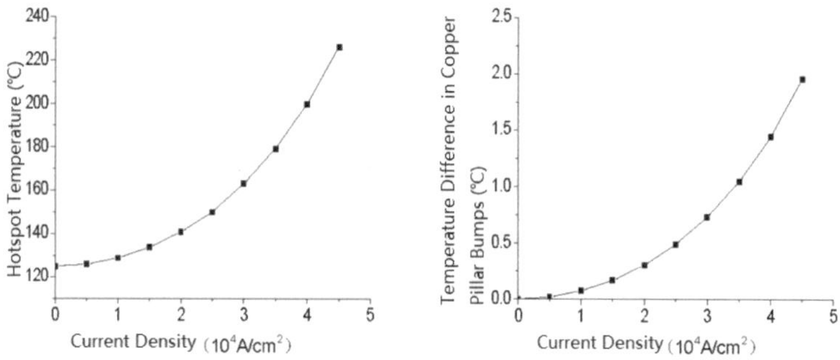

(a) Hotspot Temperature (b) Maximum Temperature Difference

Fig. 6.3 Relationship between temperature and current density

density reaches 4×10^4 A/cm^2, the hotspot temperature of the copper pillar bumps reaches 200 °C, close to the melting point of Sn-Ag or Sn-Ag-Cu lead-free solder. At this point, the failure mechanism of the interconnect structure changes from electromigration failure to solder melting and open-circuit failure caused by Joule heating. The Joule heating effect is positively correlated with the number of copper pillar bumps through which current flows. For integrated circuit packaging devices with more than 2000 pins currently in development, the number of internal bumps may exceed 7000. If current is applied to all bumps at once, there will be a significant risk of Joule heating-induced failure. Therefore, for packaging bumps under high current density, the Joule heating effect must be considered for thermal and electrical reliability. Since metal interconnect structures have good thermal conductivity, even with a high overall temperature rise, the temperature difference within a single bump remains relatively small. As shown in Fig. 6.3b, although the maximum temperature difference inside a single bump increases with current density, it stays below 2 °C.

6.2 Fundamentals of Thermal Analysis in Packaging

According to the second law of thermodynamics, heat always transfers spontaneously from a higher-temperature object to a lower-temperature one. It moves in the direction of decreasing temperature gradients. Temperature difference is the sole driving force for thermal conduction. During heat transfer, the lower-temperature object absorbs heat, causing its temperature to rise and internal energy to increase. Meanwhile, the higher-temperature object loses heat, leading to a decrease in temperature and internal energy. The higher the temperature of a material, the more intense the thermal motion of its molecules. Molecules with higher kinetic energy transfer heat to those with lower kinetic energy, enabling heat transfer.

Heat transfer occurs in three main forms: conduction, convection, and radiation. Sometimes, these mechanisms act together. Conduction occurs when a temperature gradient exists in a stationary and continuous medium. Heat is transferred through the medium. Convection happens when an object's surface is in contact with a moving fluid at a different temperature. Heat is exchanged through fluid movement. Radiation occurs when two objects with different temperatures are in space. They emit and absorb thermal radiation from each other. Conduction and convection require a medium, while radiation does not. In integrated circuit packaging, heat mainly transfers through conduction from the chip to the package shell and convection from the shell to the environment. Radiation heat transfer is minimal.

6.2.1 Thermal Conduction

Thermal conduction [6] refers to the transfer of heat within a material or between two objects in contact. This process occurs due to the thermal motion of microscopic particles, including molecules, atoms, and free electrons. Figure 6.4 shows a typical thermal conduction model in packaging.

Thermal conduction happens at the microscopic level when energy transfers between particles with different energy levels. Higher-energy particles pass their vibrational energy to adjacent lower-energy particles. This process allows heat to move through a material. Thermal conduction is the primary mode of heat transfer in solids. In insulating solids, it mainly occurs through lattice vibrations. In metals, free electrons also play a significant role in transferring energy. Liquids can conduct heat, but their thermal conductivity is much lower than that of solids. When a solid transitions to a liquid, molecular forces weaken, and structural order decreases. As a result, molecules move more freely, reducing thermal conductivity. When a liquid vaporizes, molecular bonds loosen further. Gas molecules move randomly and only interact through collisions. Due to this randomness, gases have extremely low thermal conductivity.

According to Fourier's law of thermal conduction, heat flux is proportional to the heat transfer area, thermal conductivity, and temperature gradient. It is inversely proportional to the material thickness. As shown in Fig. 6.5, consider a flat plane with a cross-sectional area *A* and thickness *L*. The temperatures on its two sides are

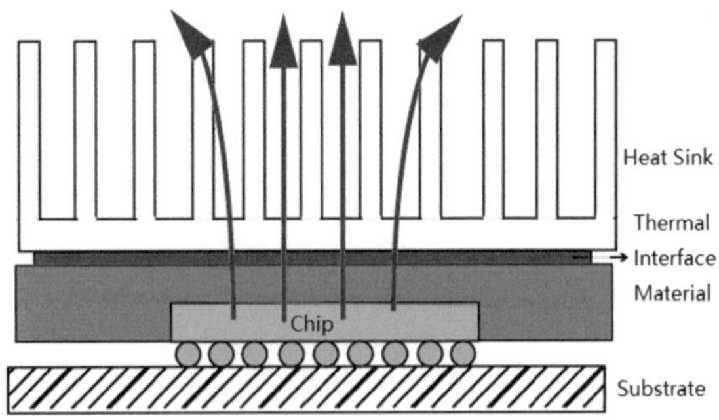

Fig. 6.4 Schematic diagram of thermal conduction in typical packaging

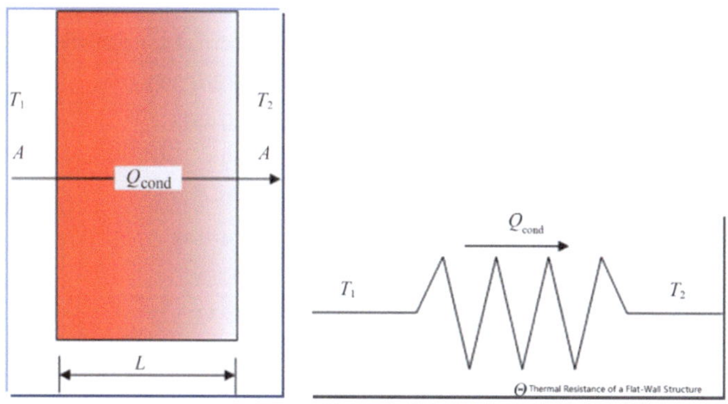

Fig. 6.5 Schematic diagram of thermal conduction in a flat-wall structure

T_1 and T_2. The heat flux Q_{cond} can be expressed as:

$$Q_{\text{cond}} = kA\frac{T_1 - T_2}{L} \tag{6.3}$$

where k is the thermal conductivity of the material. Its unit is W/(m·K). This value represents the heat transfer capability of a material.

Table 6.1 shows the thermal conductivity of common materials in electronic packaging. Diamond has the highest thermal conductivity among pure materials. Metals are widely used due to their excellent thermal conduction properties. In contrast, gases, such as air, have poor thermal conductivity.

Under normal conditions, a material's thermal conductivity is a variable. For gases, it increases as temperature rises. For most liquids and pure metals, it decreases with increasing temperature. Within the typical operating range of silicon-based devices (0–150 °C), temperature has little effect on the thermal conductivity of most materials. For example, for pure copper, the change is less than 3%. However, silicon, widely used in integrated circuits, behaves differently. Its thermal conductivity is highly sensitive to temperature and decreases as temperature rises.

6.2.2 *Convective Heat Transfer*

Convective heat transfer is different from heat convection. It refers to heat exchange between a solid surface and the surrounding fluid (liquid or gas) due to a temperature difference. This process involves three factors: thermal conduction between the fluid and the solid surface (e.g., a device's package shell), thermal conduction between fluids, and heat convection within the fluid. Convective heat transfer occurs only when there is a temperature difference. Figure 6.6 illustrates this process. It happens when there is a temperature gradient between an object's surface and a moving fluid.

Table 6.1 Thermal conductivity of common electronic packaging materials [7]

Material	Thermal conductivity (W/(m·K))
Diamond	2000~2300
Copper	380~403
Gold	319
Aluminum	210~237
Silicon	124~148
Ermanium	64
Alumina	15~33
Aluminum nitride	82~320
Silicon carbide	120~270
Solder (95Pb5Sn)	35.5
Solder (37Pb63Sn)	50.6
Epoxy resin	0.19
Polyimide	0.18
Epoxy glass	0.35
25 °C saturated water	0.613
25 °C air	0.0261

In daily life, wind results from the heat convection of cold and warm air, while air cooling in electronic devices is an example of convective heat transfer.

There are two types of convective heat transfer: natural convection and forced convection. Both involve fluid movement caused by density differences between hot and cold fluids. Natural convection: Temperature controllers work based on this principle. The heater warms the surrounding air, making it less dense than the cooler air farther away. The less densehot air rises, while cooler air moves in to maintain pressure balance. This continuous circulation causes hot and cold air to alternate,

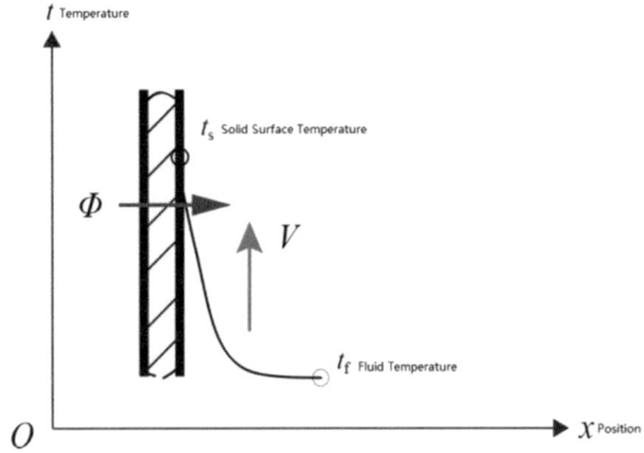

Fig. 6.6 Convective heat transfer diagram

Table 6.2 Convective heat transfer coefficients of common fluids [8]

Fluid	Flow cause	Convective heat transfer coefficient (W/(m^2·K))	
		Range	Typical value
Air	Natural convective heat transfer	3~12	5
	Forced convective heat transfer	10~100	50
Water	Natural convective heat transfer	200~1000	600
	Forced convective heat transfer	1000~15000	8000

raising the indoor temperature. Forced convection: High-power fans, blowers, and pumps increase fluid velocity to enhance heat transfer.

Both natural and forced convection involve two flow types: laminar flow and turbulent flow. Difference: laminar flow is slow, streamline, and orderly. While turbulent flow is fast, irregular, and pulsating. To improve convective heat transfer, turbulent flow should be introduced whenever possible. It enhances heat exchange efficiency.

In 1701, Newton proposed a fundamental equation for convective heat transfer, known as Newton's Law of Cooling.

$$Q_{\text{conv}} = hA(T_{\text{s}} - T_{\text{f}}) \tag{6.4}$$

where: h is the convective heat transfer coefficient (W/(m^2·K)); A is the surface area of the object (m^2); T_s is the thermodynamic temperature of the object's surface (K); T_f is the thermodynamic temperature of the fluid (K). The convective heat transfer coefficient (h) differs from thermal conductivity. Unlike thermal conductivity, which is a material property, h depends on external factors. Thus, it is not an intrinsic property of a material. Several factors affect h, including: Fluid state, Flow velocity, Temperature difference between the object and fluid, Gravitational acceleration of the fluid, Geometric characteristics of the flow path, Geometric characteristics of the object-fluid contact surface and Type of fluid flow. Table 6.2 shows the convective heat transfer coefficients of common fluids.

6.2.3 *Radiative Heat Transfer*

Process of Radiative Heat Transfer: An object emits heat through electromagnetic radiation. This process does not require a medium. The wavelength of electromagnetic radiation ranges from nanometers to kilometers, covering various scales. The Sun's heat affects Earth's temperature through radiative heat transfer. A thermos bottle has a double-layered structure with an inner silver coating to reduce heat loss from radiation. Radiative heat transfer has three main characteristics:

(1) No direct contact is needed. Heat radiation travels as electromagnetic waves. These waves can move without a medium, even in a vacuum.

(2) Energy conversion occurs during the process. Some internal energy of an object transforms into electromagnetic waves and is emitted. At the same time, another object absorbs these waves and converts them back into internal energy.
(3) It happens everywhere. As long as there is a temperature difference, radiative heat transfer occurs. A low-temperature object absorbs less radiation energy than it receives from a high-temperature object. This results in heat transfer from the high-temperature object to the low-temperature object.

Temperature, surface area, relative position, and surface radiation properties can influence radiative heat transfer. Radiative heat transfer calculations are complex, so only simple cases are considered. When an object with a uniform temperature T_s and a surface area A is exposed to a large surrounding space at temperature T_{env}, the large space refers to a closed area formed by physical structures around the object [9]. According to Stefan's Radiation Law, the net radiative heat transfer Q_{rad} between the object and the environment is as follows:

$$Q_{rad} = \varepsilon\sigma A(T_s^4 - T_{env}^4) \tag{6.5}$$

In this equation, T_s and T_{env} represent the thermodynamic temperatures of the object and its surrounding environment. Both are measured in Kelvin (K). Their relationship with Celsius (°C) is: $T(\text{K}) = T(°\text{C}) + 273.15$; σ is the Stefan-Boltzmann constant, with a value of ($\sigma = 5.67 \times 10^{-8}$ W/(m^2·K^4)). The emissivity ε depends on the surface properties of the object and ranges from 0 to 1. A is the surface area of the object.

The radiative heat transfer coefficient h_{rad} is defined as:

$$h_{rad} = \varepsilon\sigma\left(T_s^2 + T_{env}^2\right)(T_s + T_{env}) \tag{6.6}$$

Thus,

$$Q_{rad} = h_{rad}A(T_s - T_{env}) \tag{6.7}$$

Equation (6.7) is similar to Newton's Law of Cooling for convective heat transfer. However, their reference temperatures are different. The environmental thermodynamic temperature T_{env} may not be equal to the fluid temperature T_f around the object. For example, in a desktop computer, some areas inside the case may be hotter than the chassis temperature.

6.3 Failure Modes Related to Thermal-Induced Packaging

For silicon-based semiconductor devices, the junction temperature should not exceed 175 °C. However, to ensure long-term stability and reliability, it is usually kept below 125 °C. High temperatures weaken the packaging materials of microelectronic devices. The maximum temperature that molding compounds can endure is typically 115 °C, which corresponds to the deformation temperature of phenolic resin materials. Additionally, when the temperature exceeds 100 °C, the strength of solder joints gradually decreases. For microelectronic modules exposed to vibration and impact, sudden temperature changes often cause fatigue fractures, which shorten the mechanical lifespan of the package. D. Su et al. studied the displacement of flip-chip interconnect solder joints under thermal shock. They found that the failure of interconnect structures is directly related to the thermal expansion coefficient mismatch between the substrate and the silicon-based chip [10]. Microelectronic device packaging consists of materials with different coefficients of thermal expansion (CTE). During manufacturing, testing, storage, and transportation, CTE mismatches can cause thermal stress, leading to stress-induced damage. This thermal stress may result in irreversible damage to the packaging structure. The process temperature during packaging is usually higher than the operating temperature of the device. As a result, residual thermal stress may form during assembly and increase further at lower temperatures. Even without CTE mismatch, an uneven temperature distribution within the packaging structure can still create thermal stress.

6.3.1 Correlation Between Temperature and Packaging Failure

In 2019, Xilinx introduced the "Virtex UltraScale + VU19P", the world's largest Field Programmable Gate Array (FPGA) chip. It contains 35 billion transistors. Although its overall power consumption is 60% lower than the previous generation, its power density has significantly increased, raising temperature-related failure risks. The failure of integrated circuits (ICs) is closely related to temperature. The failure rate of a device follows an exponential relationship with temperature, as shown in Fig. 6.7. The failure rate rises sharply as temperature increases. If a device's operating temperature rises from 75 to 125 °C, its failure rate increases fivefold. For every 10 °C rise, the failure rate nearly doubles. Temperature is a key factor in determining the reliability of microelectronic device packaging.

According to incomplete statistics, about 55% of microelectronic product failures are caused by excessive heat or thermal-related issues [12]. The main causes of electronic product failures and their proportions are shown in Fig. 6.8. Failures in integrated circuits due to high temperatures are mainly related to packaging. These include electrical failures and mechanical failures.

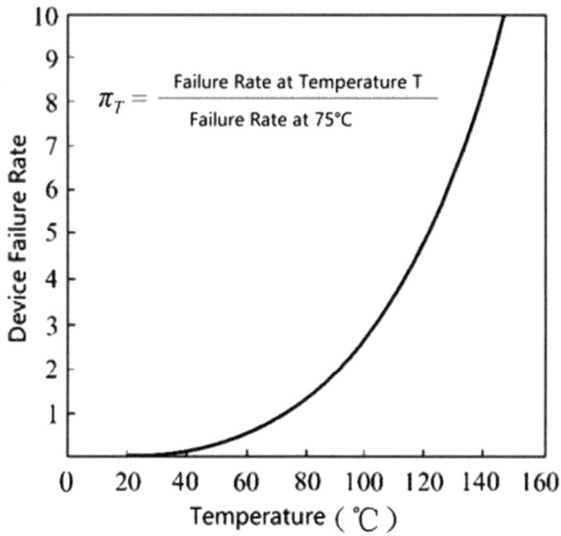

Fig. 6.7 Relationship between device failure rate and temperature [11]

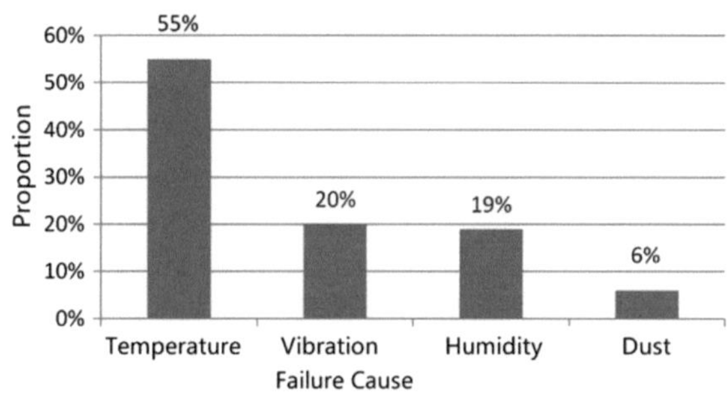

Fig. 6.8 Main causes of electronic product failures and their proportions

As microelectronics technology advances, process nodes shrink and integration density increases. This brings new reliability challenges in packaging. In high-density packaging, the coefficient of thermal expansion (CTE) of chips, solder, substrates, and underfill materials are different. During operation, temperature cycling occurs. Materials with different coefficient of thermal expansion (CTE) undergo unequal deformation. This creates long-term cyclic thermal stress, leading to material degradation and thermal stress mismatch. Currently, high-density packaged ICs have more than 2000 external pins. Flip-chip bump counts reach 8000–10,000. Under thermal stress, thermal warpage deformation worsens. This may cause substrate delamination, interconnect cracking, or solder joint electromigration failure, affecting IC packaging reliability [13]. Thermal failures in IC packaging mainly include the following aspects.

6.3.2 Cracking Failure Caused by Thermal Mismatch

In 3D packaging, thermal conduction pathways affect thermo-mechanical stress distribution. The highest stress usually appears at the top and bottom chips. Stress concentration often occurs where the chips contact the epoxy molding compound (EMC). This can cause chip separation and crack initiation. It may also lead to the popcorn effect in single-chip packaging. Due to the large difference in coefficients of coefficient of thermal expansion between molding compounds and silicon chips, CTE mismatch can occur at high temperatures. This may result in delamination and cracking. When stress or pressure in the chip reaches a critical level, vertical cracks may appear. This can cause chip failure. For example, cracks are likely to form when stress reaches 100–200 MPa or pressure exceeds 600 MPa [14]. Figure 6.9 shows thermal mismatch-induced cracking in stacked chips.

In packaging interconnects, the elastic modulus of intermetallic compounds (IMCs) is much higher than that of solder. When external thermal stress is applied, IMC deformation is smaller than solder deformation. As a result, solder near the IMC/solder interface experiences higher constraint, causing stress concentration. In contrast, solder further from the interface is less constrained. It can rotate or deform elastically to release external stress. Near the solder/IMC interface, high-stress and high-strain concentration zones form due to differential deformation. Under cyclic

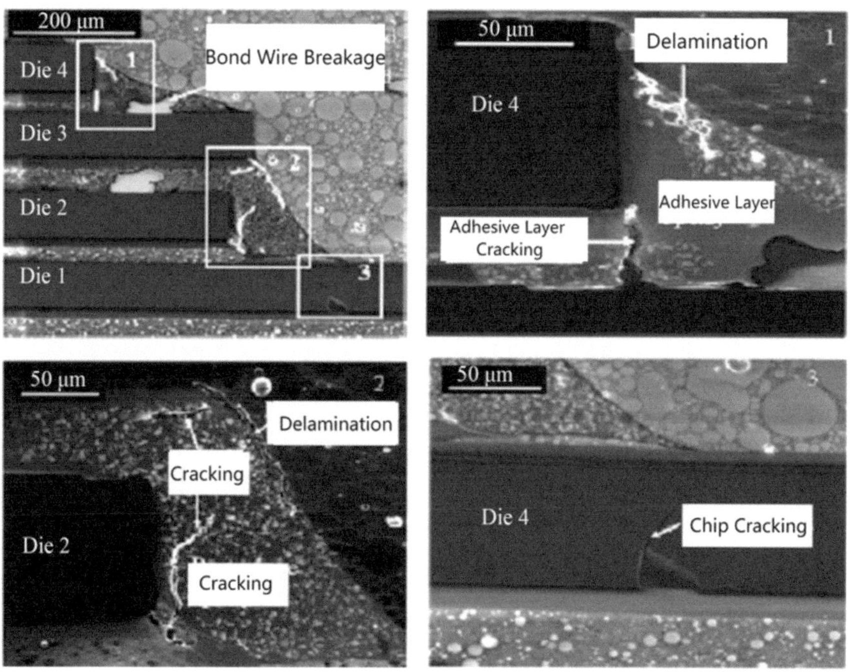

Fig. 6.9 Thermal mismatch-induced cracking in stacked chips

shear stress, the IMC boundary near the solder joint end is prone to thermal mismatch-induced cracking and failure. Figure 6.10 shows solder joint separation caused by thermal mismatch.

The coefficient of thermal expansion of silicon and molding compounds are different. Silicon has 3 ppm/°C, while molding compounds have 20 ppm/°C. In assembly processes, different adhesives required different bonding temperatures. Polyimide adhesives bonded at 270 °C, while epoxy adhesives bonded at 170 °C. During chip operation, heat causes expansion mismatch between the chip and molding compound. The chip is compressed, while the molding compound is stretched. This difference creates thermal stress, which may lead to crack formation and propagation. Figure 6.11 shows cracks in a molded chip at 260 °C caused by thermal stress [15].

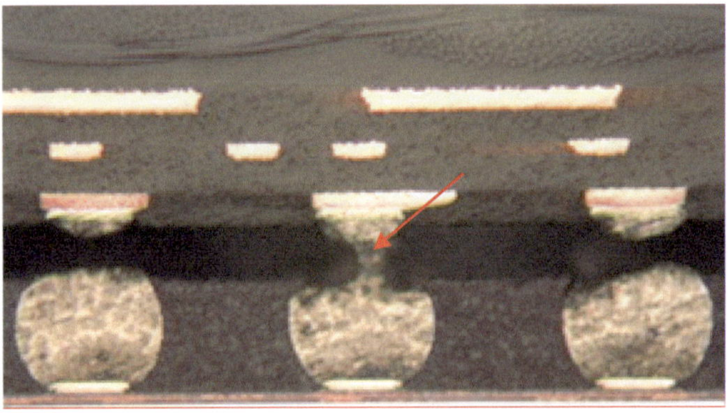

Fig. 6.10 Solder joint separation caused by thermal mismatch

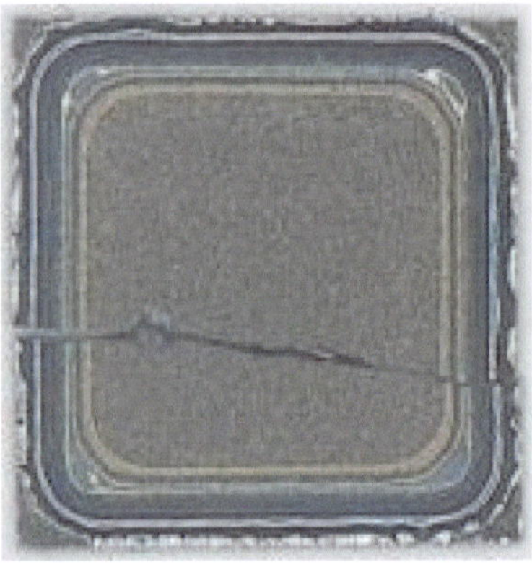

Fig. 6.11 Cracks in a molded chip at 260 °C due to thermal stress

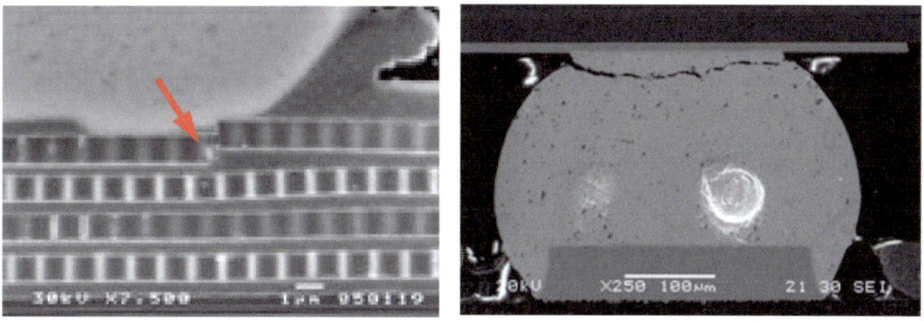

(a) Silicon Substrate Fatigue Crack (b) Solder Joint Fatigue Failure

Fig. 6.12 Typical thermal fatigue failure in integrated circuits

6.3.3 Cracking Failure Caused by Thermal Fatigue

Fatigue failure occurs when plastic deformation weakens dislocation mobility. As fatigue continues, dislocations accumulate. When their density exceeds a critical level, the crystal structure breaks down, leading to crack initiation. Further fatigue stress causes crack growth, eventually leading to fracture failure. In integrated circuits, 90% of mechanical and electrical failures result from fatigue damage.

As integration density increases and package size decreases, heat flux rises sharply. This causes higher internal temperatures. Inside the package, different materials react differently to temperature changes. High-density packaging reduces heat dissipation, leading to thermal accumulation and stress concentration. At bonding points and solder joints, temperature cycling creates cyclic shear stress. This may cause fatigue fractures at bonding points. Figure 6.12 shows a typical thermal fatigue failure in an integrated circuit.

6.3.4 Creep Failure Caused by High Temperature

Under sustained high temperatures and stress below the yield strength, plastic deformation increases over time. This process is called high-temperature creep. Creep is different from plastic deformation. Plastic deformation occurs when stress exceeds the yield strength, causing permanent damage. Creep, however, happens under low stress over a long period, eventually leading to material failure. Prolonged high temperatures can cause creep failure in solder joints and metal interconnections.

There are three types of metal creep failure: intergranular creep, transgranular creep, and diffusional creep. Intergranular creep: This is the main failure mode in metal materials. Under low stress and high temperatures ($T > 0.3T_m$),(T_m represents the melting point of the material expressed in thermodynamic temperature). Grain boundary sliding and diffusion cause voids and cracks to form and grow along grain

Fig. 6.13 Creep failure in columnar solder joints

boundaries; Transgranular creep: This occurs under high stress. Its fracture mechanism is similar to ductile fracture at room temperature; Diffusional creep: This happens at very high temperatures ($T > 0.6T_{\mathrm{m}}$). During fracture, new fine grains form inside the material. This process is called dynamic recrystallization.

The operating and storage temperatures of integrated circuits and electronic devices are usually above $0.3T_{\mathrm{m}}$. Therefore, creep in solder joints and lead materials occurs under most conditions. Figure 6.13 shows creep failure in columnar solder joints.

6.3.5 Interconnect Degradation Failure Caused by High Temperature

1. Adhesive Degradation Failure

Adhesive bonding is widely used in stacking packaging devices such as memory and System in Package (SiP). Compared to eutectic soldering, adhesives have lower high-temperature resistance and weaker thermal conductivity. After high-temperature storage, adhesion weakens. Poor bonding can cause voids, leading to thermal breakdown. Physical degradation appears as a decrease in chip shear stress. Electrical degradation is seen in a significant increase in V_{ces} (Collector-Emitter Saturation Voltage) (in transistors). Thermal degradation results in higher thermal resistance, raising junction temperature and shortening product lifespan. The poor thermal stability and outgassing properties of adhesives remain major obstacles in high-reliability electronic packaging. Figure 6.14 shows bonding failure caused by temperature changes.

2. Interconnect Bump Degradation Failure

In advanced high-density integrated circuits such as FPGAs, flip-chip bump interconnects are widely used. These interconnects include solder bumps and copper pillar bumps. Under high-temperature aging, metal atoms in the solder and substrate diffuse

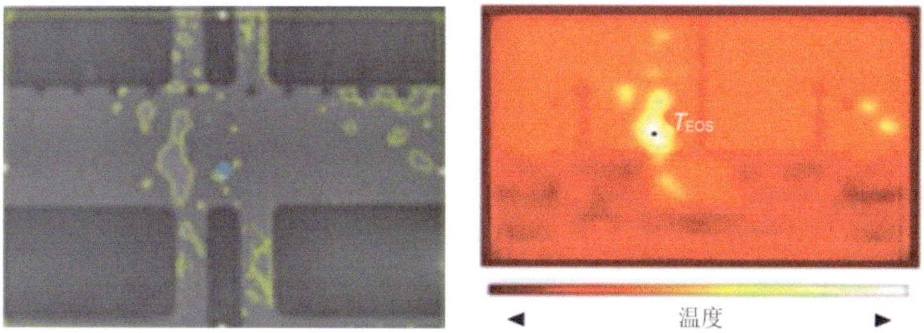

(a) Void Distribution in Adhesive (b) Device Temperature Distribution

Fig. 6.14 Bonding failure caused by temperature [16]

into each other, leading to continuous intermetallic compound (IMC) growth. Meanwhile, a large temperature gradient forms between the chip's heat dissipation structure and heat source, accelerating IMC growth on the cooler side [17]. IMC is a brittle phase. A moderate thickness (typically 1–3 μm) ensures good metallurgical bonding at the interface. However, excessive IMC growth, especially Cu_3Sn, increases brittleness, reduces interface shear strength, and causes cracking failure. Figure 6.15 shows the microstructure of copper pillar bump solder joints under different high-temperature aging conditions. As temperature rises, atomic diffusion accelerates, promoting IMC growth and structural changes. At 150 °C, Cu_6Sn_5 rapidly converts into Cu_3Sn.

3. Metal Bonding Interface Degradation Failure

Gold wire, aluminum wire, or copper wire bonding is a common method for interconnecting IC chips and substrates. The bonding interface typically forms compounds like gold-aluminum, gold-copper, or aluminum-copper. Under high temperature stress, metal elements from the different metals diffuse excessively, forming brittle IMCs and Kirkendall voids. This leads to an increase in contact resistance, a decrease in bonding strength, and even detachment failure. Figure 6.16 shows interface delamination caused by gold wire bonding degradation.

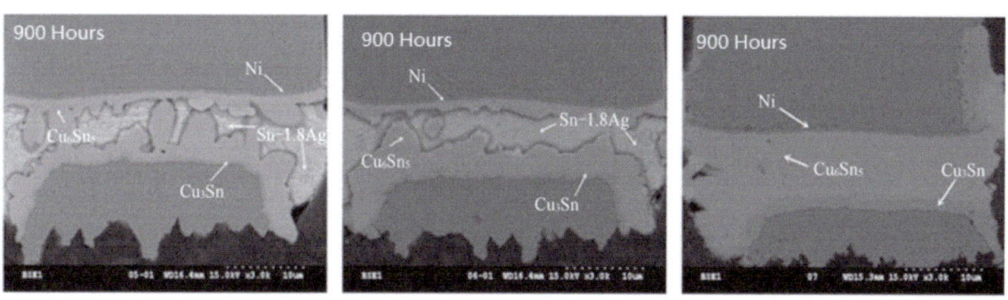

(a) 100℃ (b) 125℃ (c) 150℃

Fig. 6.15 Microstructure of copper pillar solder joint interface under different high-temperature aging conditions

Fig. 6.16 Degradation of gold wire bonding leading to interface delamination

Au-Al bonding involves bonding gold wire to aluminum chips, aluminum wire to thick-film gold conductors, or aluminum wire to gold-plated external lead pillars. Under high-temperature conditions, five types of IMCs are formed: Au_4Al, Au_5Al_2, Au_2Al, $AuAl_2$ (purple phase), and AuAl. These IMCs have low conductivity, and their lattice constants and coefficient of thermal expansion are different. This creates internal stresses at the bonding site. Au_5Al_2 is the dominant IMC. After prolonged high-temperature exposure, the bonding strength of Au-Al bonding decreases. Contact resistance increases, and open circuit failure can occur.

Cu-Al bonding refers to bonding copper wire to aluminum chips. Under aging conditions of 100 to 500 °C, interfacial voids diffuse, forming IMCs such as $CuAl_2$, CuAl, Cu_9Al_4, and CuAl. Kirkendall voids do not form in the bonding interface. However, $CuAl_2$, the primary IMC, is brittle and significantly reduces shear strength. Moreover, copper wire is prone to oxidation.

Degradation at high temperatures usually increases thermal resistance, which raises the junction temperature. This results in failure or shortened lifespan. The high-temperature degradation lifetime of interconnect structures is mainly evaluated using the Arrhenius model. The acceleration factor is calculated using the following formula:

$$A_f = e^{\left[\frac{E_a}{K_B}\left(\frac{1}{T_U}-\frac{1}{T_S}\right)\right]} \tag{6.8}$$

In the formula, A_f is the acceleration factor, E_a is the activation energy for failure, T_U is the operating environment temperature, T_S is the accelerated test environment temperature, and K_B is the Boltzmann constant. By processing multiple sets of accelerated lifetime test data, the failure activation energy and acceleration factor in the Arrhenius model can be derived. This allows the actual lifetime of the package interconnect structure to be predicted.

6.3.6 Chip Overheating and Burnout

High-density integrated circuits generate a significant amount of heat during operation. If the heat is not dissipated in time, it accumulates inside the device. This causes the junction temperature to rise. An excessive junction temperature can lead to degraded electrical performance, a shortened lifespan, and reduced reliability. In severe cases, it can even cause chip burnout. The internal working temperature of an integrated circuit package depends on power consumption, thermal resistance, and ambient temperature. Given a specific power consumption and ambient temperature, thermal resistance is the main factor that determines the chip's junction temperature. To evaluate a device's thermal performance, key parameters such as junction temperature and thermal resistance must be tested. In an overheating situation, two types of failure typically occur inside the package.

(1) Thermal Breakdown of the Chip. This can occur in two forms: thermal breakdown and secondary breakdown. Secondary breakdown is more common. Research indicates that secondary breakdown has two types. The first is forward bias secondary breakdown, which is related to the thermal properties of the device, such as doping concentration and intrinsic concentration. The second is reverse bias secondary breakdown, which is caused by avalanche multiplication of carriers in the space charge region near the collector. Both types lead to current concentration.
(2) Chip Overcurrent Burnout. In high-density packaged devices, instantaneous overcurrent pulses may occur during operation. These overcurrent pulses can cause failures, such as the melting of the top-layer IMC, especially in diodes used in the bonding process. Figure 6.17 shows a case of overheating and melting burnout in the chip's wiring area.

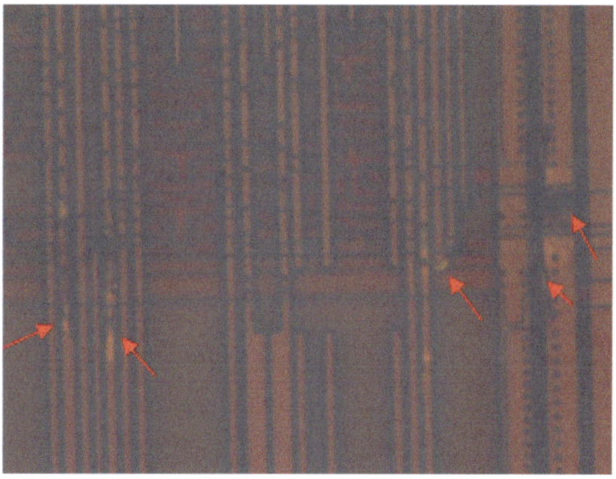

Fig. 6.17 Overheating and melting burnout in the chip wiring area

6.4 Main Thermal Performance of Integrated Circuit Packaging

Thermal performance is an important indicator for evaluating microelectronic packaging. The junction temperature (T_J) and thermal resistance ($R_{\theta JX}$ or θ_{JX}) are the two key parameters used to measure the heat dissipation performance of a package [18]. Integrated circuits have a region that generates most of the heat. This region is the hottest part of the circuit and is called the junction. The temperature of this junction is known as the junction temperature. The maximum allowable junction temperature is limited by circuit capacity, structural reliability, and the properties of the chip and packaging materials. The primary goal of thermal management in integrated circuits is to design an efficient heat dissipation method for the chip. This ensures that the junction temperature stays below the maximum allowed level. The chip package provides a path for heat transfer from the chip's hot junction to the external surface. Various cooling techniques on the package surface help improve heat dissipation. Therefore, the heat transfer performance from the inside to the outside of the integrated circuit package is usually represented as mathematical expressions. These are called the thermal parameters or thermal metrics of the package. Figure 6.18 shows typical thermal parameters and their applications.

Thermal resistance refers to how much a specific thickness of a structure resists heat flow. It measures the material's or structure's ability to transfer heat. For homogeneous materials, thermal resistance is directly proportional to the thickness of

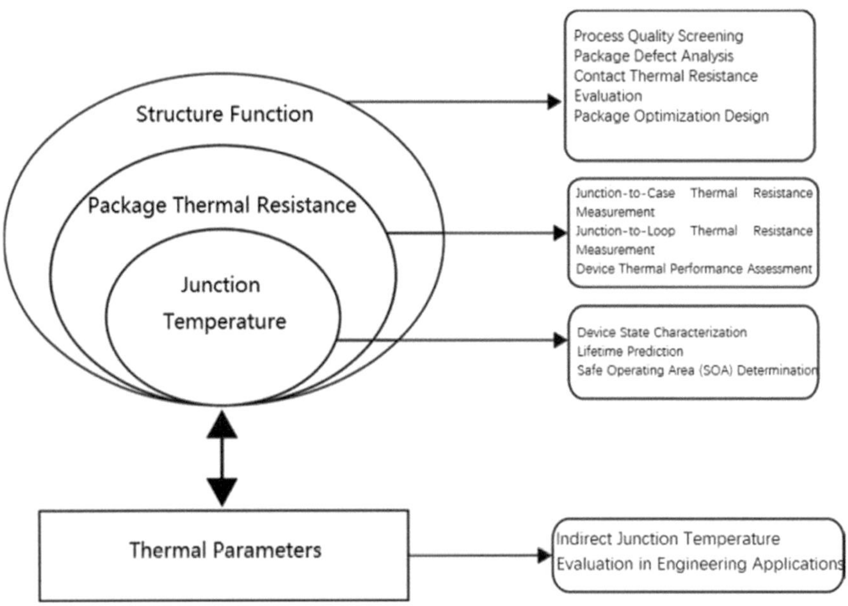

Fig. 6.18 Typical thermal parameters and application fields

the material and inversely proportional to the material's thermal conductivity and surface area. For non-homogeneous materials, thermal resistance typically increases with thickness, and the relationship may be non-linear. Thermal resistance is similar to electrical resistance. The temperature difference across the two sides of the material produces heat flow, just as a voltage difference creates current. Heat flows from high temperature to low temperature, similar to how current flows from high to low potential. Thermal conduction can also be represented by series and parallel methods in thermal resistance networks.

Thermal resistance can be divided into conduction thermal resistance, convection thermal resistance, and radiation thermal resistance. It can also be categorized as steady-state thermal resistance or transient thermal resistance, depending on whether thermal equilibrium is reached.

6.4.1 Steady-State Thermal Resistance

Steady-state thermal resistance in semiconductor devices is defined as the ratio of the temperature difference between the device junction temperature and a specific location to the dissipated power under thermal equilibrium conditions [19, 20].

$$\theta_{\mathrm{JX}} = \frac{T_{\mathrm{J}} - T_{\mathrm{X}}}{P_{\mathrm{H}}} \tag{6.9}$$

In the equation, θ_{JX} represents the thermal resistance from the device junction to the specified location or environment (°C/W). When X represents the packaging shell, it refers to the junction-to-case thermal resistance (θ_{JC}); T_{J} represents the junction temperature of the device under steady-state test conditions (°C); T_{X} represents the reference temperature in the specified environment (°C); P_{H} represents the dissipated power of the device (W). The junction temperature of the device under test conditions can be defined as:

$$T_{\mathrm{J}} = T_{\mathrm{J0}} + \Delta T_{\mathrm{J}} \tag{6.10}$$

In the equation, T_{J0} represents the initial junction temperature before applying the thermal load (°C); ΔT_{J} represents the change in junction temperature after applying the thermal load (°C).

In electrical testing methods, temperature-sensitive parameters (TSP) are commonly used to monitor the change in junction temperature of the device under test (DUT) under electrical power loading. This is expressed by the following formula:

$$\Delta T_{\mathrm{J}} = K \times \Delta\mathrm{TSP} \tag{6.11}$$

In the equation, ΔTSP represents the change in temperature-sensitive parameter (°C /(m·V)); K is a constant that defines the relationship between T_J and TSP (°C/(m·V)).

In practical testing, the steady state of the device is usually judged by the temperature change curve (heating or cooling curve). When the temperature curve becomes flat in a new phase, it can be considered that steady-state has been initially reached. Determining the thermal steady-state is important in practical testing. If the heating time is too short and thermal equilibrium is not reached, it will not reflect the actual values under specific conditions. If the heating time is too long, it will not improve the measurement accuracy, but it will reduce the testing efficiency.

If it is difficult to generate a temperature change curve, a flowchart for confirming the steady-state phase can be used (see Fig. 6.19). This flowchart helps to determine the time when steady-state occurs in three steps. When it seems that steady-state has been reached, the first step is to record the heating time and thermal resistance value. In the second step, increase the heating time by 10%, record the new thermal resistance value, and compare it with the initial thermal resistance value. If the conditions in Fig. 6.19 are met, the third step is performed. Increase the heating time by another 10%, and record the thermal resistance value again. If the conditions in Fig. 6.19 are satisfied, steady-state has been reached. The final thermal resistance value is the steady-state thermal resistance, and the third heating time is the time required to reach steady-state.

The larger the thermal resistance, the stronger the obstruction to the heat flow path. Heat that is difficult to dissipate accumulates inside the package, leading to high temperatures. Thermal resistance can help predict the heat dissipation performance of the packaging structure in actual use. Common microelectronic packaging thermal resistances include junction-to-case thermal resistance θ_{JC}, junction-to-board thermal resistance θ_{JB}, and junction-to-ambient thermal resistance θ_{JA}. These are all steady-state thermal resistances. θ_{JC} refers to the thermal resistance through which most of the heat from the chip junction is transferred to the surface of the integrated circuit package shell. θ_{JB} refers to the thermal resistance through which most of the heat from the chip junction is transferred to the PCB. θ_{JA} refers to the thermal resistance through which heat from the chip junction is transferred to the air under natural or forced convection conditions.

Junction-to-case and junction-to-board thermal resistances are conductive thermal resistances. These are inherent properties of the packaging and depend on geometric dimensions and material characteristics. During testing, the effects of convective heat transfer should be minimized. Junction-to-ambient thermal resistance is a convective and radiative thermal resistance, and its value depends on geometric dimensions, material properties, airflow speed, and environmental conditions.

1. Junction-to-Case Thermal Resistance

Junction-to-case thermal resistance θ_{JC} quantifies the heat dissipation capability of a semiconductor device. It measures the heat dissipation from the heat-generating junction surface to the package shell surface. According to JESD51-1, junction-to-case thermal resistance is defined as the thermal resistance on the package shell's surface.

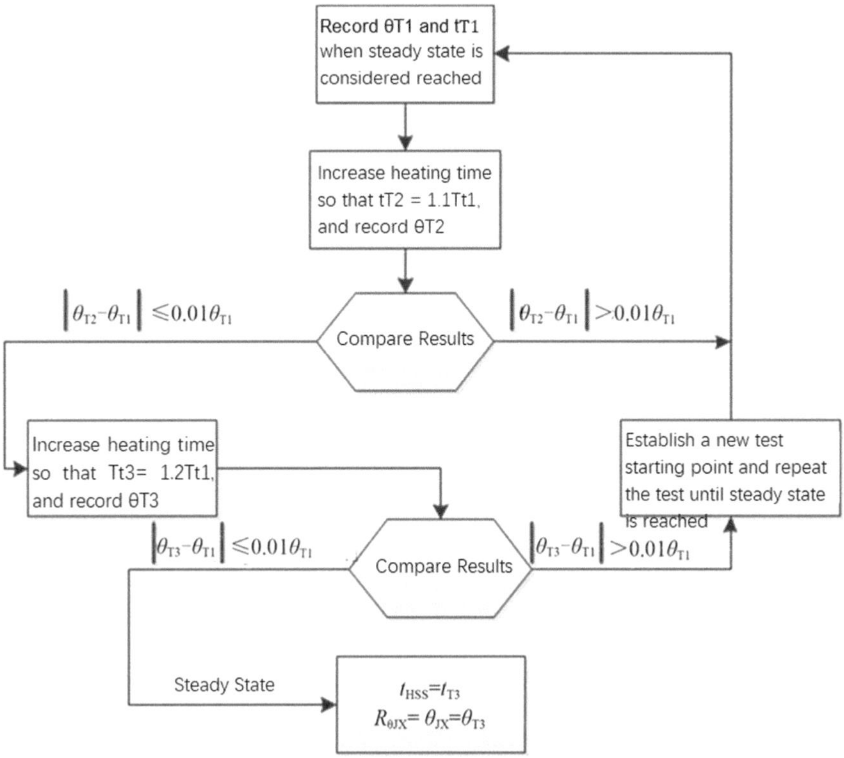

Fig. 6.19 Flowchart for confirming the thermal steady-state stage [21]

This occurs when the package shell has good heat dissipation and minimal surface temperature change. It is the main heat dissipation path closest to the chip's peak temperature area. In measuring junction-to-case thermal resistance, efforts should be made to ensure that heat dissipates only through the main heat dissipation surface. Using high thermal conductivity plates and insulating materials ensures thermal insulation on the sides.

The U.S. military standard MIL883 details the traditional thermocouple measurement method. This method requires ensuring good contact between the device package shell and the heat sink. Then, the junction temperature T_J, shell temperature T_C, and junction-to-case thermal resistance are determined. The formula for calculating junction-to-case thermal resistance is as follows:

$$\theta_{JC} = \frac{T_J - T_C}{Q_{JC}} \tag{6.12}$$

In the formula, θ_{JC} represents the junction-to-case thermal resistance, T_J and T_C represent the junction and shell temperatures, respectively, and Q_{JC} is the amount of heat transferred from the chip to the package shell. For the package structure

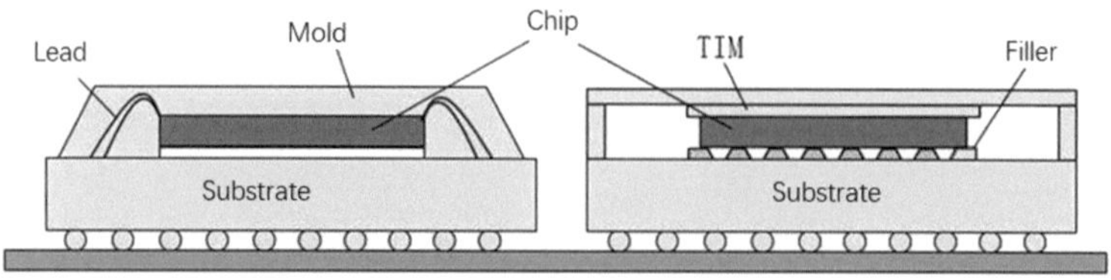

Fig. 6.20 Wire bonding molded package and flip-chip metal package

shown in Fig. 6.20, the heat from the source flows through different thermal paths. It ultimately reaches the top (or the top heat sink) and bottom (or the bottom PCB) of the chip package. This process achieves heat dissipation. Junction-to-case and junction-to-board thermal resistances represent two different heat flow paths.

For flip-chip packages, the package shell refers to the top surface of the package body. For metal packages such as TO-220, TO-263, and chip-up square flat packages (Quad Flat Package, QFP), the package shell refers to the bottom surface of the package body. Therefore, junction-to-case thermal resistance is usually represented by the symbol θ_{JCX}, where X represents either the top or bottom surface, depending on whether the package shell is the top or bottom of the package body.

Junction-to-case thermal resistance is used to measure the thermal performance of the device itself. It is mainly influenced by materials and structure. During measurement, it is important to minimize the effects of convective heat transfer. Therefore, the measurement is typically done when all heat is dissipated through the package shell. It is recommended, as shown in Fig. 6.21, to use insulating materials with a thermal conductivity less than 0.1 W/(m·°C) to insulate the non-main heat transfer paths. Additionally, efficient heat sinks or cold plates should be connected to the package shell. During thermal loading, use rated power or a power level that results in a junction temperature rise of 20 °C. A typical temperature rise is between 30–60 °C, and the shell temperature is measured at the center of the package shell. One challenge with this measurement method is that it is difficult to measure the shell temperature accurately when the package shell surface is in close contact with the heat sink. As a result, the value of θ_{JC} may vary depending on the measurement equipment used.

2. Junction-to-Board Thermal Resistance

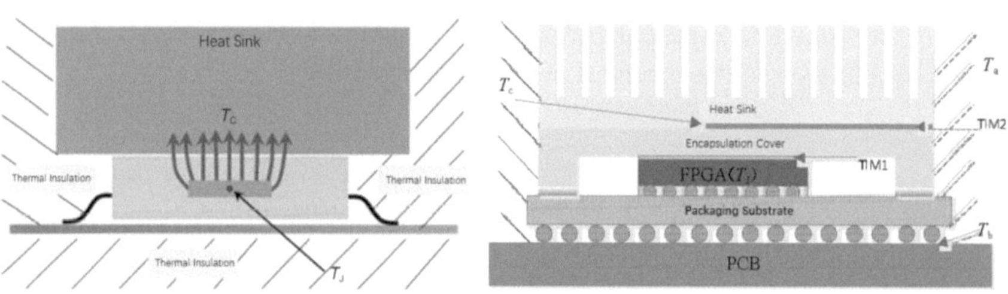

Fig. 6.21 Schematic diagram of insulation treatment for junction-to-case thermal resistance testing

According to the JESD51-8 standard, junction-to-board thermal resistance θ_{JB} can be used to compare the thermal performance of surface-mount integrated circuit packages installed on standard boards. The definition of junction-to-board thermal resistance is as follows.

$$\theta_{JB} = \frac{T_J - T_B}{Q_{JB}} \tag{6.13}$$

In the formula, T_J and T_B represent the junction and board temperatures. Q_{JB} represents the heat transferred from the junction to the board. In the Joint Electronic Device Engineering Council (JEDEC) standards, θ_{JB} refers to junction-to-board thermal resistance. The JEDEC standard JESD51-8 outlines the method for measuring it.

Junction-to-board thermal resistance is a type of conductive thermal resistance. The testing process requires insulation to reduce convective heat transfer. Typically, copper annular cold plates (thermal conductivity > 300 W/(m·K)) clamp both ends of the test board. The temperature of the cold plate should stay within room temperature + 2/− 5 °C, with fluctuations under 0.2 °C. The inner edge of the cold plate should be at least 5 mm from the package's outer edge. The cold plate directly contacts the wiring covered by the solder mask on the test board. The contact width should be at least 4 mm. Clamping force does not significantly affect the test results. Usually, 200 g of force is applied evenly to the board, and thermal grease is not needed. Insulating materials (thermal conductivity < 0.1 W/(m·°C)) should surround the package's top and bottom. This setup minimizes heat transfer in those directions, ensuring heat flows mainly to the board's two ends [22] $Q_{JB} \approx Q$. A 1–5 mm gap should be maintained between the insulating material and the package's top and bottom. Figure 6.22 shows the schematic diagram of the test fixture's cross-section.

Thermocouples measure the board's surface temperature. T-type thermocouples are recommended because they are easy to solder. For QFP packages, the thermocouple should be placed at the root of the pins along the device's centerline. For ball

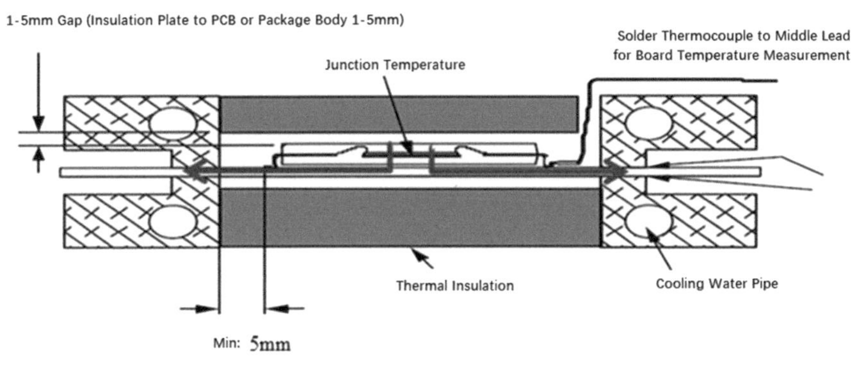

Fig. 6.22 Schematic diagram of the junction-to-board thermal resistance test fixture cross-section

grid array packages, the thermocouple should be positioned at the board's centerline, 1 mm from the package, near the wiring layer.

3. Junction-to-Ambient Thermal Resistance

Integrated circuits have various packaging types. However, heat is always dissipated into the surrounding air through the package structure. Junction-to-Ambient thermal resistance is an important indicator of a package's cooling efficiency. Its definition is as follows.

$$\theta_{JA} = \frac{T_J - T_A}{Q} \tag{6.14}$$

In the formula, θ_{JA} represents the junction-to-ambient thermal resistance under natural convection. T_J and T_A indicate the junction temperature and ambient air temperature. Q refers to the total heat dissipation of the chip. A wind speed of zero means natural convection, while a nonzero wind speed indicates forced convection.

The JESD51-2 standard provides procedures for testing junction-to-ambient thermal resistance under natural convection. If the chip size is under 27 mm, the package is mounted on a 76.2 mm × 114.3 mm test board. If the chip size is 27 to 54 mm, the package is placed on a 101.6 mm × 114.3 mm test board. The test board is 1.6 mm thick. It can be either a low thermal conductivity 1s0p board (one signal layer, no ground or power layers) or a high thermal conductivity 2s2p board. The package is positioned at the center of a 76.2 mm × 76.2 mm or 101.6 mm × 101.6 mm area at the front of the board. The test board is fixed horizontally on a vertical wall, with one edge attached. The whole setup is enclosed in a 0.0283 m^3 stationary air chamber, as shown in Fig. 6.23. This figure illustrates the JEDEC apparatus for testing junction-to-ambient thermal resistance under natural convection. The air chamber is made of polycarbonate, cardboard, wood, or other low thermal conductivity materials. A thermocouple measures the ambient air temperature. It is placed 25.4 mm below the test board and 25.4 mm from the sidewall. Junction temperature is determined by a temperature-sensitive parameter, such as the forward voltage drop of an internal diode. Power dissipation is calculated from the device's current and voltage. Once temperature and power dissipation are measured, Eq. (6.14) is used to determine junction-to-ambient thermal resistance.

The JESD51-6 standard describes the procedure for testing junction-to-ambient thermal resistance under forced convection. The test board and package placement remain the same as in the natural convection test. Figure 6.24 shows the JEDEC apparatus for testing under forced convection. The test board is placed inside a wind tunnel, where stable, uniform, one-dimensional laminar airflow is maintained. An airflow velocity probe measures wind speed upstream of the test setup. A thermocouple records airflow temperature. It is positioned 100–150 mm upstream of the test setup, 25 mm below the test board, and 25 mm from the test section wall. The methods for measuring junction temperature and power dissipation are the same as in the natural convection test.

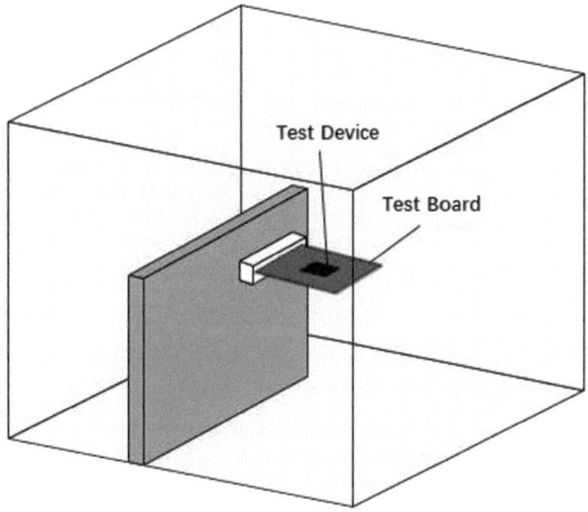

Fig. 6.23 JEDEC test setup for junction-to-ambient thermal resistance under natural convection

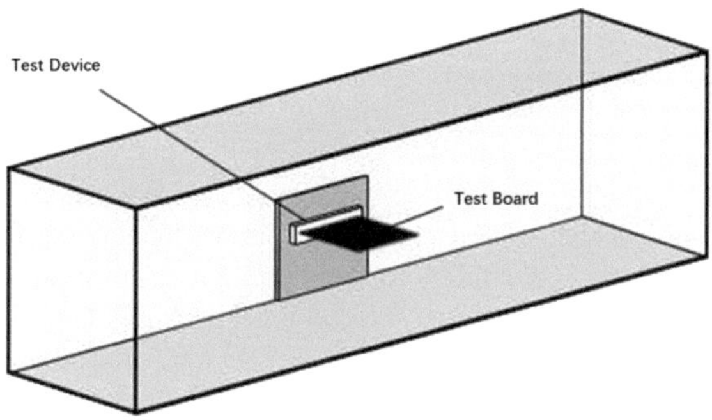

Fig. 6.24 JEDEC test setup for junction-to-ambient thermal resistance under forced convection

6.4.2 Thermal Characteristic Parameters

Junction-to-case thermal resistance is measured when almost all heat from the chip transfers to the package shell. Junction-to-board thermal resistance is measured when most heat transfers to both the package shell and the PCB. To meet these conditions, a high thermal conductivity heatsink can be installed on the package shell or board. Another option is mounting the package on a high thermal conductivity heat dissipation plate. If airflow around the package remains in natural convection, a high-efficiency heatsink is not necessary. In this case, most heat flows into the board, which meets the requirements for measuring junction-to-board thermal resistance. Only under these heat dissipation conditions does it make sense to estimate junction temperature using Junction-to-case or junction-to-board thermal resistance, package shell or board temperature, and power dissipation. In real applications, most chips dissipate only part of their heat through the package shell. The majority dissipates through the board. This makes it difficult to separately measure the heat dissipation of the package shell Q_{JC} or board Q_{JB}. Therefore, Eqs. (6.12), (6.13), and (6.14) are not suitable for junction temperature calculation. To solve this issue, thermal characteristic parameters are defined [23].

ψ_{JX} is a thermal characteristic parameter, defined as follows.

$$\psi_{JX} = \frac{T_J - T_X}{Q} \tag{6.15}$$

In the equation, ψ_{JX} represents the thermal characteristic parameter from the junction to the reference surface. T_J and T_X represent the junction temperature and the reference surface temperature, respectively. Q represents the total heat dissipation of the chip. The definitions of ψ_{JX} and θ_{JX} are similar. The difference is that ψ_{JX} refers to the condition where most heat is transferred, while θ_{JX} refers to the condition where all heat is transferred. In actual circuit operation, heat does not flow in a single direction. Instead, it dissipates in multiple directions. Therefore, ψ_{JX} is defined to meet real measurement conditions.

The junction-to-top thermal characteristic parameter ψ_{JT} is defined as follows:

$$\psi_{JT} = \frac{T_J - T_T}{Q} \tag{6.16}$$

the equation, T_J represents the junction temperature. T_T represents the temperature at the center of the package shell top surface. Q represents the total heat dissipation of the chip. ψ_{JT} refers to the thermal resistance along the heat flow path from the chip junction to the top surface of the package. This definition helps predict the junction temperature based on the surface temperature of the integrated circuit package. Compared to thermal resistance, it has greater engineering significance.

The junction-to-board thermal characteristic parameter ψ_{JB} is defined as follows.

$$\psi_{JB} = \frac{T_J - T_B}{Q} \tag{6.17}$$

In the equation, T_J and T_B represent the junction temperature and board temperature. Q represents the total heat dissipation of the chip. ψ_{JB} is measured at the same location as the junction-to-board thermal resistance θ_{JB}. However, ψ_{JB} differs due to changes in the testing environment. The chip's heat transfers from the junction to the test board. It is exposed to either natural or forced convection. The heat flow through this structure creates thermal resistance. ψ_{JB} can help predict the device's junction temperature based on the PCB temperature.

ψ_{JT} and ψ_{JB} can be measured at the same time as θ_{JA}. It is important to note that thermal characteristic parameters are similar to thermal resistance. Both depend on the package, substrate, and environmental conditions. However, they are different. Their calculations use different heat transfer values. The heat transfer amount Q used for thermal characteristic parameters is not the actual heat transfer between two points. Studies show that ψ_{JB} is less than or approximately equal to θ_{JB}, ψ_{JT} is much smaller than θ_{JT}, and ψ_{JT} is usually smaller than ψ_{JB}. Junction-to-case thermal resistance θ_{JC}, junction-to-board thermal resistance θ_{JB}, and thermal characteristic parameters are meaningful only under specific measurement conditions. In real engineering applications, devices usually have multiple heat dissipation paths. For most integrated circuits, it is difficult to measure case temperature accurately. In such cases, using thermal characteristic parameters to predict junction temperature is more practical. Therefore, thermal characteristic parameters are mainly engineering values.

6.4.3 Transient Thermal Impedance

Transient thermal impedance (Z_{th}) refers to the thermal resistance of a device when power is applied or turned off before it reaches thermal equilibrium. GB/T 2900.32 defines transient thermal impedance as the ratio of the temperature difference between two specific points (or regions) to the dissipated power. The power changes as a step function over a given time interval.

$$Z_{th}(t) = \frac{\left[T_r^*(0) - T_r^*(t)\right] - \left[T_r(0) - T_r(t)\right]}{P(0) - P(t)} \tag{6.18}$$

Transient thermal impedance is both a quantitative parameter and a physical value. It is an unsteady-state thermal resistance measured in °C/W. This means it represents thermal resistance during heating or cooling before reaching thermal equilibrium. During this process, thermal resistance changes over time. It is usually shown as a transient thermal impedance curve. When steady-state conditions are reached, heat fills the thermal capacitance of the object. The object then reaches its maximum temperature. At this point, transient thermal resistance equals steady-state thermal resistance. Researchers introduced transient thermal impedance to calculate junction temperature, power consumption, or load capacity in electronic circuits. It is useful for operations such as switching, cutoff, surges, and pulses.

Both steady-state thermal resistance and transient thermal impedance can be tested using two methods: static and dynamic test modes, as shown in Figs. 6.25 and 6.26. In the static test mode, heating power is applied to the device under test while monitoring the junction temperature with thermal-sensitive parameters. This method is highly suitable for thermal test chips and some functional integrated circuit devices. For thermal test chips, the method shown in Fig. 6.25a is used. The power chip is heated while the thermal-sensitive chip is monitored in real-time. For functional integrated circuits, the method in Fig. 6.25b is used. Power is first applied. Once thermal equilibrium is reached, the heating power is turned off, and a small test current is applied to measure the junction temperature. The dynamic test mode requires switching to thermal-sensitive parameter measurement conditions first. Then, power is applied to the device for a certain time, followed by a return to measurement conditions. This method is suitable for functional integrated circuits and most thermal test chips.

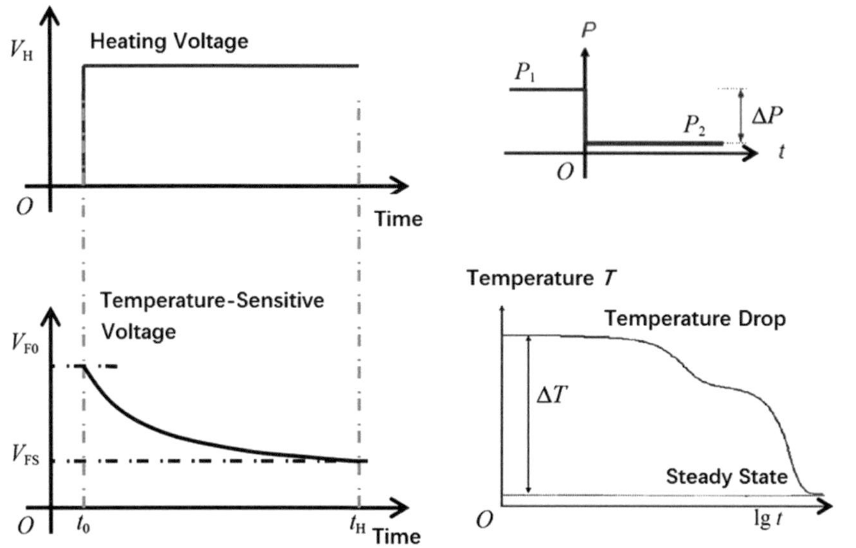

(a) Static test mode for thermal test chip (b) Static test mode for functional integrated circuits

Fig. 6.25 Static test mode

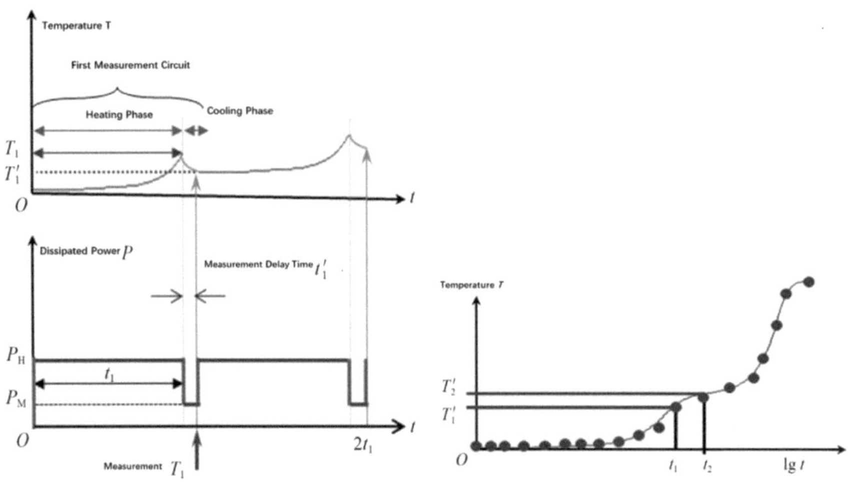

Fig. 6.26 Dynamic test mode

Compared to the dynamic test mode shown in Fig. 6.26, the static test mode offers the following advantages:

(1) The static test mode uses a single heating process with real-time monitoring. It only needs a single error correction when the heating current is cut off and the

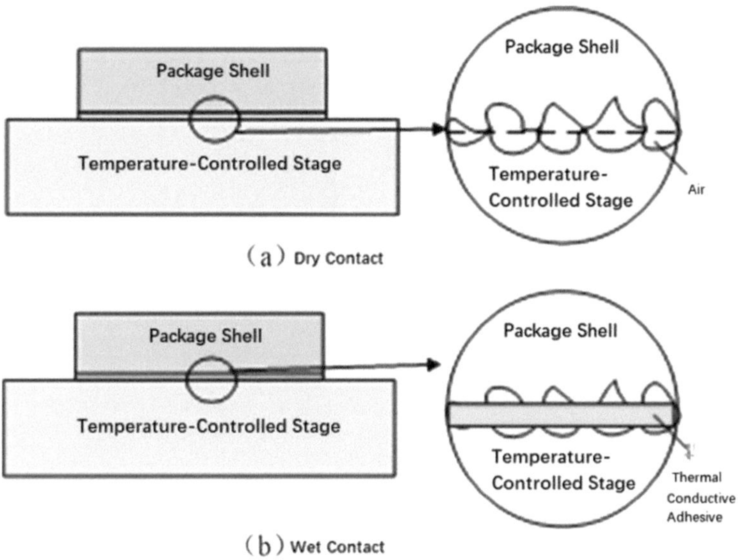

Fig. 6.27 Illustration of dry and wet contact

test current is applied. This ensures high accuracy, good consistency, and fast measurement speed.

(2) The static test mode avoids interference from minor fluctuations in heating power during testing.

(3) Traditional dynamic test mode uses thermocouples to monitor shell temperature. However, thermocouples only measure temperature at the contact point with the package, which may not reflect the actual maximum shell temperature. In addition, non-insulated contact between the thermocouple and the cooling plate, as well as loose contact between the thermocouple and the package shell, can affect accuracy. These issues introduce significant errors in thermal resistance calculations. In contrast, the static test mode based on the cooling curve uses a transient dual-interface testing method. It does not require shell temperature measurement. Instead, it calculates thermal resistance by monitoring junction temperature changes, eliminating errors from shell temperature measurement.

The JESD51-14 standard [24] defines the transient dual-interface testing method. The core of this method is to measure the transient junction temperature response of the same device twice on a controlled-temperature heat sink. In the first step, the device is in dry contact with the heat sink that has good thermal dissipation. In the second step, a thin layer of thermal interface material, such as thermal grease or oil, is applied between the device and the heat sink to form a wet contact. The dry and wet contacts are shown in Fig. 6.27. During both measurements, the difference in the dry and wet contact interfaces causes differences in the contact thermal resistance between the device and the heat sink. Therefore, when heat from the device junction

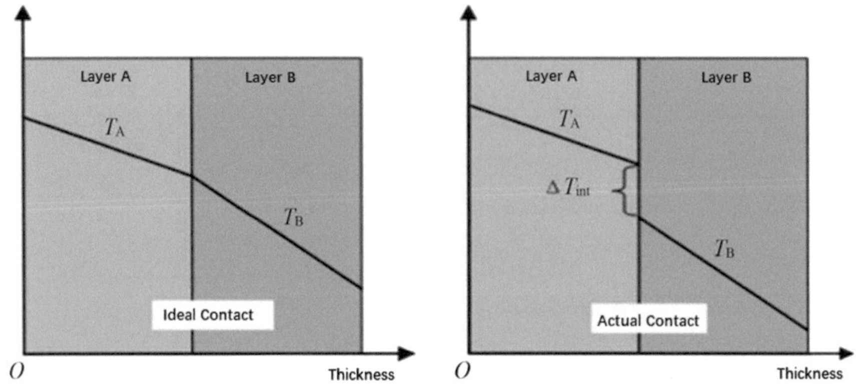

Fig. 6.28 Ideal contact and actual contact

area is transferred to the interface with the controlled temperature platform, the heat transfer path changes. This is shown by the two transient thermal impedance curves separating from that point t_s onward. The separation point indicates when the heat is transferred to the main heat dissipation interface of the device (the package shell). Typically, the contribution value of the transient thermal impedance curve $Z_{\theta JC}(t_s)$ at this point is approximately equal to the steady-state thermal resistance θ_{JC}. By examining the $Z_{\theta JC}(t_s)$ separation point of the curve, the transient junction temperature response curve θ_{JC} can be estimated. This curve contains detailed thermal information about each layer in the device's thermal conduction path. Further analysis of the thermal resistance in each package layer and diagnosis of packaging defects can be done based on the structural function in the JESD51-14 standard.

Contact thermal resistance θ_c is a key characteristic of all contact interfaces. To achieve ideal contact, the corresponding points of both surfaces must touch. However, due to manufacturing limitations, surface roughness cannot be perfectly smooth. In most cases, the actual contact area is less than 0.1% of the theoretical area. This causes a temperature difference ΔT_{int} at the contact interface. The difference between ideal and actual contact is shown in Fig. 6.28. At the microscopic level, contact occurs at small points with air gaps in between. Air is a poor thermal conductor. These air gaps block heat flow and force it through small contact points. The air gaps act as parasitic thermal resistance in the thermal conduction path. The thermal resistance per unit contact area is called contact thermal resistance. It is measured in °C $\cdot W^{-1} \cdot m^{-2}$. Its expression is given as follows.

$$\theta_c = \frac{\Delta T_{int}}{Q/A} \tag{6.19}$$

Smoother contact surfaces reduce air gaps and lower thermal resistance. Tighter compression increases actual contact area, decreases air gaps, and further reduces thermal resistance. The total thermal resistance of a structure is the sum of material resistance and contact resistance at all interfaces. Surface flatness, roughness, clamping pressure, material thickness, and compressive modulus all affect contact thermal resistance [25].

6.4.4 Specific Heat Capacity and Structural Function

Specific heat capacity is a fundamental physical quantity in thermodynamics. It refers to the amount of heat required to raise or lower the temperature of a unit mass of a substance by 1 °C. It measures a material's ability to store heat, not its ability to disspation heat. The relationship between specific heat capacity, heat transfer, and temperature change is expressed as:

$$c = \frac{\Delta Q}{m\Delta T}$$

where c is the specific heat capacity, measured in $\mathrm{J}/(kg \cdot K)$; ΔQ is the heat absorbed or released, measured in $\mathrm{J/mol}$; m is the mass of the material, measured in kg; ΔT is the temperature difference, measured in K. Specific heat capacity is not applicable during phase transitions. If a material undergoes a phase change due to heat absorption or release, this equation no longer holds. This is because the heat involved in a phase change does not cause a temperature increase or decrease.

For a simple package structure made of a single material, its thermal behavior can be described using an RC (resistor-capacitor) network model. This model consists of a single thermal resistance and a single thermal capacitance connected in parallel. When a power input of ΔP_{H} is applied, the temperature rises exponentially [26].

$$T(t) = \Delta P_{\mathrm{H}} \cdot R_{\mathrm{th}} \cdot [1 - \exp(-t/\tau)]$$

where $\tau = R_{\mathrm{th}} \cdot C_{\mathrm{th}}$ represents the thermal time constant of the package model.

The size of this model is determined by the values of τ and R_{th}. The thermal time constant model for a single package is shown in Fig. 6.29.

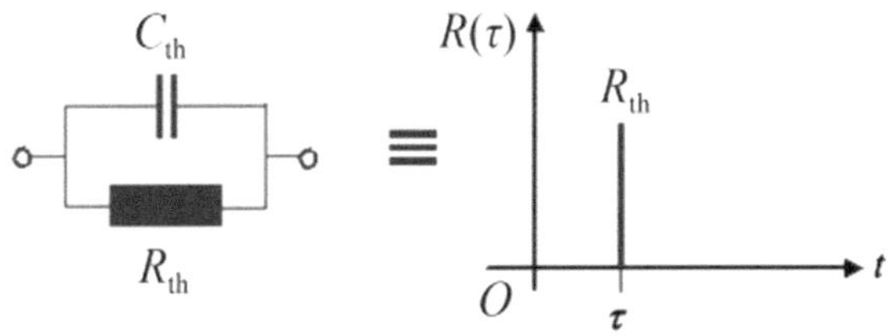

Fig. 6.29 Thermal time constant model of a single package

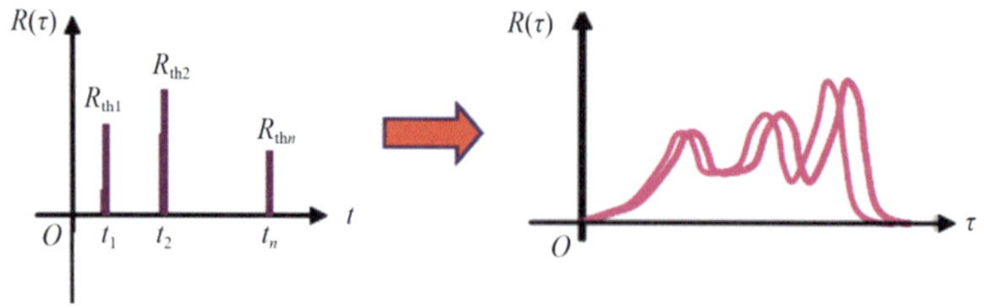

Fig. 6.30 Processing discrete thermal time constant values using a continuous thermal time constant spectrum

For a real packaged device, multiple RC models can be used to describe its thermal behavior. In this case, a continuous thermal time constant spectrum is applied to process discrete thermal time constant values, as shown in Fig. 6.30.

The unit step response function is described using an integral method. Assuming the applied power is $\Delta P_{\mathrm{H}} = 1\mathrm{W}$, the temperature response $a(t)$ can be expressed as follows:

$$a(t) = \int_{0}^{\infty} R(\tau) \cdot [1 - \exp(-t/\tau)]\mathrm{d}\tau$$

First, the thermal time constant spectrum $R(\tau)$ is obtained through deconvolution. Second, the spectrum is divided into small segments of length Δz. Each Δz corresponds to a parallel RC thermal network. Finally, the Foster thermal network model is built. However, the Foster model lacks physical significance. It connects thermal capacitances node-to-node, but real thermal capacitances should connect from known nodes to the ground. This forms a stacked structure. Therefore, the Foster model must be converted into the Cauer thermal network model, as shown in Fig. 6.31. The Cauer model is a step-like thermal network. Each thermal capacitance is connected to the ground, making it more consistent with physical reality. After obtaining the Cauer RC thermal model, a graph is plotted. The vertical axis represents cumulative thermal capacitance, while the horizontal axis represents cumulative thermal resistance. This graph approximates the device's integral structural function. Using this structural function, the thermal resistance of each packaging layer can be analyzed and extracted.

6.4.5 *Major Thermal Testing and Analysis Standards*

Thermal testing of integrated circuits involves analyzing the heat performance of the actual chip. This helps assess the heat dissipation efficiency of the package. Since chip packaging forms and real application environments vary by manufacturer, a series of standards have been established. These standards facilitate communication

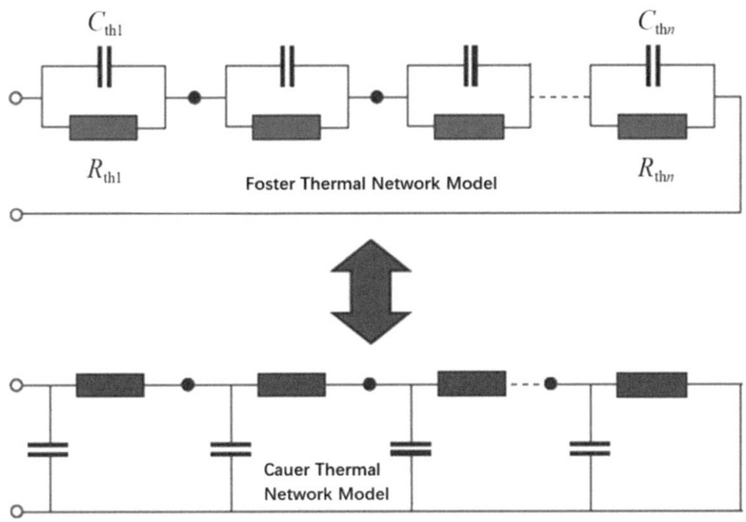

Fig. 6.31 From foster thermal network model to Cauer thermal network model

within the industry and standardize testing methods for various types of packaged chips.

There are many thermal testing standards both domestically and internationally. These standards mainly focus on chip junction temperature and thermal resistance testing under single heat source conditions. Key standards include the JESD51 (JEDEC JESD51) series, Semiconductor Equipment and Materials International (SEMI) series, International Electrotechnical Commission (IEC) series, national standards (GB), U.S. military standards MIL-STD-883J, MIL-STD-750F, and Chinese military standards GJB 548B, GJB 128A, and others. Early industrial standards for thermal resistance testing defined methods for testing integrated circuit packages under natural convection, forced convection, and near-infinite plate conditions. These belong to the SEMI series. In the 1990s, JEDEC invited packaging testing vendors and researchers to develop new industrial standards. These standards focused on thermal management. Compared to SEMI standards, JEDEC standards are more comprehensive. While the basic testing methods and principles are similar, JEDEC standards offer more detailed specifications. They cover thermal resistance definitions, single-chip thermal testing, temperature-sensitive parameter calibration, testing environment correction, and multi-heat-source thermal resistance matrices. These standards apply to discrete devices, integrated circuits, diodes, and multi-heat-source devices. JEDEC standards emphasize detailed research on testing methods, test environment definitions, and data collection and processing methods.

The JEDEC committee has developed a series of standards specifically for packaging thermal testing. These form the most comprehensive standard system in the field of thermal analysis. As packaging forms evolve rapidly, the JEDEC committee has added new definitions, such as θ_{JB}, ψ_{JT}, and ψ_{JB}, to the latest versions of the

JEDEC standards. While thermal resistance values measured with the new JEDEC standards can predict junction temperatures in real-world conditions, they do not provide major improvements in actual applications. Junction temperature predictions still have limitations. To predict the junction temperature of a package using parameters such as θ_{JA}, θ_{JC}, θ_{JB}, ψ_{JT}, and ψ_{JB}, special attention must be paid to the PCB size of the tested sample. It is crucial to clarify the differences between natural convection, forced convection, and real working conditions [27].

The JESD51-1 standard explains how to measure the thermal characteristics of single-junction semiconductor devices using electrical methods. This method uses the temperature-sensitive parameters of semiconductor devices. For example, the forward voltage of a diode under constant current shows a linear relationship with the device's junction temperature. The JESD51-1 standard applies to specialized chips (thermal test chips) and integrated circuit devices with different functions. The testing method is limited to the packaging of a single chip (specialized test chip or functional chip). The thermal characteristics of any semiconductor device vary with temperature and power dissipation. Therefore, thermal testing should be conducted in environments that closely resemble real-world conditions.

The JESD51-2 standard defines how to test junction-to-ambient thermal resistance under natural convection. The standard requires the installation of the specified SMT packaged device on a standard test board. External atmospheric interference must be prevented during testing. Thus, the test board should be placed flat in a test chamber with still air.

The JESD51-3 standard explains how to measure the junction-to-ambient thermal resistance of devices on low thermal conductivity PCBs. This standard applies to surface-mount devices with lead spacings greater than 0.35 mm and package sizes up to 48 mm. It does not apply to through-hole mounted, BGA, or socket-type devices.

The JESD51-4 standard specifies the requirements for thermal test chips with wire-bonding types. It provides specific design requirements for these test chips to minimize differences in test results between non-standard chips.

The JESD51-5 standard describes how to design test PCBs for attaching packaged chips to the PCB.

The JESD51-6 standard explains how to test junction-top and junction-to-board thermal characteristics. It defines testing methods and environmental conditions, such as forced-air cooling and related test methods.

The JESD51-7 standard is used to test junction-to-ambient thermal resistance of devices on high thermal conductivity PCBs. It specifies the dimensions, UBM quantity, and thickness of high thermal conductivity test boards. This standard does not apply to through-hole, BGA, or socket-type devices.

The JESD51-8 standard explains how to test junction-to-board thermal resistance with embedded two-layer copper test boards. It does not apply to packages with heat sinks installed on one side or those with asymmetric heat flow paths.

The JESD51-9 standard describes the PCB specifications for solder ball array and grid array packaged devices. However, it does not apply to through-hole socket-type devices.

The JESD51-10 standard outlines PCB requirements for single/dual in-line package and SiP packaged devices during testing.

The JESD51-11 standard specifies the thermal test PCB requirements for grid array packaged devices.

The JESD51-12 standard provides an overview of how to use the thermal characteristic parameters of electronic devices defined in the JESD51 series standards. Previously, θ_{JC} the surface of the device was considered the primary heat dissipation channel. This surface referred to the top or bottom surface of the package shell. The JESD51-12 standard defines the junction-to-case thermal resistance of the top surface of the package shell as $\theta_{JC\text{-}Top}$. It also explains the meanings of junction-top thermal characteristics ψ_{JT} and junction-to-board thermal characteristics ψ_{JB}.

The JESD51-13 standard provides a unified set of terms and definitions commonly used in semiconductor thermal testing. Some terms and definitions go beyond those in the JESD51 series standards. This helps describe the thermal performance of different semiconductor packages and packaged devices more accurately and comprehensively.

The JESD51-14 standard explains the heat dissipation from the device's junction to the package shell surface along a one-dimensional heat transfer path. It also provides a repeatable test method for semiconductor junction-to-case thermal resistance, called the transient double-interface testing method. Additionally, it introduces structural thermal resistance analysis methods using integral and differential structural functions.

The JESD51-31 standard adjusts the thermal test environment and methods prescribed in the JESD51 series standards for single-chip conditions, specifically for multi-chip components.

The JESD51-51 standard describes methods for testing the thermal resistance and impedance of LEDs using electrical methods. It also explains the relationships between thermal power, luminous flux, junction temperature, forward voltage, and forward current. This standard is suitable for testing the junction temperature and thermal resistance of individual LEDs and certain LED arrays in laboratory environments. However, it is not suitable for high-volume screening tests.

MIL-STD-883J and GJB 548B are essentially the same. They define methods for testing the thermal performance of microelectronic devices, including junction temperature, thermal resistance, shell temperature, package surface temperature, and thermal response time. They cover infrared and electrical methods, focusing on test procedures, circuits, equipment, sample mounting, power application, and thermal resistance calculations. These standards mainly target steady-state thermal performance testing. MIL-STD-750F 3100 series methods and GJB A128A are similar. They focus on thermal testing circuits for discrete semiconductor devices like insulated-gate power field-effect transistors (IGFETs) and insulated-gate bipolar transistors (IGBTs). Other test methods are similar to those in MIL-STD-883J. IEC 60747 and GB series standards provide detailed test methods and circuits for various semiconductor device categories, including thermal performance testing.

For high-end integrated circuits like CPUs, foreign countries have already provided specific technical requirements and testing conditions for junction temperature in their detailed CPU product specifications. In China, GJB 7704-2012 “Military CPU Testing Methods” sets performance testing standards for CPUs, but it does not address junction temperature testing requirements or methods. Currently, temperature readings are mainly taken through internal reading circuits in the CPU. However, due to the large number of CPU cores and the large area, the temperature readings from these circuits are highly correlated with the measurement location, leading to significant testing errors.

6.5 Thermal Analysis Technology

As described in Sect. 6.4, junction temperature, thermal resistance, and thermal characteristic parameters are key indicators for measuring the thermal properties of microelectronic packages. These indicators are used to assess the thermal reliability of semiconductor devices. Among the reliability issues in microelectronic packaging, thermal-related issues account for over 20%. Therefore, whether it is the self-heating effect of the device or thermal reliability issues caused by high-temperature operation, assessing thermal reliability becomes crucial when evaluating the overall reliability of device packages. Testing and analyzing key thermal parameters is very important. As devices continue to evolve toward smaller sizes, higher integration, faster speeds, and higher power, analyzing thermal properties in device packaging presents higher demands, including precise analysis accuracy, higher spatial resolution, and higher time resolution. Regardless of the thermal parameter being measured, accurate temperature testing is necessary.

6.5.1 Main Thermal Analysis Methods and Comparison

The main thermal analysis methods currently are: ① The electrical method, which uses temperature-dependent electrical parameters to characterize temperature; ② The optical method, which uses temperature-dependent optical parameters to characterize temperature; ③ The physical method, which uses materials in close contact with the device to measure temperature; ④ The simulation method, which uses simulation techniques such as finite element analysis to characterize temperature and thermal properties. A comparison of the advantages and disadvantages of these thermal analysis methods is shown in Table 6.3.

Optical methods are currently widely recognized. The basic principle is based on characterizing temperature-sensitive optical parameters on the surface of semiconductor devices. This method measures the object’s own radiation, reflected radiation, or stimulated radiation. Optical testing can be divided into two categories: emission testing and excitation testing. In emission testing, the object itself is considered the

Table 6.3 Comparison of various thermal analysis methods and their advantages and disadvantages

Category	Method	Advantages	Disadvantages
Electrical	• Junction voltage method • Threshold voltage method • Current method • Gain method	• Non–destructive measurement of packaged devices • Non–contact measurement	• The measured junction temperature reflects the average effect in the junction area • Requires specific working modes • Cannot obtain a surface temperature distribution map
Optical	• Field–induced luminescence • Infrared method • Thermal reflection method • Photoluminescence method • Raman scattering method	• Obtaining a surface temperature distribution map • Non–contact measurement • High spatial resolution	• Requires a surface view angle • Destructive measurement when measuring junction temperature of packaged devices • Expensive equipment
Physical	• Scanning probe method • Liquid crystal method • Phosphor method	• Obtaining a surface temperature distribution map • High spatial resolution	• Requires a surface view angle • Destructive measurement when measuring junction temperature of packaged devices • Direct contact may interfere with the temperature field distribution
Simulation	• ANSYS • Fluent • Flotherm • Icepak	• Obtain surface temperature distribution map and transient response curve • Flexible loading, easy structural modifications	Accuracy of the analysis needs verification

radiation source. This can involve testing fluorescence spectra excited by an external electric field (electroluminescence) or infrared radiation produced by blackbody radiation, both of which are strongly related to temperature. Excitation testing compares incident light with reflected (or scattered) light to characterize temperature. Examples include thermal reflection, photoluminescence, and Raman scattering methods. However, the problem is that only when photons interact with the device being tested can the effects of temperature be detected. These interactions are often so subtle that they are considered noise and ignored. Therefore, optical methods are considered non-contact measurement techniques. The spatial resolution of optical methods is mainly determined by the performance of the light source and optical lens. The time

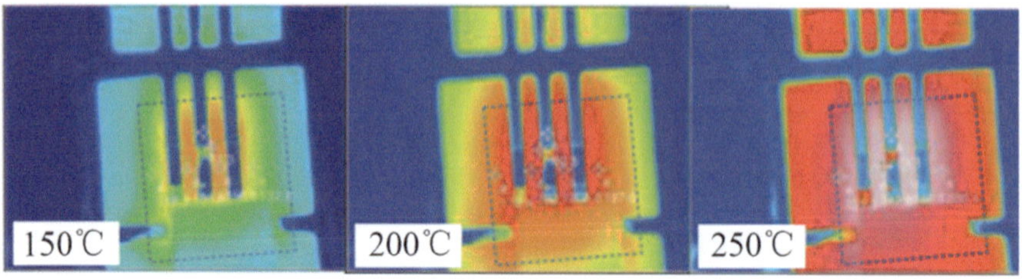

Fig. 6.32 Thermal imaging map of a MOSFET device at different temperatures after removing the surface resin

resolution is determined by how the optical phenomenon responds to temperature changes, but it is also affected by the actual response capabilities of the equipment.

Optical methods have the advantage of high spatial and temporal resolution. They can measure the temperature of very small areas on a device's surface. Optical methods can also scan the device's surface temperature point by point. This makes it easy to obtain the surface temperature distribution map. Figure 6.32 shows a thermal imaging map of a MOSFET device at different temperatures after removing the surface resin [28]. The disadvantage of optical methods is that the light beam must contact the object's surface. This makes optical methods unsuitable for packaged devices, unless the device is unsealed. Optical equipment is high-precision, making it difficult to manufacture and very expensive. Additionally, the operation process is complex.

Among the thermal analysis methods mentioned, the infrared method is the most widely used optical method. Newer optical methods include thermal reflection and Raman scattering. A comparison of the main technical indicators for optical, electrical, and thermocouple methods is shown in Table 6.4.

6.5.2 Electrical Method

In semiconductor devices, temperature-related electrical parameters are called temperature-sensitive parameters. Not all temperature-sensitive parameters can be used to measure temperature. The parameters must have a strong linear relationship with temperature, such as the forward voltage drop of the PN junction, threshold voltage drop, leakage current, and gain.

The electrical method for thermal resistance testing uses the temperature-sensitive properties of semiconductor devices. By measuring the temperature-sensitive parameters, the average junction temperature of the device can be determined, allowing the calculation of thermal resistance. After measuring the temperature-sensitive parameters, the actual temperature of the semiconductor device can be deduced using the relationship between the parameters and temperature. The electrical method for measuring junction temperature involves two steps. First, the relationship between the

Table 6.4 Comparison of main technical indicators of optical method, electrical method, and thermocouple method

Method	Maximum resolution			Imaging capability	Remarks
	Spatial resolution	Temperature resolution	Temporal resolution		
Thermocouple method	50 μm	0.01 °C	0.1~10 s	NO	Packaging shell, PCB, temperature control platform
Electrical method	–	0.01 °C	1 μs	NO	Separate devices or other temperature–sensitive components
Infrared method	3~5 μm (15 magnification, 2.7 μm/ pixel)	0.1 °C	10 ms	YES	Conventional semiconductor materials, low radiation coefficient of metals, GaN materials are transparent to infrared bands
Thermal reflection method	0.3~0.5 μm (100 magnification, 45 nm/pixel)	0.5 °C (5 min) 0.25 °C (30 min)	50 ns	YES	Can test metals and semiconductor materials
Raman scattering method	0.5~1 μm	1 °C	100 ns	YES	Semiconductor materials

temperature-sensitive parameters and temperature is established to obtain the k coefficient. Then, thermal testing equipment is used to measure the voltage variation curve of the device as it transitions from a higher stable temperature to ambient temperature. By multiplying the voltage variation curve by the k coefficient, the cooling curve of the device's junction over time is obtained. The junction cooling curve is analyzed using thermal transient measurement software to obtain the thermal characteristics of the semiconductor package (such as thermal capacity and thermal resistance). Figure 6.33 shows the process of obtaining the device junction temperature response curve using the electrical method.

When testing the thermal resistance of a semiconductor device's PN junction in a natural convection environment, the device must be placed in a standard still air box. A thermocouple amplifier is used to connect the thermocouple for measuring the ambient temperature. When testing the thermal resistance in a forced convection environment, the device must be placed in a wind tunnel. When testing the thermal resistance from the device's PN junction to the package shell, the device must be fixed to a liquid cooling plate. If the device's power exceeds the power the test equipment can provide, a power amplifier is used to supply power to the device. The electrical method thermal resistance testing system is shown in Fig. 6.34.

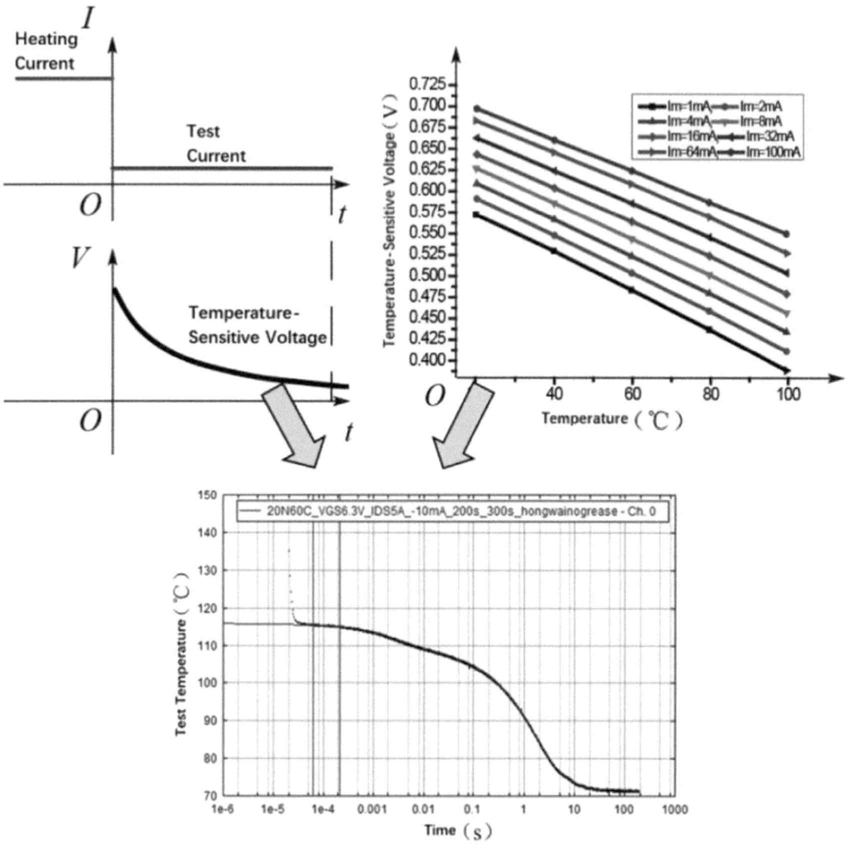

Fig. 6.33 Process of obtaining the device junction temperature response curve using the electrical method

The*k* coefficient calibration is typically conducted in an oil bath or a closed temperature-controlled chamber. This calibration assumes that the device temperature matches the temperature of the oil bath or chamber. Therefore, the test current must not cause self-heating in the device during the process. To prevent self-heating, pulse measurement calibration can be used. Alternatively, calibration can be done at low input power levels. For calibration at low input power, the temperature measurement process is as follows: First, work power is applied to raise the device temperature to saturation. Then, a small test power is applied with high-speed switching. The device temperature is monitored in real time as the temperature decreases.

Due to the nature of the electrical method, only the junction temperature of the device is obtained, not the surface temperature distribution. For instance, when the forward voltage of the PN junction is used as the temperature-sensitive parameter, it only reflects the average temperature of the heated junction. It cannot represent the

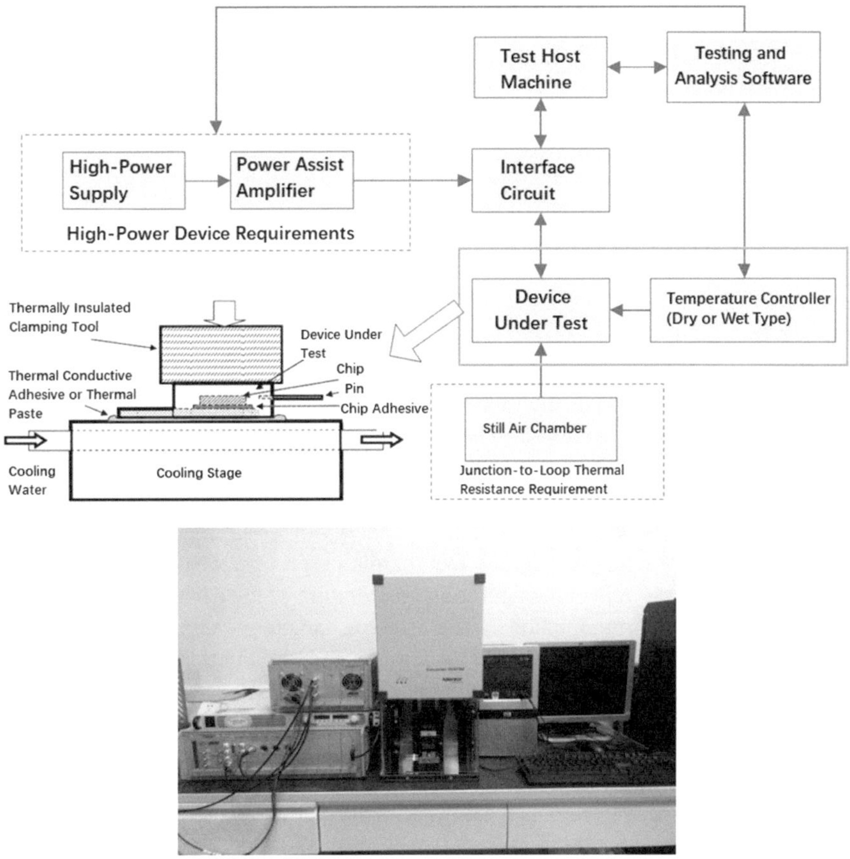

Fig. 6.34 Electrical method thermal resistance testing system

temperature of other areas of the device. In most cases, the average junction temperature can approximate the overall device temperature. However, in some cases, such as when there is a significant temperature gradient in the junction, the electrical method becomes less effective. For example, in GaN High Electron Mobility Transistors (HEMT), the electrical method only gives the average temperature between the source and drain. The lateral temperature distribution in this device is very uneven. As a result, the temperature measured by the electrical method is much lower than the actual peak temperature, reducing its accuracy.

While the electrical method sacrifices some measurement characteristics, it does not require a specific test structure. All electrical connections for this test are the same as those for the device's normal operation. This makes the electrical method a non-contact testing method. The electrical method has other advantages: it is the only method that can perform non-destructive testing on packaged devices. It can quickly test the thermal performance of devices with different processes and conduct thermal analysis. It also provides accurate information about the junction temperature, which is the highest temperature region of the device. Thus, the electrical method has become a widely used technique for studying junction temperature and conducting longitudinal thermal resistance analysis of microelectronic devices in both domestic and international research.

6.5.3 Infrared Method

The infrared method is an optical technique that can map the surface temperature distribution of a device. It works by receiving and processing infrared radiation emitted from the surface of the object. This allows for direct measurement of the surface temperature distribution. Therefore, the infrared method is widely used in surface temperature measurements. After years of development, infrared-based temperature measurement products have been available for many years. The photoluminescence method measures the bandgap energy of materials by monitoring the radiative decay of carriers during the recombination process. Changes in the material's bandgap energy reflect the operating temperature of the device [29]. Additionally, the Raman scattering method can be used to characterize temperature. It works by measuring the phonon frequency of the semiconductor. Since temperature changes are caused by the generation or disappearance of photons, the scattered photon spectrum varies with temperature [30]. The thermal reflectance method indirectly reflects the surface temperature of a device by observing changes in light reflection from the material's surface with temperature. Its advantage over the infrared method is its spatial resolution, which can reach sub-micron levels [31, 32].

The infrared method is a common technique for detecting emitted radiation from an object. It directly obtains the surface temperature distribution of the device by using the infrared radiation emitted from the object's surface. The method relies on a key feature: the peak wavelength of the radiated energy from the object is related to its temperature. The relationship between the object's radiated energy peak and temperature is shown in Fig. 6.35. By using an infrared lens module, the system can continuously measure the object's surface. It records the peak wavelength of the radiation energy emitted from each point on the surface. A computer then converts this data into temperature values for each point on the surface [33, 34].

The schematic diagram of the infrared method is shown in Fig. 6.36. The infrared lens module collects radiation signals from each point on the object's surface and focuses them onto an array detector on the infrared focal plane. The analog signals are then converted into digital signals. These signals are processed by software to create a pseudo-color image. Based on the color distribution in the image, the temperature distribution of the surface is displayed in real-time. The infrared method measures the temperature at various points on the surface simultaneously, making it possible to directly detect the temperature distribution. This enables accurate measurement of the device's surface temperature distribution.

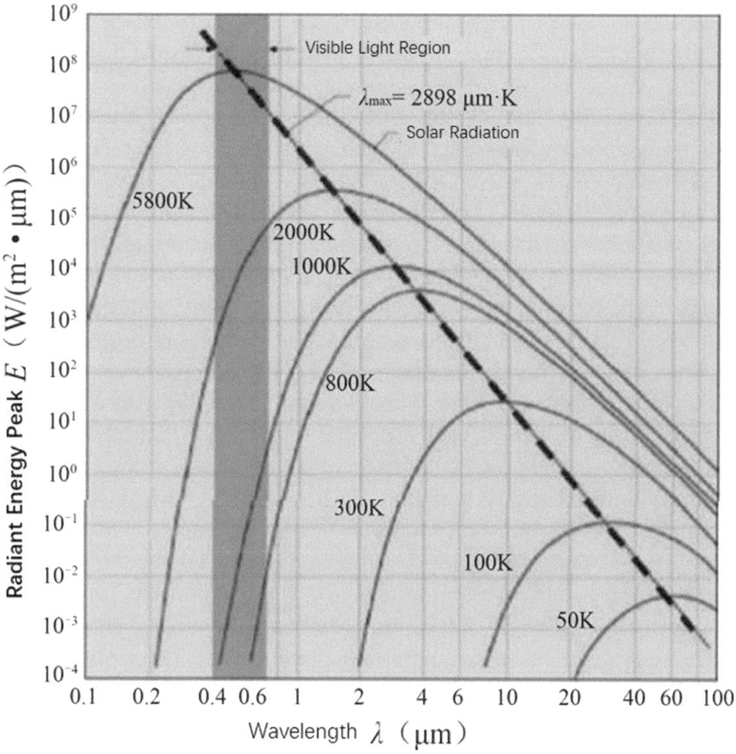

Fig. 6.35 The relationship between the object's radiated energy peak and temperature

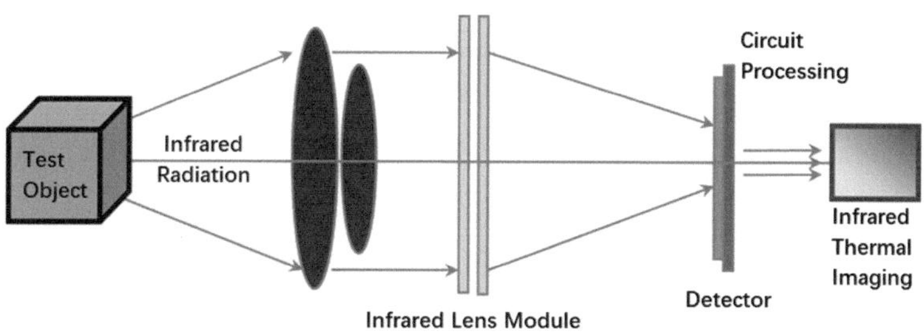

Fig. 6.36 Infrared testing principle diagram

Infrared thermography technology requires obtaining the infrared emissivity of each pixel area of the object being tested. Emissivity refers to the ratio of the energy radiated by the object's surface to the energy radiated by a blackbody at the same temperature. It ranges from 0 to 1. A blackbody is an object that absorbs all incoming electromagnetic radiation. Due to this property, its surface emissivity is 1. The emissivity of different objects depends on their material properties. At the same temperature, different objects radiate different amounts of energy, which results in different emissivities. Generally, thermal and electrical insulators, such as ceramics, rubber, and plastics, are good emitters. Metals, such as copper, iron, aluminum, nickel, and lead, are poor emitters. Their emissivity is usually below 0.25. Therefore, the appropriate method must be chosen based on the area of the integrated circuit to be tested. In their study of graphene as a thermal interface material, due to its high thermal conductivity and good electrical insulation, N. Wang and colleagues used infrared testing to measure the surface temperature of power devices with different thermal interface materials. They found that graphene has excellent electrical performance, mechanical properties, and heat dissipation capabilities [35]. Figure 6.37 shows the operating temperature of power devices when using different thermal interface materials.

A typical infrared thermography device is the QFI TM-HST MWIR-512. Figure 6.38 shows the photos of the high-speed dynamic infrared thermography device and the surface emissivity of the collected samples. This device uses a dual-detector microscope system: one for static images and one for transient images. It features a liquid nitrogen-cooled InSb sensor with a temperature resolution of 0.1 °C and a maximum spatial resolution of 2.7 μm. The dynamic measurement time constant is 3 μs, and the temperature range of the controlled platform is 25 to 145 °C. To obtain clear infrared thermography images, the controlled platform is typically heated to 70 °C or higher during measurement. The device includes a manual objective lens rotating stage that can hold five different infrared lenses.

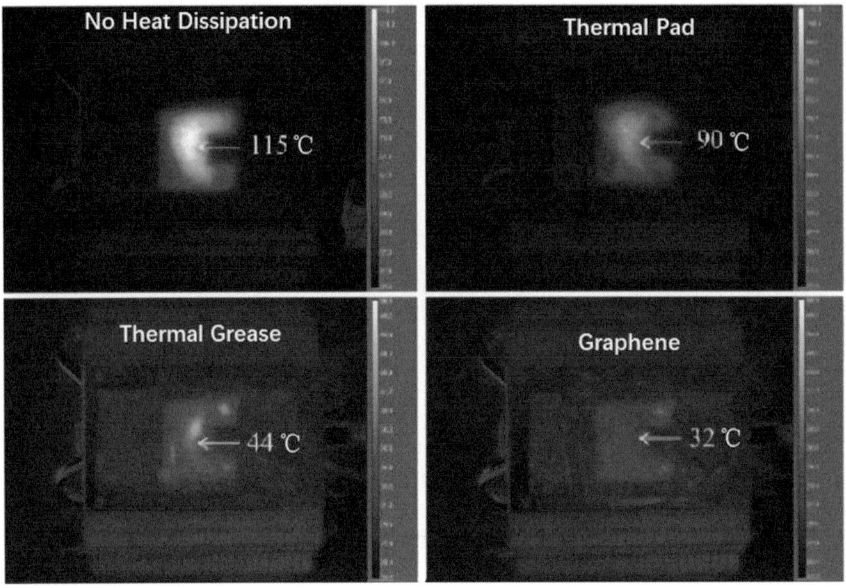

Fig. 6.37 Operating temperature of power devices with different thermal interface materials

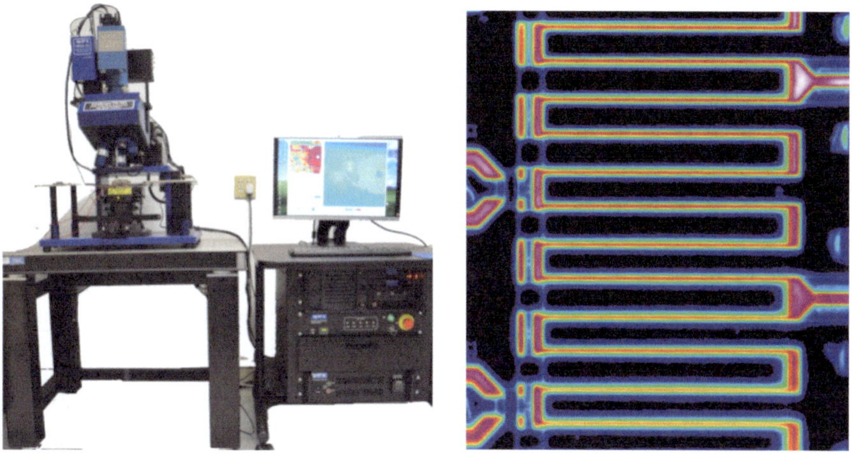

Fig. 6.38 Photos of high-speed dynamic infrared thermography equipment and surface emissivity of collected samples

6.5.4 Raman Scattering Method

The Raman scattering method is a new optical technique. It has been widely used in the study of thermal performance of power devices in recent years. The method has high spatial resolution and can provide temperature distribution information on the device surface and depth. This makes it a popular tool for studying the thermal parameters of semiconductor devices.

Raman scattering is a light scattering technique used to detect the vibrational frequency or energy of material optical phonons. It works by measuring the energy difference between incident light and scattered light. When a surface is illuminated by light of a certain frequency, most light just changes direction and undergoes elastic scattering without energy exchange. This is called Rayleigh scattering. Some of the scattered light, however, changes both direction and frequency. This is called inelastic scattering or Raman scattering. When the frequency decreases, it is called Stokes scattering. When the frequency increases, it is called anti-Stokes scattering. Stokes scattering is usually stronger than anti-Stokes scattering. The frequency difference between the scattered and incident light is called the Raman shift. The ratio of Stokes to anti-Stokes intensity is the polarization rate. Temperature changes cause variations in both the Raman shift and the polarization rate. By measuring these changes, the device temperature can be determined. Each method has its own advantages and disadvantages.

Phonon frequency depends on both temperature and lattice strain. Temperature changes can shift the phonon frequency. Mechanical stress also changes the vibration frequency. Semiconductor devices are often made of composite materials with different coefficient of thermal expansion. When the device operates, local heating causes expansion and deformation. This results in thermal coupling, which causes frequency changes. Therefore, phonon frequency changes during device operation are the result of both temperature and stress. This affects the accuracy of temperature measurement using Raman shift.

Unlike the Raman shift method, the polarization rate method can obtain the device temperature without being affected by stress. It uses the correlation between the Stokes and anti-Stokes intensity ratio and temperature. However, the anti-Stokes signal is very weak. The required collection time is long, and the holographic filters are expensive and have a short lifespan. These issues make the polarization rate method difficult to apply in practice. Therefore, the most common method for measuring device temperature using Raman scattering is the Raman shift method. Figure 6.39 shows an energy level diagram of Rayleigh scattering and Raman scattering processes.

The Raman scattering method typically monitors the side of the scattered light with lower energy than the incident light to obtain the Stokes Raman shift. This is because, at room temperature, the intensity of the Stokes signal is much stronger than the intensity of the anti-Stokes signal. Figure 6.40 shows the Raman scattering of GaN substrates at different temperatures. It can be observed that as the temperature increases, the Stokes Raman spectrum shifts horizontally, the spectral lines broaden, and the signal intensity decreases.

Andrei Sarua and his team from the University of Bristol reported using the Raman shift method to measure the junction temperature of GaN devices. Figure 6.41 compares the results of different thermal testing methods for GaN devices [36]. The test results show that the Raman scattering method has a significant advantage in spatial resolution over the infrared method. The spatial resolution of the Raman scattering method is about 1 μm, which is crucial for accurately characterizing the

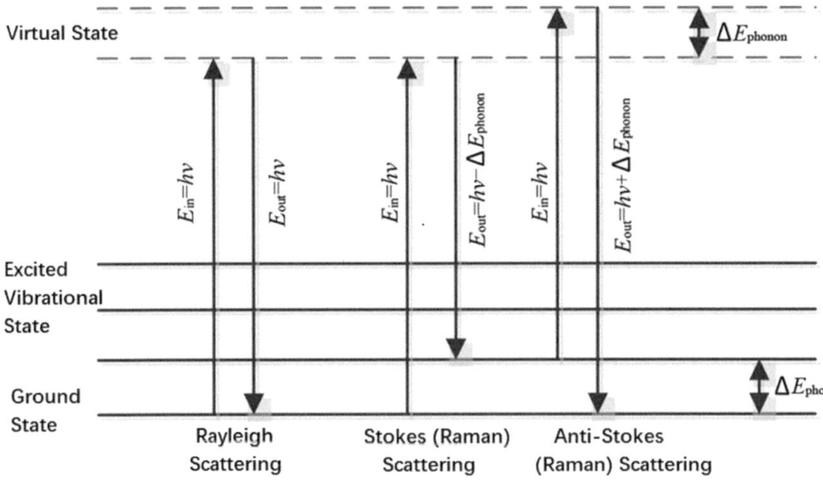

Fig. 6.39 Energy level diagram of Rayleigh scattering and Raman scattering processes

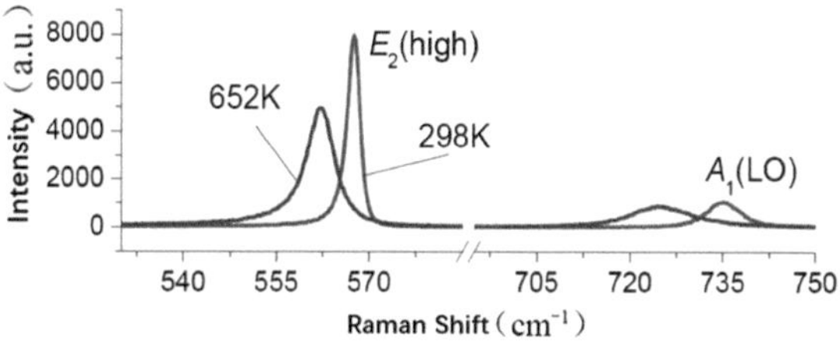

Fig. 6.40 Raman scattering of GaN substrate at different temperatures

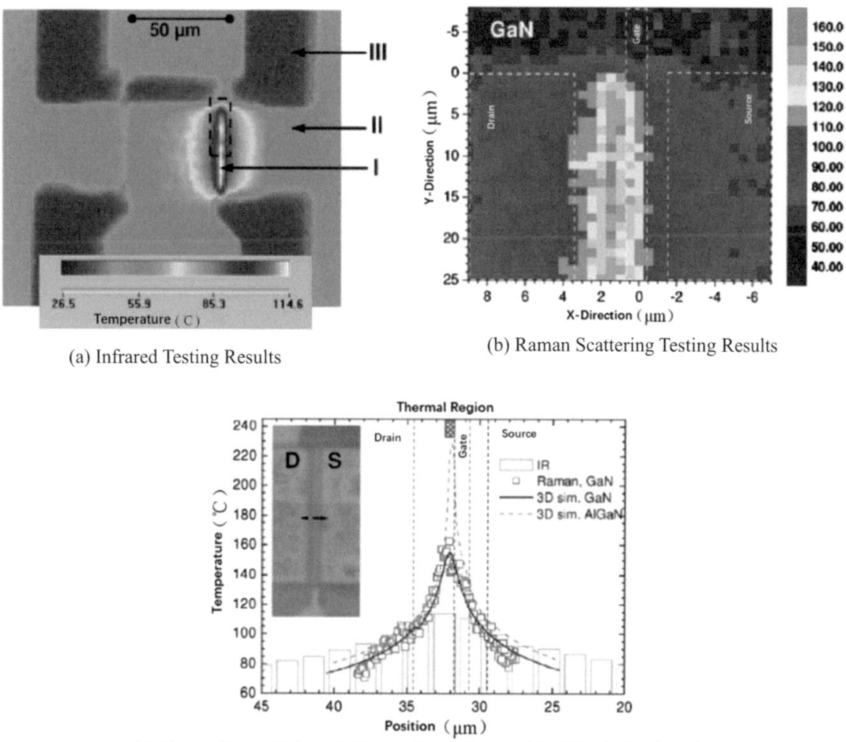

(c) Comparison of Infrared, Raman Scattering, and 3D Simulation Results

Fig. 6.41 Comparison of different thermal testing methods for GaN devices

peak temperature of the device channel. However, the drawback of the Raman scattering method is that the measurement cycle is quite long. Since the method requires point-by-point scanning, it is only suitable for localized, small-scale temperature measurements.

Additionally, the Raman scattering method can be used for vertical depth temperature distribution analysis. Figure 6.42 shows the temperature analysis results for the vertical depth of the silicon carbide (SiC) substrate in the test structure.

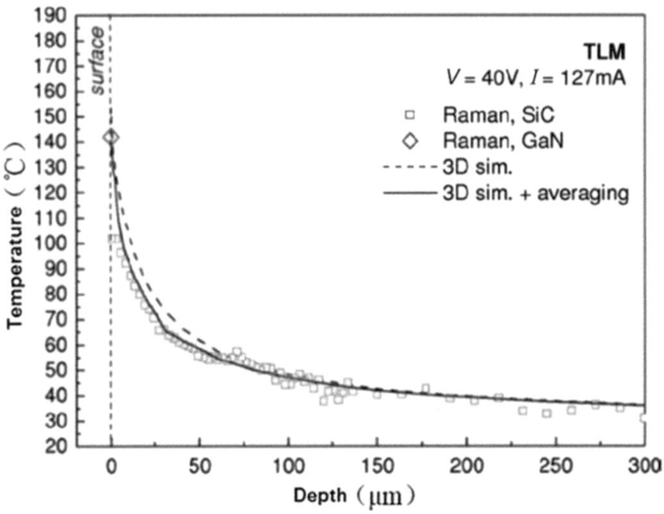

Fig. 6.42 Temperature analysis results for the vertical depth of the SiC substrate in the test structure

However, the spatial resolution decreases when performing vertical depth analysis with Raman scattering.

6.5.5 *Thermal Reflection Method*

The thermal reflection method is a new optical characterization technique. In thermal imaging technology, spatial resolution is related to wavelength. The longer the wavelength, the lower the resolution. Infrared thermal imaging uses the mid-infrared band with wavelengths ranging from 3 to 12 μm. This theoretically limits the spatial resolution to about 3 μm. To improve spatial resolution, thermal reflection imaging using shorter wavelengths, such as near-infrared light and monochromatic visible light, has been developed in recent years.

Thermal reflection imaging works by utilizing the principle that the reflectance of materials changes with temperature. This allows real-time characterization of the temperature distribution of the tested device under test. When a light beam strikes a material, it is reflected. When the material's temperature changes, the intensity of the reflected light changes. Therefore, the temperature of the device can be determined by measuring the change in the reflected light intensity from the material's surface. Unlike the infrared method, thermal reflection requires an active incident light source, calibration of the reflection coefficient, and detection of the reflected light intensity. The infrared method does not need an active incident light source and directly detects the infrared radiation emitted by the device's surface.

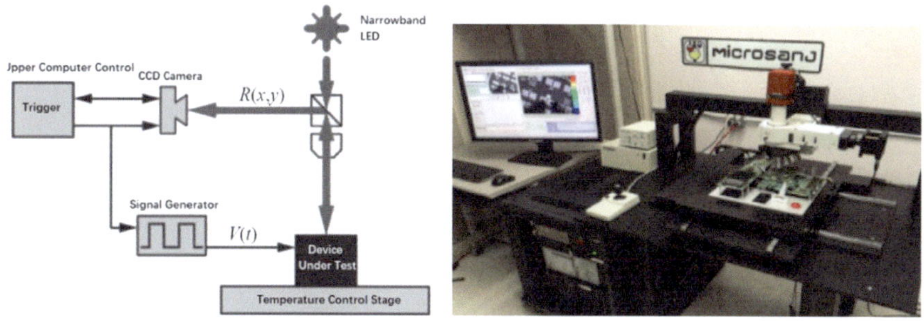

Fig. 6.43 Principle and physical image of the thermal reflection method testing system

The change in incident light reflectance (R) is linearly related to temperature, as shown in Eq. (6.20) [37].

$$\frac{\Delta R}{R} = \left(\frac{1}{R}\frac{\partial R}{\partial T}\right)\Delta T = \kappa\ \Delta T \tag{6.20}$$

In this equation, T represents temperature. $\kappa$is the thermal reflection calibration coefficient, typically ranging from 10^{-2}~10^{-5} K^{-1}. It is largely dependent on material properties, light wavelength, incident angle, surface roughness, and the composition of the sample in multi-layer structures.

Figure 6.43 Principle and physical image of the thermal reflection method testing system [38], A narrow-band LED provides a stable incident light intensity, which is directed at the device's surface. A Charge Coupled Device (CCD) or detector detects changes in the reflected light intensity with temperature. This generates a surface temperature distribution map of the device. To improve temperature resolution in the thermal reflection imaging system, phase-locking technology is used to enhance the signal-to-noise ratio, as the reflectance signal has a low signal-to-noise ratio and is easily affected by noise

According to the Sparrow criterion, thermal reflection imaging is related to the wavelength and numerical aperture of the lens:

$$D = \frac{0.5\lambda}{\text{N.A.}} \tag{6.21}$$

In this equation, D represents spatial resolution, N.A. is the numerical aperture of the lens, and λ is the wavelength of the light source. By choosing visible light with a wavelength range of 400–800 nm, the theoretical spatial resolution for temperature detection in thermal reflection imaging can reach up to 300 nm. This gives it a significant advantage in temperature imaging of small structures. This sub-micron spatial resolution is highly attractive for thermal testing of GaN integrated devices. The U.S. Air Force Research Laboratory (AFRL) introduced a thermal reflection imaging system when studying the thermal physical properties of GaN-based HEMT devices. Figure 6.44 shows the thermal reflection test results for GaN-based HEMT

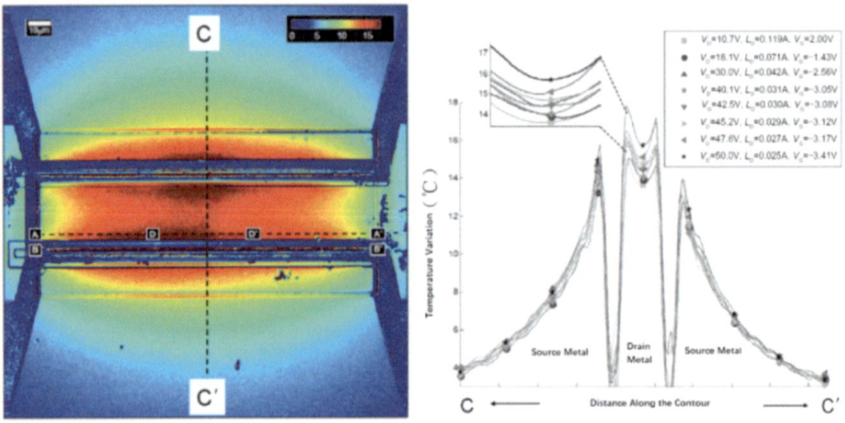

(a) Temperature Distribution Map Obtained by the Thermal Reflection Method (b) Temperature Scan of the C-C' Region [39]

Fig. 6.44 Thermal reflection test results for GaN-based HEMT devices

devices [31]. Thanks to the extremely high spatial resolution, the accuracy of the surface temperature distribution of the GaN devices is extremely high, allowing thermal reflection imaging to be more widely applied in the thermal analysis of GaN devices.

Sub-micron spatial resolution and nanosecond-level temporal resolution are unique to the thermal reflection method. Reference [43] conducted transient self-heating effect tests on GaN-based HEMT devices using the thermal reflection method. The test results are shown in Fig. 6.45. It can be seen that the high-temperature region is located in the GaN channel, where the temperature rise is the fastest. The temperature of the gate metal above the GaN channel is similar, while the low-temperature region is located in the drain metal, where the temperature rise is slower. This example shows that device-level thermal testing requires very high resolution.

Advantages of the thermal reflection method: The spatial resolution is sub-micron, and the temporal resolution is in the nanosecond range. The surface temperature distribution map is easy to interpret. It does not require knowledge of the material's emissivity and can operate at room temperature or below. These three advantages make the thermal reflection method very suitable for testing thermal parameters at the device level, such as for GaN devices.

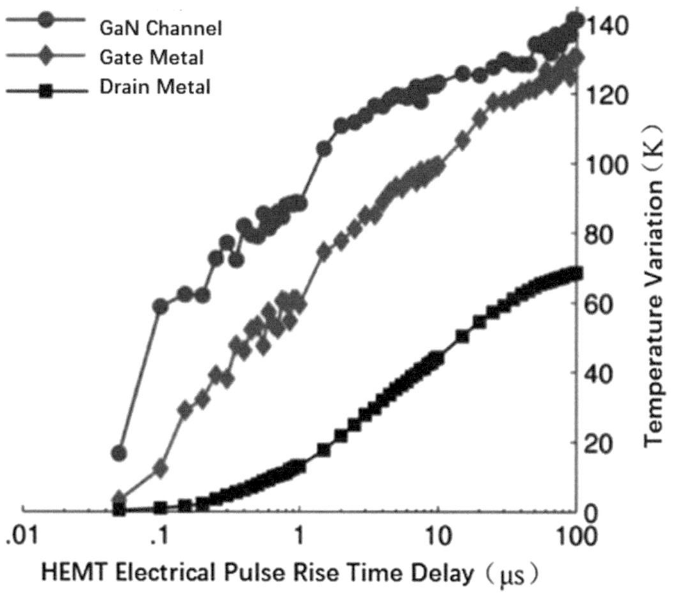

Fig. 6.45 Transient self-heating effect test results for GaN-based HEMT devices [43]

6.6 Main Factors Affecting Package Thermal Performance

6.6.1 Package Materials

Materials with high thermal conductivity, such as ceramics, metals, and high-conductivity substrates, reduce junction-to-case thermal resistance and junction-to-board thermal resistance. This lowers the overall junction-to-ambient thermal resistance. For instance, ceramic packaging has much higher thermal conductivity than plastic. As a result, ceramic packaging has lower overall thermal resistance and better thermal performance compared to plastic packaging of the same type. Figure 6.46 shows the change in junction-to-ambient thermal resistance with the number of pins in QFP packaging under natural convection.

6.6.2 Package Size

Device packaging has two types of thermal resistance. The first type is internal conduction resistance, which includes junction-to-case and junction-to-board resistance. The second type is external convection and radiation resistance, which includes case-ambient and board-ambient resistance. Both types of resistance are inversely proportional to the surface area of the device package. Therefore, the size of the

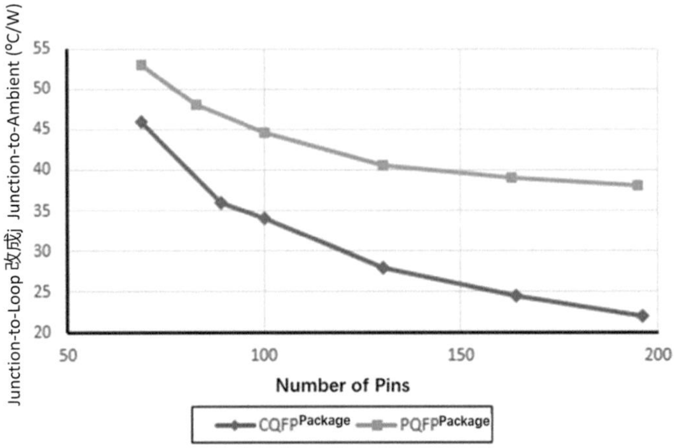

Fig. 6.46 Change in junction-to-ambient thermal resistance with the number of pins in QFP packaging under natural convection [39]

package is inversely related to thermal resistance. Figure 6.47 shows that, under natural convection, the junction-to-ambient thermal resistance of three different packages decreases as the number of pins (or package size) increases.

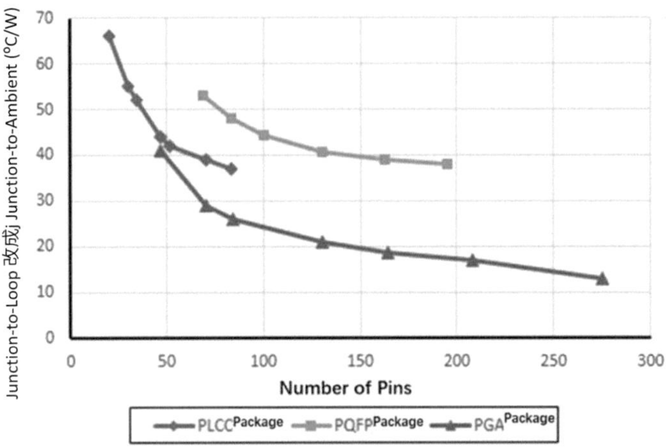

Fig. 6.47 Change in junction-to-ambientthermal resistance with the number of pins for three types of packaging under natural convection [39]

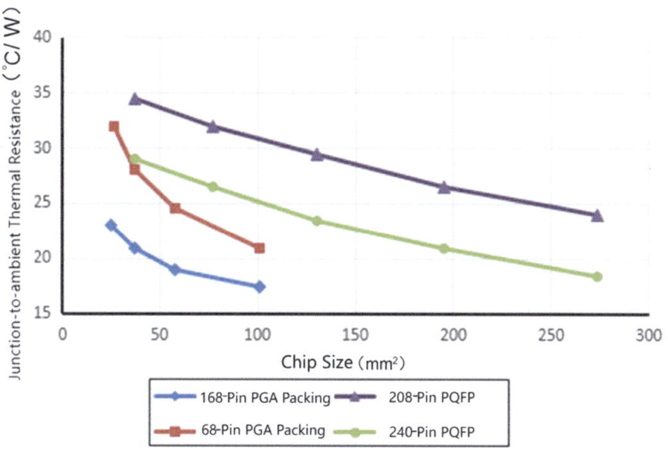

Fig. 6.48 Change in junction-to-ambient thermal resistance with chip size for different packages under natural convection [39]

6.6.3 Chip Size

The chip size is usually much smaller than the size of its substrate or cover plate. From practical experience and relevant mathematical formulas, it is known that conduction thermal resistance between the chip and substrate, as well as between the chip and cover plate, decreases as the chip size increases. Figure 6.48 shows that, under natural convection, the junction-to-ambient thermal resistance of PGA packaging with different number of pins and PQFP packaging with different number of pins decreases as chip size increases. This phenomenon occurs when the heat source area is much smaller than the actual chip area, leading to a larger temperature gradient when the chip generates heat.

6.6.4 Device Heat Dissipation

The temperature is positively correlated with the heat dissipation of electronic devices. When the temperature of an object is higher, more heat is dissipated through natural convection and radiation. Figure 6.49 shows how junction-to-ambient and junction-to-case thermal resistance change with power dissipation (heat dissipation). It can be observed that as the heat dissipation increases, junction-to-air thermal resistance decreases. However, junction-to-shell thermal resistance remains almost constant. This happens because junction-to-air thermal resistance is convective-radiative resistance, while junction-to-shell thermal resistance is conductive. Conductive resistance depends only on packaging materials and size.

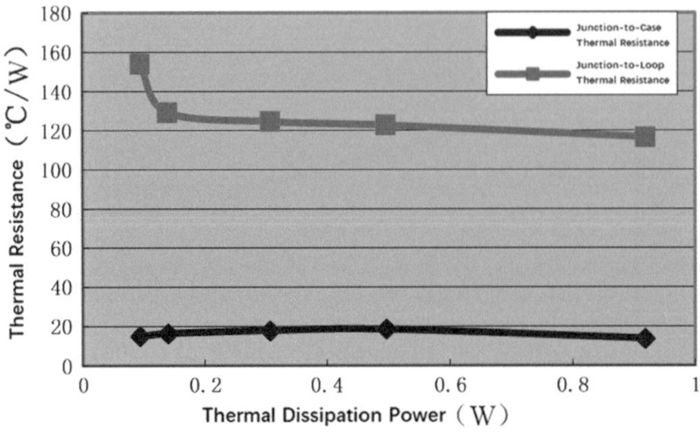

Fig. 6.49 Change in junction-to-ambient and junction-to-case thermal resistance with power dissipation

6.6.5 *Airflow Speed*

The higher the airflow speed, the more efficient the heat transfer by convection. As a result, the shell-air and board-air thermal resistance decreases. Therefore, junction-to-ambient thermal resistance is negatively correlated with airflow speed. Figure 6.50 shows how junction-to-case and junction-to-ambientthermal resistance for a 168-pin PGA package change with airflow speed. When the airflow speed reaches a certain point, the junction-to-air thermal resistance no longer decreases. Instead, it stabilizes. This suggests that increasing airflow speed further would not be useful. Figure 6.50 also indicates that airflow speed does not significantly affect the junction-to-case thermal resistance of the package.

Figure 6.51 shows the experimental setup and fitting curve for the relationship between convective heat transfer coefficient and airflow speed at a 28 °C indoor environment. As airflow speed increases, the convective heat transfer coefficient increases. This leads to a reduction in junction-to-ambient thermal resistance.

6.6.6 *Board Size and Thermal Conductivity*

A substrate with high thermal conductivity not only reduces the bulk thermal resistance but also further decreases the board-ambient thermal resistance. A larger PCB area lowers the convective-radiative thermal resistance, which also helps to reduce the board-ambient thermal resistance. Therefore, substrates with larger areas or higher thermal conductivity reduce the board-ambient thermal resistance, and in turn, decrease the junction-to-ambient thermal resistance. Figure 6.52 shows the junction-to-ambient thermal resistance for a 132-pin PQFP package with different PCB areas

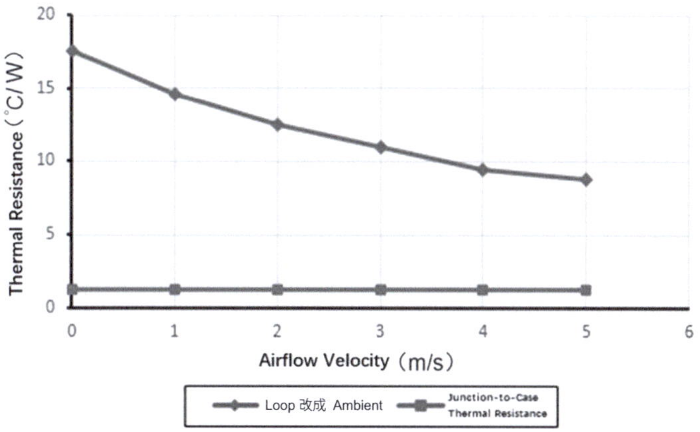

Fig. 6.50 Change in junction-to-ambient and junction-to-case thermal resistance with airflow speed for a 168-pin PGA package

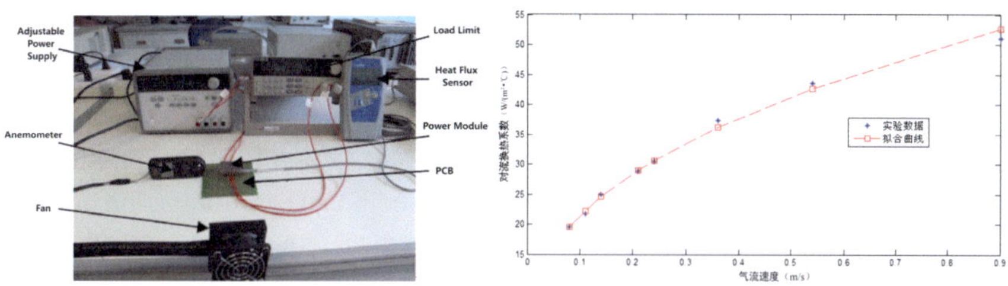

Fig. 6.51 Experimental setup and fitting curve for the relationship between convective heat transfer coefficient and airflow speed [40]

and thermal conductivities. The figure shows that when a 132-pin PQFP package is mounted on a PCB with high thermal conductivity, its junction-to-ambient thermal resistance is lower than when mounted on a PCB with low thermal conductivity. It also shows that as the PCB area increases, the junction-to-ambient thermal resistance decreases. However, experimental data show that there is a limit to the contribution of PCB area and package area in lowering thermal resistance. When the ratio of PCB area to package area exceeds 5, the junction-to-ambient thermal resistance does not decrease significantly with increasing PCB area. This is because, at positions far from the package, the temperature difference between the board and the ambient is small, so no significant convective heat transfer occurs. Therefore, the distant area does not aid in heat dissipation.

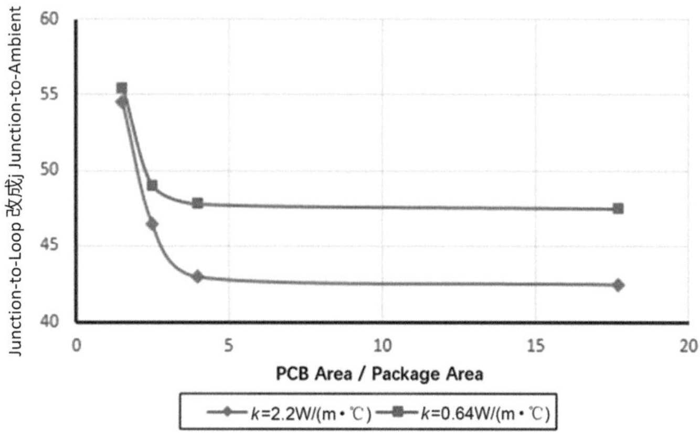

Fig. 6.52 Junction-to-ambient thermal resistance for a 132-pin PQFP package with varying PCB area and thermal conductivity [39]

6.7 Microchannel Thermal Characteristics and Thermal Management

Thermal conduction, forced air cooling, liquid cooling, and heat pipes are widely used in various packaging forms. However, these traditional heat dissipation methods are no longer sufficient to meet the thermal management requirements for chips with thermal flux densities over 100 W/cm^2 due to their low efficiency and low integration. This has created a need for new thermal management technologies. The Nanjing Electronic Technology Research Institute proposed a multi-layer microchannel design for high-power chip arrays. The design divides the cooling structure into the cooling liquid supply layer, cooling liquid return layer, and microchannel layer. By optimizing the cooling liquid path, it enhances heat transfer and ensures uniform flow in multi-chip systems [41]. Microchannels are an advanced cooling method designed for near-junction heat dissipation. Near-junction cooling offers several advantages over traditional methods, such as smaller cooling device mass, smaller size, lower thermal resistance along the heat flow path, and higher effective heat exchange rate per unit volume. These benefits support the miniaturization of high-power devices. In 2020, Remco Van Erp and other scholars from the Swiss Federal Institute of Technology in Lausanne introduced a microchannel etched manifold in a single silicon wafer. This design achieved a chip-level cooling capacity of 1723 W/cm^2, which is the highest reported chip-level cooling capability to date [42].

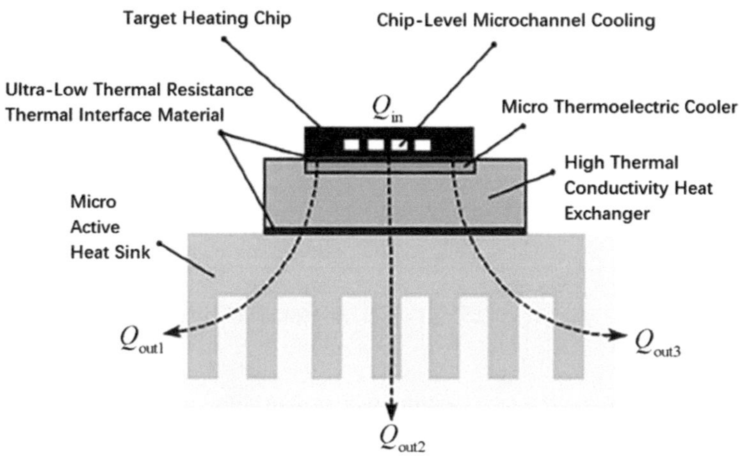

Fig. 6.53 "Thermal management technology" project

6.7.1 Microchannel Technology and Heat Transfer Efficiency

The main challenges of microchannel technology are manufacturing difficulties and the high pressure drop within the channels. In 2008, the United States launched the "Thermal Management Technology" project. This project divides thermal management into three levels and five technical approaches, as shown in Fig. 6.53 [43].

(1) Chip-level cooling technology: Chip-level microchannels, ultra-low thermal resistance thermal interface materials (CNT/GNP).
(2) Module-level cooling technology: Micro semiconductor coolers, high thermal conductivity heat exchangers (TGP).
(3) System-level cooling technology: Micro active heat sinks (heat sinks, fans).

Many experiments show that chip-level microchannels can achieve heat removal of 400 to 1000 W/cm^2 under reasonable flow channel pressure drops and pump power. The efficiency of different heat exchange methods is shown in Table 6.5.

6.7.2 Microchannel Thermal Management

Single-phase heat transfer can achieve higher heat removal rates. However, it has a higher pressure drop compared to two-phase heat transfer. This is because bubbles formed during boiling in single-phase heat transfer can block the flow channels. This creates significant resistance to the fluid flow. If the pump power is insufficient, the flow channels will clog. As a result, the chip's heat cannot be dissipated quickly. This

Table 6.5 Efficiency of different heat exchange methods

	Heat exchange method	Technology	Heat removal (W/cm^2)
Prasher et al. [44]	Single-phase	Nanostructured cooling rod	1250
Kosar et al. [45]	Two-phase	Nanostructured cooling rod	400
Weibel et al. [46]	Two-phase	Micrometric channel	500
Narayanan et al. [47]	Two-phase	Thin layer evaporation	600

leads to an increase in chip temperature, which may cause overheating and damage to the chip.

In microchannel thermal management, the flow channel shape, topology, material, coolant type, and coolant path can be optimized. The optimization has three main goals. First, to improve heat exchange efficiency. Second, to reduce the temperature gradient of hot spots and achieve uniform temperature distribution. Third, to balance the pressure drop inside the microchannel and increase its reliability. For example, Tang et al. analyzed the topology of microchannels. They optimized parameters such as nozzle diameter, length, and microchannel width, based on jet impingement cooling technology. This improved cooling efficiency [48]. T. Wei et al. optimized the cooling path between microchannels in chip arrays. They increased the number of inlets and outlets and reduced the series connection ratio between paths. This reduced the temperature gradient between chips and created a uniform flow [41]. Table 6.6 shows the design optimization of microchannels.

6.8 Thermal Analysis and Junction Temperature Prediction Case for Stacked Chip Packaging [62]

In 3D integrated circuits, chip size is decreasing, while power is increasing. As a result, the heat generated by stacked chips rises. The power dissipation per unit area increases sharply. Due to the small gap between stacked chips, sufficient cooling channels are not available for fluid flow. Thin chips in the stack are more likely to develop hotspots. The heat dissipation of chips in the stack may vary, and the hotspot temperature of inner-layer chips is difficult to detect. Therefore, thermal management becomes an important issue in 3D integrated packaging [63].

3D stacked chips resemble a sandwich structure. It is hard to directly measure the junction temperature of the inner chips. Usually, simulations are used for analysis. Alternatively, the junction temperature of the top-layer chips is measured. Based on thermal conduction theory, the junction temperature of the inner-layer chips is calculated. The theoretical methods for predicting junction temperature include the

Table 6.6 Optimization of microchannel design

Optimization direction	Category	Specific form	Diagram	Features
Heat dissipation	Channel shape	Pin-fin shape [49]	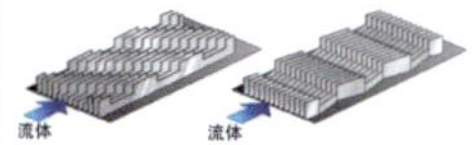	1. Generate secondary flows (such as vortices and turbulence) that superimpose on the main flow 2. Disrupt the development of the thermal boundary layer. This layer forms near the boundary due to heating or cooling and has a significant impact on thermal resistance
		Wavy shape [50]	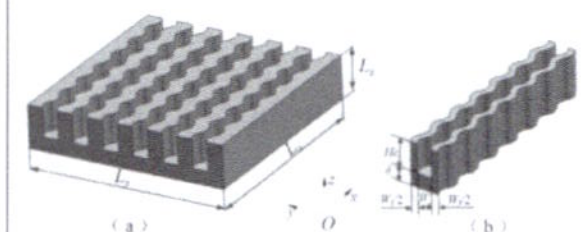	
		Serpentine shape [51]	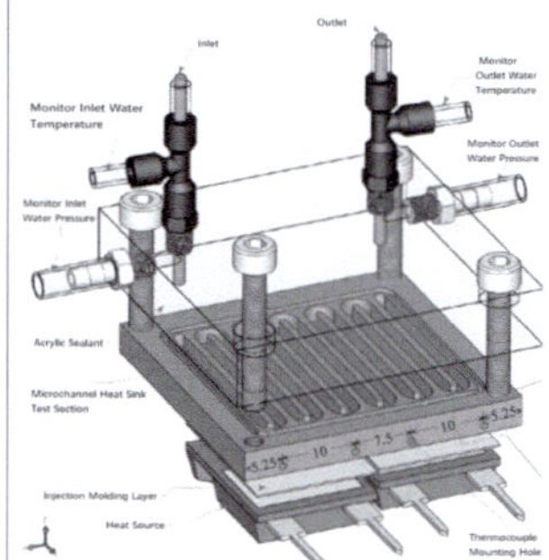	

(continued)

Table 6.6 (continued)

Optimization direction	Category	Specific form	Diagram	Features
Heat dissipation	Flow path	Copper/copper alloy [52] Aluminum/aluminum alloy [53] SiC [54] Diamond [55] Low−temperature sintered ceramic [56]	–	Increase the effective heat exchange per unit volume
	Flow path	Cyclic type [57]	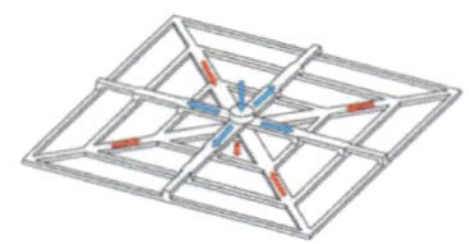	Improve the heat exchange efficiency per unit volume of the working fluid
High pressure drop issue	Topological structure	Gas–liquid two−phase heat transfer structure [58]	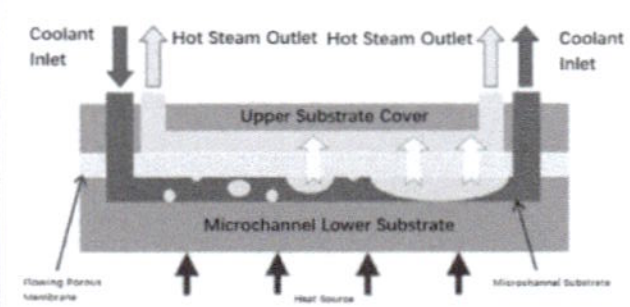	Reduce the non−uniform effects in the working fluid's boiling state. The presence of bubbles causes uneven fluid distribution Increase the number of inlets and outlets for the flow channel. This shortens the effective flow path
		Manifold−type topology structure [59]	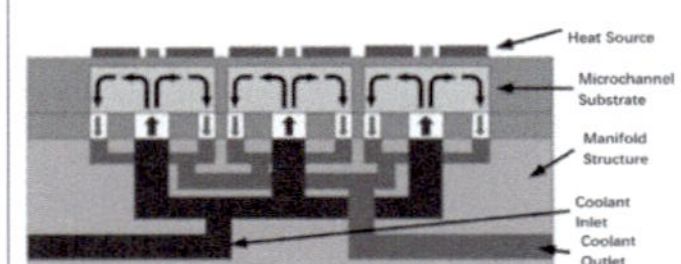	

(continued)

Table 6.6 (continued)

Optimization direction	Category	Specific form	Diagram	Features
	Cooling fluid	Gallium alloy (EGaInSn, EGaIn, GaSn, GaIn)	–	The low melting point and high boiling point characteristics address the non–uniform effects during boiling
	Flow path	Divergent type [60]	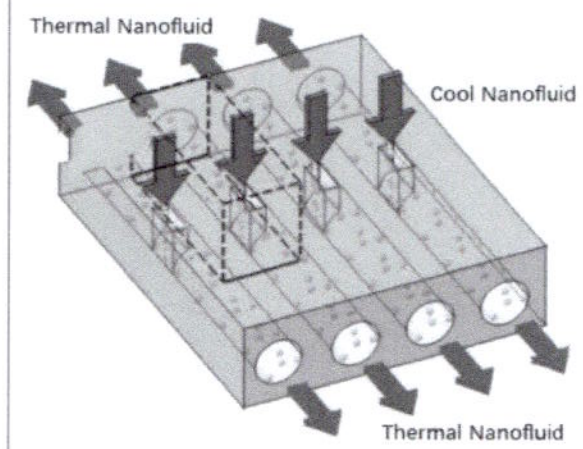	Shorten the effective flow path and allocate the cooling fluid more efficiently
		Convergent type [61]	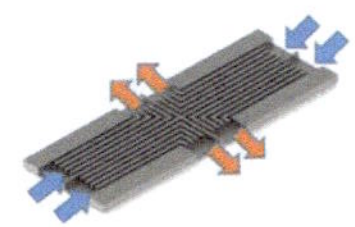	

thermal resistance matrix method based on the linear superposition principle and the thermal resistance network method based on the 45° thermal conduction rule. This section describes a case study for predicting and validating the junction temperature of a six-layer stacked chip.

6.8.1 *Thermal Testing of Stacked Chips and Test Board Design*

First, a thermal testing bare chip and test board were designed for the six-layer bonded stack. Each chip has an independent PN junction heat source, thermistor, and thermosensitive PN junction. The stacked package is created using bonding and adhesive methods. The junction temperature of each layer can be independently tested based on the thermistor and thermosensitive PN junction. Then, the junction temperature prediction for the inner layers of the stacked chip was carried out using the thermal resistance matrix method based on the linear superposition principle and the thermal resistance network method based on the 45° thermal conduction rule. The results were verified using test data.

The physical image of the thermal test chip is shown in Fig. 6.54. The central area is a comb-shaped PN junction heating zone, which provides the heat source for the bare chip. The power dissipation is calculated by multiplying the current flowing through the heating PN junction and the voltage across its ends. The thermistor's sensing area is embedded directly beneath the heating center. It can monitor the temperature at the center of the heating zone and the temperature of other surrounding heat source chips when they dissipate heat.

By measuring the resistance of the thermistor, the micro-area temperature field data can be obtained based on the relationship between resistance and temperature. The resistance measurement process uses the four-wire method. The test current should be small enough to avoid generating measurable Joule heat in the UBM. First, a stable current I_m is applied to both ends of the test line. Then, the voltage V_1 is measured. The resistance value $R_1(T_1)=V_1/I_m$ is calculated. Next, the current is reversed, and the voltage V_2 at both ends of the test line is measured. Then,

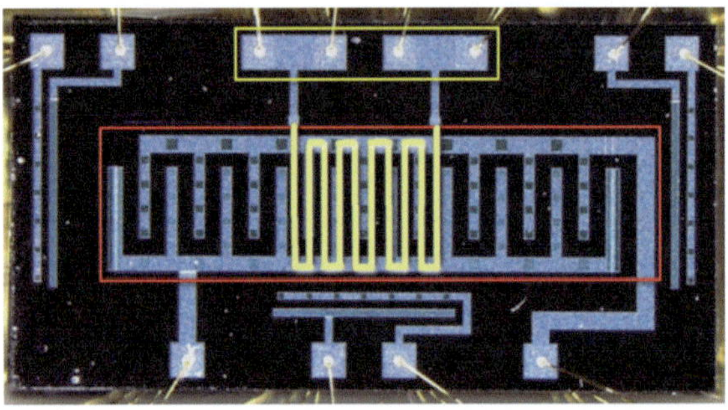

Fig. 6.54 Physical image of the thermal test chip

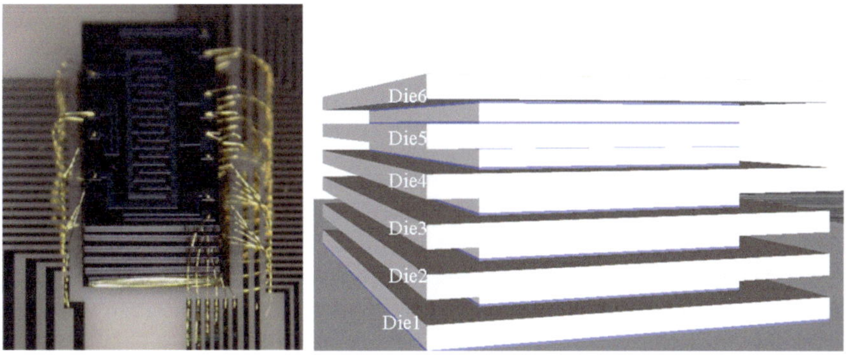

Fig. 6.55 Structural diagram of the stacked chip

the resistance value $R_2(T_2)$=V_2/I_m is calculated. The resistance value of the metal temperature sensor is obtained by averaging $R_1(T_1)$ and $R_2(T_2)$.

The relationship between resistance and temperature is as follows:

$$R(T_{\text{test}}) = R(T_{\text{ref}})\left[1 + \text{TCR}(T_{\text{ref}})'(T_{\text{test}} - T_{\text{ref}})\right] \tag{6.22}$$

In the formula, T_{test} represents the actual temperature of the thermistor test area, measured in °C; $R(T_{\text{test}})$ is the resistance at this temperature; $\text{TCR}(T_{\text{ref}})$ is the temperature coefficient, measured in °C^{-1}, a typical thermistor has a temperature coefficient of about 500 ppm/°C. It can be calibrated through actual measurements at known temperatures; $R(T_{\text{ref}})$ is the resistance at room temperature, measured in ohms Ω; T_{ref} refers to room temperature in °C. Based on Eq. (6.22), the junction temperature can be determined from the resistance measurements.

Figure 6.55 shows the structure of the stacked chip. The chip adopts a cantilever stacking method, forming a "chip-spacer-chip" structure. The spacer is made of silicon, the same material as the chip. An adhesive bonds the chips and spacers together. The chips are numbered sequentially from bottom to top as bare dies 1 (Die1) to bare dies 6 (Die6). There are no electrical connections between the bare dies in the thermal test chip. Each layer is connected individually to the pads on the substrate through wire bonding.

6.8.2 Temperature Measurement of Stacked Chips Based on Thermistors

The infrared method can only measure the surface chip temperature. It cannot detect the temperature inside the stacked chips. To solve this, thermistors in the thermal test chip are used to extract the temperature of each layer. First, a thermostatic chamber is set to temperatures between 30 and 100 °C at 10 °C intervals. A four-wire method is used to measure resistance. When the set temperature matches the actual temperature inside the chamber, and the thermistor values stabilize, the resistance of each layer is recorded. A fitted curve is then generated to show the relationship between resistance and temperature for each chip layer, as illustrated in Fig. 6.56.

To analyze temperature distribution, each bare chip received 2 W of power separately. The ambient temperature was 23 °C. The thermistor resistance of each layer was recorded. A fitted function was used to calculate temperature. The results are shown in Fig. 6.57.

When power is applied to a single chip, the top layers reach higher temperatures. The powered chip has the highest temperature. If the heat source is in the upper layers, the lower chips show little change in temperature. The chips below the heat

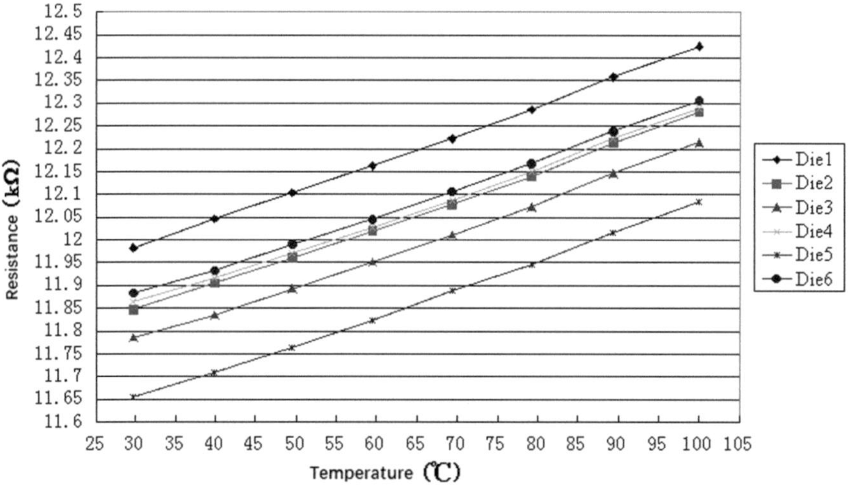

Fig. 6.56 Relationship between thermistor resistance and temperature

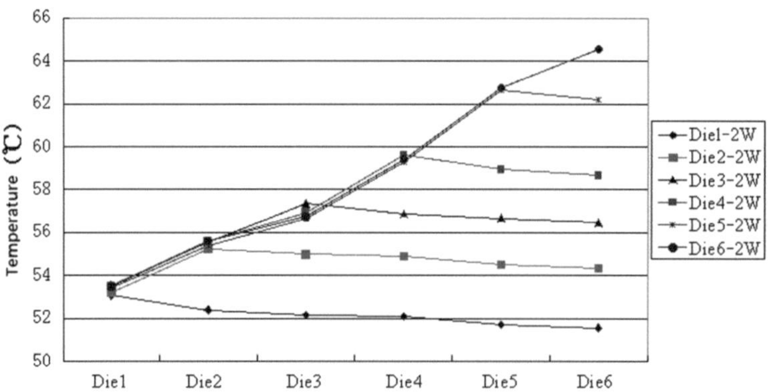

Fig. 6.57 Temperature with 2 W power applied to each chip

Table 6.7 Temperature of each layer when 0.33 W power is applied simultaneously

	Die1	Die2	Die3	Die4	Die5	Die6	Ambient temperature (°C)
Experimental results (°C)	53.94	56.45	58.10	59.18	60.94	61.67	23

source are cooler than those above. This suggests that heat mainly flows downward toward the substrate. For example, when Die4 receives power, it reaches the highest temperature. Both upper and lower layers gradually decrease in temperature. The temperatures of Die3, Die2, and Die1 are lower than those of Die5 and Die6. This means Die4 is thermally closer to the upper layers.

According to Fourier's law, heat transfer from high to low temperatures. When power is applied to a single chip, it has the highest temperature. Heat escapes through two main paths. One moves upward and dissipates into the air. The other moves downward, passing through the substrate and PCB before reaching the air. The lower chips are cooler than the upper ones. This happens because downward heat transfer faces less thermal resistance. Most heat follows the path with lower resistance.

With a total power of 2 W, each layer received 0.33 W equally. The temperature distribution is shown in Table 6.7.

As shown in Table 6.7, when 2 W of power is evenly distributed across all chip layers, the temperature decreases gradually from the top to the bottom. The top chip reaches 61.67 °C, which is lower than when it receives 2 W alone. The bottom chip has the lowest temperature at 53.94 °C, which is higher than when it receives 2 W alone. This shows that the bottom chip dissipates heat more efficiently than the top chip. It also suggests that high-power chips should be placed at the bottom. This improves heat dissipation in stacked chips.

6.8.3 Thermal Analysis of Stacked Chips Based on Finite Element Simulation

FloTHERM software was used for steady-state thermal simulation. The model included a six-layer stacked chip and a static air test chamber. The test environment followed the JESD51-2 standard. The chip was placed on a test board inside a no-airflow static air chamber. The chamber temperature was set to 23 °C. The natural convective heat transfer coefficient was 6 $W/(m^2{\cdot}K)$. The simulation model under natural convection is shown in Fig. 6.58.

To analyze the heat flow path in the stacked chip, the internal thermal distribution was studied when only Die6 received 2 W of power. Figure 6.59 shows the thermal distribution when Die6 was powered. To verify the accuracy of the thermal simulation model, both infrared measurement and thermistor measurement were used. The infrared method provides high accuracy on a 70 °C constant temperature platform. For consistency, 70 °C was set as the reference ambient temperature. During the

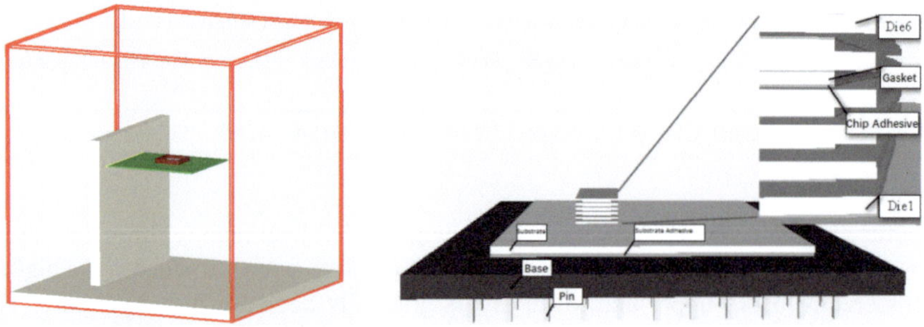

Fig. 6.58 Thermal simulation model under natural convection

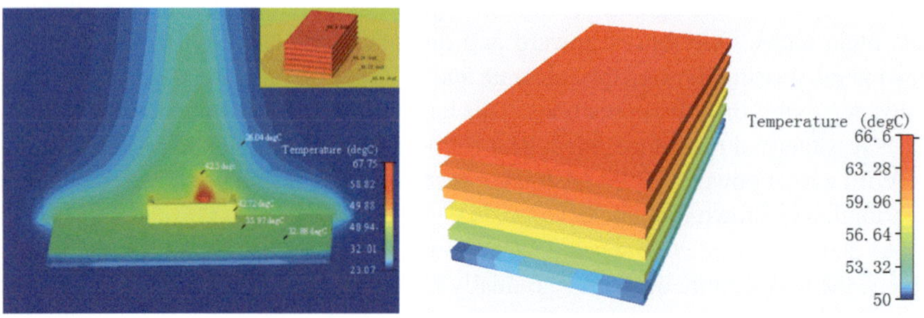

Fig. 6.59 Thermal distribution when Die6 receives 2 W of power

infrared thermal imaging test, thermistor readings were also recorded. This allowed a comparative analysis of internal temperature results.

Figure 6.60 shows the infrared thermal image of Die6 under 2 W of power. The thermistor was embedded beneath the heat source, while the simulation temperature measurement points were placed at the center of each chip layer. Table 6.8 presents the temperature distribution of each layer when Die6 alone received 2 W of power. As shown in Table 6.8, the infrared thermal test recorded an average temperature of 94.89 °C for the top chip. The highest temperature, 96.80 °C, appeared near the center. The simulated temperature for the top chip was 93.23 °C, which closely matched the infrared result. The thermistor measurement recorded 95.74 °C, differing by only 0.85 °C from the infrared test's heat source area. The maximum difference between the simulation and thermistor results across all layers was about 2 °C. These results confirm the accuracy of the simulation model in this study.

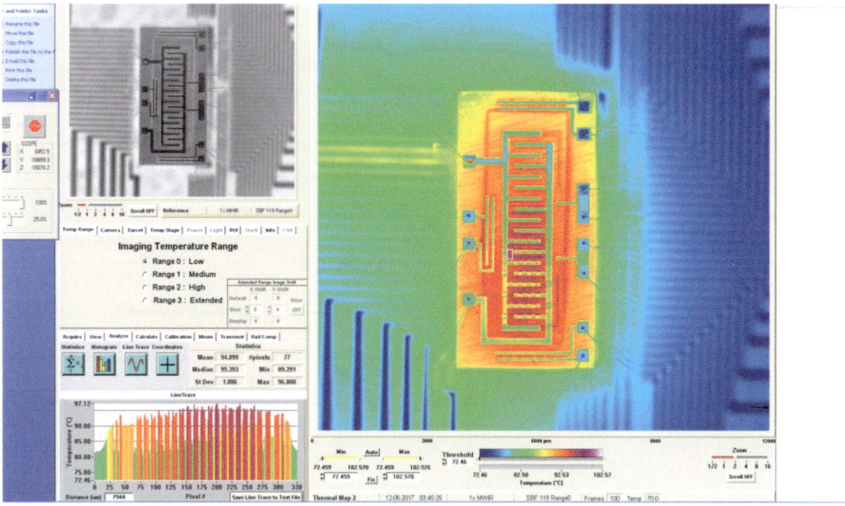

Fig. 6.60 Infrared thermal image of Die6 surface under 2 W power

Table 6.8 Temperature of each layer when only Die6 receives 2 W of power

	Die1	Die2	Die3	Die4	Die5	Die6	Infrared measurement of Die6 junction	
Simulation results (°C)	82.65	84.55	86.76	89.04	91.27	93.23	Average value	Maximum value
Thermistor measurement results (°C)	83.66	85.95	87.38	90.30	93.91	95.74	94.89	96.80

6.8.4 Temperature Prediction Model and Validation for Stacked Chips

For single heat source chip, many thermal resistance network models have been developed. However, there are few studies on thermal resistance networks for stacked chips with multiple heat sources. This section builds junction temperature prediction models for stacked chips based on heat flow paths. Two methods are used: thermal resistance matrix and thermal resistance network.

1. Chip Temperature Prediction Based on Thermal Resistance Matrix

   (1) Thermal Resistance Matrix Theory

The thermal resistance matrix is a mathematical model based on the linear superposition principle of heat coupling. It describes how each heat source in a multi-source chip affects other chips when generating heat independently. Figure 6.61 shows the linear heat coupling superposition principle (the horizontal axis represents position,

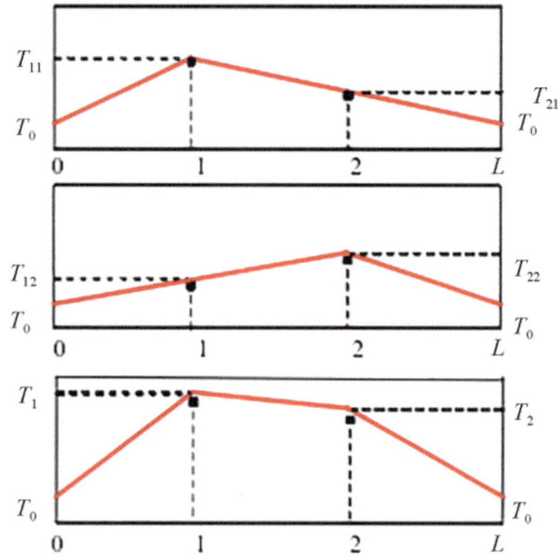

Fig. 6.61 Illustration of the linear heat coupling superposition principle

and the vertical axis represents temperature). A metal rod with internal heat sources is considered. If only its two ends conduct heat while all other surfaces are perfectly insulated, the temperature decreases linearly from the heat source toward both ends.

When only Position 1 is heated, the temperatures at Position 1 and Position 2 are T11 and T21, with T0 as the ambient temperature; when only Position 2 is heated, the temperatures at Position 1 and Position 2 are T_{12} and T_{22}. When both Position 1 and Position 2 are heated together, the total temperature rise at Position 1 equals the sum of the rise caused by Position 1 alone and the rise caused by Position 2 alone at Position 1. The same superposition rule applies to Position 2. This relationship can be written as:

$$\begin{aligned} T_1 - T_0 &= (T_{11} - T_0) + (T_{12} - T_0) \\ T_2 - T_0 &= (T_{21} - T_0) + (T_{22} - T_0) \end{aligned} \tag{6.23}$$

The linear superposition principle above is represented as a thermal resistance matrix:

$$\begin{bmatrix} T_1 - T_0 \\ T_2 - T_0 \end{bmatrix} = \begin{bmatrix} \theta_{11} \ \theta_{12} \\ \theta_{21} \ \theta_{22} \end{bmatrix} \cdot \begin{bmatrix} P_1 \\ P_2 \end{bmatrix} \tag{6.24}$$

Here, θ_{ij} represents the thermal resistance at position i when power is applied only at position j. This can be expressed by Eq. (6.25).

$$\theta_{ij} = \frac{(T_{ij} - T_0)}{P_j} \quad i, j = 1, 2 \tag{6.25}$$

The number of heat sources determines the size of the thermal resistance matrix. N heat sources are represented by an $N \times N$ matrix. When any bare chip in a stacked chip generates heat, it affects other layers through coupled heating. The temperature rise of each bare chip depends not only on its own power consumption but also on the heat contribution from other heated bare dies. Therefore, when calculating the thermal resistance of a specific bare chip in a stacked chip, the coupling effects from other bare dies must be considered. According to the linear superposition principle, for a stacked chip package with six layers, a 6 × 6 thermal resistance matrix can be used to describe the heat dissipation characteristics of the package. Equation (6.26) provides a method to calculate the temperature of each bare chip layer using the environmental temperature and the corresponding thermal resistance matrix.

$$[T_i] = \begin{bmatrix} T_1 \\ T_2 \\ T_3 \\ T_4 \\ T_5 \\ T_6 \end{bmatrix} = \left[\theta_{ij}\right] \cdot [P_k] + [T_A] = \begin{bmatrix} \theta_{11} & \theta_{12} & \theta_{13} & \theta_{14} & \theta_{15} & \theta_{16} \\ \theta_{21} & \theta_{22} & \theta_{23} & \theta_{24} & \theta_{25} & \theta_{26} \\ \theta_{31} & \theta_{32} & \theta_{33} & \theta_{34} & \theta_{35} & \theta_{36} \\ \theta_{41} & \theta_{42} & \theta_{43} & \theta_{44} & \theta_{45} & \theta_{46} \\ \theta_{51} & \theta_{52} & \theta_{53} & \theta_{54} & \theta_{55} & \theta_{56} \\ \theta_{61} & \theta_{62} & \theta_{63} & \theta_{64} & \theta_{65} & \theta_{66} \end{bmatrix} \cdot \begin{bmatrix} P_1 \\ P_2 \\ P_3 \\ P_4 \\ P_5 \\ P_6 \end{bmatrix} + [T_A] \tag{6.26}$$

In the equation, T_i represents the temperature of the i layer bare chip, and P_k represents the power of the k layer bare chip. In the thermal resistance matrix, θ_{11} refers to the temperature difference between the bare chip and the environment when only the first layer bare chip dissipates unit power. This is the chip's self-thermal resistance; $\theta_{\text{i}1}$ denotes the temperature difference between the i layer bare chip and the environment due to the coupled heating effect from the first layer bare chip when it dissipates unit power. This is the coupling thermal resistance from the first layer chip to the i layer chip. Equation (6.26) can be used to calculate the temperature of any layer in a six-layer stacked chip structure, as shown in Eq. (6.27).

$$T_i = \theta_{i1}P_1 + \theta_{i2}P_2 + \theta_{i3}P_3 + \theta_{i4}P_4 + \theta_{i5}P_5 + \theta_{i6}P_6 + T_A \tag{6.27}$$

$$\theta_{ik} = \frac{T_i - T_{Ak}}{P_k} \tag{6.28}$$

Any element θ_{ik} in the thermal resistance matrix can be obtained using Eq. (6.28). In Eq. (6.28), P_k represents the power applied only to the k layer chip. T_i and T_{Ak} are the temperatures of the k layer chip and the ambient temperature when only the i layer chip consumes power. The thermal resistance matrix shows the relationship between self-heating and coupling heating.

(2) Verification of the Thermal Resistance Matrix

Table 6.9 shows the temperatures of each bare chip when 1.2 W power is applied individually. Using the thermal resistance matrix method and Eq. (6.28), the self-thermal resistance and coupling thermal resistance can be calculated. The same method applies when 0.8 and 1.6 W power are used, forming thermal resistance matrices $[\theta]_{0.8}$ and $[\theta]_{1.6}$.

The values in the thermal resistance matrix are not constant. As power increases, the thermal resistance from the chip to the environment decreases. The average values of elements in matrices $[\theta]_{0.8}$, $[\theta]_{1.2}$ and $[\theta]_{1.6}$ form the final matrix $[\theta]$ below:

$$[\theta] = \begin{bmatrix} 16.62 & 16.26 & 16.34 & 16.21 & 16.42 & 16.38 \\ 16.22 & 17.35 & 17.61 & 17.17 & 17.16 & 16.91 \\ 16.08 & 17.24 & 18.68 & 17.72 & 18.31 & 18.19 \\ 16.05 & 17.30 & 18.59 & 20.04 & 19.68 & 19.28 \\ 15.30 & 16.55 & 17.71 & 18.45 & 21.46 & 21.34 \\ 15.56 & 16.49 & 17.76 & 18.65 & 20.78 & 22.17 \end{bmatrix}$$

Table 6.9 Temperatures of each bare chip when 1.2 W power is applied

	Die1-1.2 W	Die2-1.2 W	Die3-1.2 W	Die4-2 W	Die5-1.2 W	Die6-1.2 W
Die1 temperature (°C)	43.12	42.04	42.85	42.25	42.73	42.74
Die2 temperature (°C)	42.90	43.63	43.63	43.66	43.45	43.51
Die3 temperature (°C)	42.68	42.80	44.91	43.74	44.96	44.61
Die4 temperature (°C)	42.54	42.84	45.00	47.10	46.75	46.46
Die5 temperature (°C)	42.04	42.18	44.36	44.65	49.12	48.49
Die6 temperature (°C)	41.95	42.01	44.01	44.86	47.40	49.46
Ambient temperature (°C)	22.6	22.4	22.5	22.8	22.7	22.5

Given the ambient temperature, Eq. (6.27) and the thermal resistance matrix $[\theta]$ are used to predict chip temperature. The total power is 1.2 W, with each bare chip consuming 0.2 W. The results are then compared for verification. Table 6.10 shows the comparison between thermistor test results and thermal resistance matrix calculations.

From Table 6.10, the maximum error between the two methods is − 7.60%, with a temperature difference of about 4 °C. This shows that the thermal resistance matrix provides a reasonable temperature prediction for stacked chips. The error may be due to the power assumption in the thermal resistance matrix. The actual applied power could be lower. Since higher power reduces thermal resistance, the calculated values may be lower than the test results.

2. Chip Temperature Prediction Based on Thermal Resistance Network Topology Model

   (1) Stacked Chip Thermal Resistance Network Topology Model

When heat transfers from a narrow chip to a wide substrate, it diffuses at a 45° angle. The thermal resistance 45° calculation method can be used to solve this. Figure 6.62 shows this method. If A is the length of the heat-generating chip, then D = A + X is the average length after 45° diffusion. C = $A + 2X$ is the maximum length after diffusion. The heat dissipation area is based on this maximum length.

Table 6.10 Comparison of thermistor test results and thermal resistance matrix calculations

	Die1	Die2	Die3	Die4	Die5	Die6	Ambient temperature (°C)
Thermal resistance matrix calculation (°C)	44.65	45.48	46.24	47.19	46.15	47.28	25
Thermistor test results (°C)	46.68	47.94	49.07	50.51	50.98	51.17	
Error (%)	− 4.36	− 5.12	− 5.76	− 6.58	− 7.50	− 7.60	

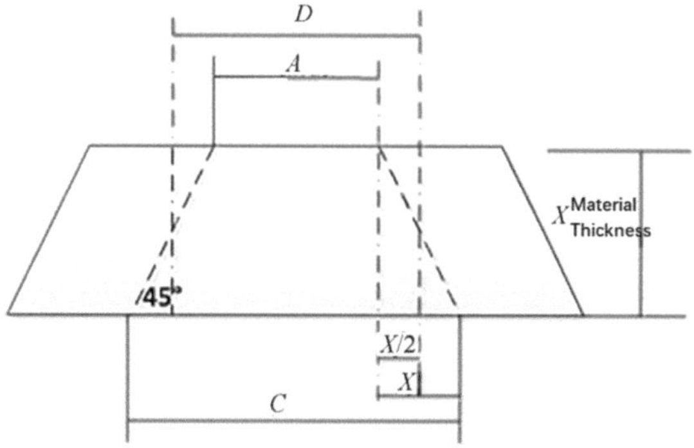

Fig. 6.62 Thermal resistance 45° calculation method

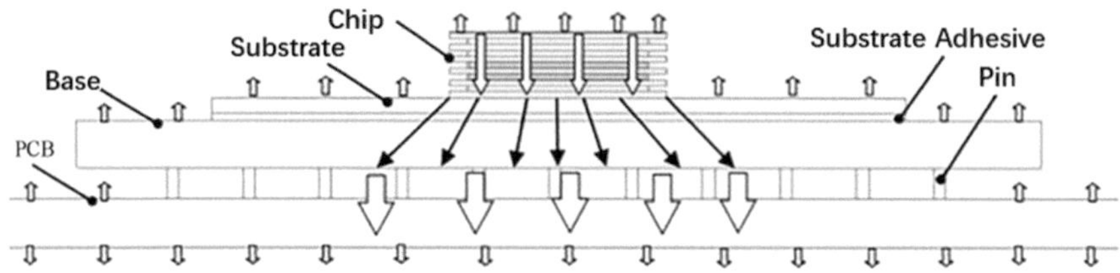

Fig. 6.63 Heat dissipation path distribution of stacked chips

Simulation results show that most of the chip's heat flows through the chip adhesive and substrate, then passes through the pins, and finally dissipates across the PCB. Stacked chips mainly release heat through thermal conduction and heat convection. Figure 6.63 illustrates the heat dissipation paths.

The thermal resistance for Die6 transferring heat to the air through convection is \theta_{air-Die6}$R_{\text{air-Die6}}$. The thermal resistance of the sixth-layer chip is R6. Other layers follow the same naming rule. Each chip layer consists of two adhesives (e.g., adh61, adh62), a silicon spacer (e.g., spacer6), and a bare chip (e.g., Die6). The total thermal resistance of each layer includes these four parts. Based on the heat transfer paths and internal heat sources, a thermal resistance network model is proposed for single power dissipation, as shown in Fig. 6.64. (To simplify the diagram, points A and B are connected, and the ground symbols are also linked.)

To simplify notation, define:

$$\theta_6 = \theta_{\text{Die6}} + \theta_{\text{adh62}} + \theta_{\text{spacer6}} + \theta_{\text{adh61}} \tag{6.29}$$

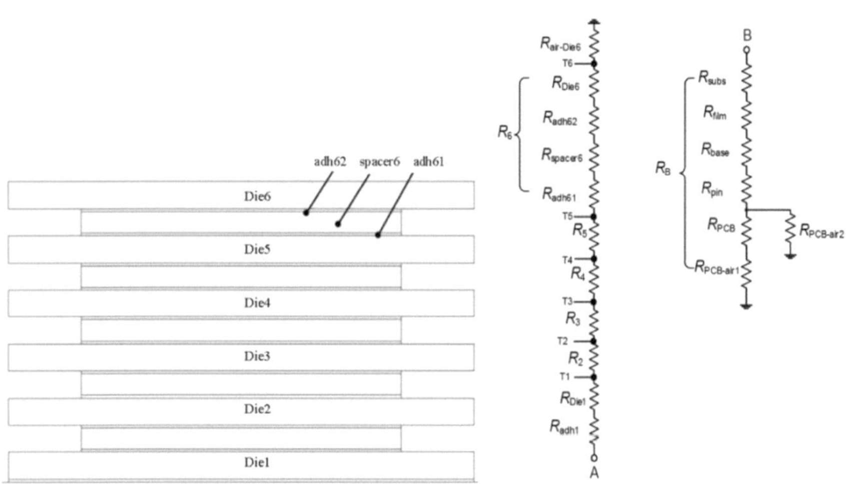

Fig. 6.64 Thermal resistance network model for single power dissipation

$$\theta_{\text{B}} = \theta_{\text{subs}} + \theta_{\text{film}} + \theta_{\text{base}} + \theta_{\text{pin}} + (\theta_{\text{PCB}} + \theta_{\text{PCB-air1}})//\theta_{\text{PCB-air2}} \tag{6.30}$$

In the model shown in Fig. 6.64, thermal resistances \theta_{2}, R_2, θ_3, θ_4 and θ_5 have similar meanings. They represent the thermal resistance of chips from layer 2 to layer 5, respectively. A bare chip has two main heat transfer paths. The thermal resistances of each component along the paths are connected in series. The heat flowing to the PCB is partly dissipated through the top surface to the air and partly through the bottom surface to the air. In the network model, these two heat transfer paths at the PCB exist in parallel, as shown in Eq. (6.31). The top-layer chip (Die6) dissipates power through two paths, which transfer heat simultaneously. In the thermal resistance network model, these two thermal resistances are connected in parallel, representing the thermal resistance from Die6 (measured at T6) to the air.

$$\theta_{\text{T6-air}} = \theta_{\text{air-Die6}}//(\theta_6 + \theta_5 + \theta_4 + \theta_2 + \theta_{\text{Die1}} + \theta_{\text{adh1}} + \theta_{\text{B}}) \tag{6.31}$$

Similarly, when power is applied individually to Die1 to Die5, the thermal resistance from each layer to the air can be determined using the thermal resistance network topology in Fig. 6.64.

In the stacked chip thermal resistance network, heat flow between chips is sequential. Most heat transfers downward. The heat passing through a specific chip equals the sum of the heat dissipated by that chip and all chips above it. The temperature of each chip is the sum of its temperature difference from the bottom chip and the bottom chip's temperature relative to the ambient environment. The mathematical expression is as follows:

$$T_i = T_1 + (T_2 - T_1) + \cdots + (T_i - T_{i-1}) \tag{6.32}$$

$$T_i = \left(\sum_{k=1}^{6} P_k\right) \cdot \theta_{\text{T1-air}} + \left(\sum_{k=2}^{6} P_k\right) \cdot \theta_{\text{T2-air}} + \cdots + \left(\sum_{k=i}^{6} P_k\right) \cdot \theta_{\text{T}i\text{-air}} + T_{\text{air}} \tag{6.33}$$

(2) Validation of the Thermal Resistance Network Model

The thermal resistance network model is a topological network based on the positions of heat sources within the chip. It calculates the temperature at the heat source region using the thermal resistance of each chip component and their network relationships.

The one-dimensional thermal conduction resistance for each chip layer is determined using Eq. (6.34). The thermal conduction resistance of the substrate, substrate adhesive, and base follows the 45° heat flow diffusion theory. The convective heat transfer resistance of the top and bottom surfaces of the PCB is calculated using Eq. (6.35). The detailed calculation process for some components is shown below.

$$\theta_{\text{cond}} = \frac{T_1 - T_2}{Q_{\text{cond}}} = \frac{L}{\lambda A} \tag{6.34}$$

$$\theta_{\text{conv}} = \frac{1}{hA} = \frac{T_{\text{s}} - T_{\text{f}}}{Q_{\text{conv}}} \tag{6.35}$$

$$\theta_{\text{Die6}} = \frac{0.19 \times 10^{-3}}{7.22 \times 10^{-3} \times 3.65 \times 10^{-3} \times 145} = 0.0497\,^{\circ}\text{C/W}$$

$$\theta_{\text{LTCC}} = \frac{0.5 \times 10^{-3}}{7.72 \times 10^{-3} \times 4.15 \times 10^{-3} \times 20} = 0.7803\,^{\circ}\text{C/W}$$

$$\theta_{\text{PCB-air1}} = \frac{1}{114.3 \times 10^{-3} \times 101.6 \times 10^{-3} \times 6} = 14.3519\,^{\circ}\text{C/W}$$

Using Eqs. (6.34) and (6.35), along with the simulated dimensions and thermal conductivity of each component, the thermal resistances in the thermal resistance network model are obtained. By applying the corresponding data into the thermal resistance network relationships in Fig. 6.64 and Eq. (6.31), the thermal resistance from each heat source measurement point to the environment can be calculated when power is applied individually to each bare chip. Based on these results, the temperature of each bare chip under 2 W power dissipation can be determined. The comparison between the calculated chip temperatures and the experimental results is shown in Table 6.11.

From Table 6.11, when power is applied individually to each layer of the stacked chip, the maximum difference between the calculated results based on the thermal resistance network model and the experimental results is about 2 °C. The relative error is less than 4%. To further verify the model's accuracy under multiple heat sources, the temperature is calculated using Eqs. (6.32) and (6.33). Each bare chip layer is subjected to 0.33 W power, with a total power of about 2 W. Table 6.12 shows the comparison of calculated and experimental results.

From Table 6.12, when each chip layer receives 0.33 W power simultaneously, the maximum temperature difference between calculated and experimental results is about 2 °C. The relative error remains below 4%.

These cases show that for both single-chip and multi-chip power dissipation, the calculated results based on the thermal resistance network model stay within a 5%

Table 6.11 Comparison of calculated and experimental temperatures for bare dies

Power applied (W)	Measurement point	Calculated result (°C)	Experimental result (°C)	Relative error (%)
2	T6	62.79	64.56	– 2.75
2	T5	60.57	62.52	– 3.33
2	T4	58.36	59.62	– 2.11
2	T3	56.15	57.35	– 2.09
2	T2	53.94	55.22	– 2.32
2	T1	51.72	53.06	– 2.52

Table 6.12 Comparison of calculated and experimental results

Total power (W)	Measurement point	Calculated result (°C)	Experimental result (°C)	Relative error (%)
2	T6	59.22	61.07	– 3.97
	T5	58.85	60.64	– 3.43
	T4	58.11	59.18	– 1.81
	T3	57.01	58.10	– 1.88
	T2	55.54	56.45	– 1.61
	T1	53.69	53.94	– 0.46

error range. This confirms that the model is valid and reliable for stacked chip temperature prediction. The calculated results are slightly lower than the experimental data. This difference may come from size variations in chip manufacturing and contact thermal resistance between components. These factors were not considered in the thermal resistance model when calculating the heat flow path resistance.

References

1. Zhou LZ (2006) Microelectronic device packaging: packaging materials and packaging technology (in Chinese). Chemical Industry Press, Beijing, China
2. Hannemann RJ (2003) Thermal control of electronics: perspectives and prospects
3. Fu ZW (2017) Reliability study of copper pillar bump interconnects in flip-chip under thermoelectric stress. South China University of Technology, Guangdong, China
4. Li Y (2010) Numerical study of atomic migration of cylindrical copper bumps in coupled fields in integrated circuit packaging. University of Electronic Science and Technology of China, Sichuan, China
5. Frear DR, Burchett SN, Morgan HS (1994) The mechanics of solder alloy interconnects. van Nostrand Reinold, New York
6. Shabany Y (2013) Heat transfer: thermal management of power electronic devices (translated by XL Yu, WF Wu, and FL Liu). China Machine Press, Beijing, China
7. Pecht MG, Dishongh TJ, Mahajan R et al (1998) Electronic packaging materials and their properties. Plastics 31(2):203–209
8. Li B (2014) Fundamentals and application examples of FloTHERM software. China Water & Power Press, Beijing, China
9. Shabany Y (2014) Casual discussion on thermal design (translated by B Li). China Machine Press, Beijing, China
10. Su D, Zhao D, Zhang L et al (2020) Reliability assessment of flip chip interconnect electronic packaging under thermal shocks. In: 2020 21st International conference on electronic packaging technology (ICEPT). IEEE, pp 1–4
11. Lou WZ, Sun YQ (2007) Fundamentals of MEMS Integration and Packaging Technology. China Machine Press, Beijing, China
12. Ren H, Liu WJ, Huang J et al (2015) Research on chassis thermal design based on icepak. Electron Sci Technol 2(6):639–644
13. NENG L, SONG T, SHAO G et al (2021) An Accurate Simulation Method of Package Warpage Experimental Results Based on FEM[C]. 2021 22nd International Conference on Electronic Packaging Technology (ICEPT). IEEE, pp 1-4.

14. Guo D (2008) Thermo-mechanical reliability analysis and parameter optimization of stacked CSP. Guilin University of Electronic Technology, Guangxi, China
15. Vitello D, Albertinetti A, Rovitto M (2019) Die thickness optimization for preventing electro-thermal fails induced by solder voids in power devices. In: 2019 IEEE 69th electronic components and technology conference (ECTC). IEEE, pp 2091–2096
16. Peng Y, Gao W, Guo Q et al (2020) Thermal stress failure analysis of power diode SMBF package. In: 2020 21st International conference on electronic packaging technology (ICEPT). IEEE, pp 1–3
17. Tang C, Zhu W, Chen Z (2021) Effect of thermomigration on evolution of interfacial intermetallic compounds in Cu/Ni/Sn-3.5 Ag microsolder joints for 3D interconnection. In: 2021 22nd International conference on electronic packaging technology (ICEPT). IEEE, pp 1–6
18. EIA/JESD51 (1995) Methodology for the thermal measurement of component packages. JEDEC Standard, USA
19. GB/T 2900.32-94 (1994) Electrotechnical terminology—power semiconductor devices. National Standard of the People's Republic of China
20. EIA/JESD51-1 (1995) Integrated circuits thermal measurement method-electrical test method. JEDEC Standard, USA
21. JESD51-1 (1995) Integrated circuits thermal measurement method—electrical test method (single semiconductor device). JEDEC Standard, USA
22. JESD51-8 (1999) Integrated circuit thermal test method environmental conditions-junction-to-board. JEDEC Standard, USA
23. Liu Y, Liang LH, Qu JM (2010) Modeling and simulation of microelectronic devices and packaging. Science Press, Beijing, China
24. JESD51-14 (2010) Transient dual interface test method for the measurement of the thermal resistance junction to case of semiconductor devices with heat flow trough a single path. JEDEC Standard, USA
25. Zhang P, Xuan YM, Li Q (2012) Research progress on interfacial thermal contact resistance. J Chem Ind Eng 63(2):335–349
26. Li RG, Zhou B, Xu W et al (2016) Analysis of internal thermal resistance of high-power rectifier packaging based on structure function. Electron Compon Mater 35(1):40–43
27. JEDEC Solid State Technology Association (1999) JESD51. JEDEC Standard, USA
28. Tatsumi K, Morisako I, Wada K et al (2019) High temperature resistant packaging technology for SiC power module by using Ni micro-plating bonding. In: 2019 IEEE 69th electronic components and technology conference (ECTC). IEEE, pp 1451–1456
29. Landesman JP, Floriot D, Martin E et al (2000) Temperature distributions in III-V microwave power transistors using spatially resolved photoluminescence mapping. In: Proceedings of the 3rd IEEE Caracas conferences on devices, circuits and systems, Cancun, Mexico. IEEE, New York, pp D1114/1-D1114/8
30. Kuball M, Hayes JM, Uren MJ et al (2002) Measurement of temperature in active high-power AlGaN/GaN HFETs using Raman spectroscopy. IEEE Electron Device Lett 23(1):7–9
31. Farzaneh M, Maize K, Luerben D et al (2009) CCD-based thermoreflectance microscopy: principles and applications. J Phys D Appl Phys 42:143001
32. Maize K, Ziabari A, French WD et al (2014) Thermoreflectance CCD imaging of self-heating in power MOSFET arrays. IEEE Trans Electron Devices 61(9):3047–3053
33. Maize K, Heller E, Dorsey D et al (2012) Thermoreflectance CCD imaging of self heating in AlGaN/GaN high electron mobility power transistors at high drain voltage. In: 28th IEEE SEMI-THERM symposium. IEEE, San Jose, USA. New York, pp 173–181
34. Li RG, Liao XY, Yao B et al (2017) Progress in thermal testing technology and applications for GaN-based HEMTs devices. Electron Compon Mater 36(9):1–9
35. Wang N, Liu Y, Chen S et al (2019) Highly thermal conductive and electrically insulated graphene based thermal interface material with long term reliability. In: 2019 IEEE 69th electronic components and technology conference (ECTC). IEEE, pp 1564–1568
36. Sarua A, Ji H, Kuball M et al (2006) Integrated micro-Raman/infrared thermography probe for monitoring of self-heating in AlGaN/GaN transistor structures. IEEE Trans Electron Devices 53(10):2438–2447

37. Kuball M, Pomeroy JW (2016) A review of Raman thermography for electronic and opto-electronic device measurement with sub-micron spatial and nanosecond temporal resolution. IEEE Trans Electron Mater Reliab 16(4):1–19
38. Tadjer M, Raad P, Komarov P et al (2018) Electrothermal evaluation of AlGaN/GaN Membrane high electron mobility transistorsby transient thermoreflectance. J Electron Devices Soc 6:922–930
39. Tong XC (2016) Advanced materials for thermal management of electronic packaging (translated by B An, WW Lü, YP Wu). National Defense Industry Press, Beijing, China
40. Liu GG (2012) Thermal simulation and reliability study of DC/DC power module based on convective heat transfer coefficient. Guilin University of Electronic Technology, Guangxi, China
41. Wei T, Huang H, Ma Y et al (2021) Design and fabrication of multi-layer silicone microchannel cooler for high-power chip array. In: 2021 22nd International conference on electronic packaging technology (ICEPT)
42. van Erp R, Soleimanzadeh R, Nela L et al (2020) Co-designing electronics with microfluidics for more sustainable cooling. Nature 585:211–216
43. Yu HQ, Tang GQ, Gui JL et al (2018) New developments in thermal management technology for microsystems. Piezoelectrics Acoustooptics 40(6):931–935
44. Prasher RS, Chang JY, Sauciuc I et al (2005) Nano and micro technology-based next-generation package-level cooling solutions. Intel Technol J 9(4):285–296
45. Kosar A, Kuo CJ, Peles Y (2005) Boiling heat transfer in rectangular microchannels with reentrant cavities. Int J Heat Mass Transf 48(23/24):4867–4886
46. Weibel JA, Garimella SV, Murthy JY et al (2011) Design of integrated nanostructured wicks for high-performance vapor chambers. IEEE Trans Compon Packag Manuf Technol 1(6):859–867
47. Narayanan S, Fedorov AG, Joshi YK (2010) Experimental characterization of a micro-scale thin film evaporative cooling device. In: IEEE intersociety conference on thermal and thermomechanical phenomena in electronic systems. IEEE, pp 1–10
48. Tang G, Wai LC, Boon Lim S et al (2020) Thermal analysis, characterization and and material selection for SiC device based intelligent power module (IPM). In: 2020 IEEE 70th electronic components and technology conference (ECTC), pp 2078–2085
49. Peles Y, Kosar A, Mishra C et al (2005) Forced convective heat transfer across a pin fin micro heat sink. Int J Heat Mass Transf 48(17):3615–3627
50. Lin L, Zhao J, Lu G et al (2017) Heat transfer enhancement in microchannel heat sink by wavy channel with changing wavelength/amplitude. Int J Therm Sci 118:423–434
51. Al-Neama AF, Kapur N, Summers J et al (2018) Thermal management of GaN HEMT devices using serpentine minichannel heat sinks. Appl Therm Eng 140:622–636
52. Lee YJ, Lee PS, Chou SK (2012) Enhanced thermal transport in microchannel using Oblique Fins. J Heat Transfer 134(10):101901
53. Muhammad A, Selvakumar D, Iranzo A et al (2020) Comparison of pressure drop and heat transfer performance for liquid metal cooled mini-channel with different coolants and heat sink materials. J Therm Anal Calorim 141:289–300
54. Won Y, Houshmand F, Agonafer D et al (2014) Microfluidic heat exchangers for high power density GaN on SiC. In: Compound semiconductor integrated circuit symposium. IEEE, pp 1–5
55. Palko JW, Lee H, Chi Z et al (2017) Extreme two-phase cooling from laser-etched diamond and conformal, template-fabricated microporous copper. Adv Func Mater 27(45):1703265
56. Hu DW, Miao M, Ma SL et al (2012) Investigation of cooling performance of micro-channel structure embedded in LTCC substrate for 3D micro-system. In: 2012 IEEE 11th international conference on solid-state and integrated circuit technology (ICSICT). IEEE, pp 1–3
57. Farzaneh M, Salimpour MR, Tavakoli MR (2016) Design of bifurcating microchannels with/without loops for cooling of square-shaped electronic components. Appl Therm Eng 108:581–595
58. David MP, Miler J, Steinbrenner JE et al (2011) Hydraulic and thermal characteristics of a vapor venting two-phase microchannel heat exchanger. Int J Heat Mass Transf 54(25–26):5504–5516

59. Stevanovic LD, Beaupre RA, Gowda AV et al (2010) Integral micro-channel liquid cooling for power electronics. In: Applied power electronics conference and exposition (APEC), 2010 twenty-fifth annual IEEE. IEEE, pp 1591–1597
60. Shalchi-Tabrizi A, Seyf HR (2012) Analysis of entropy generation and convective heat transfer of Al_2O_3 nanofluid flow in a tangential micro heat sink. Int J Heat Mass Transf 55(15–16):4366–4375
61. Bahiraei M, Heshmatian S (2018) Thermal performance and second law characteristics of two new microchannel heat sinks operated with hybrid nanofluid containing graphene–silver nanoparticles. Energy Convers Manage 168:357–370
62. Zhang JY (2018) Thermal analysis and temperature prediction model for 3D stacked chips. South China University of Technology, Guangdong, China
63. Harper CA (2006) Electronic packaging materials and processes. Chemical Industry Press, Beijing, China

Chapter 7
Mechanical Characteristics and Testing of Integrated Circuit Packaging

Integrated circuit (IC) packaging structures face a variety of mechanical issues during production, transportation, and usage. These issues span multiple fields of mechanics, including material mechanics, plasticity mechanics, and fracture mechanics. With the advent of very large-scale integrated circuits and the increasing density of electronic packaging, mechanical problems have become a significant cause of IC packaging failures. Therefore, clarifying the failure modes and mechanisms of packaging mechanics is crucial for addressing the mechanical reliability issues in IC packaging and provides valuable references for structural optimization design.

This chapter begins with an analysis of the key mechanical characteristics in IC packaging, discusses the failure modes and mechanisms, proposes solutions for failure prevention, introduces traditional and microstructural mechanical testing methods for IC packaging, and finally provides methods for mechanical performance evaluation and case studies.

7.1 Mechanical Characteristics of Integrated Circuit Packaging

7.1.1 Mechanical Issues in Packaging

1. Material Mechanics Issues in Packaging

Integrated circuit packaging faces numerous material mechanics issues. Its structural design has evolved from one-dimensional to two-dimensional (2D) and three-dimensional (3D) structures, and from simple to multi-level fractal structures, with increasingly demanding mechanical performance requirements. Material mechanics focuses on analyzing deformation problems of packaging structures under external forces. Electronic packaging materials provide mechanical support, environmental

B. Zhou et al., *Reliability Technology for Integrated Circuit Packaging*,
https://doi.org/10.1007/978-981-95-3885-0_7

sealing, signal transmission, heat dissipation, and shielding. Their mechanical characteristics directly affect the lifespan of integrated circuits. During packaging, structures are often subjected to bending, folding, tension-compression, vibration, impact, and other external forces, leading to various problems such as fatigue failure of packaging structures under alternating stress, delamination between multi-layer materials, cracking of packaging cover plates under external mechanical stress, wire breakage of high-density bonding leads in vibration environments, and damage and cracking of packaging structures under impact loads. These issues are related to the packaging structures being subjected to forces beyond their strength or fatigue limits [1].

2. Plasticity Mechanics Issues in Packaging

Plastic deformation is common in IC packaging, extending beyond the realm of elastic mechanics and involving both elastoplastic and viscoplastic aspects. Elastoplasticity studies the behavior of packaging structures that undergo complete deformation under external forces, with only a small portion of the deformation disappearing immediately upon removal of the force, while the rest remains permanent. Viscoplasticity considers the viscoplastic behavior of packaging structures, where viscosity refers to time-dependent deformation characteristics. Viscosity-related mechanical phenomena include creep and stress relaxation. For describing the propagation of stress waves and the dynamic characteristics of structures under short-term high loads, both plasticity and viscosity of the packaging structures need to be considered [2]. Plasticity mechanics issues in IC packaging are prominent, such as the "popcorn" rupture phenomenon of electronic packaging materials during reflow soldering due to moisture ingress and elastoplastic problems. Other examples include the viscoplastic mechanical characteristics of plastic ball grid array (PBGA) materials and the significant stress concentration at interlayer interfaces in multi-layer integrated circuits due to different material characteristics and structural loading forms. With increasing cycles, plastic deformation accumulates, leading to ratcheting effects, which significantly shorten the fatigue life of chip interconnect solder layers under stress-controlled low-cycle fatigue loading [3].

3. Fracture Mechanics Issues in Packaging

During the design and fabrication of packaging structures, interfaces caused by defects such as cracks and voids exist. These interfaces can significantly degrade the performance of the packaging structures during service. For such multi-cracked and interfaced structures, assessing their fracture characteristics, especially thermal fracture characteristics, and even altering the material's fracture toughness to enhance its resistance to fracture, can have a significant impact on the design of packaging structures. Common fracture mechanics issues in IC packaging include stress corrosion cracking, where packaging structures develop cracks and eventually fail under static stress in contact with corrosive media. Another issue is solder joint failure due to defects introduced during the packaging process, which makes them more susceptible to solder joint cracking failures under external mechanical stress and thermal cycling.

7.1.2 *Main Mechanical Characteristics of Packaging*

The main mechanical characteristics of packaging are reflected in modal characteristics, vibration and impact response characteristics, and interfacial mechanical characteristics.

1. Modal Characteristics

Modal analysis is based on vibration theory and focuses on modal parameters such as modal shapes, damping ratios, and natural frequencies. It is an interdisciplinary field combining computer technology, testing techniques, signal analysis methods, data processing, and vibration theory. Modal analysis reached maturity during the period from 1970 to 1985 and has since been applied in various engineering fields, leading to the development of various tools, including modal analysis software and hardware. Modal analysis is used to study the vibration characteristics of actual structures. It first fits the transfer function curve of the structure, then identifies the modal parameters, and finally establishes a structural modal model.

In the theory of vibration mechanics, a mode is defined as the vibration form exhibited by a multi (or single) degree-of-freedom system structure vibrating at a certain natural frequency. Modal analysis is often applied in structural design to verify the rationality of the design through modal calculations, avoiding passive control measures [4].

Integrated circuits, subject to multiple excitations during operation, are complex mechanical systems. Their natural frequencies and modes are generally determined using the finite element method. Through finite element discretization and variation of the integrated circuit, the following dynamic equation can be obtained:

$$\boldsymbol{M}\ddot{\boldsymbol{X}} + \boldsymbol{C}\dot{\boldsymbol{X}} + \boldsymbol{K}\boldsymbol{X} = \boldsymbol{F} \tag{7.1}$$

In the equation, $\boldsymbol{M}$, $\boldsymbol{C}$ and $\boldsymbol{K}$ are the mass matrix, damping matrix, and stiffness matrix, respectively; $\boldsymbol{F}$ is the force vector; $\ddot{\boldsymbol{X}}$ is the acceleration of the structural vibration; $\dot{\boldsymbol{X}}$ is the velocity of the structural vibration; X is the displacement vector of the structural vibration. Modal characteristics are inherent attributes of the structure and are independent of external loads.

2. Vibration and Impact Response Characteristics

Vibration and impact are the primary mechanical stresses that integrated circuit packaging experiences during production, transportation, and application. Their resistance to vibration and impact is determined by their inherent dynamic characteristics and the features of the excitation signals. If vibration is classified based on the statistical characteristics of the vibration excitation signals, the classification is shown in Fig. 7.1.

Periodic vibration is characterized by repetitive motion, where the same variation process occurs successively in multiple cycles. It consists of simple harmonic vibration and complex periodic vibration.

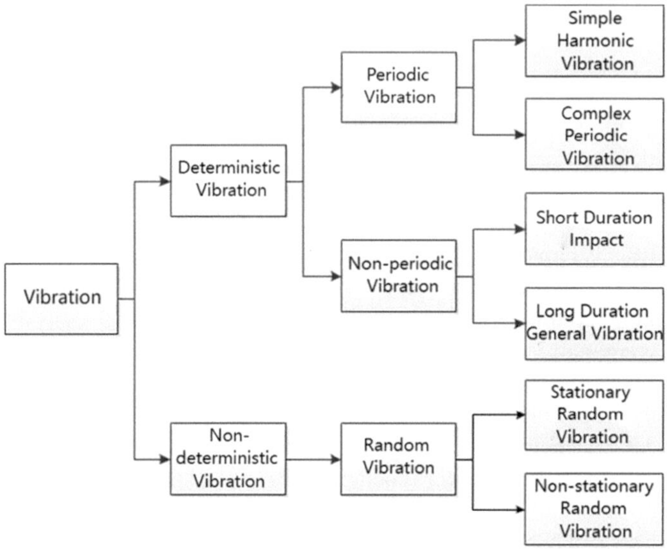

Fig. 7.1 Classification of vibration

Non-periodic vibration includes short-duration impacts and long-duration general vibrations, with mechanical impacts and earthquakes belonging to these two types, respectively.

Non-deterministic vibration refers to vibration forms that change over time, with random vibration being a typical example.

Random vibration can be classified into two categories on the stationarity of the vibration: stationary random vibration and non-stationary. Non-stationary random vibration refers to the type of vibration where the statistical characteristics change over time during the random vibration process. Conversely, stationary random vibration is defined as the type where the statistical characteristics remain constant over time. A stationary random process is one in which the statistical parameters such as the ensemble mean, variance, mean square value, and autocovariance do not change with time. An ergodic random process is a type of random process where the time average of a single sample function is equal to the ensemble average. In other words, all the statistical properties of the process are contained within any single sample function.

In general, random vibration is treated and analyzed as an ergodic random process in practical applications. The statistical characteristics of random vibration are typically reflected in the time domain, frequency domain, and amplitude domain. Random vibration, as a typical non-deterministic form of vibration, more accurately represents the real vibration environment for electronic products such as integrated circuits. This includes vibrations encountered in aircraft, missiles, automobiles, trucks, trains, as well as those in chemical processing facilities and CNC milling machines [5–8]. Under the excitation of a random vibration environment, the packaging structure of an integrated circuit will undergo random vibration. The vibration response cannot be expressed by a single function and can only be described through the statistical characteristics of the response. Modal analysis must first be used to solve for the modes and obtain modal parameters before random vibration analysis can be conducted. Based on vibration theory, modal analysis is used to study the dynamic characteristics of integrated circuits, identify their natural frequency characteristics, and thereby guide fault diagnosis, performance analysis, and design of the integrated circuits themselves.

Impact is generally considered as a rapid transfer of energy in a mechanical system, resulting in a significant increase in stress, velocity, acceleration, or displacement within the system. The duration required for this energy transfer is related to the characteristics of the system itself. Impact can cause large displacements in the packaging structure and bond wires of integrated circuits, leading to structural cracking of the packaging and short-circuit failures in the bond wires [6].

The sources of mechanical impact vary, and so do the forms of action and their effects. The most commonly used descriptions of impact include three types: impact pulses, velocity impacts, and shock response spectra. Common impact pulses include square wave pulses, half-sine pulses, and various triangular pulses. Impact pulses are easy to implement due to their simple form, but they rarely fully represent the actual impact environment. Nevertheless, using impact pulses is effective for assessing the impact resistance of product structures. Velocity impacts refer to impacts where the velocity changes abruptly. For example, when a falling package hits the ground and its velocity suddenly drops to zero, this is referred to as a drop shock. Another type of velocity impact involves using a heavy hammer to strike a fixture holding a sample. The hammer transfers its velocity to the fixture and the sample. The US Navy uses this type of velocity impact to simulate the effects of explosions on ship and submarine equipment.

Both impact pulses and velocity impacts are characterized by the time histories of acceleration, velocity, and displacement to describe the impact environment. However, the purpose of impact research is not the impact itself, but rather the destructive effects of the impact on the product. The inherent characteristics of the product influence its performance after being subjected to impact, and this can be observed through the response generated after the impact. Components may experience impact during transportation and handling, and impact pulses are commonly used to test their adaptability in these processes. For example, standards such as GJB 548 and GJB 360 specify various types of impact pulses and detailed methods for conducting tests with these pulses.

The half-sine impact pulse is almost the m ost commonly used impact pulse for testing various commercial and industrial products. This is because the half-sine impact pulse is the easiest to generate, analyze, and evaluate. However, this does not mean that the half-sine impact pulse accurately represents the actual impact that products experience in real environments.

When an integrated circuit is subjected to impact excitation, its response process is the zero-state response of the circuit system to the impact signal excitation. If it is simplified as a single-degree-of-freedom system [9, 10], as shown in Fig. 7.2, then the vibration equation of the integrated circuit under impact pulse excitation is:

$$m\ddot{x}(t) + c\dot{x}(t) + kx(t) = f(t) \tag{7.2}$$

In the equation, $f(t)$ represents the impact pulse excitation function. If a half-sine impact pulse is used, as shown in Fig. 7.3, then we have:

$$f(t) = \begin{cases} F\sin\frac{\pi}{D}t & 0 \le t \le D \\ 0 & t > D \end{cases} \tag{7.3}$$

In the equation, F is the amplitude and D is the duration of the half-sine impact pulse. Under the action of a half-sine impact pulse, the response of a single-degree-of-freedom system is divided into two stages: the first stage is the forced vibration produced during the load application time, and the second stage is the free vibration

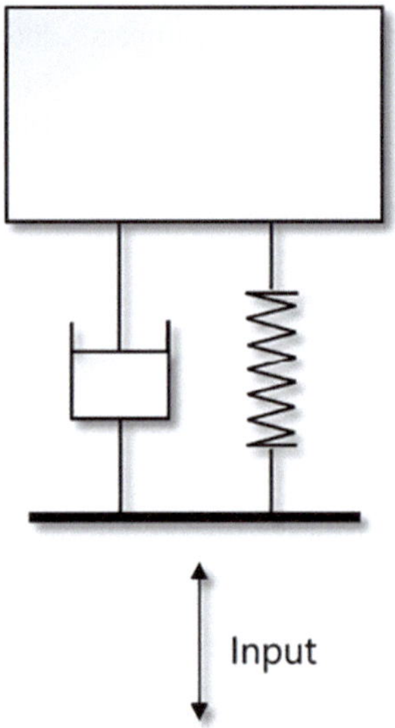

Fig. 7.2 Single-degree-of-freedom system

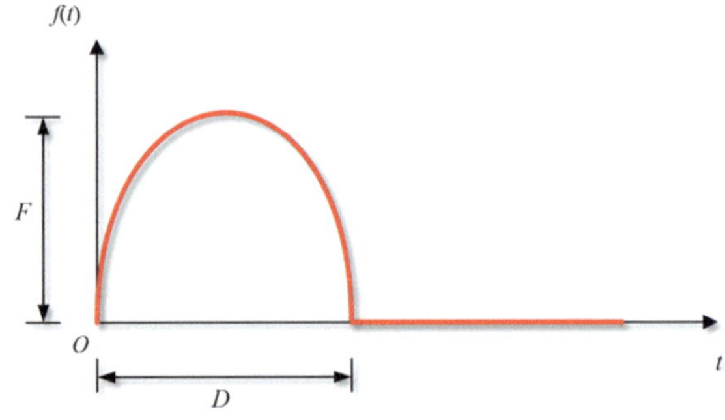

Fig. 7.3 Half-sine impact pulse

produced after the forced vibration. The solution to Eq. (7.2) is:

$$x(t) = \frac{\frac{F\omega}{\omega_n} e^{-2\xi t}}{\sqrt{\left(\omega_n^2 - \omega^2\right) + (2\xi\omega)^2}} \sin \omega_d t + \frac{F}{\sqrt{\left(\omega_n^2 - \omega^2\right) + (2\xi\omega)^2}} \sin \omega t \tag{7.4}$$

In the equation, $\xi = c/2m\omega_n$ is the damping ratio coefficient; $\omega = 2\pi/D$ is the excitation base frequency; $\omega_d = \omega_n\sqrt{1 - \xi^2}$ is the damped natural frequency of the system; and $\omega_n = \sqrt{k/m}$ is the undamped natural frequency of the system.

It can be seen from Eq. (7.4) that the response of a single-degree-of-freedom system to impact excitation is related to the amplitude and duration of the excitation pulse, as well as the system's natural frequency and damping.

3. Interfacial Mechanical Characteristics

There are many joints in the interfaces between the solder joints and the soldering materials, and many interfaces exist at these joints. Interfacial stress singularity fields arise due to the different material properties on either side of the interface, which significantly affect the failure of solder joints.

Macroscopically, the junctions of various materials and structures have a profound impact on the overall functionality and mechanical behavior of the structure, such as the interlaminar interfaces in composite laminates, metal-ceramic interfaces, and coating interfaces in thin-film materials. Microscopically, the design and development of new advanced materials primarily focus on the mechanical behavior of grain boundaries and the interfaces between matrix materials and fibers, which have significant implications for the precise evaluation of structural lifetimes [11, 12].

Macroscopic defects, such as poorly bonded areas and large voids, must be considered in modeling. However, the impact of microscopic defects in the interfacial phase on the overall model can be neglected. From a mechanical perspective, interfaces can be categorized into three types, as shown in Fig. 7.4.

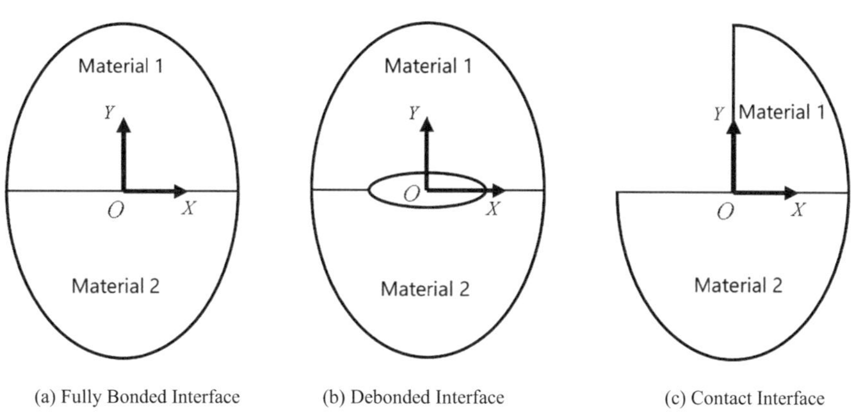

Fig. 7.4 Classification of interfaces

(1) Fully bonded interface

A fully bonded interface is one that lacks macroscopic defects such as poorly bonded areas and large voids. It satisfies the continuity of displacement and force across the interface and is also known as an ideal interface. This condition holds true for all surfaces. Taking Fig. 7.4a as an example, it can be described as:

$$\tau_{xy1} = \tau_{xy2}, \sigma_{y1} = \sigma_{y2}, u_1 = u_2, v_1 = v_2 \tag{7.5}$$

In the equation, τ_{xy} is the shear stress in the XY plane; σ_{y1} and σ_{y2} are the stress components in the Y direction for materials 1 and 2, respectively; u is the displacement; and v is the magnitude of the interfacial force.

According to the relevant theories of elasticity mechanics, there are two cases where the normal stress is discontinuous when it is parallel to the interface: when the normal strain is perpendicular to the interface and when the shear strain is located on either side of the interface.

(2) Debonded interface

As shown in Fig. 7.4b, a debonded interface refers to the boundary where the materials on either side of the interface are separated at the same location, meaning there are voids or unbonded areas at the interface. The primary manifestation is an interface crack with an opening. The normal and shear stresses at the debonded interface should satisfy:

$$\sigma_{y1} = \tau_{xy1} = 0, \sigma_{y2} = \tau_{xy2} = 0 \tag{7.6}$$

In the equation, σ_{y1} and τ_{xy1} are the normal stress in the Y direction and the shear stress in the XY plane for material 1, respectively.

(3) Contact interface

A contact interface refers to the interface where two materials are in contact but not bonded under the action of residual stress or external forces. The main regions after deformation of a contact interface are the opening zone, slip zone, and adhesion zone. Under the premise of small deformation, as shown in Fig. 7.4c for a contact interface, the following should be satisfied in the slip zone:

$$\sigma_{y1} = \sigma_{y2} \leq 0, v_1 = v_2, \tau_{xy1} = \tau_{xy2} = \pm f \sigma_y \tag{7.7}$$

In the equation, f is the dynamic friction coefficient on the contact interface. The opening zone, slip zone, and adhesion zone that appear after deformation of the contact interface can be solved based on the actual loading conditions and multiple iterations.

7.1.3 Mechanical Failure and Prevention of Packaging

Mechanical failure of integrated circuit packaging structures is primarily caused by external vibration or impact stresses. The failure modes can be categorized into two types based on the mechanisms of mechanical failure: displacement-type failure and stress-type failure. Displacement-type failure refers to the failure that occurs when the vibration deformation of the packaging structure exceeds the allowable value under vibration or impact conditions. This type of failure often leads to instantaneous short-circuiting of integrated circuit devices and, in severe cases, can cause device burnout [13]. Figure 7.5 illustrates a board-mounted integrated circuit experiencing bond wire short-circuiting during mechanical impact. Bond wires undergo vibration deformation in a vibrating environment, especially when the natural frequency of the bond wires falls within the environmental vibration frequency range. Resonance occurs in the bond wires, and once it does, the maximum vibration amplitude of the bond wires exceeds the distance between them, causing the bond wires to come into contact with each other (wire-to-wire contact) and resulting in a short circuit [14, 15]. As shown in Fig. 7.6, the bond wires return to normal after the vibration ceases.

Stress-type failure refers to the failure phenomenon that occurs when the dynamic stress generated by the packaging structure under vibration or impact conditions exceeds the material's endurance limit. When the dynamic stress surpasses the material's strength limit, cracks or fractures may appear in the regions of maximum stress within the packaging structure. This type of failure often occurs early in the device's usage, especially when manufacturing defects are present, such as instantaneous fractures caused by early cracks or solder joint detachment [16]. When the dynamic stress exceeds the material's fatigue limit, long-term dynamic stress can lead to fatigue fracture failure of the packaging structure. Common failure modes include solder joint cracking, bond wire breakage (see Fig. 7.7), hermetic package lid cracking (see Fig. 7.8), and packaging structure fracture (see Fig. 7.9).

Currently, in packaging and testing companies, two main interconnection forms have been developed and widely applied in secondary packaging for high-reliability

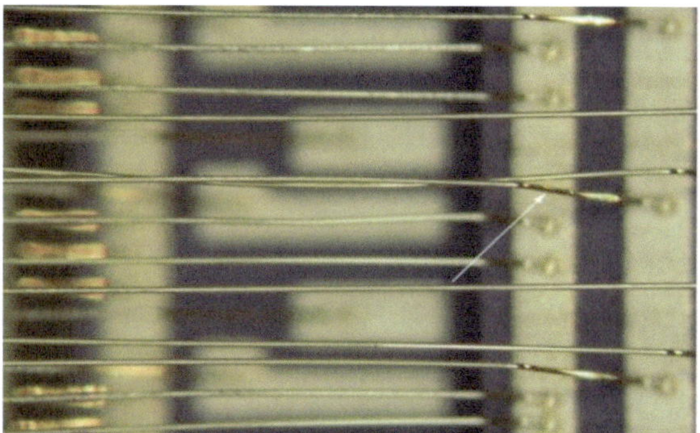

Fig. 7.5 Bonding wire shorting in board-mounted integrated circuits during mechanical impact

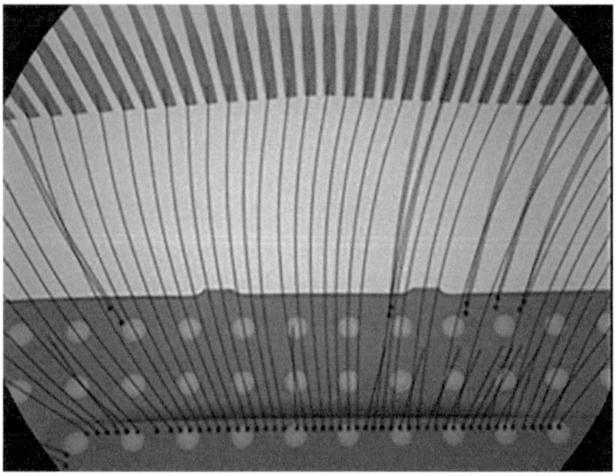

Fig. 7.6 Bonding wire shorting failure due to wire-to-wire contact

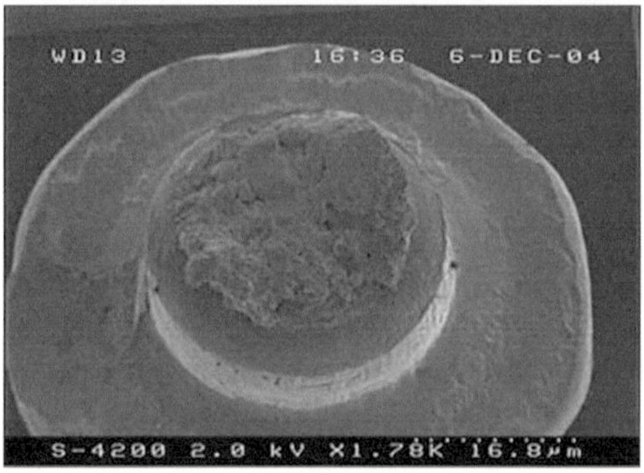

Fig. 7.7 Bonding wire fracture

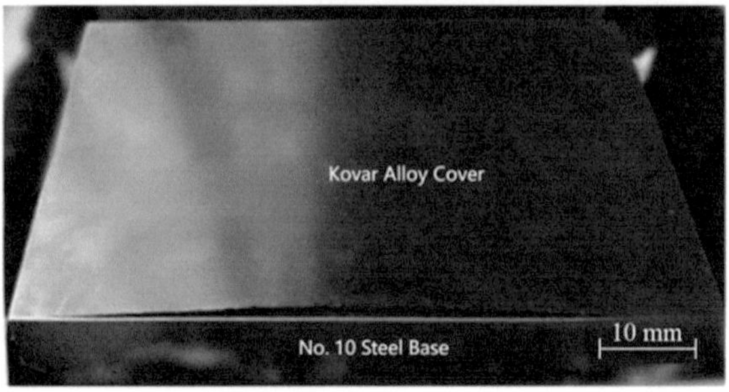

Fig. 7.8 Hermetic package lid cracking

electronic products [17]. Figure 7.10 shows the failure modes of CBGA (Ceramic Ball Grid Array) packaging under different loading conditions. As shown in Fig. 7.10a, under thermal cycling conditions, CBGA packaging primarily undergoes two types of deformation: creep deformation and plastic deformation. As shown in Fig. 7.10b,

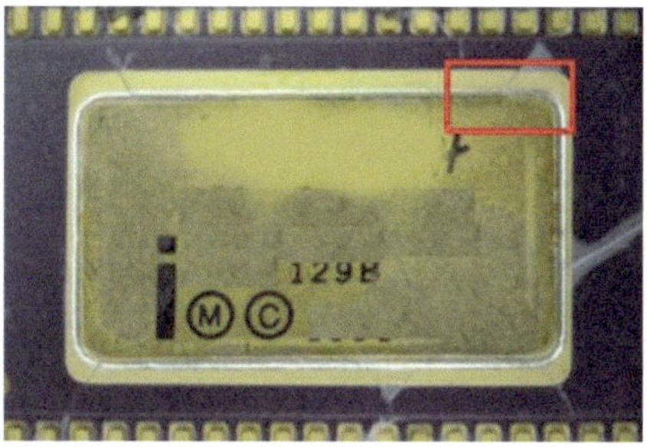

(a) Ceramic Body Crack

(b) Metal Lid Seal Separation

Fig. 7.9 Packaging structure fracture

under vibration impact conditions, CBGA packaging mainly experiences brittle fracture along one end of the PCB. Comparing Fig. 7.10a with b, it is evident that different loading conditions lead to two completely different failure modes of CBGA packaging.

Figure 7.11 illustrates the failure modes of CCGA (Ceramic Column Grid Array) packaging under different loading conditions. In Fig. 7.11a, under thermal cycling

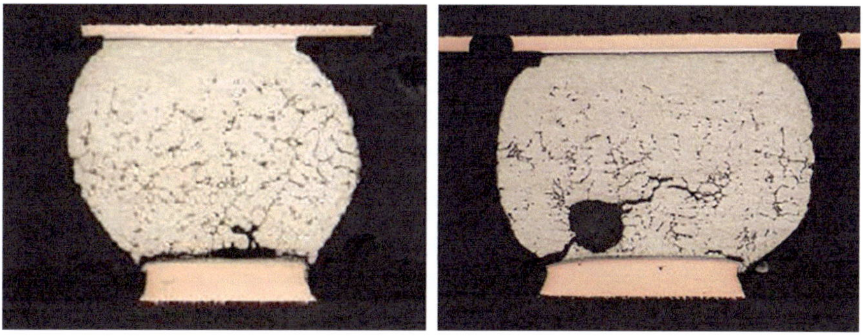

(a) Cracking Failure Under Temperature Cycling Conditions

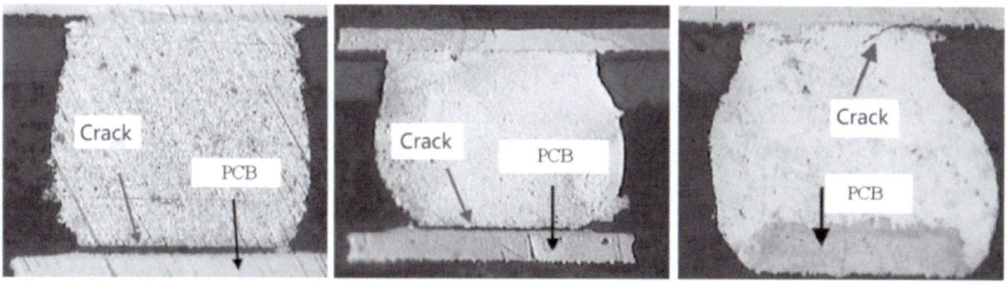

(b) Cracking Failure Under Vibration and Impact Conditions

Fig. 7.10 Failure modes of CBGA packaging under different loading conditions

(a) Cracking Failure under Temperature Cycling Conditions (b) Cracking Failure under Vibration and Impact Conditions

Fig. 7.11 Failure modes of CCGA packaging under different loading conditions

conditions, CCGA packaging effectively accommodates plastic and creep deformation due to thermal mismatch. In Fig. 7.11b, CCGA packaging is too weak to withstand high levels of vibration impact stress, resulting in brittle fracture at the base.

In summary, CBGA and CCGA packaging structures have their respective advantages and disadvantages in terms of manufacturing process, space occupancy within the package, electrical performance, and service conditions. The comparison of their pros and cons is shown in Table 7.1.

In Table 7.1, CCGA has higher thermo-mechanical reliability, while CBGA has higher dynamic mechanical reliability. This is because, in ceramic packaging, the chip can better match the thermal expansion coefficient (TEC) of the substrate. In contrast, the TEC of epoxy PCBs is 18×10^{-6}~20×10^{-6}/°C, which is 3 to 4 times that of ceramic substrates. As a result, the basic size of CCGA must be limited to within 32.5 mm. When this size is exceeded, the thermal fatigue life of the solder joints is significantly affected, thereby reducing the reliability of the device. However, the slender solder columns of CCGA offer certain advantages in this regard. When the solder column height increases, CCGA can release stress through column bending. Compared with CBGA, under thermal fatigue, the flexible solder columns can have a longer life, but they are more susceptible to mechanical and physical damage.

Table 7.1 Comparison of advantages and disadvantages of CBGA and CCGA

	CBGA	CCGA
Advantages	High dynamic mechanical reliability Standardized manufacturing process Occupies less space Low inductance and capacitance Good self-alignment characteristics	High thermo-mechanical reliability The number of I/O > 1657
Disadvantages	Low thermo-mechanical reliability The number of I/O < 1657	Low dynamic mechanical reliability No standardized manufacturing process Requires more vertical space

In practical applications, when devices are subjected to loads during service, high temperature and vibration often act simultaneously. Therefore, the selection between CBGA and CCGA must be based on systematic testing and theoretical research to reach a conclusion [18].

The prevention of vibration-induced failure in packaging structures essentially involves "vibration isolation and damping." This can be divided into active prevention and passive prevention. Active prevention refers to improving the vibration resistance of the structure from the perspective of product structural design, such as increasing structural stiffness and reducing structural mass to minimize structural response, including stress and displacement response. Passive prevention refers to reducing structural response through auxiliary strategies without altering the product structure, such as increasing damping or installing vibration dampers. For stress-induced failure, active prevention mainly focuses on reducing stress and increasing material strength. Stress reduction is primarily achieved through optimization using simulation tools during the design phase. For displacement-induced failure, active prevention mainly involves threshold optimization and displacement optimization. Threshold optimization refers to increasing the threshold through local structural adjustments, such as increasing the spacing between bonding wires to address wire-to-wire contact issues. Displacement optimization involves quantitative structural optimization using simulation tools during the design phase to reduce the displacement response of the packaging structure under vibration conditions. Passive prevention involves using damping effects to reduce vibration, such as applying adhesive. Among all preventive measures, optimization through simulation tools is the most complex. The principles and methods for optimization design are introduced below.

The prevention of vibration-induced failure in packaging structures needs to be considered from the design stage during the development phase. The three key elements of vibration testing include frequency range, vibration amplitude or spectrum shape, and duration. Considering these three elements, vibration-resistant design of packaging structures can be divided into three aspects: avoiding resonance, reducing response, and fatigue life design.

1. Avoiding Resonance

Firstly, the natural frequency of the product must be considered. The natural frequency should be designed to avoid the high spectral density frequency bands of the environmental vibration. These high spectral density bands have high energy, and if the natural frequency range is close to them, the required vibration resistance of the product will be extremely high. This not only significantly increases the development cost of the product but also extends the testing time. In some cases, it may even be impractical from an engineering perspective. If it is impossible to avoid the high spectral density frequency bands, especially the peak frequencies, the natural frequency should at least be separated from the peak frequency by a factor of $\sqrt{2}$, and preferably more than $\sqrt{2}$. For tests involving fixed-frequency sinusoidal loads, the natural frequency of the product must be far from the sinusoidal fixed-frequency test frequency by a factor of $\sqrt{2}$.. For sinusoidal sweep frequency tests, the natural frequency of the product must be outside the input frequency range. For example, the sweep frequency vibration (Method 2007) specified in the GJB 548B—2005 standard has a frequency range of 20–2000 Hz, while the random vibration (Method 2026) has a frequency range of 50–2000 Hz. The natural frequency of the packaging structure is generally above 50 Hz. To avoid resonance during sweep frequency testing, the first-order natural frequency of the packaging structure should be greater than 2000 Hz, and preferably more than $\sqrt{2}$ times 2000 Hz, i.e., greater than 2828 Hz.

2. Reducing Response

The specific frequency conditions of the environment are one of the critical factors for product failure, and they are related to the intensity of the vibration. In vibration testing, the vibration amplitude or spectrum shape determines the vibration intensity. Not all integrated circuit packaging structures can meet the requirement of having a first-order natural frequency greater than 2828 Hz to avoid resonance. When the natural frequency is within the range of 20–2000 Hz, the packaging structure may resonate under vibration excitation. When the vibration intensity reaches a certain level, the packaging structure may fail due to excessive stress or displacement response. The sweep frequency vibration (Method 2007) specified in the GJB 548B—2005 standard defines three test conditions, A, B, and C, as shown in Table 7.2.

The GJB 548B—2005 standard specifies two test conditions for random vibration (Method 2026), each with 10 test levels, as shown in Fig. 7.12 and Table 7.3. The technical conditions of the product define the test conditions, and the product should be designed for vibration resistance based on these conditions. Simulation tools should be used to calculate the vibration response of the product under the specified test conditions to ensure that the response meets the requirements for strength and stiffness.

The test conditions specified in MIL-STD-883K for random vibration are similar to those in GJB 548B—2005, with two sets of conditions, each comprising 10 test levels (Table 7.4).

3. Fatigue Life Design

Alternating stress is one of the factors that cause material failure. In practical engineering applications, materials may experience fatigue failure long before reaching their yield strength or ultimate tensile strength. Throughout their lifecycle, integrated circuits are subjected to a variety of complex environmental conditions. Under vibration conditions, the magnitude and frequency of alternating stresses experienced by packaging materials are highly complex, and fatigue failure may occur after a certain period of time. Therefore, ensuring that the vibration response of the packaging

Table 7.2 Sweep frequency vibration test conditions

Test condition	Peak acceleration (g)
A	20
B	50
C	70

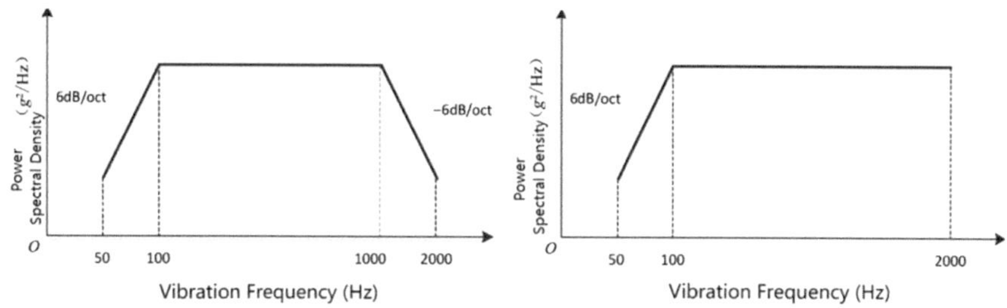

(a) Test Condition 1 (b) Test Condition 2

Fig. 7.12 Random Vibration Spectrum

Table 7.3 Random vibration test conditions

Test condition 1			Test condition 2		
Test condition no	Power spectral density (g^2/Hz)	Acceleration total RMS (g)	Test condition no.	Power spectral density (g^2/Hz)	Acceleration total RMS (g)
A	2	53.5	A	2	62.1
B	4	75.6	B	4	87.8
C	6	92.6	C	6	107.6
D	10	119.5	D	10	138.9
E	20	169.1	E	20	196.4
F	30	207.1	F	30	240.6
G	40	239.1	G	40	277.8
H	60	292.8	H	60	340.2
J	100	378	J	100	439.2
K	150	463	K	150	537.9

structure meets strength and stiffness requirements can only guarantee short-term vibration reliability. In contrast, considering vibration fatigue life design can ensure long-term vibration reliability. The concept of stress history is illustrated in Fig. 7.13.

The fatigue life assessment of integrated circuits is typically conducted through sweep frequency vibration fatigue tests. For example, GJB 548B—2005 (Method 2007) specifies a frequency range of 60 ± 20 Hz. The vibration test conditions are as follows: the test duration is 32 ± 8 h for each axis, with a total time of no less than 96 h. GJB 548B—2005 (Method 2026) specifies that the random vibration fatigue test time is 15 min for each axis, totaling 45 min. For random vibration fatigue life, the three-zone method proposed by Stenberg can be used for estimation.

The nominal stress method is a fatigue design approach that originated from the *S–N* (stress-life) curve. It references the nominal stress of the material or structure of the test specimen and the stress concentration factor at the fatigue-critical location.

Table 7.4 Random vibration test conditions

Test condition 1			Test condition 2		
Test condition no	Power spectral density (g^2/Hz)	Acceleration total RMS (g)	Test condition no	Power spectral density (g^2/Hz)	Acceleration total RMS (g)
A	0.02	5.2	A	0.02	5.9
B	0.04	7.3	B	0.04	8.3
C	0.06	9.0	C	0.06	10.2
D	0.10	11.6	D	0.10	13.2
E	0.20	16.4	E	0.20	18.7
F	0.30	20.0	F	0.30	22.8
G	0.40	23.1	G	0.40	26.4
H	0.60	28.4	H	0.60	32.3
J	1.00	36.6	J	1.00	41.7
K	1.50	44.8	K	1.50	51.1

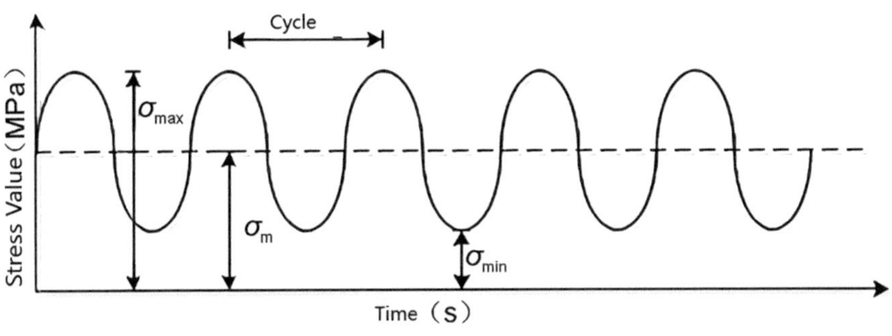

Fig. 7.13 Concept of stress history

Based on the cumulative damage theory of fatigue, the fatigue life is calculated. In the *S–N* curve, the maximum and minimum stress values are related to the load time-domain signal. According to the definition of statistical distribution, fatigue load cycles are primarily characterized by mean stress and stress amplitude.

The indicators used to basically describe the characteristics of a the stress cycle are as follows:

Stress Amplitude:

$$\sigma_a = \frac{\sigma_{\max} - \sigma_{\min}}{2} \tag{7.8}$$

Mean Stress:

$$\sigma_m = \frac{\sigma_{\max} + \sigma_{\min}}{2} \tag{7.9}$$

Ratio:

$$R = \frac{\sigma_{\min}}{\sigma_{\max}} \tag{7.10}$$

In these equations, $\sigma_{\max}$ is the maximum stress value, and $\sigma_{\min}$ is the minimum stress value.

Based on the mission profile, perform a force analysis to obtain the stress cycle characteristics of the critical components or areas where fatigue is likely to occur. This includes the maximum stress $\sigma_{\max}$, minimum stress $\sigma_{\min}$, and then calculate the corresponding *R*, σ_{a} and σ_{m}. For uniaxial 1 fatigue problems, $\sigma_{\max} = \sigma_1$ (the first principal stress at the critical point), and $\sigma_{\min} = \sigma_3$ (the third principal stress at the critical point). For multiaxial fatigue problems, $\sigma_{\max} = \sigma_{\mathrm{Von-Mises\ max}}$ and $\sigma_{\min} = \sigma_{\mathrm{Von-Mises\ min}}$. Among the above five parameters, only two are independent. To determine the stress cycle characteristics, knowing any two of them is sufficient.

The relationship between stress amplitude and the number of cycles to failure is:

$$N = N_0 \left(\frac{\sigma_a}{\sigma_{-1A}} \right)^{\frac{1}{b}} \tag{7.11}$$

In the equation, *N* represents the number of cycles to failure under the stress amplitude σ_a; N_0 is the number of cycles corresponding to the fatigue limit σ_{-1A}, σ_{-1A} is the fatigue limit, and *b* is the slope of the *S–N* curve. The slope *b* can be calculated from two points on the curve using the formula $b = \frac{\lg\sigma - \lg\sigma_{-1A}}{\lg N - \lg N_0}$. This exponent is related to the material's fatigue strength and can also be obtained from relevant handbooks.

For most materials, constant amplitude loading ($\sigma_m = 0$) is commonly used to obtain their fatigue characteristic data. In the laboratory, constant amplitude loading is applied to standard fatigue test specimens, and the material's fatigue characteristics are obtained through standard testing methods. However, in actual engineering applications, materials or structures are often subjected to random variable amplitude loading, which significantly affects the estimation of the material or structure's fatigue life. Under random alternating stress and equal stress amplitude conditions, the degree of damage caused by alternating stress increases with the increase in mean stress. Typically, tensile mean stress shortens and compressive mean stress extends the fatigue life of materials or structures. The *S–N* curve provides better predictions when considering mean stress.

Under a given stress amplitude, as the mean stress increases, the fatigue life decreases. The Haigh curve (see Fig. 7.14) can be used to correct the stress amplitude while ensuring that the fatigue life remains unchanged.

The famous correction methods are the Goodman and Gerber stress corrections.

Goodman Formula:

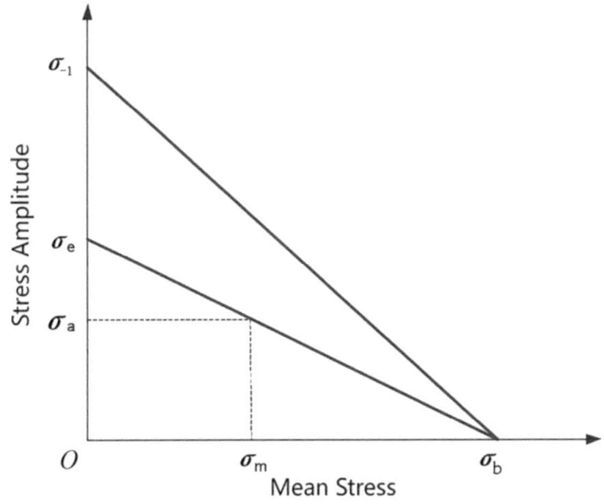

Fig. 7.14 Haigh curve (Goodman curve)

$$\sigma_e = \frac{\sigma_b}{\sigma_b - \sigma_m}\sigma_a \tag{7.12}$$

Gerber Formula

$$\sigma_e = \frac{\sigma_b^2}{\sigma_b^2 - \sigma_m^2}\sigma_a \tag{7.13}$$

In these equations, σ_e is the equivalent stress amplitude with a mean stress of zero. When the mean stress is not zero, σ_e is obtained using Eqs. (7.12) or (7.13) and is used instead of σ_a in Eq. (7.8) for the calculation.

One of the methods for estimating random vibration fatigue life is the three-zone method. The three-zone method is based on the damage law proposed by Stenberg and the Gaussian distribution. It is assumed that random vibration follows a Gaussian distribution, then for 1σ level instantaneous acceleration, the time spent between -1σ and $+1\sigma$ is 68.3%; for 2σ level instantaneous acceleration, the time spent between -2σ and $+2\sigma$ is (95.4 − 68.3%), which is 27.1%; for 3σ level instantaneous acceleration, the time spent between -3σ and $+3\sigma$ is (99.73 − 95.4%), which is 4.33%. Fatigue failure is assumed to occur when the cumulative damage reaches 1. The random vibration fatigue life can be estimated using Eq. (7.14):

$$T = \frac{1}{f_0\left(\frac{0.683}{N_{1\sigma}} + \frac{0.271}{N_{2\sigma}} + \frac{0.0433}{N_{3\sigma}}\right)} \tag{7.14}$$

In the equation, f_0 is the statistical mean frequency, usually taken as the ratio of 1σ velocity to 1σ displacement. $N_{1\sigma}$, $N_{2\sigma}$, and $N_{3\sigma}$ are the fatigue lives corresponding to 1σ, 2σ, and 3σ stress levels, respectively, which are calculated based on the *S*–*N* curve. The final vibration resistance design requirements are only met when the random vibration fatigue life is longer than the expected vibration time or the desired service life.

7.2 Mechanical Testing of Integrated Circuit Packaging

7.2.1 Conventional Mechanical Testing of Packaging

As the reliability of materials and components continues to improve, the probability of product failure is gradually decreasing, and the test cycle for reliability assessment of materials or components is lengthening. There is a desire to conduct an assessment test in a shorter period to obtain effective results, namely accelerated testing. Accelerated testing is not merely about intensifying stress to shorten the time, nor is it a comprehensive test19. The effective application of this technique can lead to different reliability assessment results and significantly different reliability test contents. Environmental reliability testing is characterized by its short duration and low cost, simulating the service environment of products, including temperature and humidity, pressure, vibration, etc.

Considering the conditions of use, the characteristics of the product should also be taken into account to infer potential failure phenomena and develop a framework for environmental testing based on the reproduction mechanism of such failures.

With the development of semiconductor technology, surface-mount technology, and LCD technology, high density, miniaturization, reliability, full functionality, low cost, and low power consumption have become the development requirements for electronic products. These requirements pose two demands on reliability testing:

① As market demand increases, the safety and reliability standards for products are further enhanced.

② The increase in market demand leads to a shortened product life cycle and accelerated R&D cycles, necessitating reliability assessment within a shorter timeframe. Currently, mature reliability theory and technology enable the study of product failure mechanisms and reliability.

1. Tension and Bending

In testing, tensile testing is commonly used to measure the mechanical properties of materials. Tensile testing refers to the axial tension applied to a component under room temperature and static load (strain rate ≤ 0.1/s). Through tensile testing, comprehensive measurements of material properties such as strength, plasticity, deformation, and fracture are obtained. These characteristics are essential for material analysis, structural design, and new material development in engineering applications. The component is stretched to fracture under uniaxial tension in the effective section. Tensile testing yields the load-deformation curve and deformation characteristics of the material. Based on specific calculations and criteria, mechanical indicators reflecting the results of the tensile test can be obtained and used to determine material performance.

2. Vibration and Shock

Vibration testing is categorized as follows:

(1) Environmental Adaptability Testing: By understanding the future vibration environment of the test object, an impact vibration tester is used to simulate its adaptability to the environment.
(2) Dynamic Strength Testing: The primary test objects are structural components. Specific test conditions are applied to assess the fatigue life and dynamic strength of the components.
(3) Dynamic Characterization Testing: This involves testing the modal parameters of an object, such as mode shapes, damping, and frequency.
(4) Other Tests: For example, low-quality products are often caused by issues in the manufacturing process. Vibration screening tests are used to screen components and assemblies on the production line to improve overall product reliability.

4. Hardness and Fatigue

The fatigue strength of materials under alternating loads can be predicted through fatigue strength testing. These tests generally have long durations and require complex equipment. The fatigue strength of materials subjected to alternating loads cannot be determined by general mechanical tests (such as impact vibration, hardness, or tensile tests). For critical components, fatigue strength testing is essential.

The ability of a material to resist deformation and failure is often measured by hardness, which reflects the material's toughness and elasticity. The commonly used test method in laboratories is the static load indentation hardness test, with related indicators including Vickers hardness (HV), Rockwell hardness (HRA/HRB/HRC), and Brinell hardness (HB). These indicators represent the material's ability to resist indentation by a hard object. Vickers hardness (HV) and Shore hardness (HS) are used for rebound testing, representing the amount of elastic deformation work in metals.

5. Fracture Strength

Fracture strength is the ratio of the tensile force at material fracture to the cross-sectional area. Depending on the material's properties and thickness, the measuring equipment may vary, but the overall principle remains the same.

When it is necessary to measure elongation simultaneously, the length of the specimen will be specified. For example, in the semiconductor industry, the fracture strength and elongation of bonding wires are important evaluation indicators, and the selection of specimen length will differ based on metric and imperial units.

When using the metric system, the specimen length is 10 cm, and a uniform descent rate of 10 mm/min is applied until the specimen breaks. The strength measured at this point is the fracture strength, which is expressed in units of force, such as g, kg, mN, cN, or N. The ratio of the elongation length to the specimen length (10 cm) is the elongation percentage (%).

When using the imperial system, the specimen length is 25.4 mm, and a uniform descent rate of 25.4 mm/min is applied. The conversion method is the same as above.

7.2.2 New Mechanical Testing Methods for Packaging

1. Laser Vibrometry

Laser vibrometry is a non-contact measurement method based on the Doppler effect. The instrument emits a laser beam (with a wavelength of λ) onto the surface of a vibrating object. The frequency of the reflected laser beam changes due to the vibration (Doppler frequency shift f_d). According to the Doppler effect, the velocity of the object is proportional to the frequency shift, i.e., $f_d = 2v/\lambda$. The frequency shift f_d is measured by the device, and after signal demodulation, the waveform of vibration velocity over time can be obtained.

(1) 3D scanning laser vibrometer

For small, lightweight, and thin electronic components, traditional contact measurement methods often introduce additional mass and stiffness to the specimen, which affects the accuracy and precision of the measurement. The 3D scanning laser vibrometer can be used for vibration testing of electronic products that are not suitable for traditional contact sensors (such as strain gauges). It is primarily applicable for dynamic vibration testing of lightweight and thin structures in electronic products.

The 3D scanning laser vibrometer is equipped with three high-precision cameras, as shown in Fig. 7.15. Using a non-contact measurement approach, it can obtain information on the dynamic response and modal parameters of the specimen.

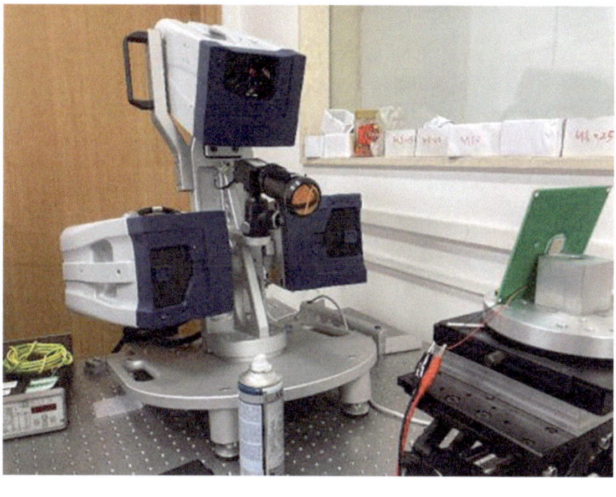

Fig. 7.15 3D scanning laser vibrometer

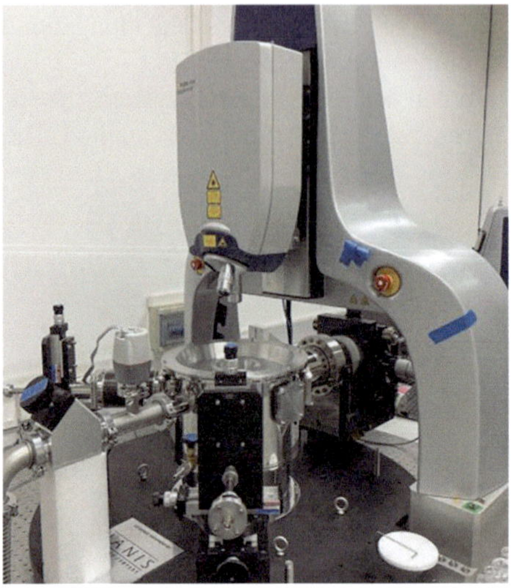

Fig. 7.16 Microscopic laser vibrometer

(2) Microscopic laser vibrometer

The measurement of vibration parameters of micro-components in electronic devices is crucial for verifying the vibration resistance design of their structures. The microscopic laser vibrometer addresses the challenge of measuring vibration parameters of microstructures within components and is primarily used for dynamic vibration testing of micro-components in electronic products under ambient and extreme temperature vacuum conditions.

As shown in Fig. 7.16, the microscopic laser vibrometer is equipped with a vacuum probe station, an *XY*-axis positioning stage, and a temperature control device. It employs a non-contact measurement method, which can meet the requirements for vibration testing of microstructures under various environmental temperatures and obtain information on the dynamic response and modal parameters of the structure.

2. Measurement of Residual Stress in Microregions

The common methods for measuring residual stress in microregions are as follows.

(1) X-ray diffraction method

X-ray diffraction is one of the most commonly used methods for measuring residual stress and is performed using specialized equipment. Figure 7.17 shows the LXRD residual stress analyzer from PROTO company.

X-ray diffraction is the most widely used method. However, the penetration depth of X-rays is approximately 30 μm, which is relatively shallow. For internal residual stress measurement, the internal structure must be exposed. For some materials, such as coarse-grained materials, diffraction is not possible, making measurement more challenging. X-ray diffraction also has high requirements for the material itself [20].

Fig. 7.17 LXRD residual stress analyzer

X-ray diffraction relies on the interference phenomenon in the crystal lattice to determine the lattice plane spacing and residual stress. This method directly measures the atomic spacing of crystals to obtain deformation information of the component, achieving high precision. Due to its non-destructive nature, X-ray diffraction has been widely applied in the measurement of residual stress in welded structures.

(2) Neutron diffraction method

Neutron diffraction offers stronger penetration capabilities compared to X-ray diffraction and can measure internal residual stress without exposing the internal structure. However, it is limited by neutron source accessibility, and it requires prior knowledge of the atomic plane spacing or Bragg angle in the free state of the lattice. As an emerging measurement technique, neutron diffraction poses certain challenges in practical application and is worthy of further research and development by researchers.

(3) Magnetic measurement method

The magnetic measurement method is a stress measurement technique that utilizes the magnetostrictive effect. Changes in strain lead to deformation and expansion, which in turn cause variations in the magnetic flux within a coil. Stress is measured by detecting changes in the induced current within the coil. The magnetic residual stress measuring device is shown in Fig. 7.18.

The magnetic measurement method is a stress measurement technique with distinct advantages and disadvantages. Its advantages include the ability to perform rapid, non-contact measurements on-site. However, the measurement results are

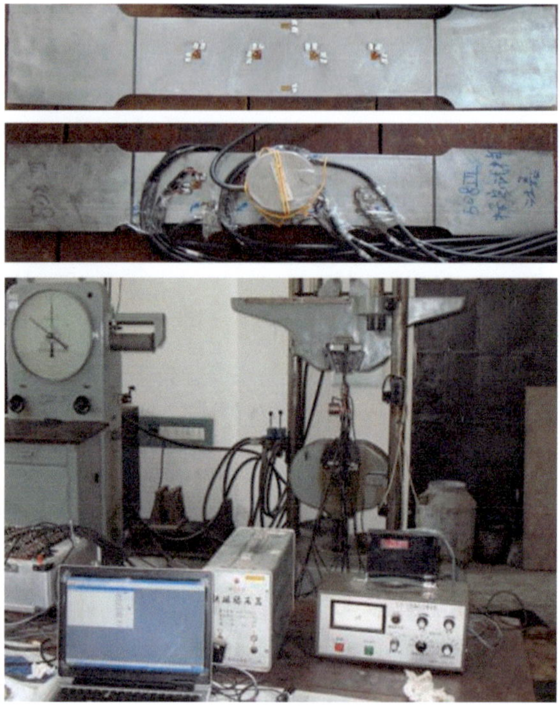

Fig. 7.18 Magnetic residual stress measuring device

influenced by multiple factors. Compared to other methods, it is challenging to quantify the values accurately, and its precision is relatively low. Additionally, the method is only applicable to ferromagnetic materials. The dedicated equipment for magnetic measurement consumes a significant amount of energy, is bulky, and the external magnetic field it generates can cause magnetic pollution and residual magnetism. The sensitivity of the material also significantly reduces the accuracy of the measurement, and the primary objects of measurement are large-scale components.

(4) Raman spectroscopy method

Raman spectroscopy is a measurement method based on the Raman scattering effect, which utilizes the sensitivity of spectral bands to stress for stress measurement. The JYT 64000 Micro-laser Raman system is shown in Fig. 7.19.

Raman spectroscopy offers high resolution and is capable of non-contact, non-destructive detection of samples at different depths with minimal sample requirements. This method can perform measurements on small samples at various depths. However, the applicability of materials is affected by Raman activity, and Raman shifts can be influenced by many external factors. In the absence of effective calibration methods, the measurement accuracy is not high.

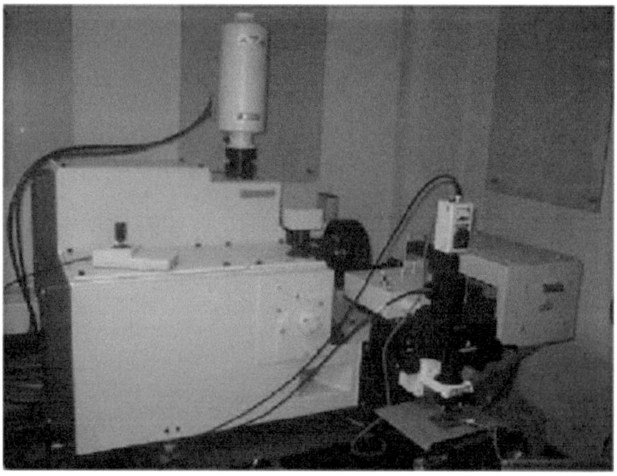

Fig. 7.19 JYT 64000 micro-laser Raman system

(5) Ultrasonic method

The ultrasonic method is a stress measurement technique based on the acoustoelastic effect. Since ultrasonic waves undergo birefringence when passing through stressed materials, their propagation speed differs between stressed and unstressed objects. There is a relationship between ultrasonic wave velocity and stress magnitude, which is utilized to measure residual stress.

Ultrasonic measurement devices are portable and capable of on-site and in-service inspections. They can non-destructively measure surface and internal stresses in components, with good directionality that allows for directional emission. Ultrasonic methods are safe and pollution-free but have relatively low measurement precision. They are particularly suitable for measuring residual stresses in large components, as well as stresses in welded joints and bolts.

(6) Nanoindentation method

As a novel measurement technique, nanoindentation utilizes the principle of stress field interference to measure residual stress. Figure 7.20 shows the DUH-211S Ultra-micro Dynamic Hardness Tester.

The primary measurement objects for nanoindentation are metals and metal-based thin films. This method has extremely high resolution in displacement and force, enabling the measurement of mechanical properties over small ranges. It can continuously record changes in displacement and force during the application or removal of load.

3. Non-contact Strain Measurement Methods

In contrast to traditional contact strain measurement methods, non-contact strain measurement methods have emerged. Contact strain measurement involves attaching strain gauges (strain rosettes) to the area of interest and using the Wheatstone bridge principle to obtain strain values before and after deformation. This method achieves high precision, up to 0.1 $\mu\varepsilon$, but its accuracy is influenced by the quality of strain

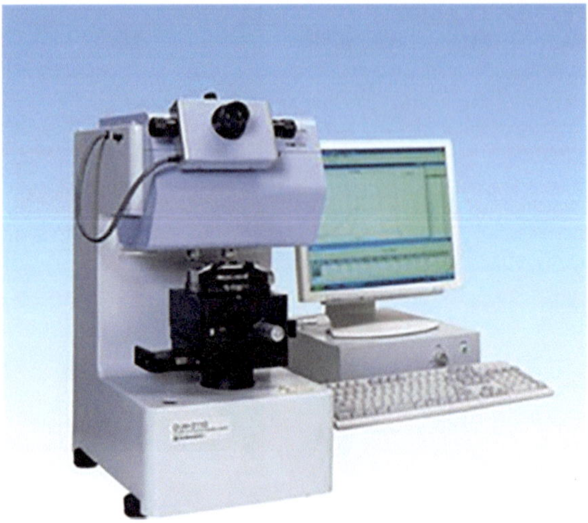

Fig. 7.20 DUH-211S dynamic ultra micro hardness tester

gauge attachment and the relative size of the gauge to the specimen. For small-scale specimens, the relatively large size of the strain gauge means that the measurement result only represents the average strain in the attached area, resulting in low spatial resolution. For example, in the case of a hermetically sealed lid for memory devices, only one strain gauge can be attached, and the thin, easily deformable lid may introduce artificial strain during the attachment process.

Non-contact strain measurement methods include Digital Image Correlation (DIC) and Electronic Speckle Pattern Interferometry (ESPI) techniques. DIC uses optical imaging principles, as shown in Fig. 7.21 with the PCC (high-speed camera software) capture site. The surface of the specimen is imaged to obtain a random gray-scale distribution. If the imaging parameters of each camera and their relative positions are known, the coordinates of surface points can be calculated using algorithms.

ESPI uses a CCD (Charge-Coupled Device) camera to capture images of an object illuminated by a laser, generating random interference patterns. These patterns change when the object is deformed under load. By calculating the phase map from these changes, displacement values can be directly obtained. ESPI can measure surface displacement, shape, and stress-strain using a single sensor.

DIC and ESPI are complementary techniques. DIC can measure large or plastic deformations but has lower sensitivity, while ESPI offers higher sensitivity and precision for small deformations.

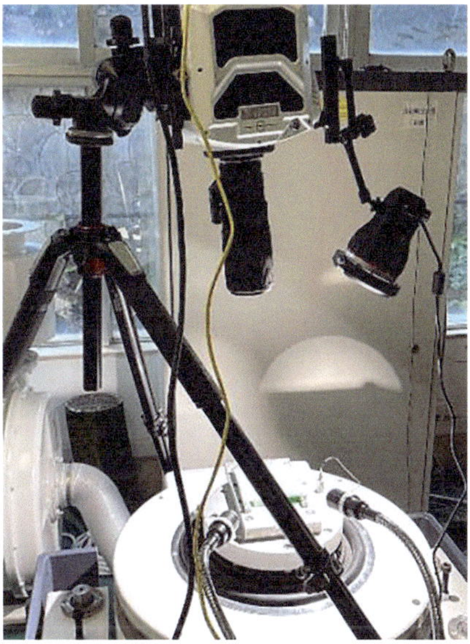

Fig. 7.21 PCC capture site

7.3 Integrated Circuit Packaging Mechanical Case Studies

7.3.1 Case Study on Vibration Characteristics of Packaging Lid

To ensure and extend the fatigue life of the packaging lid, it is necessary to obtain its dynamic mechanical properties. By using finite element analysis (FEA) software, the stress distribution map can be extracted to identify the weak regions of the packaging lid [21].

1. Modal Finite Element Analysis of Packaging Lid

Before the experiment, a 3D CAD model of a typical device packaging lid is established, as shown in Fig. 7.22. Based on the material parameters and boundary constraints given in the design drawings, the mesh is divided, and the corresponding 3D finite element model (FEM) is established, as shown in Fig. 7.23. Using FEM, modal analysis is performed on the packaging lid to obtain the modal parameters of each order. Understanding the natural frequencies, modal density, and mode shapes of the packaging lid helps determine the measurement point locations, excitation methods, and sampling frequency settings. The first eight mode shapes are particularly useful for these purposes.

2. Modal Analysis Experiment of Packaging Lid

To pre-estimate the dynamic characteristics of the packaging lid, 234 measurement points were arranged on the packaging shell. Comparative tests were conducted

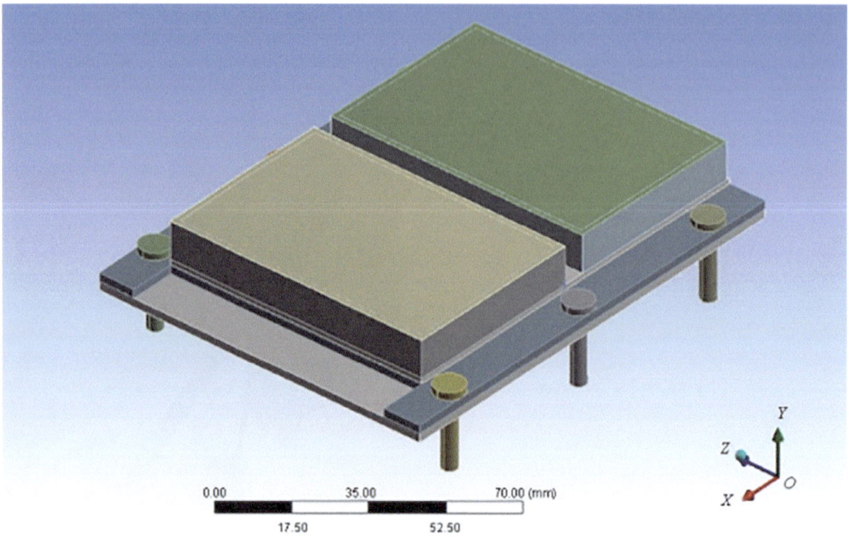

Fig. 7.22 3D CAD model of the packaging lid

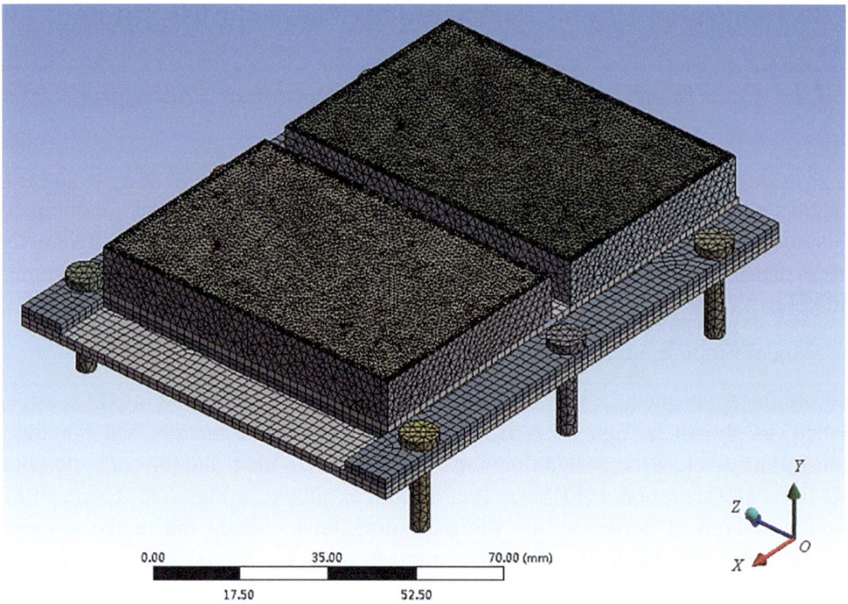

Fig. 7.23 3D FEM of the packaging lid

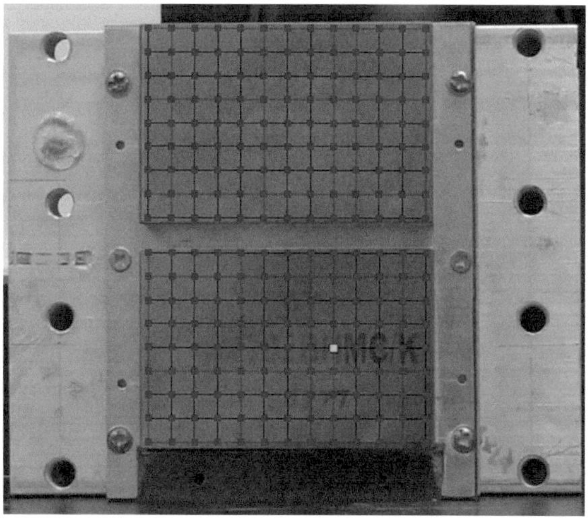

Fig. 7.24 Measurement point arrangement

under different experimental conditions, with various excitation positions and types of excitation signals. The optimal excitation position was determined through preliminary tests. Additionally, the symmetry of the frequency response function matrix and the non-linear effects of the packaging lid were tested. Finally, a non-contact scanning laser vibrometer was used for measurement, with sinusoidal swept-frequency acoustic excitation as the excitation source.

(1) Dynamic testing

The packaging lid was fixed onto a PCB, which was then secured in its actual installation orientation using fixtures. A total of 234 measurement points were selected, with their arrangement shown in Fig. 7.24. A swept-frequency acoustic excitation method was employed, and the laser vibrometer captured the vibration information at each measurement point to obtain the frequency response functions.

(2) Data post-processing

The measurement signals were analyzed and processed using the data acquisition software of the laser vibrometer. Since there were two packaging lids of the same specification, the frequencies corresponding to the same mode shapes should theoretically be identical. However, due to manufacturing process variations, some deviations occurred, resulting in closely spaced frequencies. Therefore, it was necessary to enhance the frequency resolution of the acquired data. An analysis frequency of 5000 Hz was selected, with 6400 spectral lines for the frequency response function, yielding a frequency resolution of 0.783 Hz. The frequency response functions were calculated based on the excitation and measured signals. Figures 7.25 and 7.26 present typical frequency response function curves and typical phase function curves, respectively .

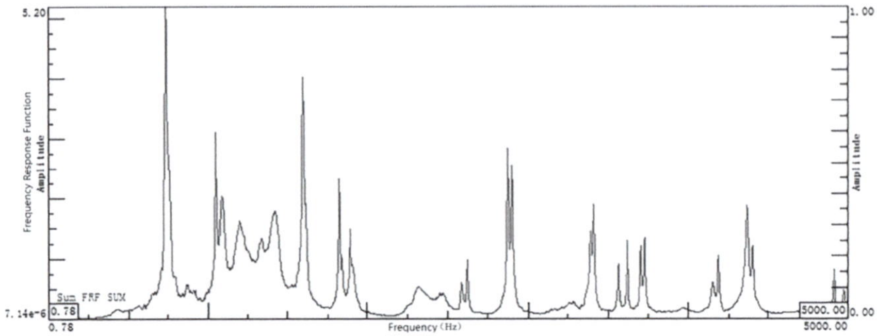

Fig. 7.25 Typical frequency response function curves

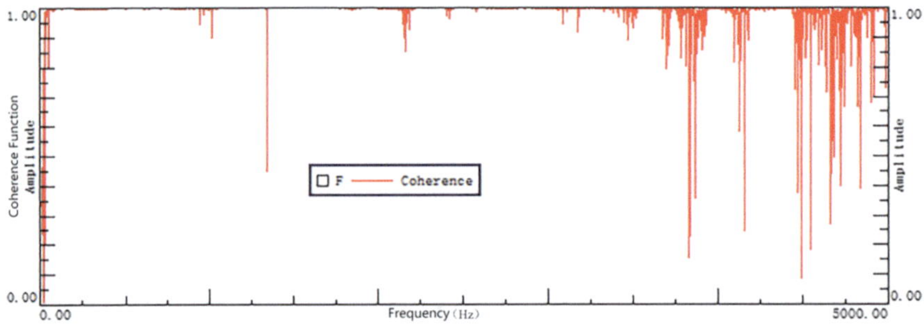

Fig. 7.26 Typical phase function curves

(3) Modal test results and analysis

The frequency response functions (FRFs) obtained from the tests were imported into the LMS Test.Lab software, and the PolyMAX method in LMS Test.Lab was used for modal identification. By summing the FRFs and using modal indicator functions, spurious modes were eliminated. A total of 26 natural frequencies and mode shapes within 5000 Hz were identified, as shown in the modal assurance criterion (MAC) plot in Fig. 7.27, which indicates that the mode shapes are mutually orthogonal. The FRFs were then correlated with the measured data for comparison, with some FRF comparisons at selected measurement points shown in Fig. 7.28. Table 7.5 presents the first eight modal parameters of the packaging lid.

The modal analysis results show that the packaging lid has a rich modal density below 2000 Hz. Given that the operational vibration bandwidth of the packaging lid is 2000 Hz, resonance is inevitable under this environmental vibration condition. This resonance will amplify the vibration level, potentially leading to fatigue failure of the lid.

3. Identification of Structural Weak Points Based on Random Vibration Analysis Results

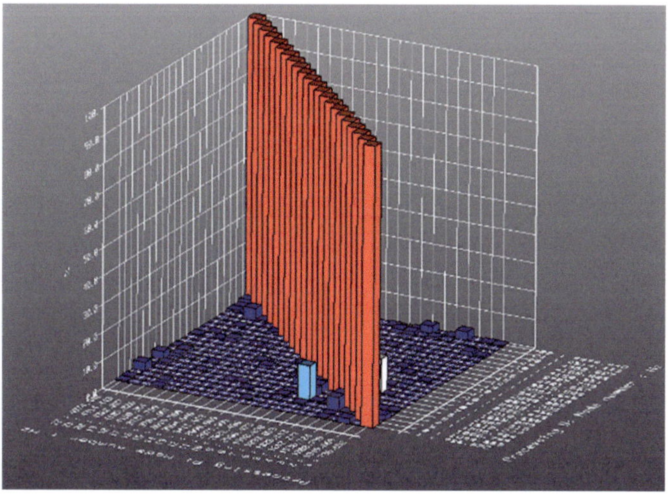

Fig. 7.27 Modal assurance criterion

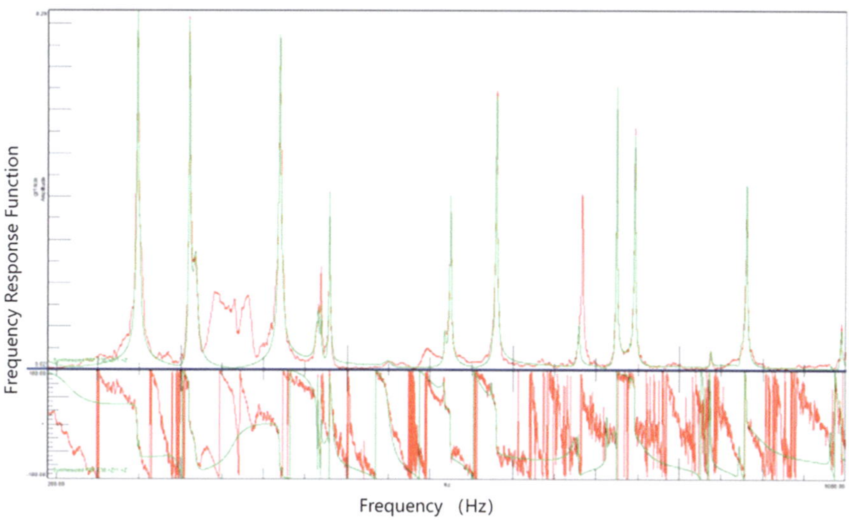

Fig. 7.28 FRF comparisons at selected measurement points

Modal analysis can provide a qualitative assessment of structural performance. However, to determine the causes of structural failure and identify weak points under vibration conditions, further in-depth calculations and analyses are required. Building on the previous section, this section introduces the identification of structural weak points based on random vibration analysis results, using the packaging lid as an example.

Table 7.5 Modal parameters of the first eight orders of the packaging lid

Modal order	Natural frequency (Hz)	Damping ratio (%)
1	736.925	0.28
2	749.436	0.34
3	1049.613	0.13
4	1088.950	0.91
5	1595.733	0.18
6	1611.995	0.14
7	1825.434	0.08
8	1843.682	0.04

(1) Model correction based on modal analysis results

By comparing the results of simulation and testing, the material parameters and boundary conditions in the finite element model (FEM) can be corrected. The comparison of natural frequencies between the simulation and testing of the packaging lid is shown in Table 7.6, with the relative error between the two being within 10%.

(2) Random vibration testing of the packaging lid

To further eliminate the influence of damping, random vibration tests were conducted to measure the acceleration response levels at critical points. The testing site for the random vibration experiment is shown in Fig. 7.29. The power spectral density (PSD) applied during the random vibration test is illustrated in Fig. 7.30, with a root mean square (RMS) value of 8*g*. The layout scheme for the measurement points during the test is depicted in Fig. 7.31, and the time-domain response of the measurement points on the packaging lid is shown in Fig. 7.32.

(3) Model testing based on random vibration testing

Table 7.6 Comparison of natural frequencies of the packaging lid

Modal order	Modal test results (Hz)	Corrected FEM simulation results (Hz)	Relative error (%)
1	736.925	734.96	0.27
2	749.436	735.72	1.83
3	1049.613	1074.30	− 2.35
4	1088.950	1074.50	1.33
5	1595.733	1476.40	7.48
6	1611.995	1479.90	8.19
7	1825.434	1811.20	0.78
8	1843.682	1815.40	1.53

Fig. 7.29 Testing site for the random vibration experiment

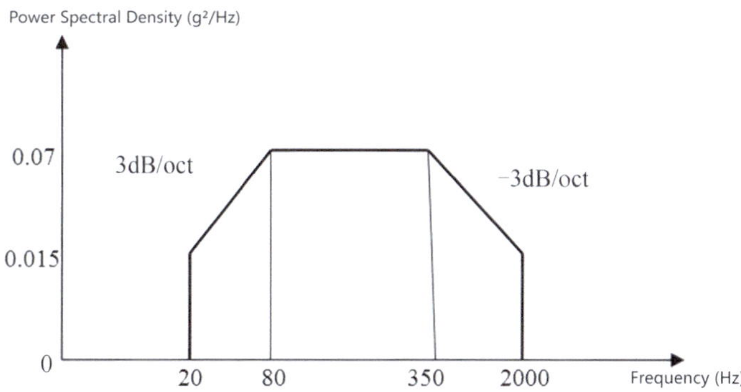

Fig. 7.30 Power spectral density applied in random vibration

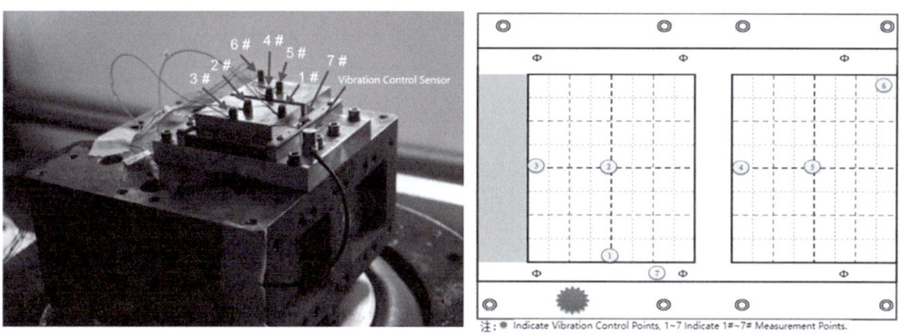

Fig. 7.31 Layout scheme for the measurement points

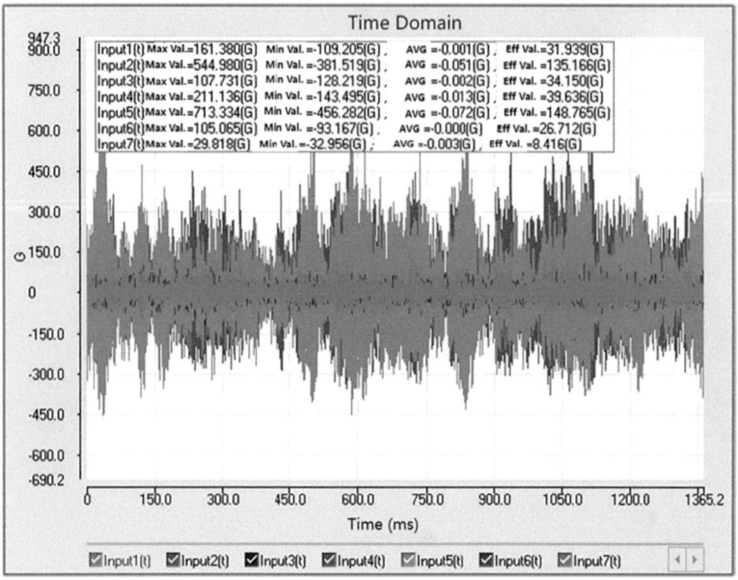

Fig. 7.32 Time-domain response of the measurement points on the packaging lid

The comparison between random vibration simulation and test results is shown in Table 7.7. By adjusting the damping ratio to 8%, the relative error between the simulation and test results was reduced to less than 5%.

(4) Extraction of structural weak points based on random vibration analysis results

Based on the verified FEM, the equivalent stress distribution map of the packaging lid under random vibration was extracted, as shown in Fig. 7.33. The results indicate that the highest stress concentration occurs at the long edge of the metal packaging

Table 7.7 Comparison of random vibration simulation and test results (measurement point 7# test data is invalid)

Measurement point	Test results of random vibration (Hz)	Verified FEM simulation results (Hz, damping ratio 2%)	Relative error (%)	Verified FEM simulation results (Hz, damping ratio 8%)	Relative error (%)
1#	31.94	35.54	11.2	32.52	1.82
2#	135.16	146.36	8.28	139.65	3.32
3#	34.15	39.24	14.9	35.31	3.40
4#	39.63	40.64	10.9	38.72	2.30
5#	148.76	149.21	3.1	142.37	4.30
6#	26.71	29.24	9.5	27.34	2.36

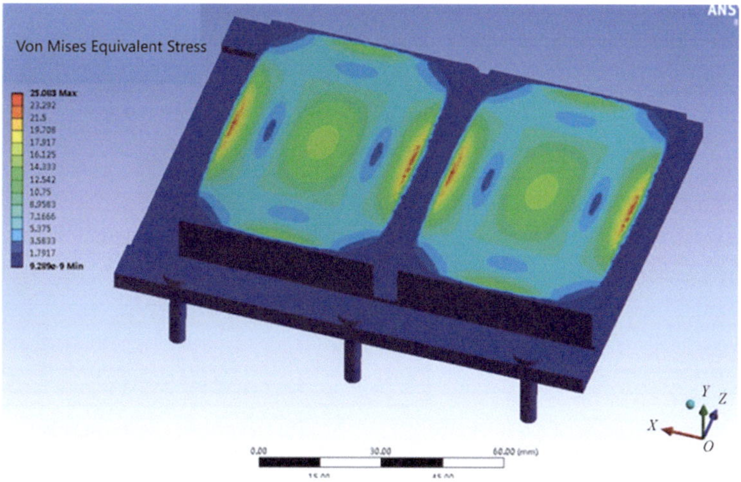

Fig. 7.33 Equivalent stress distribution map of the packaging lid under random vibration

of the packaging lid, which is identified as the weak region. This finding is consistent with the actual failure phenomena observed.

7.3.2 *Case Study on High-Density Bonding Wire Shorting*

1. Bonding Wire Shorting Based on Optical Measurement

Due to the unique structure of stacked chips, bonding wires are required to connect them to external circuits. In the service environment of electronic products, which involves external vibration, impact, or cyclic acceleration, bonding wires may experience deformation, twisting, and fracture. Additionally, short circuits due to wire-to-wire contact can occur if the spacing between adjacent bonding wires is too small under vibration or impact conditions. These issues severely threaten the safe and reliable operation of electronic products. Therefore, improving the reliability of wire bonding is key to extending the lifespan of electronic products.

The natural frequency of a device and the frequency, direction, and acceleration of resonance can all affect the deformation of bonding wires. Different components have different structures, materials, and bonding methods, leading to different natural frequencies. For the same type of device, variations in wire spacing, loop height, and wire length can result in different outcomes under the same load. Therefore, when studying the shorting behavior of device wires, it is necessary to consider not only the device's natural frequency but also the vibration results caused by key vibration parameters such as excitation frequency and acceleration.

(a) Complete Packaging of the Device (b) Device after De-capsulation

Fig. 7.34 Appearance and internal structure of the specimen

1) High-speed camera shorting test

   (1) Specimen

The appearance and internal structure of the specimen are shown in Fig. 7.34. Figure 7.34a shows the complete packaging of the device, while Fig. 7.34b shows the device after de-capsulation. In Fig. 7.34, a chip (3 mm × 4 mm × 0.1 mm) is mounted on a perforated ceramic substrate (14 mm × 14 mm × 2 mm) and connected to 100 pads via 121 bonding wires (gold wires). One hundred copper wires are externally connected, with a wire diameter of 25 μm. The minimum distance between wires is 125 μm, and the maximum distance is 325 μm.

(2) Test process

Longitudinal and transverse excitations were applied to the specimen separately (excitation perpendicular to the plane of the chip is defined as longitudinal excitation, and excitation parallel to the plane of the chip is defined as transverse excitation). The vibration tests are shown in Fig. 7.35.

(3) Test results and analysis

   ① Longitudinal excitation can induce resonance but does not cause wire-to-wire contact (shorting). In contrast, transverse excitation leads to shorting of the bonding wires. The frequency range for resonance induced by longitudinal excitation is 2000–3000 Hz. Under longitudinal excitation with an acceleration of 250 m/s^2, the vibration behavior of the bonding wires across different frequency bands is shown in Fig. 7.36. For transverse excitation, resonance occurs within the frequency range of 500–2500 Hz.

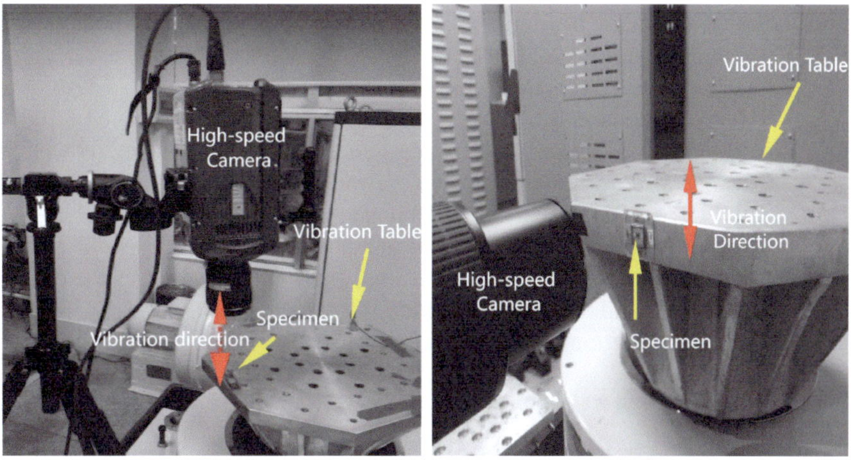

(a) Longitudinal Excitation (b) Transverse Excitation

Fig. 7.35 Vibration tests

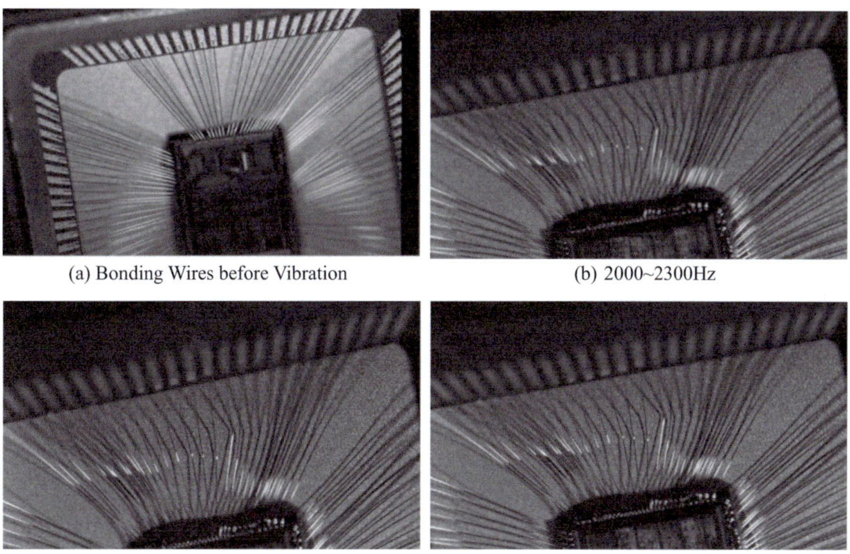

(a) Bonding Wires before Vibration (b) 2000~2300Hz

(c) 2300~2500Hz (d) 2500~3000Hz

Fig. 7.36 Vibration of bonding wires under longitudinal excitation with an acceleration of 250 m/ s^2 at different frequencies

② As the excitation acceleration increases, the vibration of the bonding wires becomes more severe. Under transverse excitation, when the acceleration reaches 250 m/s^2, shorting of the bonding wires occurs.

③ At an acceleration of 250 m/s^2 and within the frequency range of 2000–2300 Hz, shorting occurs between bonding wires 91 and 92, as well as between wires 96 and 97. When the frequency is in the range of 2300–2500 Hz, shorting occurs between bonding wires 94 and 95. This shorting leads to partial short-circuiting of the chip's circuitry, causing functional defects and raising reliability concerns. These findings identify the critical frequencies and locations for shorting, providing a basis for experiments aimed at monitoring electrical performance to detect shorting.

2) Wire bond shorting test using a microscopic laser vibrometer

(1) Specimen

The specimen is shown in Fig. 7.37a. After the lid is removed, the bonding wires to be tested are shown in Fig. 7.37b.

(2) Test process

The bonding wire modal test uses an optical microscope for observation. The MSA-600 Microscopic Laser Vibrometer is shown in Fig. 7.38. Two bonding wires were tested in this experiment, referred to as Bonding Wire 1 and Bonding Wire 2. Under a 5x magnification lens, the bonding wires appear as shown in Fig. 7.39. Only about half the length of the bonding wire can be observed within the field of view of the 5x lens, as shown in Fig. 7.39a, labeled as Part 1. Therefore, the test was conducted in segments. The remaining part of the bonding wire is shown in Fig. 7.39b, labeled as Part 2. The modal shapes of the bonding wires were then stitched together.

The two segments of the bonding wires were tested separately. A random swept-frequency signal was applied to the specimen through a piezoceramic actuator, which

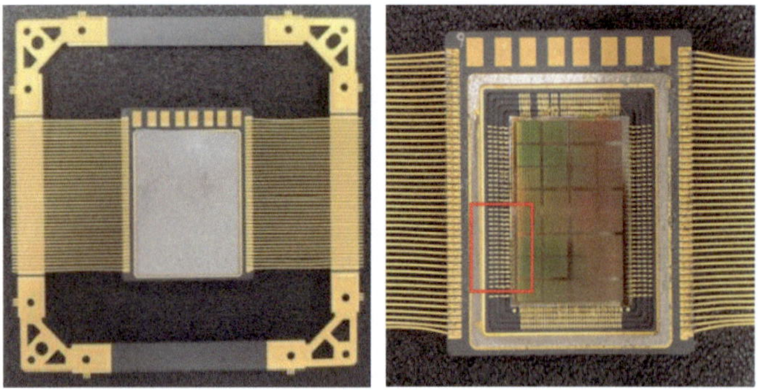

(a) Specimen (b) Bonding Wire to be Tested

Fig. 7.37 High-capacity SRAM memory

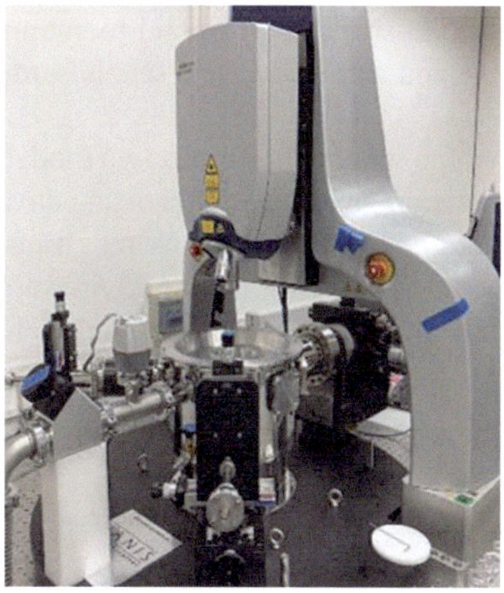

Fig. 7.38 MSA-600 microscope-based laser vibrometer

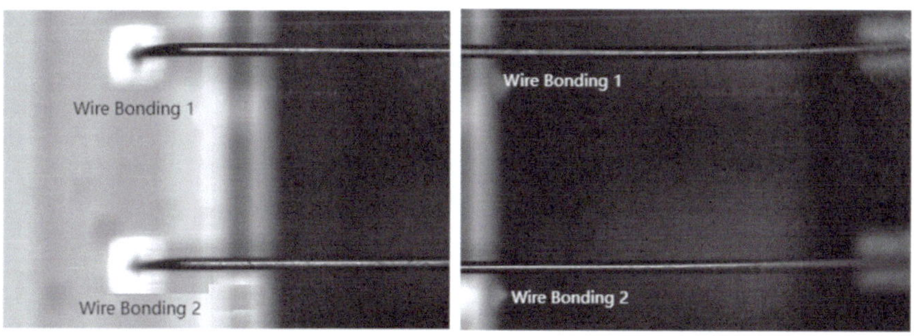

(a) Part 1 (b) Part 2

Fig. 7.39 Bonding wires under 5x magnification

provided the vibrational excitation. The signal generated by the signal source was used as the reference signal. The frequency response functions and response spectra relative to the reference signal were calculated for all measurement points. The files from the two segments were then stitched together to obtain the final response spectrum.

(3) Test results and analysis

The vibration velocity signals in the frequency domain were measured at each measurement point. The transfer functions of each measurement point relative to the reference point were calculated using the PSV software and exported as unv format files. Figure 7.40 shows the final frequency response function. The specimen exhibits two modal shapes within 125 kHz. The material properties of the two bonding wires may differ, resulting in different natural frequencies. Bonding Wire 1 has natural frequencies of 43.5 and 90.8 kHz, while Bonding Wire 2 has natural frequencies of

41.7 and 91.4 kHz. The mode shapes correspond to symmetric bending about the center of the bonding wire and anti-symmetric second-order bending, as shown in Figs. 7.41 and 7.42.

It can be concluded that the specimen exhibits two modal shapes within 125 kHz. Excitation frequencies in the range of 40–45 kHz can activate the first-order mode. Under longitudinal excitation of the bonding wires, transverse vibrations within the device plane are minimal. Therefore, bonding wires are unlikely to experience shorting under longitudinal excitation but may undergo deformation.

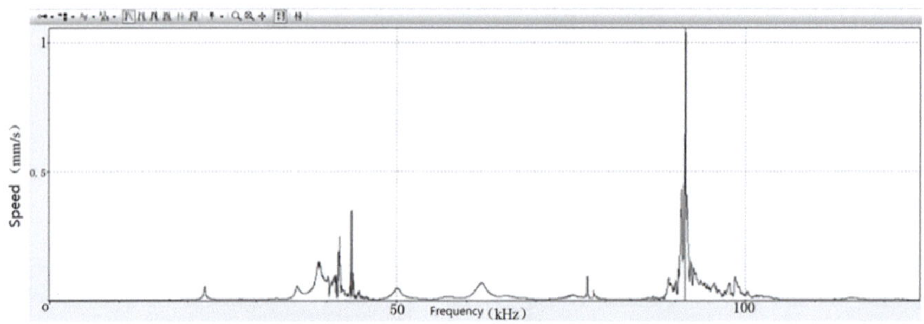

Fig. 7.40 Final frequency response function plot

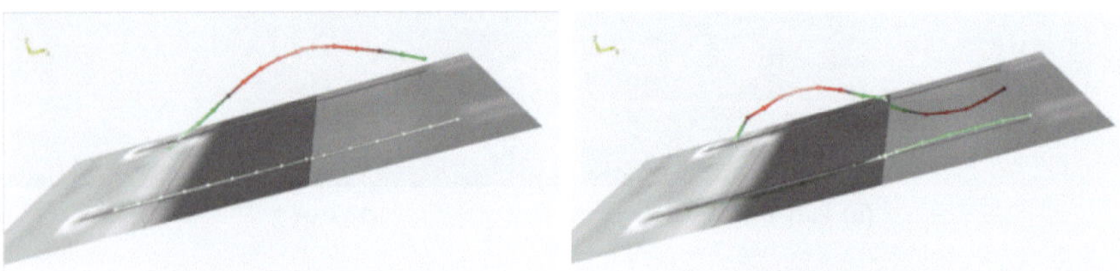

(a) First-order Modal of Bonding Wire 1 (b) Second-order Modal of Bonding Wire 1

Fig. 7.41 Modal of bonding wire 1

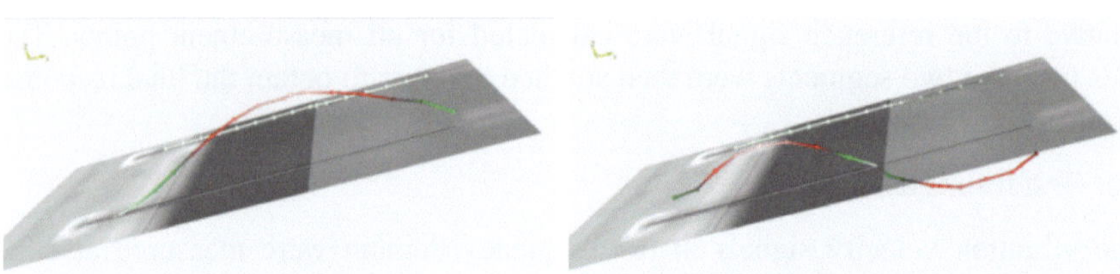

(a) First-order Modal of Bonding Wire 2 (b) Second-order Modal of Bonding Wire 2

Fig. 7.42 Modal of bonding wire 2

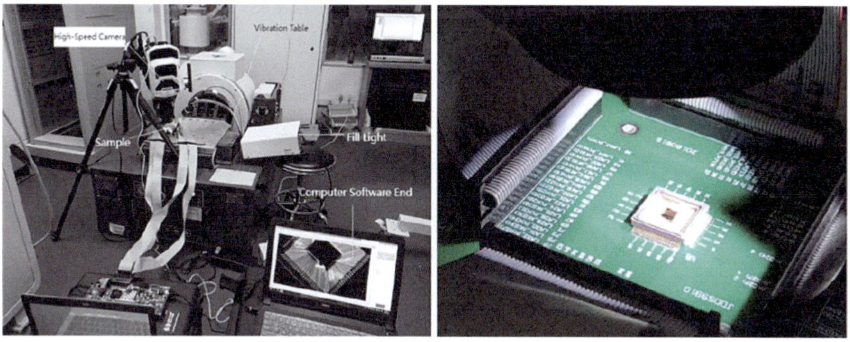

Fig. 7.43 Vibration shorting in-situ monitoring system

2. Bonding Wire Shorting Based on Electrical Measurement

To address the issue of missed detections, a new online monitoring method for bonding wire shorting based on electrical measurement has been proposed. By designing a monitoring circuit and developing a driver program, the electrical method for online monitoring was realized, and high-speed camera technology was used to complete the experimental validation.

The Xilinx KC705 FPGA kit was used as the development board. A test circuit board was designed according to the specifications of the integrated circuit pins. The FPGA development board and the test circuit board were connected through an FMC adapter board. An automatic signal acquisition program for the FPGA was developed and programmed. Communication between the test chip and the host computer was realized through a serial port debugging assistant. The vibration shorting in-situ monitoring system based on the proposed electrical method was built on a vibration test platform, as shown in Fig. 7.43.

By comparing the electrical and optical methods for shorting detection, it was found that the proposed electrical method for online monitoring can achieve in-situ detection of shorting events within 1 μs during vibration testing. This provides an effective means for vibration reliability testing and qualification of high-density bonding devices such as SiP, filling the gap in the in-situ online monitoring technology for instantaneous shorting under vibration and impact conditions.

Under vibration conditions of 2000–3000 Hz and 20 g, the static and dynamic data were analyzed and displayed using MATLAB code, as shown in Fig. 7.44. The dynamic data includes four detection cycles. Columns marked with vertical lines indicate differences between static and dynamic data. (Under normal conditions, shorting can only occur between adjacent pins. To facilitate analysis, only the detection results of two adjacent pins driven by the pins are retained.) The detection results show that pins 3 and 4, pins 10 and 11, and pins 18 and 19 experienced shorting. When shorting is detected, the two columns of the shorting pins are marked with vertical

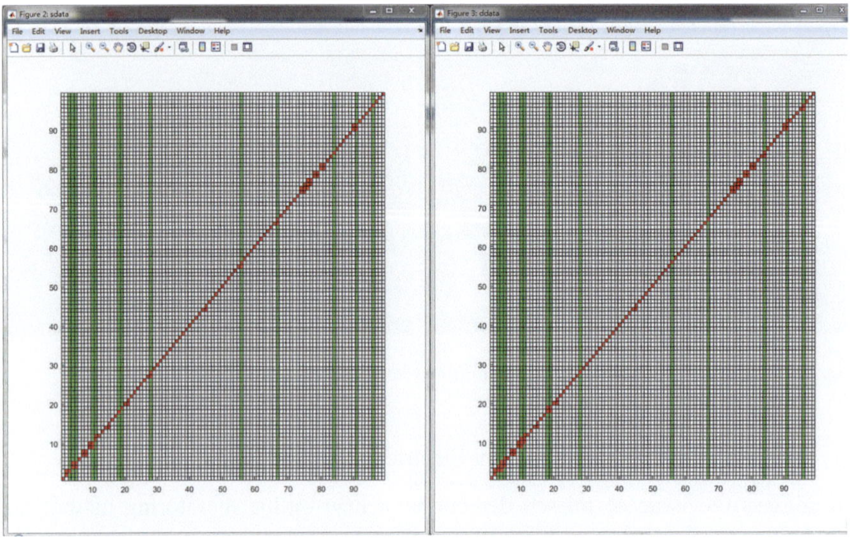

Fig. 7.44 Comparison of static and dynamic data

lines, and the corresponding rows are marked with diagonal lines. For example, if pins 3 and 4 short, the detection results show vertical lines in columns 3 and 4, and diagonal lines in squares (3,3), (3,4), (4,3), and (4,4). Other isolated vertical lines do not meet the shorting conditions and may be due to insufficient detection cycles, resulting in missed detections, or internal chip functions causing false detections.

To verify the effectiveness of the proposed electrical method for online monitoring, the project used a high-speed camera to perform synchronous high-speed video analysis of the samples. By slowing down the playback of the recorded videos, further observation and analysis were conducted. The results showed that shorting occurred between pins 3 and 4, pins 10 and 11, and pins 18 and 19, which is completely consistent with the results obtained using the electrical method. This validates the rationality and accuracy of the proposed electrical method for online monitoring. Figure 7.45 shows the shorting image of pins 3 and 4 captured after reducing the playback frame rate of the recorded video.

Tests have shown that for high-density bonding wires, shorting mainly occurs in the *XY* direction (horizontal direction) under vibration and impact conditions, with the minimum spacing being 125 μm. The frequency range for shorting is primarily concentrated between 2000 and 3000 Hz. The project team further narrowed the excitation frequency range and found that shorting occurs when the frequency reaches 2000–2300 Hz. Therefore, based on the above analysis, control of shorting failure in high-density bonding wires can be optimized from three main aspects: increasing the first-order natural frequency of the bonding wires, reducing the vibration excitation response in the *XY* direction, and increasing the minimum spacing between bonding wires.

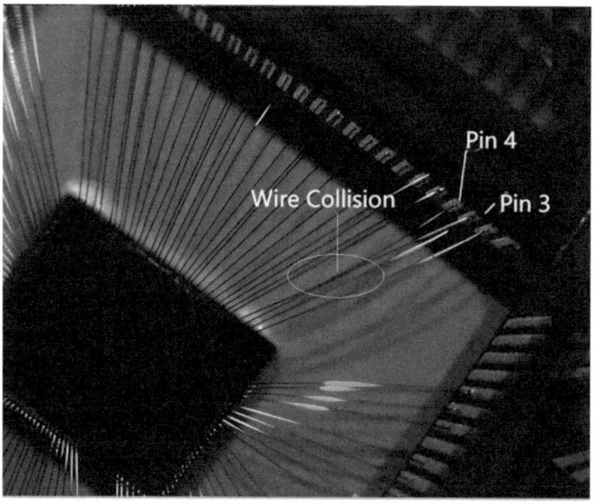

Fig. 7.45 Shorting image of bonding wires between pin 3 and pin 4

References

1. Sun X, Fang X, Guan L (2009) Mechanics of materials, 5th edn. Higher Education Press, Beijing
2. Yang G (2013) Introduction to elasto-plastic mechanics. Beijing University of Aeronautics and Astronautics Press, Beijing
3. Li Z (2012) Applied fracture mechanics. Beijing University of Aeronautics and Astronautics Press, Beijing
4. Heylen W, Lammens S, Sas P (1997) Modal analysis theory and testing. Katholieke Universiteit Leuven, Leuven, Belgium
5. Su W, Luo W, Dong X et al (2020) Ultimate strength analysis of local thinning tee pipe considering plastic strengthening effect. J Phys Conf Ser 1650(2):022024
6. Wolter A, Baumeister H, Yeni C et al (2020) A study of the board level reliability of large 16FF wafer level package for RF transceivers. In: 2020 IEEE 70th electronic components and technology conference (ECTC). IEEE, pp 1684–1690
7. Su W, Luo W, Nie Z et al (2020) A wideband folded reflectarray antenna based on single-layered polarization rotating metasurface. IEEE Access 8:158579–158584
8. Luo W, Su W, Nie Z et al (2021) Vibration characteristic measurement method of MEMS gyroscopes in vacuum, high and low temperature environment and verification of excitation method. IEEE Access 9:129582–129593
9. Li D (2012) Advanced structural dynamics. Science Press, Beijing
10. Qiu J (2015) Structural dynamics and its applications in aerospace engineering. University of Science and Technology of China Press, Hefei
11. Ndip I, Andersson K, Kosmider S et al (2020) A novel packaging and system-integration platform with integrated antennas for scalable, low-cost and high-performance 5G mm wave systems. In: 2020 IEEE 70th electronic components and technology conference (ECTC). IEEE, pp 101–107
12. Hou IC, Tseng PH, Jou JJ et al (2020) Design of 4-channel 25 Gbaud/s PAM-4 optical transmitter module for short reach applications. In: 2020 IEEE 70th electronic components and technology conference (ECTC). IEEE, pp 2272–2277
13. Jouve A, Lagoutte E, Crochemore R et al (2020) A reliable copper-free wafer level hybrid bonding technology for high-performance medical imaging sensors. In: 2020 IEEE 70th electronic components and technology conference (ECTC). IEEE, pp 201–209

14. Wang S (2007) Reliability analysis and structural parameter optimization of stacked chip packaging. Harbin University of Science and Technology, Harbin
15. Cao W (2008) Study on the performance of monocrystalline copper bonding wires. Lanzhou University of Technology, Lanzhou
16. Kuo KH, Chiang CY, Chen KH et al (2020) The board level reliability performance and process challenges of ultra-thin WLP (package height < 250 μm). In: 2020 IEEE 70th electronic components and technology conference (ECTC). IEEE, pp 2118–2123
17. Wang X, Yong G, Zhang Y et al (2019) Reliability and simulation of composite BGA solder joint connecting LTCC substrates. In: 2019 20th International conference on electronic packaging technology (ICEPT), pp 1–4
18. Jinfeng G, Chunyue H, Maolin L et al (2021) Sop welding joint bending stress finite element analysis and optimization. In: 2021 22nd International conference on electronic packaging technology (ICEPT), pp 1–4
19. Su W, Luo W, Dong X et al (2021) Residual stress analysis of laser welding between output needle and helix. J Phys Conf Ser 1802(2):022088
20. Su W, Luo W, Dong X et al (2021) Finite element analysis of subsurface damage of optical glass after grinding. J Phys Conf Ser 1802(2):022087
21. Su W, Luo W, Dong X et al (2021) Harmonic response analysis for ball grid array package using computer finite element simulation. J Phys Conf Ser 2033(1):012012

Chapter 8
Failure Analysis Techniques for Integrated Circuit Packaging

Failure analysis examines the structure, process, and materials of integrated circuit packaging. It identifies weak points, analyzes defect formation and evolution, and determines failure mechanisms. This helps recognize defects, classify failure modes, and confirm root causes. Based on these findings, engineers can propose improvements to enhance reliability [1, 2]. This chapter introduces key aspects of failure analysis, including procedures and techniques.

8.1 Key Aspects of Packaging Failure Analysis

Failure analysis uses various testing and analytical methods to identify failure modes, mechanisms, and potential root causes. Engineers then suggest improvements to prevent recurrence. For example, Fan-Out packaging often experiences process-related and stress-induced failures. Table 8.1 [3–8] lists common failure locations, along with their corresponding failure modes.

Different failure locations have different failure modes and mechanisms. Failure analysis helps confirm them. The main steps include identifying the failure object, determining the failure mode, investigating the failure mechanism, and inferring the cause. Engineers then suggest design and process improvements.

8.2 Failure Analysis Procedures for Packaging

Failures in integrated circuit packaging can occur in many forms. They may appear in external packaging, internal packaging, the chip surface, the chip underside, or the interface. A structured, step-by-step procedure is essential. This ensures no important

B. Zhou et al., *Reliability Technology for Integrated Circuit Packaging*,
https://doi.org/10.1007/978-981-95-3885-0_8

Table 8.1 Failures related to integrated circuit packaging process and stress (taking fan-out packaging as an example)

Component	Process-related failure mode	Failure mode under stress
Chip	Ultra-thin wafers are fragile and prone to cracking Improper thinning stress can cause the chip to deform or warp Chips may crack during dicing Poor chip adhesion layer misalignment Filling voids in the adhesive layer may cause delamination, etc	Under thermal stress, chips may deform or warp Under thermal stress, molded plastic may damage the chip Mechanical stress may cause the chip to crack Thermal stress lead to delamination failures Mismatch of thermal expansion coefficients can cause delamination Electrostatic damage Overvoltage damage etc
Component	Process-related failure mode	Failure mode under stress
Solder ball	Solder ball voids Solder ball cold soldering Solder ball size issues Pad detachment	Under high-temperature reflow soldering, solder ball melting causes short circuits Under high-temperature reflow soldering, solder ball cracking Mismatch of thermal expansion coefficient of filling material causes cracks and delamination, leading to solder ball cracking
TSV	Insufficient or excessive etching depth of TSV Top diameter error Bottom diameter error Sidewall verticality non-compliance and excessive roughness Copper diffusion caused by incomplete barrier layer TSV filling integrity TSV voids	Deformation caused by stress during TSV filling Cracks caused by copper precipitation / delamination of dielectric layer Thermal expansion coefficient mismatch causing plastic deformation of TSV Electromigration Stress migration Sidewall cracking Cracks between TSV bottom and metallization interface Oxidation

(continued)

Table 8.1 (continued)

Component	Process-related failure mode	Failure mode under stress
Microbumps	Small bump size leading to cold soldering Large bump size causing short-circuit failure between adjacent bumps Holes in the bump Cushion effect	Electromigration Deformation under thermal stress Cracks Cracking
Redistribution layer (RDL)	RDL voids RDL size deviation	Copper electromigration causing short circuits between RDL lines Copper electromigration causing RDL voids Thermo-mechanical stress causing RDL cracks

information is lost. Design, assembly, and manufacturing require specific defect localization and analysis techniques.

New packaging technologies have emerged in recent years. These include ultra-precision surface mounting, flip-chip (FC), packaging-on-packaging (PoP), packaging-in-packaging (PiP), ultra-thin wafer processing, chip stacking, and through-silicon via (TSV) technology [9–11]. These technologies create new challenges for failure analysis. High integration density, complex structures, and diverse materials make traditional methods less effective. Improvements are needed to meet the demands of 3D system failure analysis. Table 8.2 lists failure causes, mechanisms, and corresponding analysis methods for integrated circuit packaging.

Failure analysis follows a specific order. External analysis comes first, then internal analysis. Non-destructive analysis is done before destructive analysis. Non-destructive techniques do not affect device performance. They do not change existing defects or failures in the packaging. They also do not introduce new ones. Before non-destructive analysis, electrical methods identify the failure mode of the defective device. X-ray inspection is a non-destructive technique. However, X-rays may degrade the electrical properties of some devices. So, X-ray inspection must be done after electrical evaluation. Destructive techniques, such as decapsulation, permanently change the packaging. These methods may also alter existing defects. For example, decapsulation may destroy evidence of failures inside the packaging. A well-structured failure analysis procedure is essential. Non-destructive techniques have resolution limits. Defects smaller than 1 μm are hard to detect using standard X-ray or acoustic microscopes. High-resolution microscopes, such as C-SAM, are needed for analysis. Figure 8.1 shows the failure analysis procedure for packaging.

Table 8.2 Causes, mechanisms, and corresponding failure analysis methods of IC packaging failures [12–30]

Failure mode	No	Failure causes and mechanisms		Failure analysis methods
A Cracks	1	Thermal–mechanical mismatch	Chip solder fatigue BGA solder ball fatigue Embedded passive component damage Spacer cracking between chips Underfill cracking Chip metal wire open circuit	Stress analysis Thermal deformation interferometer, speckle interferometer Deformation analysis Image correlation, X-ray diffraction Defect isolation Magnetic microscopy localization, time-domain reflectometry, LIT, TIVA, OBIRCH Crack Detection SAM, optical microscope cross-section analysis, SEM or FIB/SEM
	2	Mechanical impact	Chip insulation layer cracking Organic substrate cracking Solder ball cracking (drop impact)	
	3	Moisture-induced expansion	Encapsulation material cracking Chip cracking Substrate cracking	
	4	Volume shrinkage or expansion due to reactions (e.g., curing)	Encapsulation material cracking Chip cracking	
	5	Internal stress (e.g., moisture evaporation at high temperature)	Encapsulation material cracking Chip cracking	
B Interface delamination	1 ~ 5	Same failure mechanisms as A cracks 1 ~ 5	Chip insulation layer delamination Underfill delamination Stacked chip delamination Organic substrate delamination Encapsulation material delamination	Stress analysis Thermal deformation interferometer, speckle interferometer Deformation analysis Image correlation, X-ray diffraction Crack detection SAM, optical microscope cross-section analysis, or SEM, FIB/SEM, FIB/TEM

(continued)

Table 8.2 (continued)

Failure mode	No	Failure causes and mechanisms		Failure analysis methods
	6	Interface reactions causing adhesive detachment (e.g., humidity, oxidation, contamination, etc.)	Underfill delamination Encapsulation material delamination Organic substrate delamination	Crack detection SAM, optical microscope cross-section analysis, or SEM, FIB/SEM, FIB/TEM Surface analysis TOF–SIMS、XPS、AES、TEM + EDX、TEM + EELS
Failure mode	No	Failure causes and mechanisms		Failure analysis methods
C Voids and porosity	1	Mechanical creep	Chip interconnect voids BGA solder ball voids	Defect isolation Magnetic microscopy localization, time-domain reflectometry, LIT, TIVA, OBIRCH Void detection X-ray microscopic analysis, optical microscope cross-section analysis, SEM or FIB/SEM (EDX, WDX, EBSD, metal interconnect X-ray diffraction analysis)
	2	Diffusion	Chip UBM lift-off Chip interconnect or via voids Bonding wire lift-off BGA solder ball lift-off	
	3	Electromigration	Chip metal wire and solder ball voids Substrate solder ball, metal wiring, and via voids Via voids	
	4	Thermal migration	Chip metal wire and solder ball voids Substrate solder ball, metal wiring, and via voids	

(continued)

Table 8.2 (continued)

Failure mode	No	Failure causes and mechanisms		Failure analysis methods
D Material corrosion and cracking	1	Chemical corrosion	Bonding wire lift-off	Defect Isolation Magnetic microscopy localization, time-domain reflectometry, LIT, TIVA, OBIRCH Mechanism Analysis Optical microscope cross-section analysis or FIB/SEM (EDX/WDX, TEM, TOF–SIMS, XPS, FTIR spectrum analysis), mechanical testing, TGA, DMA, DSC (aging), EBSD (grain analysis)
	2	Electrochemical corrosion	Bonding wire lift-off	
	3	Aging	Organic substrate cracking or delamination Underfill cracking or delamination	
	4	Grain coarsening, phase separation	Bonding wire cracking Chip solder ball fatigue BGA solder ball fatigue	

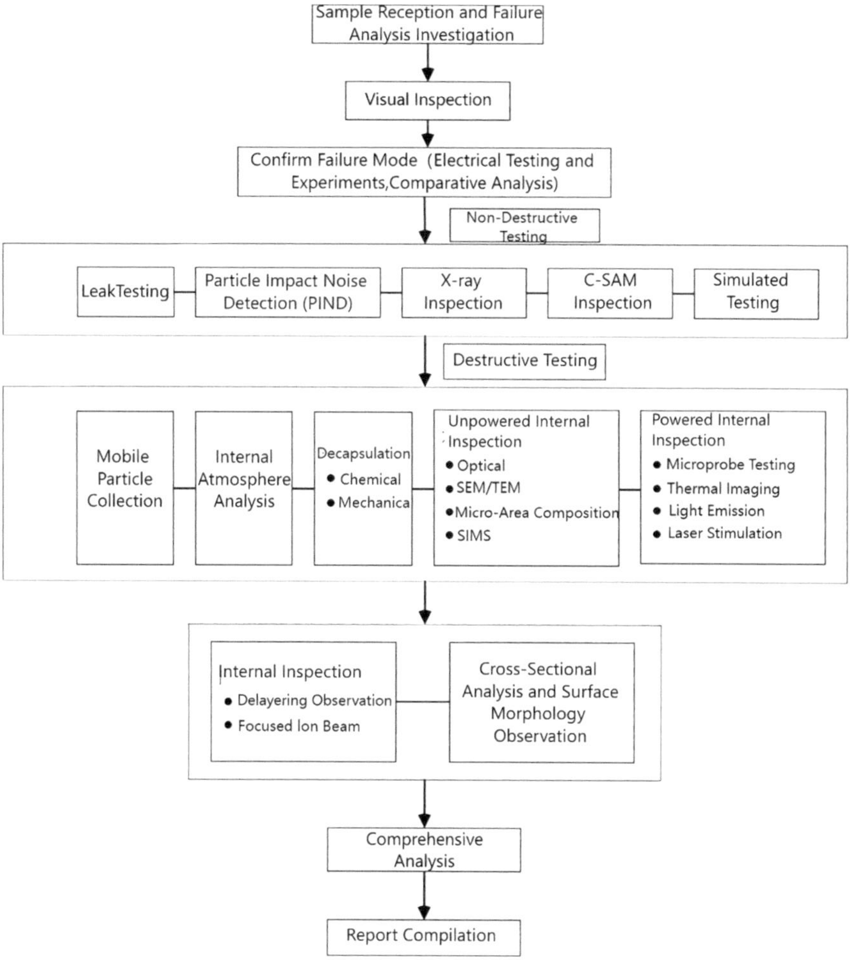

Fig. 8.1 Failure analysis procedure for packaging

8.3 Non-destructive Failure Analysis Techniques

8.3.1 Appearance Analysis Techniques

Appearance analysis provides key clues for further investigation. Optical microscopes are commonly used. They help detect surface contamination, foreign particles, pin discoloration, wire breakage, and tin whiskers.

1. Basic Principles and Key Performance Indicators

Optical microscopes are widely used for appearance analysis. They use optical lenses to magnify and observe surface defects in packagings. These microscopes are easy

Fig. 8.2 Metallurgical Microscope

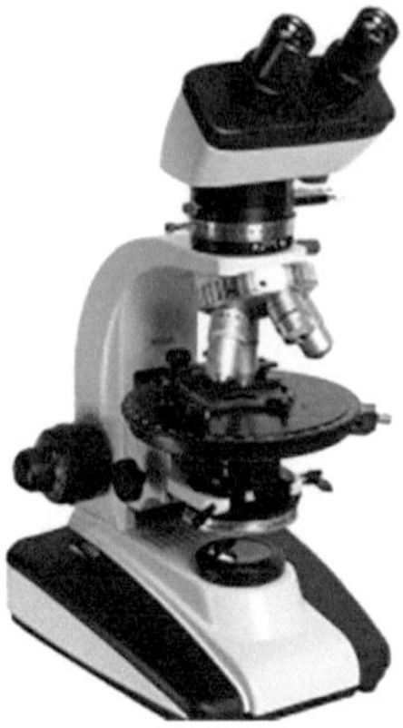

to operate but have limited magnification. The maximum magnification is 2000x, and the resolution is usually above the micrometer level. There are two main types of optical microscopes: stereo microscopes and metallurgical microscopes. Stereo microscopes have lower magnification but provide a greater depth of field.

Both types use the same imaging principle. They form images through a combination of eyepieces and objective lenses. The total magnification is the product of the eyepiece and objective lens magnifications.

The illumination system affects key optical properties. It determines objective lens magnification, numerical aperture, resolution, depth of field, and field curvature. The eyepiece system allows users to select different magnifications. Figure 8.2 shows a metallurgical microscope. Samples can be examined using reflected light. If the sample is transparent, transmitted light can also be used.

Unlike electron microscopes, optical microscopes do not require a high vacuum or conductive samples. However, their resolution is limited by the wavelength of visible light. The resolution d is related to the wavelength λ, as shown in the following equation:

$$d = \frac{\lambda}{2A} \tag{8.1}$$

where A represents the numerical aperture.

2. Case Analysis

Figures 8.3 and 8.4 show images captured by a stereo microscope. Figure 8.3 presents the external appearance of an integrated circuit. Figure 8.4 displays contamination on IC pins. Contaminated pins can cause poor contact after soldering.

Fig. 8.3 Stereo microscope image of an integrated circuit

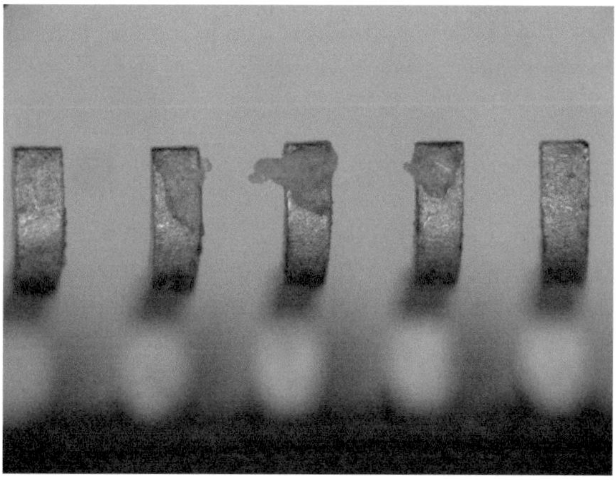

Fig. 8.4 Contamination observed on IC pins under a stereo microscope

8.3.2 X-ray Microscopy Analysis Technology

X-ray microscopy allows non-destructive analysis of internal defects in packages. It can detect voids in solder joints, wire bonding shorts, pin fractures, and sealing defects in package shells. X-ray microscopy is classified into 2D X-ray and 3D X-ray analysis. This section introduces 2D X-ray analysis. Section 8.5 will discuss 3D X-ray analysis.

1. Basic Principles and Key Performance Indicators

X-ray microscopy detects internal defects by analyzing differences in X-ray absorption and transmission across materials. It measures X-ray intensity after passing through different parts of the package [31–33]. When X-rays penetrate a package, their intensity decreases. The attenuation level depends on the material's attenuation coefficient and the sample's thickness. The relationship follows:

$$I = I_0 \cdot e^{-\mu \cdot L} \tag{8.2}$$

where I represent the intensity of the X-ray after passing through the packaging and undergoing attenuation. I_0 is the initial intensity of the X-ray before passing through the packaging. μ is the attenuation coefficient of the packaging material.

By performing a line integral on the function $\mu(x, y)$, we can obtain the X-ray intensity I passing through the 2D object.

$$I = I_0 \cdot e^{-\int_L \mu(x,y)dL} \tag{8.3}$$

Figure 8.5 shows the physical factors that affect X-ray attenuation.

2D X-ray analysis technology can detect packaging defects in real-time. It can identify the type of hidden defects and achieve accurate localization. It is also easy to operate. The resolution of 2D X-ray analysis depends on the magnification. The following equation explains the simple relationship between magnification and geometric distance:

$$M = \frac{L_1 + L_2}{L_1} \tag{8.4}$$

where M is the magnification. $L_1 + L_2$ is the distance between the X-ray source and the detector. L_1 is the distance between the X-ray source and the sample. Equation (8.4) shows that the geometric magnification of a sample closer to the X-ray source is higher than that of a sample closer to the detector.

Another important factor affecting magnification, especially at high magnifications, is the geometric size of the X-ray source (focal point). Figure 8.6 shows the relationship between the focal point size and image clarity. The larger the focal point size, the more blurred the image at the object's edge. This relationship can be expressed by the following equation:

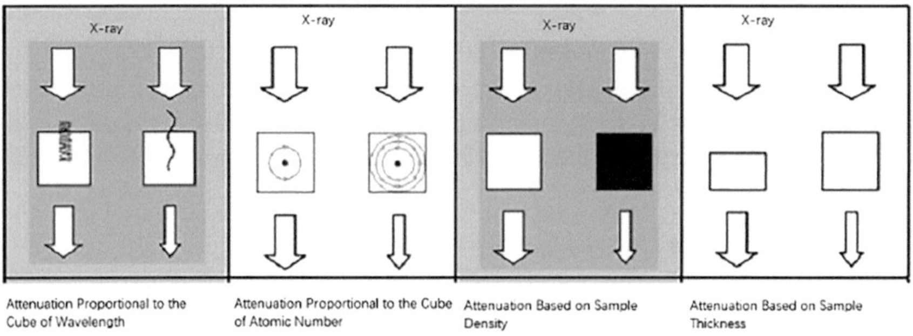

Fig. 8.5 The physical factors that affect X-ray attenuation

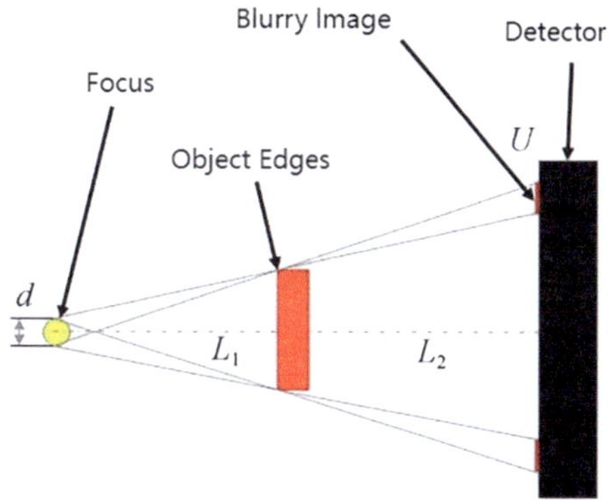

Fig. 8.6 The relationship between focal point size and image clarity

$$U = d_{\text{X-Ray}} \frac{L_2}{L_1} \tag{8.5}$$

Using a micro-focus tube can reduce the diameter of the X-ray source to the nanoscale. However, only low-power X-ray tubes can achieve this. In high-power modes, the diameter of the X-ray source is greater than 2 μm, depending on the type and manufacturer of the tube.

With the miniaturization and increased complexity of integrated circuit packaging, 2D X-ray analysis faces technical challenges. There are many obstructions and interference factors in the field of view. This makes it difficult, or even impossible, to detect small defects. In such cases, slicing preparation is often needed. However, this can lead to human-induced damage. The low resolution of 2D X-ray imaging is a common bottleneck. This has led to the development of 3D X-ray analysis technology, which has become an important method for failure analysis of microelectronic packaging.

Currently, X-ray micro-CT systems for integrated circuit packaging operate at voltage levels between 150 and 250 kV, with current levels in the tens of milliamps. The spatial resolution reaches the micron level. By adjusting the current and voltage, the penetration strength can be controlled. By adjusting the distance between the sample and the X-ray source, the magnification can be adjusted. This allows for the analysis of different packaging products.

2. Analysis Case

Figure 8.7 shows the 2D X-ray radiograph of chip bonding voids. These voids can cause poor heat dissipation in integrated circuits, which affects their long-term reliability.

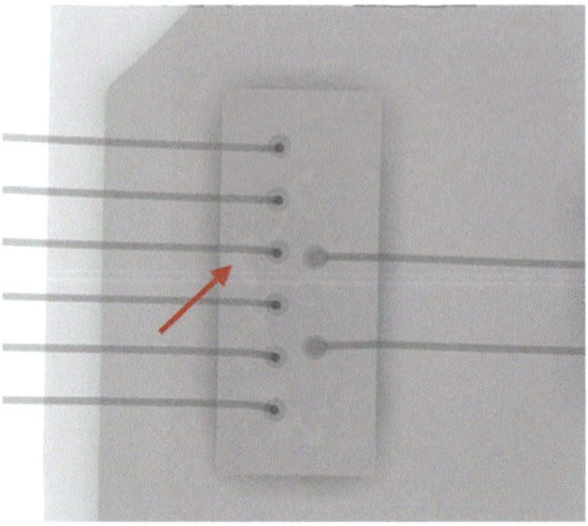

Fig. 8.7 2D X-ray radiograph of chip bonding voids

8.3.3 Scanning Acoustic Microscopy (SAM) Technology

Scanning Acoustic Microscopy (SAM) is a non-destructive analysis technique for integrated circuit packaging. It uses ultrasound to detect internal voids and other defects in the packaging. SAM is mainly used to observe defects in chip bonding layers. It is especially effective for analyzing delamination in plastic packaging.

1. Basic Principles and Key Performance Indicators

Scanning Acoustic Microscopy (SAM) technology uses an ultrasonic transducer to focus at a certain depth inside the sample. The transducer moves in various scanning modes, generating corresponding acoustic microscopic images. The basic principle is shown in Fig. 8.8. The common scanning modes are C-mode, B-mode, and Tru-mode. Both C-mode and B-mode use reflected sound waves for imaging. In C-mode, the ultrasonic transducer moves parallel to the sample's XY plane in a scanning manner. This generates acoustic microscopic images of the XY plane. In B-mode, the ultrasonic transducer moves along a straight line in the X or Y direction. It collects information at each point and depth along that line to generate XZ or YZ cross-sectional images [34]. Tru-mode uses transmitted sound waves for imaging. The transducer moves along the XY plane, collecting transmitted sound wave information at each point to create a transmitted acoustic image of the sample. The most commonly used mode is C-mode scanning. In C-mode, the focused ultrasonic transducer scans the region of interest on the plane. The lens is focused at a specific depth, and the depth selection gate determines the scanning area [35]. For example, in plastic-encapsulated integrated circuits, a narrow gate selects imaging only on the chip surface. A wider gate can image the chip surface, substrate edges, and lead frame simultaneously.

Currently, the frequency range of Scanning Acoustic Microscopy (SAM) is from 1 to 500 MHz. The spatial resolution can reach 0.1 μm. The higher the frequency of the transducer, the higher the resolution. The scanning area can cover several hundred square millimeters.

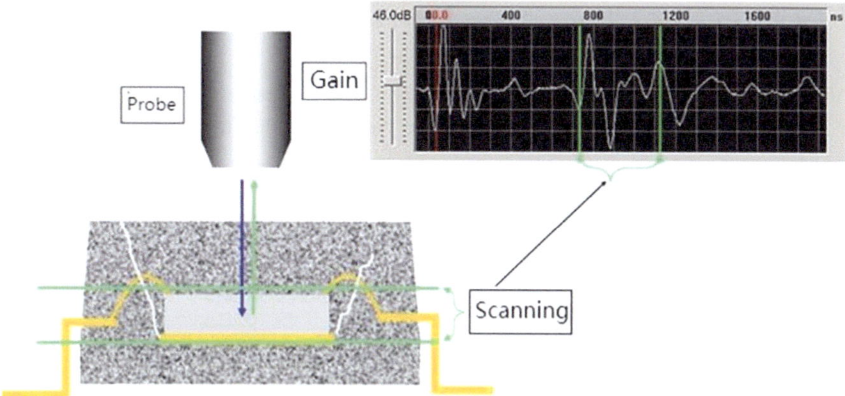

Fig. 8.8 Basic principle of scanning acoustic microscopy

2. Case Analysis

Figure 8.9 shows an image of delamination at the interface between the DSP integrated circuit molding plastic and the lead frame. Delamination in plastic-encapsulated integrated circuits may be pre-existing or can occur after moisture absorption during the reflow soldering process. For samples with existing delamination, the delamination can expand during soldering, storage, and use. This can eventually lead to failure. In failed samples, delamination or cracking at multiple internal interfaces is caused by the "popcorn effect," which results from high-stress conditions.

Fig. 8.9 Image of delamination at the interface between the DSP integrated circuit molding plastic and the lead frame

8.3.4 Particle Collision Noise Detection Technology

Particle collision noise detection technology is a non-destructive analysis method. It is used to examine movable debris inside the sealed integrated circuit packaging cavity.

1. Basic Principle and Key Performance Indicators

Particle Impact Noise Detection (PIND) is a non-destructive testing method. It checks for loose particles inside the cavity of an integrated circuit packaging to assess its integrity. When subjected to impact and vibration, these particles move faster and collide with the inner walls. Their kinetic energy converts into sound energy. A piezoelectric crystal inside the ultrasonic detector transforms this sound energy into electrical energy. After amplification and filtering, one signal is displayed on an oscilloscope, while another is converted into a low-frequency signal and sent to a sound system. This helps detect unwanted particles. Figure 8.10 shows the PIND detector.

Currently, commercial PIND detectors operate within a vibration frequency range of several tens to hundreds of hertz. The impact acceleration can reach thousands of *g*. The noise sensor sensitivity is within several tens of decibels.

2. Case Analysis

After particle collision noise detection on a sealed integrated circuit, noise bursts were observed. Decapsulation revealed movable foreign objects on the bonding wires, as shown in Fig. 8.11. SEM and EDS analysis confirmed that these objects were composed of Al, the same material as the bonding wires. This suggests that the contamination occurred during the bonding process.

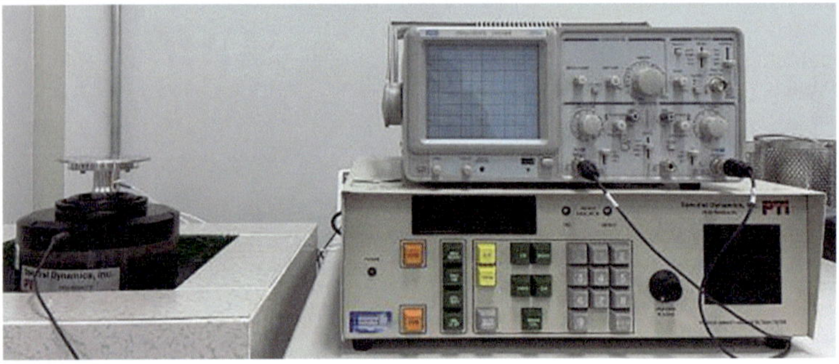

Fig. 8.10 PIND detector

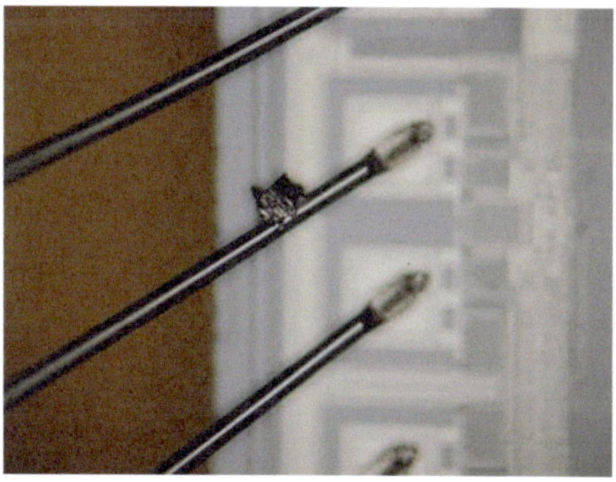

Fig. 8.11 Noise bursts caused by movable foreign objects in a sealed integrated circuit

8.3.5 Helium Mass Spectrometry Leak Detection

Helium mass spectrometry leak detection is a non-destructive method for testing the airtightness of sealed integrated circuit packaging.

1. Basic Principles and Key Performance Indicators

Helium mass spectrometry leak detection places the sealed integrated circuit inside a sealed chamber. The chamber is then pressurized with 100% helium at a fixed pressure. After a set time, the pressure is released. The sample is removed, and surface helium is cleared. It is then tested using a helium mass spectrometer to measure the leak rate. The leak rate depends on pressurization time, pressure level, and cavity volume. Figure 8.12 shows the helium mass spectrometer.

Today, advanced helium mass spectrometers can detect leak rates as low as 10^{-10} Pa·cm ^{3}·s^{-1}.

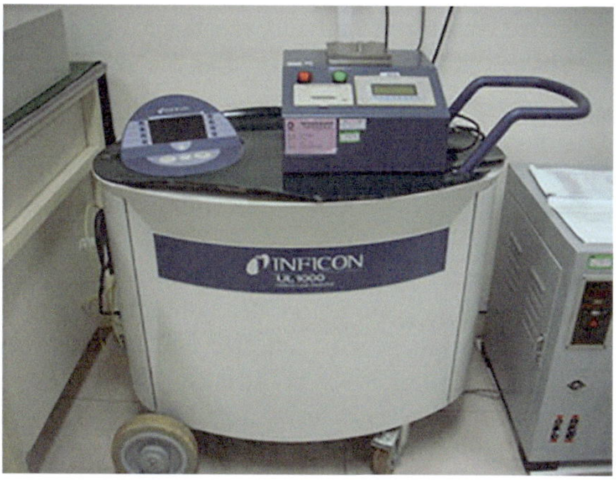

Fig. 8.12 Helium mass spectrometer

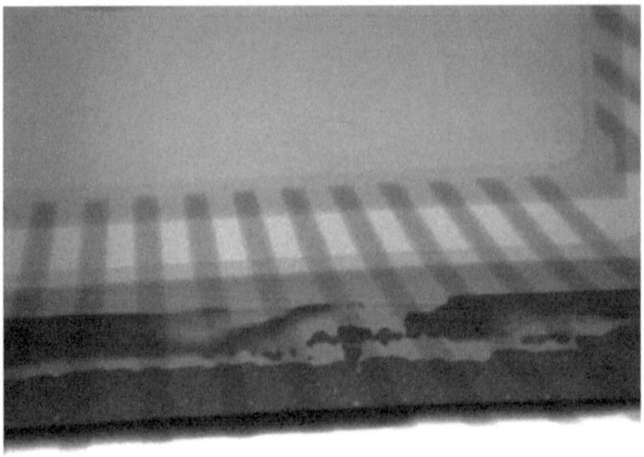

Fig. 8.13 Excessive leak rate due to insufficient sealing width

2. Case Analysis

A sealed integrated circuit was tested using a helium mass spectrometer. The leak rate was 6.2×10^{-3} Pa•cm^3•s^{-1}, which exceeded the standard limit of 5.0×10^{-3} Pa•cm^3•s^{-1}. X-ray inspection found holes in the sealing area. The sealing width did not meet the requirement, as shown in Fig. 8.13.

8.4 Destructive Failure Analysis Techniques

8.4.1 Decapsulation and Microscopic Sample Preparation

To analyze the internal structure of an integrated circuit packaging, decapsulation is necessary. Different packaging types require different methods. Non-plastic-sealed integrated circuits typically use mechanical decapsulation. Plastic-sealed integrated circuits use chemical etching or laser decapsulation methods.

1. Basic Principles and Key Performance Indicators

The principle of mechanical decapsulation is simple, so it will not be discussed in detail here. Below, chemical etching and laser decapsulation methods will be introduced.

The jet etching decapsulation machine (see Fig. 8.14) is used for decapsulating plastic-sealed devices. Before decapsulation, X-ray inspection is required to determine the chip's shape, position, size, and bonding wire height. The sample should be baked to remove moisture in the encapsulation layer. This helps prevent acid from corroding the metal and causing additional defects. When using chemical etching, the external bonding wires should not be exposed, as the chemical solvent can degrade them. After the encapsulation layer has been completely removed, the sample is rinsed with anhydrous acetone and isopropyl alcohol.

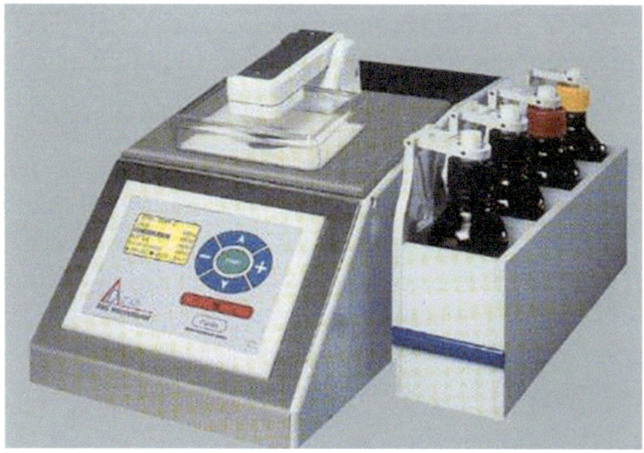

Fig. 8.14 Jet etching decapsulation machine

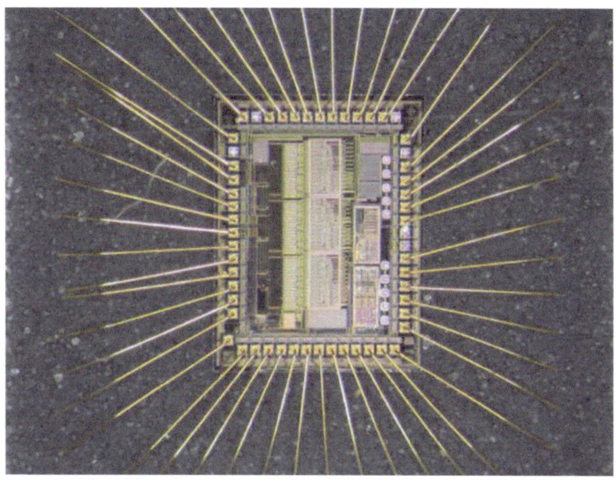

Fig. 8.15 Internal morphology of plastic-sealed integrated circuit after chemical etching decapsulation

For integrated circuits with thick molding plastic or copper wire bonding, the mold plastic on the package device can be removed by laser decapsulation method. This prevents corrosion damage to the copper wire or chip caused by prolonged chemical etching.

2. Case Analysis

Figure 8.15 shows the internal structure of a plastic-sealed integrated circuit after chemical etching decapsulation.

8.4.2 Internal Atmosphere Analysis Technology

Internal atmosphere analysis requires drilling holes in the integrated circuit packaging to extract the internal atmosphere. This is a destructive analysis and must be done before decapsulation. Internal atmosphere analysis is used to quantify various gases (including water vapor) inside the airtight packaging.

1. Basic Principles and Key Performance Indicators

Internal atmosphere analysis is an important technology for evaluating the packaging conditions, internal materials, and processing methods of integrated circuits. The analysis method used is partial pressure mass spectrometry. The process involves extracting a sample from the sealed device, ionizing it, and then using a mass spectrometer to separate and count the ions. Finally, the partial pressure ratios of various gases are calculated. The typical packaging atmosphere in sealed integrated circuits includes several types of gases (Fig. 8.16).

1. Dry Air: Dry air, which removes moisture and other harmful gases, can be used as a protective gas.
2. Dry Nitrogen: Nitrogen with 99.9% purity.
3. Dry Nitrogen/Helium Mixture.

The internal atmosphere analyzer mainly includes a vacuum system, sampling system, analysis system, data processing system, and sample holder. The internal atmosphere analyzers imported to China currently operate at an atomic level (1–512 atoms). The accuracy for major gases (such as water vapor, nitrogen, oxygen, argon, hydrogen, carbon dioxide, etc.) is within 5%. The sensitivity for detecting water vapor is 100 ppm, and for other gases, it is 10 ppm.

2. Case Analysis

Table 8.3 shows the typical internal atmosphere detection results of sealed integrated circuits. The internal water vapor content of both samples is higher than 5000 ppm. Excessive water vapor can cause chip corrosion and other failure issues.

Fig. 8.16 A typical internal atmosphere analyzer

Table 8.3 Detection results of internal atmosphere in typical sealed integrated circuits

Partial pressure and internal atmosphere		Sample No.	
		1	2
Partial pressure	torr	2.5	2.5
Nitrogen	%	86.0	86.5
Oxygen	%	6.00	5.16
Argon	%	1.10	1.22
Moisture	%	2.95	3.69
Hydrogen	ppm	632	248
Hydrocarbon	ppm	104	104
Carbon-Dioxide	%	3.83	3.37

8.4.3 Scanning Electron Microscopic Analysis

Scanning electron microscopy (SEM) is mainly used for high-resolution physical characterization in integrated circuit packaging failure analysis. The magnification of SEM can reach several hundred thousand times.

1. Basic Principles and Key Performance Indicators

Scanning electron microscopy analysis uses an electron lens to focus and accelerate the electron beam. The beam then hits the sample surface, generating secondary electrons, backscattered electrons, and other signals. These signals are collected, amplified, and displayed as an image, showing the sample's surface morphology. The principle of scanning electron microscopy is shown in Fig. 8.17.

The resolution of SEM can reach below 50 Å, with magnification ranging from a few times to several hundred thousand times. It has a large depth of field and strong stereoscopic effect. When combined with an Energy Dispersive X-ray Spectrometer (EDS), it can also perform compositional analysis. Before using SEM for analysis, preliminary defect localization and sample preparation are required. To prevent the sample surface from charging under high voltage, which could cause unclear images, the sample may need to be gold-coated. Figure 8.18 shows the appearance of an SEM.

2. Case Analysis

Figure 8.19 shows an SEM image of TSV deformation after thermal cycling. The TSV has changed from a uniform thickness (see Fig. 8.20) to a shape with a thin middle and thicker ends. This deformation occurred because the thermal stress was greatest at the middle of the TSV during the thermal cycling test, causing displacement in different parts of the copper pillar, with the ends being pushed toward the center.

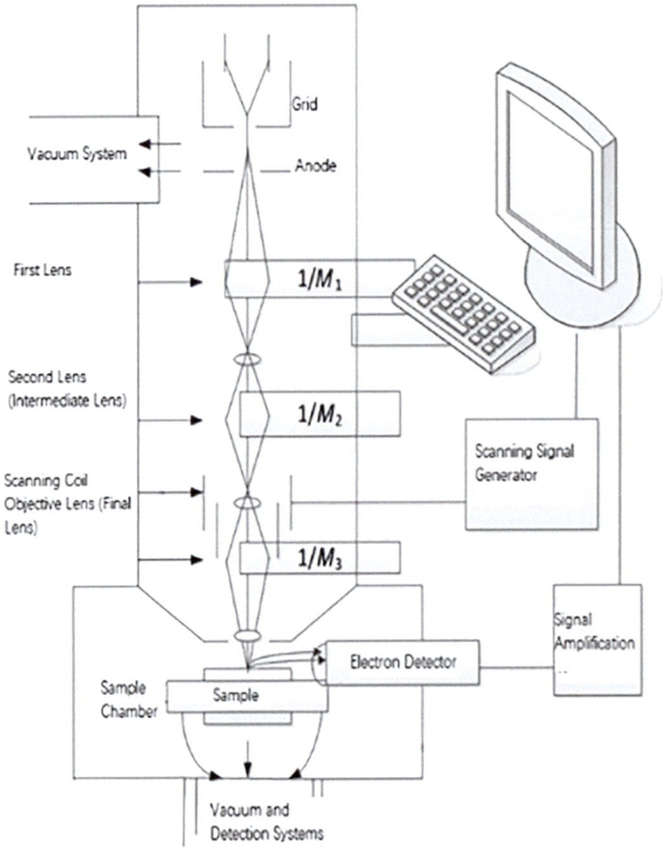

Fig. 8.17 Scanning electron microscopic analysis principle

Fig. 8.18 SEM appearance image

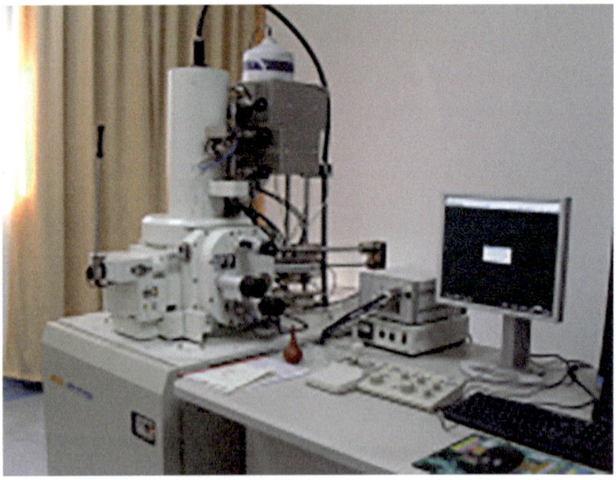

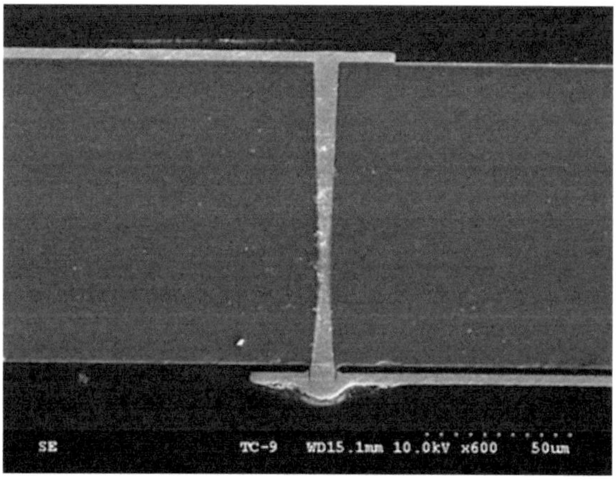

Fig. 8.19 SEM image of TSV deformation

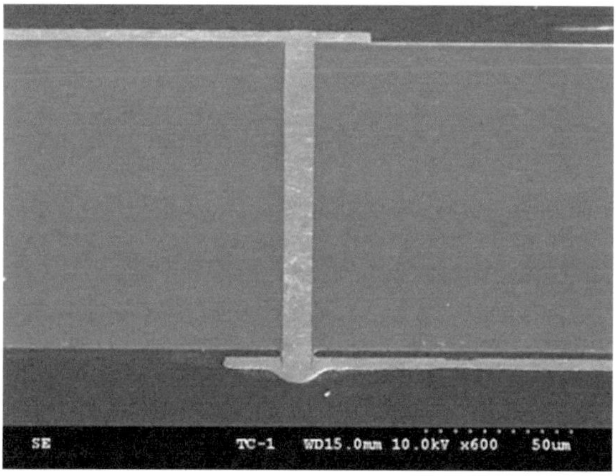

Fig. 8.20 SEM image of a normal TSV

8.4.4 Transmission Electron Microscopic Analysis Technology

Transmission electron microscopic (TEM) analysis technology is mainly used for high-resolution 3D microstructure analysis and structural analysis, as well as element composition analysis, in integrated circuit packaging failure analysis.

1. Basic Principles and Key Performance Indicators

This technology uses the wave properties of electrons to observe material and structural defects, achieving a resolution at the atomic level. Similar to SEM, it uses an electron beam as the source. The beam bends in the presence of an external magnetic or electric field. This bending creates a phenomenon similar to the refraction of visible light through glass. This principle is used to create an "electron lens," which leads to the development of TEM.

Through specific sample preparation techniques, TEM can provide ultra-high-resolution 3D images. It also allows high spatial resolution (nanoscale) analysis of structure and composition. For integrated circuits with submicron features, TEM can precisely reveal the morphological characteristics of materials used in large-scale integrated circuit chip manufacturing. Since the sample preparation process has minimal impact on the device's internal structure, the images observed are generally considered to be the original morphology of the sample.

Currently, TEM research and development in China are still in the early stages. There are no mature products available in the market. Leading companies worldwide include Thermo Fisher (formerly FEI) in the U.S., JEOL in Japan, and ZEISS in Germany. In China, universities, research institutes, and other organizations mainly use TEMs from these three brands. Advanced TEMs can achieve a point resolution of up to 0.24 nm, a line resolution of up to 0.14 nm, and scanning transmission electron microscopy (STEM) with a resolution of less than 1 nm. Figure 8.21 shows the appearance of a TEM.

2. Case Analysis

After 1000 thermal cycles (−55 ~ 125 °C), the TSV (Through-Silicon Via) interposer failed with an open circuit. Defect localization and FIB (Focused Ion Beam) cross-section analysis confirmed that the failure occurred at the TSV and top metal (TOP-M1) interface. To investigate the crack initiation mechanism further, TEM (Transmission Electron Microscopy) was used to analyze the interface region of the TSV sample. Based on the STEM morphology (Fig. 8.22) and TEM morphology (Fig. 8.23) of the TSV and TOP-M1 interface, it was confirmed that under thermal

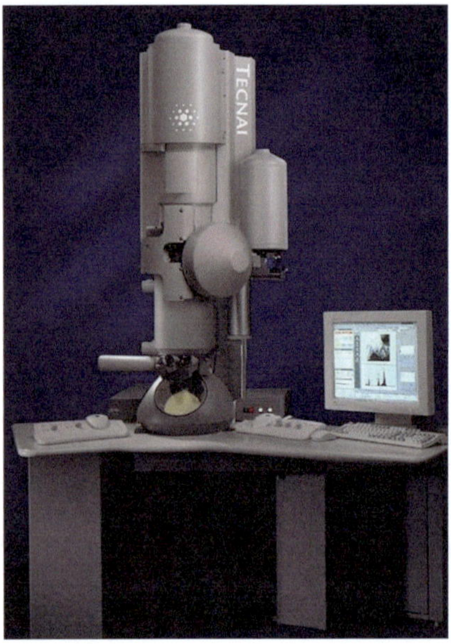

Fig. 8.21 TEM appearance image

cycling stress, cracks first appeared at the stress concentration point at the TSV top corner. These cracks then propagated inward toward the center of the TSV [9].

Using the EDS (Energy Dispersive X-ray Spectroscopy) analysis function, the distribution and diffusion behavior of elements were analyzed. This provided important evidence to understand the failure mechanism. The element analysis of the TSV interface is shown in Fig. 8.24. From Fig. 8.24, it can be seen that the TSV interface consists of Si, SiO_2, Ti, and Cu. The distribution of Ti elements indicates that after cracking at the stress concentration point at the TSV top corner, Ti diffused along the crack into the Cu substrate. This confirms the physical mechanism that cracks first appear at the stress concentration point at the TSV top corner, and then propagate inward toward the TSV center.

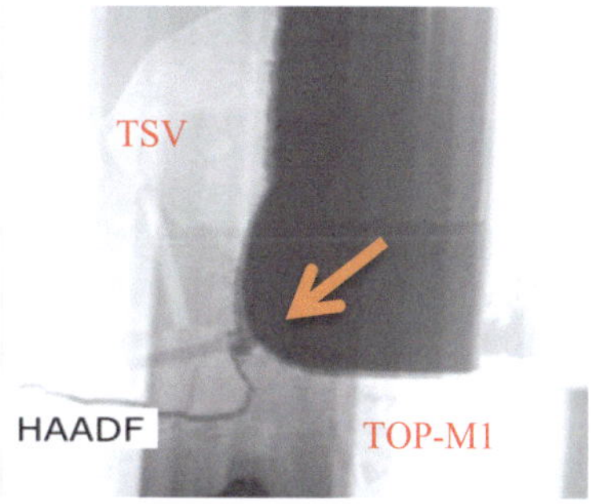

Fig. 8.22 STEM morphology of the TSV and TOP-M1 interface

Fig. 8.23 TEM morphology of the TSV and TOP-M1 interface

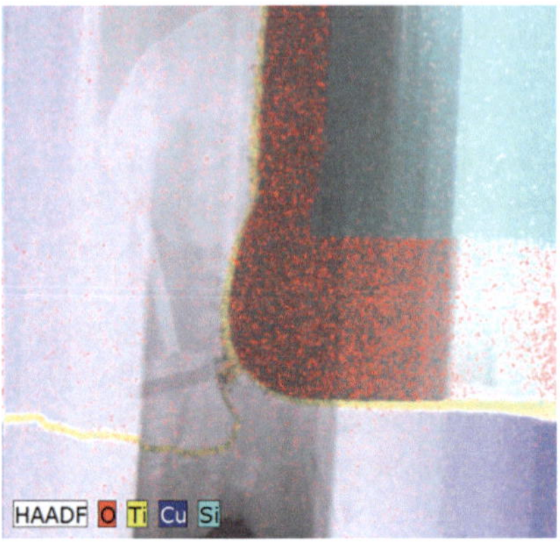

Fig. 8.24 Elemental analysis of the TSV interface

8.4.5 Focused Ion Beam Defect Analysis Technology

Focused Ion Beam (FIB) defect analysis technology is mainly used in integrated circuit failure analysis. It is used for cross-sectional analysis of failure sites or specific locations, local delayering observations, cutting isolation for auxiliary testing, and key growth testing, etc. [36].

1. Basic Principles and Main Performance Indicators

In many failure analysis cases, metallographic cross-sections are prepared to analyze the failure mechanism and identify the root cause of device failure. For failure analysis of 3D integrated circuit modules, cross-section preparation is key. This is because these modules have diverse components, materials, and complex structures.

Mechanical grinding can remove different cover materials. However, this method often causes surface contamination and roughness. Ion polishing and FIB technology improve surface quality. However, traditional FIB sample preparation is slow and has a low etching rate. This makes it suitable for chip-level samples, but not for packaging-level samples. To shorten packaging-level sample preparation time, High Current FIB (HC-FIB) technology can be used. HC-FIB uses an inductively coupled plasma (ICP) source instead of a liquid metal gallium ion source [37]. A Gas-Injection System (GIS) can inject reactive gases into the FIB, increasing the etching rate by one or two orders of magnitude. Combining FIB technology with laser etching is another way to reduce preparation time. Infineon uses HC-FIB for SiP defect analysis, and Philips Semiconductor uses laser decapsulation to expose defects in 3D integrated circuits [38].

1) Plasma FIB Technology

Traditional FIB technology uses a liquid metal gallium (Ga) ion source (LMIS) for the ion beam. The FIB system applies an electric field to the ion source, extracting the ion beam. The electron lens then focuses the ion beam, which bombards the sample surface. This process uses physical collisions to achieve cutting. A significant problem with LMIS is the limited types of metal ions available. These ions also react easily with other elements, causing metal ion contamination and sputtering.

Plasma FIB technology uses inductively coupled plasma (ICP). ICP has a low temperature, high density, and good uniformity. It has significant advantages in large-area etching applications. At low beam currents, the ion beam from ICP (about 50 nm) is less fine than that of LMIS (less than 5 nm). However, when the beam current reaches several microamps, ICP can remove materials much faster. Its performance is much better than LMIS. ICP can use various plasma sources, such as Xe, O_2, Ar, and He. The plasma source is chosen based on the material being removed. Xe is commonly used as the etching material because of its high concentration and good ion source performance. Compared to traditional FIB systems, plasma FIB systems remove materials faster. They also maintain high resolution at low beam currents, which is crucial for further analysis. Plasma FIB systems increase the beam current to microamp levels, achieving etching rates of up to 20,000 μm^3/min.

2) Combination of FIB Technology and Laser Etching

Laser etching involves focusing a laser beam on the target area of the sample surface. The high temperature melts the material. Laser etching is a non-contact process. It avoids mechanical impacts that could cause deformation or cracks. During the process, the laser is concentrated into a very small beam. The energy density is high, allowing for fast operation and a small heat-affected zone.

Table 8.4 shows several typical FIB technologies combined with laser etching for Si etching rates. It also shows the time required to remove 0.3 mm^3 of material using these methods. The diode-pumped solid-state laser (DPSS) has a wavelength of 355 nm. When using laser etching to prepare cross-sections of stacked silicon wafers, choosing the correct laser wavelength is crucial to avoid introducing new damage.

From Table 8.4, we can see that the laser etching rate reaches the order of 10^6 μm^3/s. The process first uses laser etching for rapid material removal, followed polishing to obtain a smooth metallographic cross-section by FIB. The combination of laser etching and FIB technology allows for high-quality, fast preparation of large-sized samples

However, material changes caused by laser etching may affect the feasibility of this technology. Laser etching can cause localized melting, chemical reactions, phase formation, recrystallization, surface contamination, and increased diffusion rates in

Table 8.4 Etching rate and removal time using different sample preparation methods

Method	Si etching rate(μm^3/s)	Removal time(0.3 mm^3)
FIB	2.7	3.5 Years
HC-FIB	30	116 Days
FIBwith GIS	250	14 Days
ICP ion source	2000	1.7 Days
355 nm DPSS laser	10^6	5 min

the Heat-Affected Zone (HAZ). It may also introduce cracks, surface roughness, and other damage during the sample preparation process. Therefore, it is crucial to analyze the HAZ characteristics formed during laser etching. This helps determine whether visible cracks or delamination are caused by thermal reactions introduced by the laser etching or are actual defects in the device. In failure mechanism studies, it is important to avoid removing or altering the original defects. Thus, laser etching should be done carefully to avoid introducing damage to true defects.

2. Case Analysis

HC-FIB technology can analyze failures in 3D integrated systems caused by TSV and bonding. Figure 8.25 shows how plasma FIB analysis is applied in the reliability analysis of a three-layer stacked chip. Figure 8.25a is the cross-sectional view of the region of interest (IMC bonding, TSV). Figure 8.25b shows the bottom layer structure after the sample is opened. The XeF_2 injection system accelerates the delayering and opening process. This, combined with fast selective Si etching and Xe polishing, makes the procedure more flexible and feasible. Figure 8.26 is an enlarged view of the TSV and bottom structure. It shows that plasma FIB still maintains high resolution (around 50 nm at a beam current of 60 pA) even at low beam currents, providing accurate information about the relevant structures.

(a) (b)

Fig. 8.25 Plasma FIB analysis technology in the reliability analysis of three-layer stacked chips

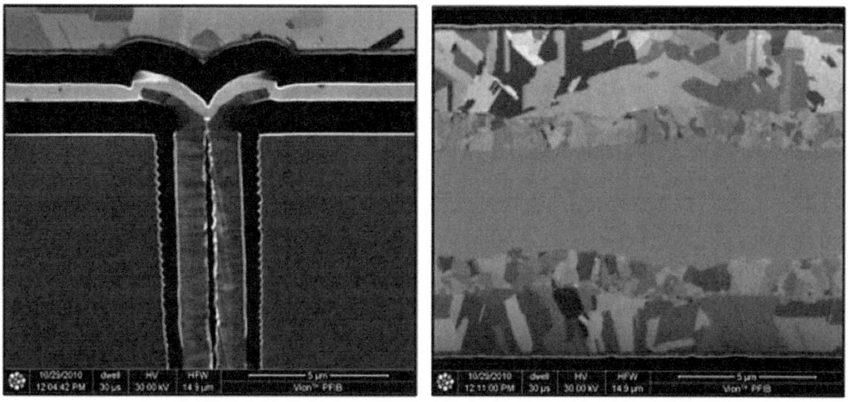

(a) (b)

Fig. 8.26 Enlarged view of TSV and bottom structure

8.5 New Technologies in 3D Packaging Failure Analysis

8.5.1 3D X-ray Analysis Technology

3D X-ray analysis technology can produce 3D images of the internal structures of integrated circuit packaging. It allows for non-destructive analysis of internal defects in complex packaging. This technology enables submicron-level 3D X-ray non-destructive detection of various packaging structures, especially advanced ones such as TSV, stacked bare-chip packaging, PiP, and PoP.[39, 40]

1. Basic Principles and Key Performance Indicators

3D X-ray analysis technology, also known as Computed Tomography (CT) technology, collects X-ray projection information from different angles of the packaging. Special computer algorithms are used to reconstruct the data into a 3D stereoscopic image. This method can solve the issue of information overlap seen in traditional 2D X-ray analysis technology [41].

With the development of advanced 3D packaging technology, complex structures like TSV, stacked bare-chip packaging, PiP, and PoP are being widely used. These complex structures challenge traditional 2D X-ray analysis technology. In 2D X-ray analysis, occlusion and interference in the field of view make it difficult to detect small defects in complex packaging or devices. In a typical 3D packaging, 2D projection results in overlapping multilayer chips, solder balls, and bonding wires in one image. This makes it hard to see each part's features. Current 2D X-ray analysis projects the internal structure of the sample onto a real-time detector. However, since the projection is 2D, depth information cannot be obtained. Slight tilting of the sample or detector may provide a clearer view, but the fundamental problem is not solved [42].

2D X-ray analysis faces limitations in resolution and imaging contrast due to basic technical factors. One difficulty often encountered is detecting defects related to solder balls, such as chip cracks, through-hole cracks, or delamination. Solder ball cracks, micron-sized voids inside the balls, non-wetting solder balls, and small metal protrusions from solder balls onto the packaging surface are usually in the range of a few micrometers to sub-micrometers. These defects are hard to detect using existing 2D X-ray analysis. Additionally, 2D X-ray analysis cannot effectively distinguish low atomic number materials, such as aluminum wire bonds, packaging resins, and silicon chips. These materials have fewer electrons in their atomic nuclei. As a result, they interact very little with X-rays. They absorb almost no X-rays, causing them to pass through as if they do not exist, which results in no image contrast on the detector.

Key indicators of 3D X-ray analysis include resolution and analysis time [43].

1) Resolution

The resolution of 3D X-ray images depends on several factors. It is influenced by geometric magnification, the focal spot size of the X-ray source, and the sample size [44, 45]. For 2D X-ray analysis, geometric magnification is expressed as:

$$M = \mathrm{DD}/\mathrm{OD} \tag{8.6}$$

where OD is the distance from the sample to the X-ray source, and DD is the distance from the X-ray source to the detector. The principle of 3D X-ray analysis is shown in Fig. 8.27. This method requires the sample to rotate 360° to capture images from different angles. To prevent collisions with the X-ray source, the sample must be positioned at a safe distance. However, increasing OD reduces magnification, which limits the maximum resolution. As a result, larger samples have lower resolution.

Maximum resolution is a key limitation of 3D X-ray analysis [46]. Most existing systems have a resolution above 10 μm [47]. Recently, some manufacturers have developed submicron 3D X-ray analysis equipment. These systems use a scintillator-coupled visible light detector and a two-stage magnification technique (see Fig. 8.28) [48]. First, the system magnifies the sample image using geometric magnification, similar to conventional 3D X-ray analysis. Then, X-rays are converted into visible light by the detector's scintillator screen. The visible light system further magnifies the image. This secondary optical magnification greatly enhances resolution. Even for large samples and long working distances, the system maintains high resolution. The maximum resolution can reach 0.6 μm [49].

2) Analysis Time

The analysis time of 3D X-ray technology depends on image acquisition and reconstruction. This method requires a 360° scan to capture images from different angles. A large amount of data must be collected. Processing and reconstructing the 3D image take much longer than acquiring the images. In integrated circuit packaging,

Fig. 8.27 Principle of 3D X-ray analysis technology

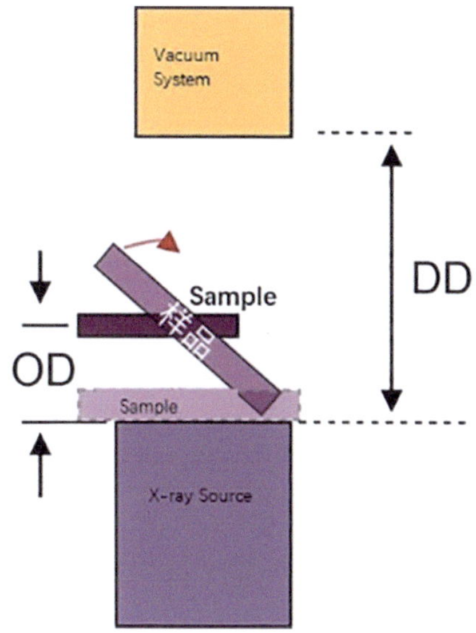

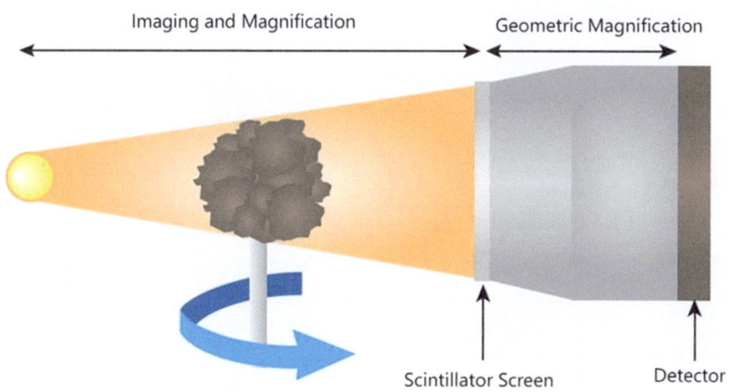

Fig. 8.28 Two-stage magnification structure

reducing analysis time is crucial. Effective methods are needed to keep image acquisition and reconstruction under 10 min. Faster analysis and higher efficiency drive the continuous improvement of 3D X-ray systems [40].

2. Case Analysis

3D X-ray technology was used to analyze a PoP sample. Figure 8.29 shows the virtual cross-sectional scan of one row of solder joints. Voids and head-in-pillow defects are clearly visible. The exact locations of defective solder joints can be identified [39].

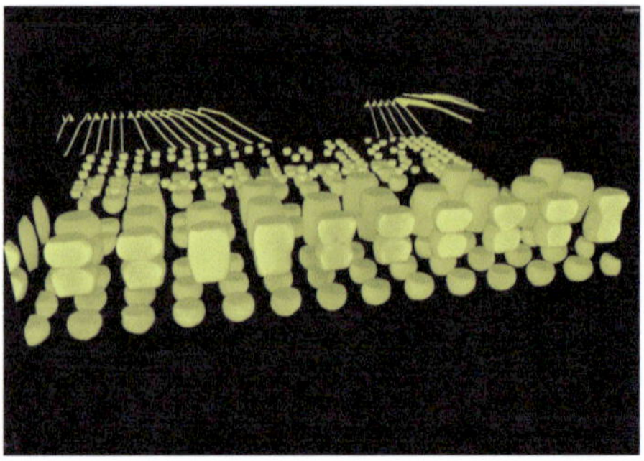

Fig. 8.29 Virtual tomographic scanning results of PoP Sample solder joints

The results of 3D X-ray analysis match well with metallographic cross-sectioning. Figure 8.30 compares the 3D X-ray morphology with the metallographic cross-section. For 3D SiP devices or multi-layer flip-chip bumps, 3D X-ray technology provides detailed defect location information.

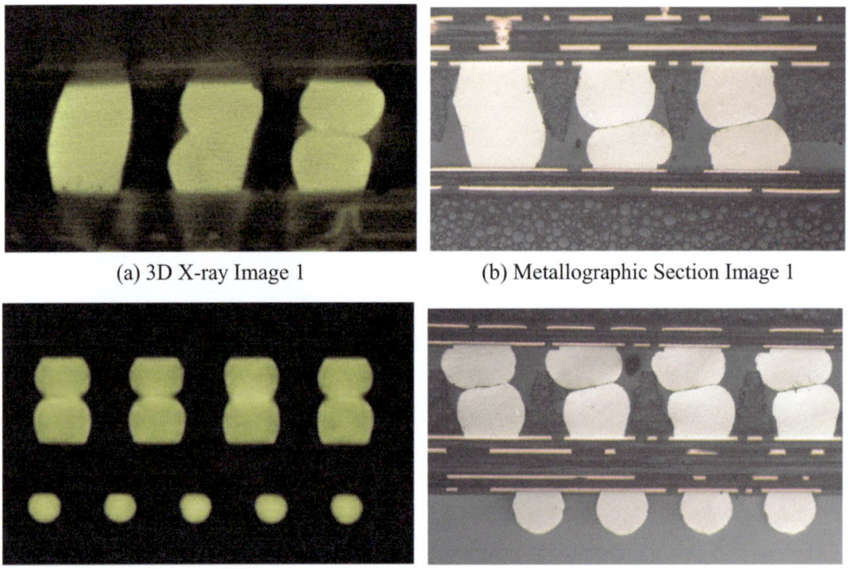

(a) 3D X-ray Image 1 (b) Metallographic Section Image 1

(c) 3D X-ray Image 2 (d) Metallographic Section Image 2

Fig. 8.30 Comparison of 3D X-ray images and metallographic section images

8.5.2 Magnetic Microscopy Analysis

Magnetic microscopy is a non-destructive, non-contact defect localization method. Unlike thermal, optical, ion, or electron beam techniques, it is not affected by packaging materials or structures. Current images can pass through multiple metal layers, chips, or packaging materials from both sides of the device.

Magnetic induction imaging does not require sample preparation, such as decapsulation or passivation removal. Common failure analysis methods, like optical and thermal emission, detect only leakage and short circuits. Traditional TDR technology for open-circuit detection has a resolution of 1 mm. In contrast, magnetic microscopy can locate short circuits, open circuits, and leakage defects with much higher accuracy.

1. Basic Principles and Key Performance Indicators

Magnetic microscopy analysis uses the relationship between current and magnetic fields to locate defects. According to Biot-Savart's law, when current flows through a conductor, it generates a magnetic field. Figure 8.31 shows this relationship. When current flows inside a device, a magnetic probe detects the internal magnetic signal. A computer processes the magnetic field image and converts it into a corresponding current image. This process is called Magnetic Current Imaging (MCI). By comparing the CAD or X-ray images of the device, the exact failure location can be identified.

Most materials used in integrated circuits and packaging are non-magnetic. As a result, the magnetic field remains unaffected by them. In complex packaging structures, especially system-in-Packaging packaging (SiP), stacked chips are densely packed with different materials, making defect localization difficult. Thermal, optical, and electron signals can be easily blocked by complex structures. However, magnetic signals are less affected. Even if the material has low magnetic permeability, the signal can still be detected.

Magnetic microscopy analysis can locate failures at the chip, interconnection, and packaging levels. It detects short circuits, leakage, and high-resistance defects, such as broken traces, poor wetting, C4 solder bump cracks, and via delamination. For defect analysis in 3D stacked packaging, it is an effective method.

By detecting the magnetic field in the Z-axis direction and using iterative algorithms, defects in 3D structures can also be located. The closer the current path is to the magnetic sensor, the stronger the detected magnetic field. If the metal trace

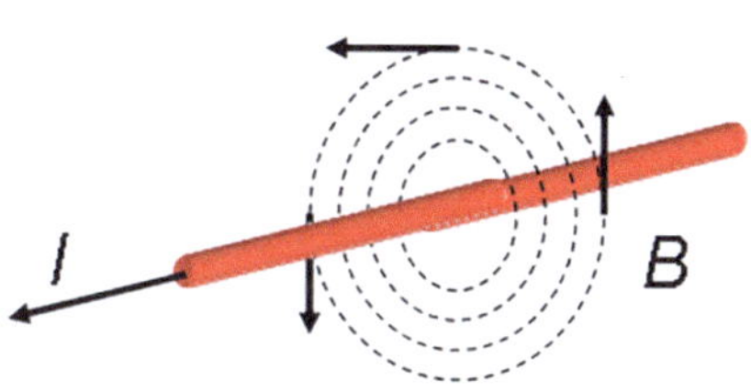

Fig. 8.31 Relationship between current and magnetic field

width remains the same in each layer, and the two metal traces from different layers are not directly connected, different metal layers can be displayed in a 3D image. When current flows through metal in multiple planes or stacked structures, the chip's internal layers and their physical conditions can be visualized.

Magnetic microscopy defect localization uses two main detection technologies: Superconducting Quantum Interference Device (SQUID) detection and Giant Magnetoresistance (GMR) detection. The primary difference between them is in sensitivity and resolution.

SQUID has extremely high sensitivity and resolution. It can measure not only magnetic flux, but also other magnetic field quantities, such as magnetic induction intensity, magnetic field gradient, and magnetic susceptibility, which can be converted into magnetic flux. The magnetic field resolution of SQUID is less than 10^{-10} T, and its current resolution is less than 10^{-9} A. GMR has a larger detection range, over 700 mm, but its resolution is lower.

1. Case Analysis

Magnetic microscopy was used to locate defects in multi-layer chip integrated circuits. The system was equipped with a SQUID magnetic detector. The generated current paths were clearly visible. Short-circuit failures usually show as concentrated current. The magnetic microscopy results showed that the defect was located at a point with higher current concentration along the failure path, and it provided the exact coordinates. A typical image of magnetic induction defect localization in the sample is shown in Fig. 8.32 [50].

The sample was locally opened using a jet etching machine, exposing the chip surface. Based on the magnetic microscopy localization results, observations were made using a stereo microscope. Measurements of the exposed sample were taken with NIS Elements F software.

The failure location of sample D2 is shown in Fig. 8.33a. The failure point is on the bottom layer of the chip. The local magnified morphology of the failure site is shown in Fig. 8.33b. First, the top and bottom chips were separated. The bottom bare chip was placed into a reactive ion etching machine to remove the passivation

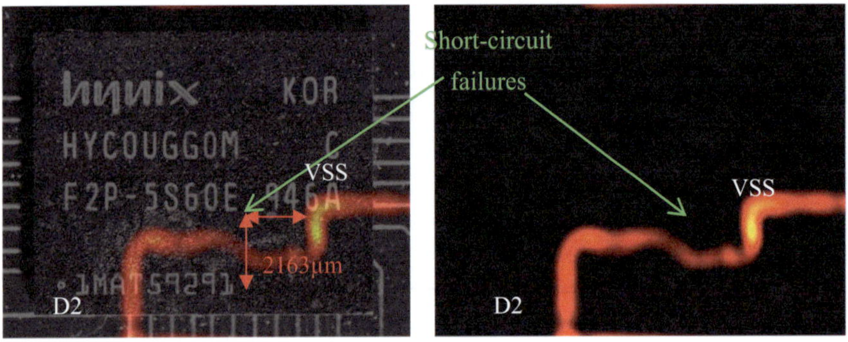

Fig. 8.32 Typical image of magnetic induction defect localization in sample

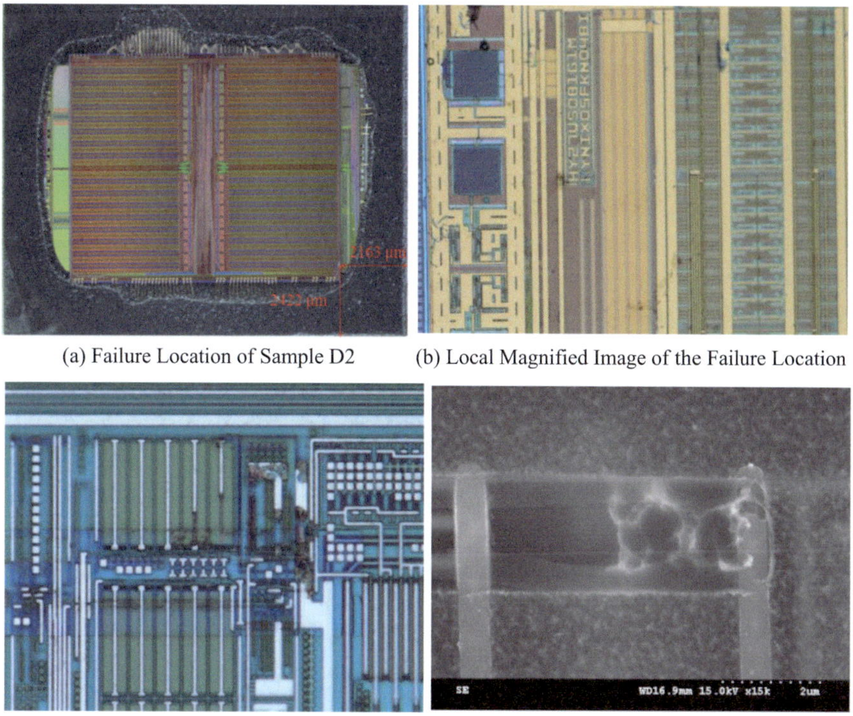

(a) Failure Location of Sample D2

(b) Local Magnified Image of the Failure Location

(c) SEM Image of the Failure Location

(d) Locally Magnified SEM Image

Fig. 8.33 Physical analysis process corresponding to the failure location of sample D2

layer. After removal, the failure location was observed with a metallurgical microscope, but no obvious anomalies were found. Next, the metal layer was removed. No failure morphology was observed under an optical microscope. Then, the sample was placed back into the reactive ion etching machine to remove the dielectric and metal layers. Under a metallurgical microscope, the SEM morphology of the failure location is shown in Fig. 8.33c. Anomalous morphology was found at the failure location identified in Fig. 8.33a. Layer removal continued until the silicon substrate was exposed. SEM observation revealed molten deformation at the failure location. The SEM local magnified morphology is shown in Fig. 8.33d.

8.5.3 Lock-In Thermography (LIT) Analysis Technology

Lock-In Thermography (LIT) analysis technology detects hotspots to analyze failures, such as metal layer short circuits, oxide layer breakdown, ESD damage, and latch-up. Compared to traditional infrared thermal imaging, LIT offers deeper detection. Its phase images provide more information than amplitude images. LIT is an emerging non-destructive testing technique. It is suitable for defect localization in 3D integrated circuits [51].

1. Basic Principle and Key Performance Indicators

With the development of microelectronic packaging technology, many common defect localization techniques are no longer suitable for 3D packaging defect analysis. Techniques such as photonic emission microscopy, liquid crystal hotspot detection, and OBIRCH are less effective for 3D devices. 3D device packaging structures are more complex than planar ones. Defects can appear anywhere, such as in the bottom chip or substrate. The multi-layer stacking of materials increases the difficulty of detection and localization [52].

LIT analysis, also known as phase-locked thermal imaging, is a type of infrared thermal imaging technology [53]. Under external excitation at a specific frequency, the defect-induced thermal signals within the device under test (DUT) propagate through various material layers and reach the surface. This causes temperature changes on the surface, oscillating at that frequency. Although the materials above the defect absorb infrared wavelengths, the thermal waves from the defect can pass through these materials and reach the surface, causing periodic temperature increases. High-sensitivity infrared thermal imaging equipment captures the thermal response on the surface. The collected signals are processed using phase correlation methods (in-phase signal S0° and out-of-phase signal S90°). The results depend on the material's thermal properties. Additionally, the silicon substrate is transparent to infrared light, allowing LIT detection on the backside of SiP devices [54].

Figure 8.34 shows the principle of using LIT technology for non-destructive detection of packaging defects. Researchers have studied the relationship between defect depth, phase shift, and frequency in depth. They have applied LIT to detect and locate various 3D packaging defects. The results confirm that LIT is effective for failure analysis in flip-chip, TSV, stacked-chip packaging, and other SiP packaging structures [55].

LIT phase images can separate weak hotspots from nearby stronger interference hotspots. The phase images also contain depth information about the defects. This removes the need for multiple thermal images and device delayering, making it suitable for defect analysis of 3D SiP devices.

The modulation frequency is the most important factor affecting the accuracy of LIT detection. There is an optimal detection frequency when using LIT for non-destructive testing. This optimal frequency depends on the material properties of the device and the depth of the defect. In practice, a higher modulation frequency should be set first, then gradually reduced to the optimal detection frequency for accurate localization [56].

One challenge of LIT is improving spatial resolution. The higher the spatial resolution, the more precise the hotspot localization will be. The smaller the defect's cross-sectional area, the more important it is for failure analysis. However, the heat from the defect spreads to the surface and diffuses through the molding material.

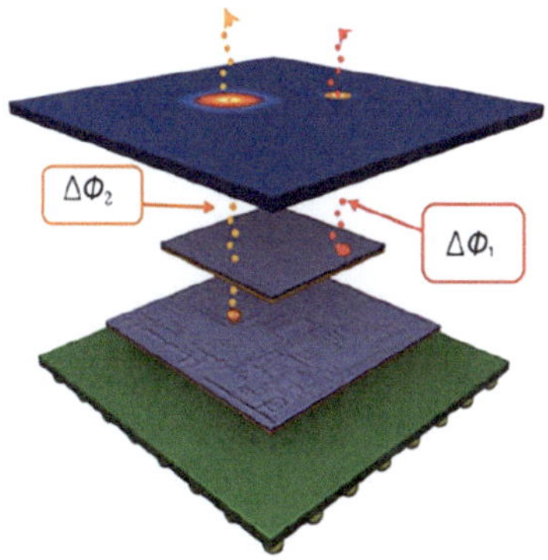

(a) Defects in Different Layers Cause Different Phase Shifts

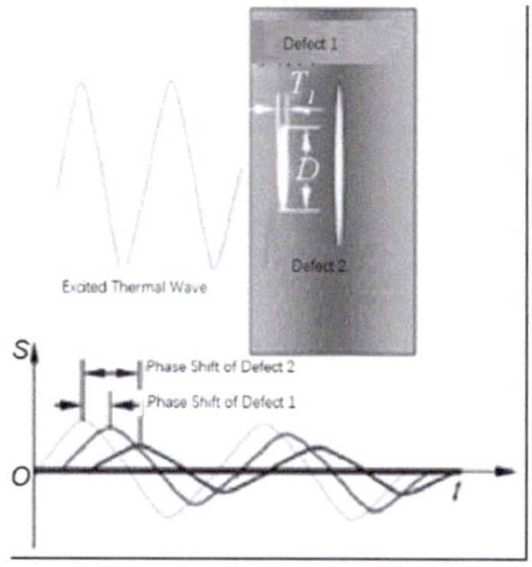

(b) Defect Localization Through Phase Information

Fig. 8.34 Principle of non-destructive detection of packaging defects using LIT technology

This reduces the spatial resolution and complicates detection. Increasing the detection frequency can improve spatial resolution. Currently, advanced LIT equipment can achieve a spatial resolution of up to 1 μm.

2. Case Analysis

Defect localization inside stacked-die packagings was carried out using Lock-in Thermography (LIT). By obtaining reference and actual phase shift values under different frequency conditions and comparing them, Z-axis information was obtained, indicating that the defect was located in the upper die, as shown in Fig. 8.35. Combined with X-ray analysis, the thermal image was overlaid to determine the exact defect location, as shown in Fig. 8.36. After decapsulation and physical analysis, the defect area identified by thermal imaging was confirmed, as shown in Fig. 8.37. SEM analysis of the defect site revealed signs of breakdown and burnout, as shown in Fig. 8.38.

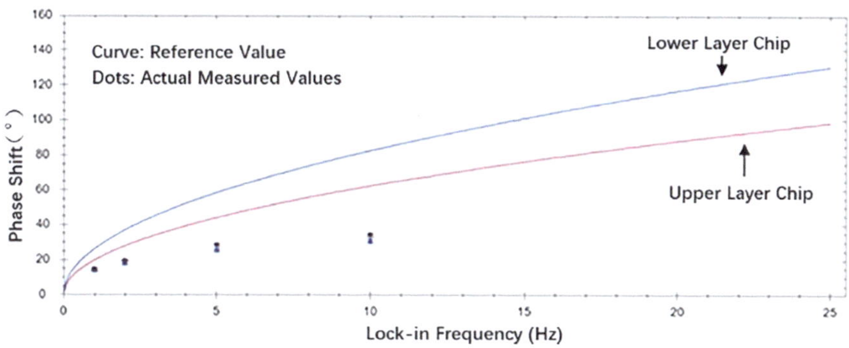

Fig. 8.35 Determining that the defect is located in the upper die chip

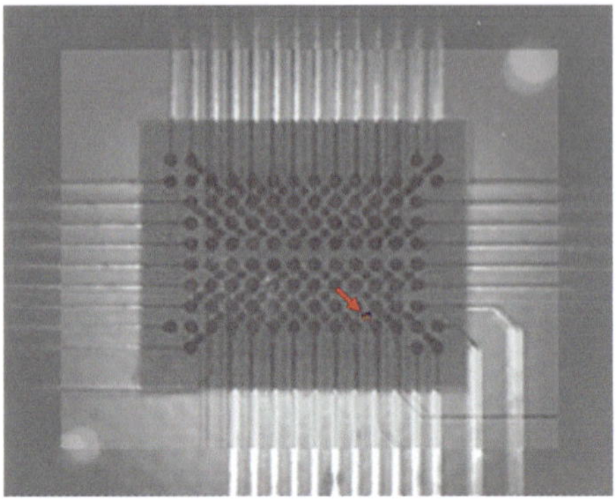

Fig. 8.36 Determining the specific location of the defect

Fig. 8.37 Defect range locked based on thermal imaging detection

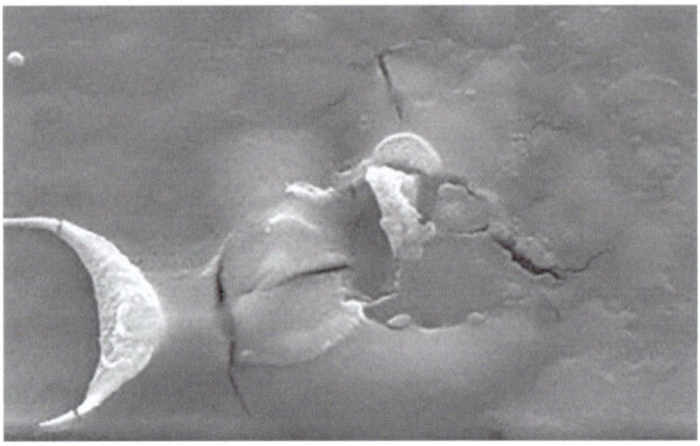

Fig. 8.38 Defect area showing breakdown and burn marks under SEM

8.6 Fault Tree Analysis of Integrated Circuit Packaging

The traditional Fault Tree Analysis (FTA) method builds fault trees based on functional logic relationships. It is used to analyze failure causes in complete systems. Basic fault events occur at the component level. This section introduces a fault tree method based on failure physics logic. It is suitable for analyzing failures in electronic components. Basic fault events occur within the internal physical structure. This method allows fault analysis, failure localization, and failure mechanism study at the physical level. Fault tree analysis of integrated circuit packaging focuses on component-level failures in the packaging.

Fault tree analysis helps identify failure modes and locations in integrated circuit packaging. It determines failure mechanisms and key influencing factors. It also assesses potential failure modes and calculates their importance. This process identifies weak points in reliability and safety limits. The findings help improve design and enhance reliability.

8.6.1 Fault Tree Analysis Method for Integrated Circuit Packaging

1. Purpose of Fault Tree Analysis for Integrated Circuit Packaging

The goal of fault tree analysis for integrated circuit packaging is to identify failure mechanisms and failure paths related to packaging. It also examines factors that influence these failures. This analysis helps eliminate failures and improves the reliability of integrated circuit packaging.

2. Fault Tree Construction Method Based on Failure Physics

Fault trees are built separately for different types of integrated circuit packaging. Each tree follows six levels of failure physics and their logical relationships. To simplify the structure, the method integrates common-cause failure mechanism subtrees and common-cause failure module subtrees. The final fault tree consists of six failure physics levels and n-level events ($n \geq 6$).

Figure 8.39 defines the event levels in the fault tree. The "fault object" event comes from its direct lower-level cause, the "failure mode" event. The "failure mode" event comes from the "failure location" event. The "failure location" event comes from the "failure mechanism" event. The "failure mechanism" event comes from the "mechanism factor" event. The "mechanism factor" event comes from the "influencing factor" event [57].

Fault Tree Construction for Integrated Circuit Packaging: The fault tree for integrated circuit packaging has six levels.

(1) Top Event (First Level)—Fault Object, It is built separately for different types of integrated circuit packaging. The top event represents the faulty component in the system, such as failures in plastic or non-plastic packaged integrated circuits.

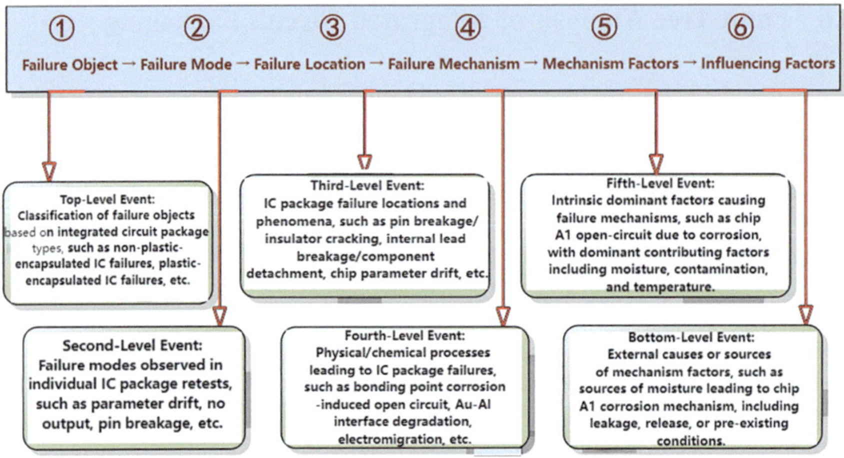

Fig. 8.39 Definition of event levels in the fault tree for integrated circuit packaging

(2) Second-Level Event—Failure Mode shows failure modes detected in independent tests. Examples include parameter drift, no output, and pin breakage.
(3) Third-Level Event—Failure Location further identifies the failure category and describes the failure phenomenon. Examples include pin breakage, insulator cracking, internal lead breakage, component detachment, and chip parameter drift.
(4) Fourth-Level Event—Failure Mechanism defines the physical or chemical process causing the failure. Examples include bond corrosion, Au-Al interface degradation, and electromigration.
(5) Fifth-Level Event—Mechanism Factor identifies the internal factors that lead to the failure mechanism. For example, in chip Al corrosion-induced open circuits, key factors include moisture, contamination, and temperature.
(6) Bottom Event (Sixth Level)—Influencing Factors represents external factors that trigger the mechanism factor. For example, moisture in chip Al corrosion may come from leakage, release, or pre-existing conditions.

The selection of the bottom event depends on the analysis goal. For mechanism research, the bottom event is set at the failure mechanism or influencing factor level. For failure prevention, it is set at the external cause level.

A fault tree is constructed for each type of component based on failure physics and logical relationships from the fault database.

8.6.2 Application of Fault Tree Analysis in Integrated Circuit Packaging

Plastic-Packaged FPGA: Fault Tree Analysis of the "Popcorn Effect".

【Fault Issue】 During system debugging, a plastic-packaged FPGA failed to load the program.

【Failure Mode】 Multiple ports became open circuits.

【Fault Tree Analysis】 The open-circuit failure in the plastic-packaged FPGA might result from failures in pins, chips, or bond points. Fault tree analysis can identify the failure mechanisms, key factors, and external influences. Figure 8.40 shows the fault tree for this failure.

【Analysis Verification】 SAM results revealed severe delamination at internal interfaces (see Figs. 8.41, 8.42, 8.43 and 8.44). Multiple ports were open circuits to GND. However, applying pressure restored the characteristic curve to match a normal sample. Cross-sectional analysis showed delamination between the molding compound and PCB. The solder mask also separated from the substrate, with some areas cracking. Additionally, copper foil lifted from the substrate. These findings confirm that internal delamination caused the loss of electrical connections.

【Failure Mechanism】 Moisture inside the plastic package expands under high temperatures. This expansion causes delamination at the package interface. The stress from thermal expansion breaks the bonding wires.

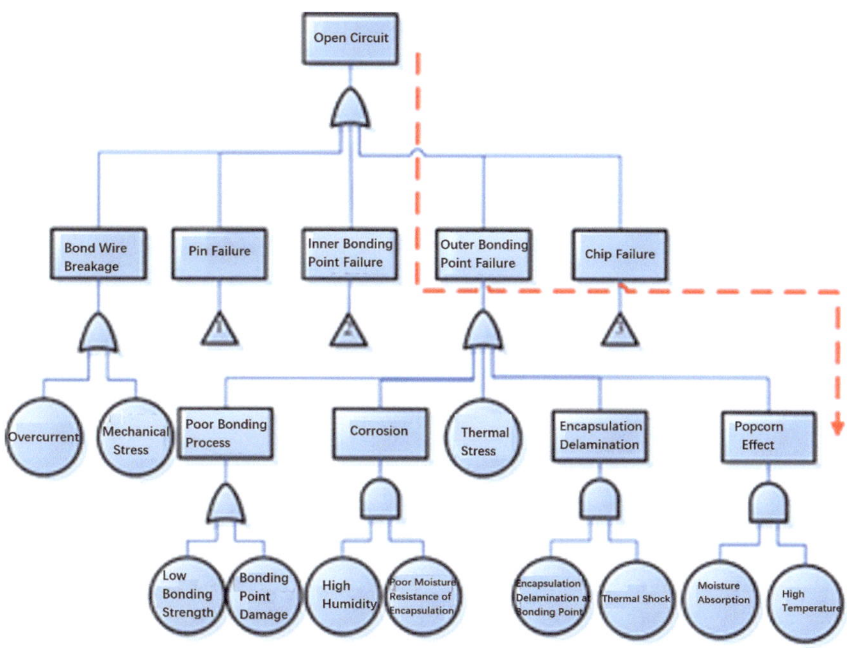

Fig. 8.40 Fault tree of open circuit failure in plastic encapsulated FPGA

Fig. 8.41 SAM image of the failed sample

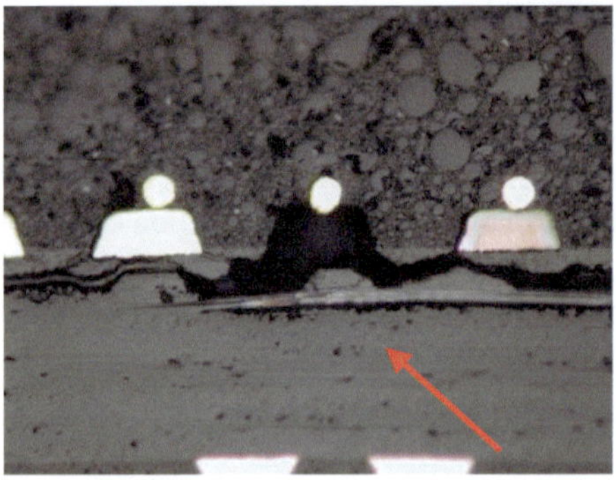

Fig. 8.42 Missing appearance of fractured copper foil after polishing

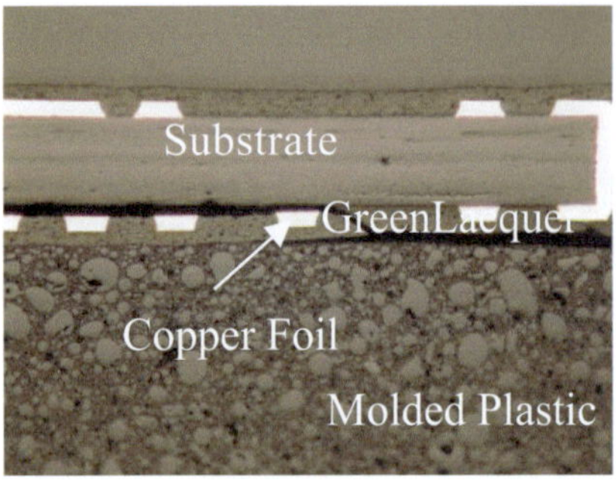

Fig. 8.43 Delamination between molded plastic and PCB, and between PCB surface green lacquer and substrate

【Analysis Conclusion】 Plastic-packaged integrated circuits absorb moisture when exposed to air for long periods. During soldering, trapped moisture can lead to internal delamination. This delamination is often hidden. If moisture accumulates, the issue may worsen during soldering and operation. It poses a serious reliability risk.

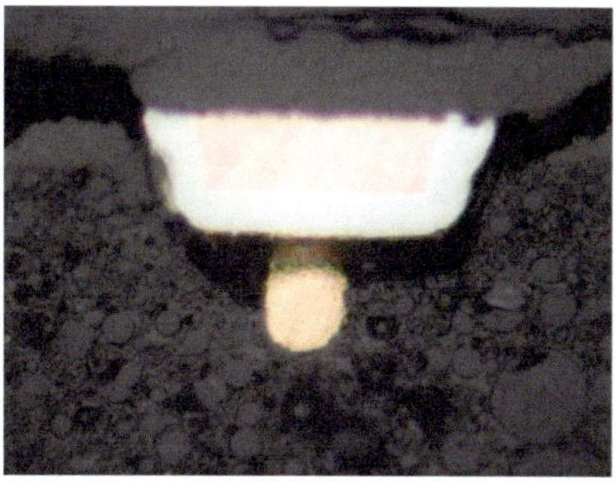

Fig. 8.44 Delamination between molded plastic and PCB at the bonding wire location

【Corrective Measures】 Baking the sample after programming and before soldering is recommended. This step prevents moisture-induced internal delamination. For mounted components, 100% SAM screening is advised. It helps remove components with potential reliability risks.

References

1. Kong X, En Y (2006) Failure analysis and typical cases of electronic components. National Defense Industry Press, Beijing
2. En Y, Lai P, Li S (2015) Failure analysis techniques for electronic components. Electronics Industry Press, Beijing
3. Chen Y, Zhang P, Xia K (2018) Diagnosis and analysis of TSV process defects. Semicond Technol 43(6):473–479
4. Premachandran C, Tran-Quinn T, Burrell L et al (2019) A comprehensive wafer level reliability study on 65nm silicon interposer. In: 2019 IEEE international reliability physics symposium (IRPS), pp 1–8
5. Chan JM, Lee K C, Tan CS (2018) Effects of copper migration on the reliability of through-silicon via (TSV). IEEE Trans Device Mater Reliab 18 (4):520–528
6. Premachandran CS, Choi S, Cimino S et al (2018) Reliability challenges for 2.5D/3D integration: an overview. In: 2018 IEEE international reliability physics symposium (IRPS), 5B.4-1-5B.4-5
7. Shen WW, Chen KN (2017) Three-dimensional integrated circuit (3D IC) key technology: through-silicon via (TSV). Nanoscale Res Lett 12(1):1–9
8. Daniel HJ, Youngwoo K, Jonghoon J et al (2017) Through silicon via (TSV) defect modeling, measurement, and analysis. IEEE Trans Compon Packag Manuf Technol 7(1):138–152
9. Chen Y, Su W, Huang HZ et al (2020) Stress evolution mechanism and thermo-mechanical reliability analysis of copper-filled TSV interposer. Eksploatacja i Niezawodnosc–Maintenance Reliability 22(4):705–714
10. Chen Y, Lai P, Huang H-Z et al (2021) Open localization in 3D package with TSV daisy chain using magnetic field imaging and high-resolution three-dimensional X-ray microscopy. Appl Sci 11(17):1–11
11. Chen Y, Zhang P, Su W et al (2019) Boundary layers defect diagnosis and analysis of through silicon via (TSV). Int J Perform Eng 15(1):97–106
12. Mees F, Swennen R, Geet MV et al (2003) Applications of X-ray computed tomography in the geosciences. J Geol Soc Lond Spec Publ 215(1):1–6
13. Vallett MD (2011) A comparison of lock-in thermography and magnetic current imaging for localizing buried short-circuits. In: ISTFA proceedings, pp 146–152

14. Gourikutty SBN, Chow YM, Alton J et al (2020) Defect localization in through-Si-interposer based 2.5D ICs. In: 2020 IEEE 70th electronic components and technology conference (ECTC), pp 1180–1185
15. Wang Z, Wu Z, Sun J et al (2019) Defect location and physical analysis in chip-on-chip device. IN: IEEE 26th international symposium on physical and failure analysis of integrated circuits (IPFA), pp 1–3
16. Jakubas A (2019) Diagnostics of the Fe-based composites using a magnetic field camera. In: 2019 progress in applied electrical engineering (PAEE), IEEE, pp 1–4
17. Wolf ID, Croes K, Beyne E (2018) Expected failures in 3-D technology and related failure analysis challenges. IEEE Trans Compon, Packag, Manuf Technol, pp 1–8
18. Altmann F, Brand S, Petzold M (2018) Failure analysis techniques for 3D packages. In: 2018 IEEE international symposium on the physical and failure analysis of integrated circuits (IPFA), IEEE, pp 1–8
19. Xu X, Huang W, Chen Y (2022) Failure localization method in open mode failure analysis of advanced package. In: 2020 21st international conference on electronic packaging technology (ICEPT), IEEE, pp 1–5
20. Hartfield C, Schmidt C, Gu A et al (2018) From PCB to BEOL: 3D X-ray microscopy for advanced semiconductor packaging. In: 2018 IEEE international symposium on the physical and failure analysis of integrated circuits (IPFA), IEEE, pp 1–7
21. Liu CY, Kuo PS, Chu CH et al (2016) High resolution 3D X-ray microscopy for streamlined failure analysis workflow. In: 2016 IEEE 23rd international symposium on the physical and failure analysis of integrated circuits (IPFA), IEEE, pp 216–219
22. Mohammad-Zulkifli S, Zee B, Thomas G et al (2019) High-resolution 3D X-ray microscopy for structural inspection and measurement of semiconductor package interconnects. In: 2019 IEEE 26th international symposium on the physical and failure analysis of integrated circuits (IPFA), IEEE, pp 1–4
23. Oravec M, Pacaiova H, Izarikova G et al (2019) Magnetic field image—source of information for action causality description. In: 2019IEEE 17th world symposium on applied machine intelligence and informatics (SAMI), IEEE, pp 101–106
24. Schmidt C, Lechner L, Wolf ID et al (2018) Novel failure analysis techniques for 1.8 m pitch wafer-to-wafer bonding. In: 2018 IEEE 68th electronic components and technology conference (ECTC), IEEE, pp 92–96
25. Khosravani MR, Reinicke T (2020) On the use of X-ray computed tomography in assessment of 3D-printed components. J Nondestruct Eval 39(4):1–6
26. Jacobs K, Li Y, Stucchi M et al (2020) Optical beam-based defect localization methodologies for open and short failures in micrometer-scale 3-D TSV interconnects. IEEE Trans Compon, Packag, Manuf Technol 99:1–1
27. Gaudestad J, Talanov V, Gagliolo N et al (2012) Space domain reflectometry for open failure localization. In: Physical and failure analysis of integrated circuits (IPFA), 19th IEEE international symposium on the 2012, pp 11–15
28. Gaudestad J, Talanov V, Huang PCJMR (2012) Space domain reflectometry for opens detection location in microbumps. Microelectron Reliab 52(9–10):2123–2126
29. Spinella L, Jiang T, Tamura N et al (2019) Synchrotron X-ray microdiffraction investigation of scaling effects on reliability for through-silicon vias for 3-D integration. IEEE Trans Device Mater Reliab 19(3):568–571
30. Murugesan M, Takeuchi A, Fukushima T et al (2019) X-ray computed tomography studies on directed self-assembly formed vertical nanocylinders containing metals for 3D LSI applications-characterization technique-dependent reliability issues. Jpn J Appl Phys 58(SB):SBBC05
31. Yang L (2009) Research on X-ray image sensing and digital imaging experiment, Ph.D. dissertation, Chongqing University, Chongqing, China
32. Cheng Y (2003) Research on industrial radiographic real-time imaging technology and development of high-performance digital imaging system, Ph.D. dissertation, Nanjing University of Science and Technology, Nanjing, China

33. Li Y (2006) Review of the current status of digital radiography. Nondestruct Test 28(4):198–203
34. Semmens JE, Kessler LW (1988) Nondestructive evaluation of thermally shocked plastic integrated circuit packages using acoustic microscopy. Int Symp Test Fail Anal, pp 211–215
35. Cichanski FJ (1989) Method and system for dual phase scanning acoustic microscopy: US, US4866986A
36. Shang Y, Zhao H (2011) Ion source for focused ion beam systems. Nucl Phys Rev 28(4):439–443
37. Teng L, Yu Z, Yang M et al (2012) Optimization and damage analysis of GaN inductively coupled plasma etching. Micro Nano Electron Technol 49(3):181–186
38. Kho WF, Leow JL, Cheah YC et al (2012) Alternative FIB cross section and laser ablation methods to improve failure analysis throughput of copper wire moisture related reliability failures. In: 2012 19th IEEE international symposium on the physical and failure analysis of integrated circuits (IPFA), IEEE, pp 1–4
39. Chen Y, Lin N, Lai P (2014) Three-dimensional X-ray laminography as a tool for detection and characterization of package on package (PoP) defects. In: 201410th international conference on reliability, maintainability and safety (ICRMS), IEEE, pp 275–278
40. Chen Y, Lin N (2013) Three-dimensional X-ray laminography and application in failure analysis for system in package (SiP). In: 14th international conference on electronic packaging technology, pp 746–749
41. Li Y, Cai Y, Pacheco M et al (2012) Non destructive failure analysis of 3D electronic packages using both electro optical terahertz pulse reflectometry and 3D X-ray computed tomography. In: International symposium for testing and failure analysis American society for metals, pp 95–99
42. Mario P, Deepak G (2011) Detection and characterization of defects in microelectronic packages and boards by means of high-resolution X-ray computed tomography (CT). In: Electronic components and technology conference, pp 1263–1268
43. Pacheco M, Goyal D (2010) X-ray computed tomography for non-destructive failure analysis in microelectronics. In: Reliability physics symposium, pp 252–258
44. Pacheco M, Goyal D (2008) New developments in high-resolution X-ray computed tomography for non-destructive defect detection in next generation package technologies. In: International symposium for testing and failure analysis, pp 30–35
45. Oppermann M, Zerna T, Wolter KJ (2009) X-ray computed tomography on miniaturized solder joints for nano packaging. In: Electronics packaging technology conference, pp 70–75
46. Scott D, Duewer F, Kamath S et al (2004) A novel X-ray microtomography system with high resolution and throughput for non-destructive 3D imaging of advanced packages. In: International symposium for testing and failure analysis American society for metals, pp 94–98
47. Kalukin AR, Sankaran VJ (1997) Three-dimensional visualization of multilayered assemblies using X-ray laminography. IEEE Trans Compon Packag Manuf Technol Part A 20(3):361–366
48. Sankaran V, Kalukin AR (1998) Improvements to X-ray laminography for automated inspection of solder joints. IEEE Trans Compon Packag Manuf Technol Part C 21(2):148–154
49. Gondrom S, Zhou J, Maisl M et al (1999) X-ray computed laminography: an approach of computed tomography for applications with limited access. Nucl Eng Des 190(1–2):141–147
50. Yuan C, Na L (2014) Magnetic microscopy for 3D devices failure localization and analysis. In: 2014 IEEE international conference on electron devices and solid-state circuits (EDSSC), IEEE, pp 1–2
51. Liu J, Dai J, Wang Y (2010) Infrared lock-in thermal wave detection technology and defect depth measurement. Opt Precis Eng 18(1):37–44
52. Schmidt C, Grosse C, Altmann F (2010) Localization of electrical defects in system in package devices using Lock-in Thermography. In: 2010 IEEE 3rd electronic system-integration technology conference (ESTC), IEEE, pp 1–5
53. Tay MY, Tan MC, QIU W et al (2011) Lock-in thermography application in flip-chip packaging for short defect localization. In: 2011 IEEE 13th electronics packaging technology conference (EPTC), IEEE, pp 642–646

54. Schmidt C, Altmann F, Schlangen R et al (2010) Non-destructive defect depth determination at fully packaged and stacked die devices using Lock-in Thermography. In: 17th IEEE international symposium on the physical and failure analysis of integrated circuits (IPFA), IEEE, pp 1–5
55. Huo Y, Zhao Y, Li Y et al (2009) Comparative study of pulsed and lock-in infrared thermal imaging detection techniques. Laser Infrared 39(6):602–604
56. Zhao Y (2012) Nondestructive testing and fatigue performance research based on lock-in infrared thermography theory, Ph.D. dissertation, Dalian University of Technology, Dalian, China
57. Chen Y, He XQ, Lai P (2013) The application of fault tree analysis method in electrical component. In: 20th international symposium on the physical and failure analysis of integrated circuits (IPFA), pp 666–669

Chapter 9
Quality and Reliability Assurance in Integrated Circuit Packaging

The quality and reliability of integrated circuit (IC) packaging play a crucial role in product performance. Ensuring reliability involves multiple aspects, from structural design and manufacturing processes to product screening, qualification, and lifespan assessment. Various inspection and testing methods help verify that the packaging meets or exceeds design and usage requirements.

To maintain high standards, regulatory bodies worldwide have established general specifications and standards for IC products. These standards define the necessary tests and verification procedures. However, requirements vary depending on application fields and operating environments. Some standards impose stricter regulations than others. This chapter compares different standards related to packaging quality and reliability, covering physical properties and environmental adaptability. It includes industrial, military, and automotive electronics standards. This chapter also introduces key analysis techniques for assessing IC packaging quality and reliability. These include Destructive Physical Analysis (DPA), Construction Analysis (CA), and counterfeit detection methods. Such techniques help identify design flaws, manufacturing defects, and potential failure risks. They also detect counterfeit or refurbished products. Finally, this chapter discusses accelerated life testing and common acceleration models used in IC packaging. By increasing stress levels, these tests generate failure data more quickly. This allows researchers to analyze lifespan distribution and evaluate reliability in a shorter time frame.

B. Zhou et al., *Reliability Technology for Integrated Circuit Packaging*,
https://doi.org/10.1007/978-981-95-3885-0_9

9.1 Packaging Quality Inspection and Environmental Adaptability Requirements

9.1.1 Packaging Quality Inspection Requirements

To ensure packaging quality, various tests are required. Common tests include external and internal visual inspection, Scanning Acoustic Microscope inspection, wire integrity testing, bond strength testing, flip-chip pull-off testing, die shear strength testing, hermeticity testing, internal moisture content testing, solderability testing, dimensional inspection, lid torque testing, and solder ball shear testing.

1. Non-Destructive Physical Testing Requirements

Non-destructive physical testing does not damage components during the process. These tests, also called non-destructive tests, include external visual inspection, dimensional inspection, hermeticity testing, and X-ray inspection. They help detect internal defects in packaging, such as delamination, wire bond issues, adhesion defects, or voids in glass-sealed packages. Table 9.1 compares non-destructive testing requirements in different standards. The requirements are mostly the same, with minor differences in sample size.

Table 9.1 Requirements for two types of non-destructive physical tests in different standards

Test item	Sample qty	Acceptance criteria	Standard
Dimensional test	5	A sample fails if its size exceeds the specified tolerance or limit. 0 failures allowed	National military standard
	5	A sample fails if its size exceeds the specified tolerance or limit. 0 failures allowed	U.S. military standard
	10/ Lot;3Lots	*CPK* > 1.67; A sample fails if its size exceeds the specified tolerance or limit. Zero failures allowed	Automotive electronics industry standard
	30/Lot;1Lot	A sample fails if its size exceeds the specified tolerance or limit. 0 failures allowed	Industrial standard
Sealing test	15	Fine leak test first, then gross leak test. 0 failures allowed	National military standard
	15	Fine leak test first, then gross leak test. 0 failures allowed	U.S. military standard
	15/Lot;1Lot	Fine leak test first, then gross leak test. 0 failures allowed	Automotive electronics industry standard
	–	Fine leak test first, then gross leak test	Industrial standard

Note: C_{PK} refers to the Process Capability Index. “Lot” means batch.

Dimensional inspection ensures that manufacturing follows design specifications. It also guarantees proper assembly in later stages. Different packaging structures follow industry standards that define specific dimensions and tolerances.

Hermeticity testing checks whether packages with internal cavities are airtight. It involves two steps: fine leak testing and gross leak testing. Airtightness directly affects the reliability of integrated circuits. Moisture or acidic gases entering the cavity can corrode the chip, metal interconnections, and non-metallic structures. This may cause performance degradation or even failure. Therefore, hermetic packages must meet the leakage rate limits set by standards.

The standards listed in this chapter refer to the following:

(1) Chinese Military Standards: GJB 597B-2012 General Specification for Semiconductor Integrated Circuits, GJB 548B-2005 Test Methods and Procedures for Microelectronic Devices, and GJB 2438B-2017 General Specification for Hybrid Integrated Circuits.
(2) U.S. Military Standards: MIL-STD-883 Test Method Standard for Microelectronic Devices and MIL-PRF-38535 General Specification for Integrated Circuit Manufacturing.
(3) Automotive Electronics Standards: AEC-Q100 Failure Mechanism Based Stress Test Qualification for Integrated Circuits and AEC-Q104 Failure Mechanism Based Stress Test Qualification for Multi-Chip Modules (MCMs) in Automotive Applications.
(4) Industrial Standards: JESD47 Stress-Test-Driven Qualification of Integrated Circuits, JESD22 series Reliability Test Methods for Packaged Devices, and IPC/JEDEC J-STD series standards.

2. Destructive Physical Testing Requirements

Destructive physical testing damages the product's structure using physical or chemical methods. It verifies whether the materials, processes, and design meet the intended use or relevant standards. Common tests include solderability testing, lid torque testing, internal moisture content testing, solvent resistance testing, bond strength testing, and die shear strength testing. Table 9.2 lists the requirements for these tests. Chinese and U.S. military standards have similar test conditions and acceptance criteria. However, military standards differ significantly from automotive and industrial standards in test conditions, sample sizes, and acceptance requirements.

Wire bond pull testing evaluates whether the bond strength meets the required standards. Common wire bond failures include wire breakage, bond interface failure, metallization lift-off, chip cracking, and substrate cracking. Military, automotive, and industrial standards have the same minimum wire pull strength requirements. However, automotive standards specify that gold wires thicker than 1 mil must have a pull strength greater than 0.03N after thermal cycling.

The wire integrity test checks the strength of leads (terminals) and their soldering. It includes tensile testing, bending testing, wire fatigue testing, wire torque testing, bolt torque testing, and pad adhesion testing for leadless chip carriers and similar

Table 9.2 Requirements for destructive physical tests

Test item	Test condition	Sample qty	Acceptance criteria	Standard
Wire bond pull test	Use GJB 548B–2005, Method 2011, Fig. 9.1 to determine the minimum tensile strength	At least 15 wires from a minimum of 4 devices	0 failures	National military standard
	Use MIL-STD-883, Method 2011, Fig. 9.1 to determine the minimum tensile strength	At least 15 wires from a minimum of 4 devices	0 failures	U.S. military standard
	For wires > 1 mil, the minimum tensile strength after thermal cycling is 0.03N. For wires < 1 mil, use MIL-STD-883, Method 2011, Fig. 9.1	At least 30 wires from a minimum of 5 devices	$C_{PK} > 1.67$ or 0 failures after thermal cycling	Automotive electronics industry standard
	Use MIL-STD-883, Method 2011, Fig. 9.1 to determine the minimum tensile strength	At least 30 wires from a minimum of 5 devices	$C_{PK} \geq 1.33$ or $P_{PK} \geq 1.66$	Industrial standard
Bond strength test	Minimum bond strength: 0.05N × number of bonds	22 chips	0 failures	National military standard
	Minimum bond strength: 0.05N × number of bonds	22 chips	0 failures	U.S. military standard
	–	At least 10 devices with 5 solder balls each; 3 lots	$C_{PK} > 1.67$	Automotive electronics industry standard
	–	10 solder balls per device; 5 devices	–	Industrial standard

(continued)

Table 9.2 (continued)

Test item	Test condition	Sample qty	Acceptance criteria	Standard
Wire integrity test	Test condition: Wire fatigue	45 wires, taken from at least 3 devices	0 failures	National military standard
	Test condition: Wire fatigue	45 wires, taken from at least 3 devices	0 failures	U.S. military standard
	–	10 wires; 5 units; 1 lot	No wire breakage or cracks	Automotive electronics industry standard
	–	45 wires; at least 5 units	Based on manufacturer specifications or applicable procurement documents	Industrial standard
Solderability test	Solder temperature: 245°C ± 5°C	22 chips	>95% coverage, 0 failures	National military standard
	Solder temperature: 245 °C ± 5°C	22 chips	>95% coverage, 0 failures	U.S. military standard
Solderability test	Solder temperature: 245°C ± 5°C	15/Lot;1Lot	>95% coverage, 0 failures	Automotive electronics industry standard
	Solder temperature: 245°C ± 5°C	22/Lot;3Lots	>95% coverage, 0 failures	Industrial standard
Die shear strength test	Use GJB 548B−2005, Method 2019, Fig. 9.4 to determine values for different chip sizes	3 chips	0 failures	National military standard
	Use MIL-STD-883, Method 2019, Fig. 9.4 to determine values for different chip sizes	3 chips	0 failures	U.S. military standard

(continued)

Table 9.2 (continued)

Test item	Test condition	Sample qty	Acceptance criteria	Standard
	Use MIL-STD-883, Method 2019, Fig. 9.4 to determine values for different chip sizes	5/Lot;1Lot	0 failures	Automotive electronics industry standard
	–	–	–	Industrial standard
Lid torque test	Applicable only to glass-sealed packages	5 chips	0 failures	National military standard
	Applicable only to glass-sealed packages	5 chips	0 failures	U.S. military standard
	Applicable only to ceramic cavity packages	5/Lot;1Lot	0 failures	Automotive electronics industry standard
	–	5/Lot;1Lot	Based on manufacturer specifications or applicable procurement documents	Industrial standard
Internal moisture content test	Not exceed 5000 ppm	3 or 5 chips	0 failures or up to 20% failure rate	National military standard
	Not exceed 5000 ppm	3 or 5 chips	0 failures or up to 20% failure rate	U.S. military standard
	–	5/Lot;1Lot	0 failures	Automotive electronics industry standard
	–	1/Lot;3Lots	0 failures	Industrial standard

Note: "Lot" refers to a batch.

packages. Military standards clearly state that wire fatigue testing is required to assess resistance to bending fatigue. However, automotive and industrial standards do not specify which wire integrity test is necessary.

The solderability test examines whether the surface treatment of package leads, solderable terminals, solder balls, and solder columns supports reliable soldering. It ensures that integrated circuits do not face reliability issues during assembly. Poor wettability can cause defects such as cold solder joints, weak bonds, bridging, and poor solder strength. These defects increase quality control workload and repair costs, leading to wasted resources. Standards usually require that the wetted area in the solderability test must exceed 95% coverage.

The die shear strength test verifies the reliability of the materials and processes used to mount chips onto substrates or package enclosures. It also assesses the quality of die attachment. Military, automotive, and industrial standards follow the same shear strength requirements.

The lid torque test evaluates the shear strength of integrated circuit package lids. All four standards in Table 9.2 include this test. The sample size is typically five units. Military and automotive standards require zero failures, while industrial standards follow manufacturer specifications or applicable procurement documents.

The internal moisture content test measures the moisture level inside hermetically sealed integrated circuits with metal or ceramic packages. Excessive moisture can cause metallic corrosion, electrochemical corrosion, and Au-Al bonding degradation. Military standards usually require the internal moisture content to be below 5000 ppm.

9.1.2 Environmental Adaptability Requirements for Packaging

Temperature, humidity, smoke, vibration, and shock can degrade packaging performance. They also affect the reliability of integrated circuits. Standards and specifications set environmental adaptability requirements. These ensure product quality and reliability during transportation, storage, and use. Packaging is tested under specific conditions to find structural defects. This helps provide information about product quality.

1. Temperature Stress

Temperature is a key factor in packaging reliability. Different thermal expansion coefficients of materials can cause issues. These include interface fatigue, delamination, and residual stress release from manufacturing [1–3]. Table 9.3 shows temperature adaptability requirements in different standards. The thermal cycling test measures the impact of repeated high and low temperatures. The thermal shock test evaluates reliability under sudden temperature changes. It also identifies related issues. The high-temperature storage test examines how time and temperature affect packaging. High temperatures can cause cracks, Au-Al interdiffusion, IMC formation, voids, and increased thermal resistance [4].

By comparing, it can be observed that the temperature environmental adaptability requirements in the Chinese military standards and U.S. military standards are consistent. The automotive electronics industry standards require more samples and higher cycles for thermal cycling tests. They also demand higher temperature reliability.

2. Mechanical Stress

Table 9.3 Temperature environmental adaptability requirements for integrated circuit products in different standards

Test item	Test condition	Sample qty	Acceptance criteria	Standard
Thermal cycling test	−65°C~150°C	15	100 cycles/0 failures	National military standard
	−65°C~150°C	15	100 cycles/0 failures	U.S. military standard
	Minimum Operating Temperature ~ Maximum Operating Temperature	30/ Lot;3Lots	1000 cycles/0 failures	Automotive electronics industry standard
	−55°C ~ 125°C	25/ Lot;3Lots	700 cycles/0 failures	Industrial standard
Thermal shock test	−65°C~150°C	15	15 cycles/0 failures	National military standard
	−65°C~150°C	15	15 cycles/0 failures	U.S. military standard
	–	–	–	Automotive electronics industry standard
	−55°C ~ 125°C	25/ Lot;3Lots	700 cycles/0 failures	Industrial standard
High-temperature storage test	–	–	–	National military standard
	–	–	–	U.S. military standard
	150°C	30/Lot;1Lot	1000 h/0 failures	Automotive electronics industry standard
	150°C	25/ Lot;3Lots	1000 h/0 failures	Industrial standard

Note: "Lot" refers to a batch.

Mechanical stress affects the quality and reliability of the packaging structure in the following ways:

(1) Structural Damage: This refers to deformation, bending, cracks, fractures, and fatigue damage caused by long-term alternating stress.
(2) Operational Failure: This is mainly due to poor contact between components, resulting in performance issues such as irregular operation or instability.

(3) Functional Failure: This type of failure typically involves issues like solder joint failure, which often happens in a short period.

Mechanical stress tests include sweep frequency vibration tests, mechanical shock tests, and constant acceleration tests. Table 9.4 provides the mechanical environmental adaptability requirements.

The sweep frequency vibration test measures the impact of vibration within a specific frequency range on integrated circuit products. For example, strong vibrations occur during aerospace launches. It is necessary to test the reliability of products in such an environment. Common failure mechanisms caused by vibration include bond interface failure and interference from foreign objects inside the package (contaminants from the manufacturing process). The frequency requirement for this test is 20–2000 Hz.

The mechanical shock test determines whether integrated circuit products can work under moderate mechanical shock conditions. Such shocks occur when the product experiences sudden external forces or changes in motion during transportation or operation. These shocks can affect product performance or even cause damage [5]. As the shock pulse repeats, the damage becomes more severe. The military standard requires pulses in the X_1, X_2, Y_1, Y_2, Z_1, and Z_2 directions. The automotive electronics industry and industrial standards only require pulses in the Y1 direction.

The constant acceleration test identifies defects not found in the sweep frequency vibration or mechanical shock tests. It can detect performance degradation or failure in the leads, chips, and substrate welds under constant acceleration.

1. Humidity and Heat Stress

Moisture in the air is a significant factor causing package failure. When moisture enters the package structure, it can cause corrosion of pads and metals. At room temperature, moisture enters slowly. Therefore, a high-accelerated humidity and heat stress test is usually used. This test combines high temperature, high humidity, and electrical stress to assess the package's adaptability to humidity environments. It can lead to failures in metals and pads, crack formation, delamination, electrochemical corrosion, IMC formation, and insulation failure [6–8].

Table 9.5 lists the humidity environment adaptability requirements. The Chinese military and U.S. military standards both use humidity resistance tests. The test conditions and pass criteria are the same for both. This test involves thermal cycling to cause moisture inside the package to condense and dry out. This "breathing" effect accelerates corrosion. In high-temperature environments, degradation from moisture becomes more severe due to faster chemical reactions. The test also includes low-temperature sub-cycles. In low temperatures, moisture condensation creates significant stress, speeding up the appearance of defects like cracks and delamination.

The automotive electronics industry and industrial standards also use the high-accelerated humidity and heat stress test. The test conditions are the same, but the sample size requirements differ. The automotive electronics industry standard requires more samples to ensure higher quality and reliability.

Table 9.4 Mechanical environmental adaptability requirements

Test item	Test condition	Sample qty	Acceptance criteria	Standard
Sweep frequency vibration test	Peak acceleration of 196 m/s^2 (20 g), frequency 20–2000 Hz	15	0 failures	National military standard
	Peak acceleration of 20 g, frequency 20–2000 Hz	15	0 failures	U.S. military standard
	Peak acceleration of 50 g, frequency 20–2000 Hz	15/ Lot;1Lot	0 failures	Automotive electronics industry standard
	Peak acceleration of 50 g, frequency 20–2000 Hz	39/ Lot;3Lots	0 failures	Industrial standard
Mechanical shock test	5 pulses in X_1, X_2, Y_1, Y_2, Z_1, and Z_2 directions; pulse width 0.5 ms; peak acceleration of 1500 g	15	0 failures	National military standard
	5 pulses in X_1, X_2, Y_1, Y_2, Z_1, and Z_2 directions; pulse width 0.5 ms; peak acceleration of 1500 g	15	0 failures	U.S. military standard
	5 pulses in Y_1 direction; pulse width 0.5 ms; peak acceleration of 1500 g	15/ Lot;1Lot	0 failures	Automotive electronics industry standard
	5 pulses in Y_1 direction; pulse width 0.5 ms; peak acceleration of 1500 g	39/ Lot;3Lots	0 failures	Industrial standard
Constant acceleration test	Y_1 direction, 294,000 m/s2 (30,000 g)	15	0 failures	National military standard
	Y_1 direction, 30,000 g	15	0 failures	U.S. military standard
	Y_1 direction: for packages with fewer than 40 pins, 30,000 g; for packages with more than 40 pins, 20,000 g	15/ Lot;1Lot	0 failures	Automotive electronics industry standard
	Y_1 direction: for packages with fewer than 40 pins, 30,000 g; for packages with more than 40 pins, 20,000 g	39/ Lot;3Lots	0 failures	Industrial standard

Note: ① "Lot" refers to a batch.; ② After the test, perform a visual inspection of the markings and inspect the leads or sealing condition of the package to determine whether the product is qualified

Table 9.5 Humidity and heat environmental adaptability requirements

Test item	Test condition	Sample qty	Acceptance criteria	Standard
Moisture resistance test	25 °C/(80%~100%) RH to 65 °C/(90%~100%) RH, 10 cycles (at least 5 cycles at low temperature sub−cycles, −10 °C/uncontrolled humidity)	15	0 failures	National Military Standard
	25 °C/(80%~100%) RH to 65 °C/(90%~100%) RH, 10 cycles (at least 5 cycles at low temperature sub−cycles, −10 °C/uncontrolled humidity)	15	0 failures	U.S. Military Standard
	–	–	–	Automotive electronics industry standard
	–	–	–	Industrial standard
High accelerated moisture stress test	–	–	–	National military standard
	–	–	–	U.S. Military Standard
	High-humidity test130 °C/85% RH, 96 h or 110 °C/85% RH, 264 h, maximum VCC	30/ Lot;3Lots	0 failures	Automotive electronics industry standard
	130 °C/85% RH, 96 h or 110 °C/85% RH, 264 h, maximum VCC	25/ Lot;3Lots	0 failures	Industrial standard

Note: "Lot" refers to a batch.

9.2 Quality and Reliability Analysis Techniques

9.2.1 Destructive Physical Analysis

Destructive physical analysis provides important information about integrated circuit design and packaging process defects. It is a key technique for quality and reliability analysis. This method can identify batch quality issues and help customize selection and identification test plans.

Destructive physical analysis related to integrated circuit packaging mainly includes nine items: external visual inspection, X-ray inspection, sealing test, internal moisture content test, internal visual inspection, bond strength test, baseline structure inspection, chip shear test, and SEM inspection. Details of the procedures for each item can be found in GJB 548B—2005. For plastic-sealed circuits, GJB 4027A—2006 adds two more items: acoustic SEM inspection and glass passivation layer integrity inspection.

The destructive physical analysis process for sealed packaging is shown in Fig. 9.1. The sampling requirements are specified in GJB 4027A—2006. Specific requirements and criteria for each analysis project are provided in the operational procedures. Among these, external visual inspection, X-ray inspection, sealing test, internal moisture content test, internal visual inspection, bonding strength test, and chip shear test all involve packaging inspections.

If the product is applied to critical and harsh fields, such as aerospace and military applications, it is best to add the following additional analyses or request suppliers to provide the following data:

- Mold plastic filler particle analysis: particle diameter and particle sharpness.
- Physical properties analysis of mold plastic (2 products per batch): Glass transition temperature (T_g).
- Electron Dispersive Spectroscopy (EDS) analysis: pin coating materials.
- Volatility of packaging materials: suitable for aerospace applications.
- Moisture absorption characteristics of packaging: water vapor diffusion constant and moisture expansion coefficient.

(1) External Visual Inspection

External visual inspection checks the packaging quality of integrated circuit products and identifies any damage caused during transportation and use. The main areas inspected include:

- Packaging deformation: unevenness, distortion, etc.
- Impurities in the packaging, voids, or cracks in the mold plastic.
- Pin deformation, peeling, bubbling, or coating corrosion.
- Condition of pins and coatings.
- Correctness and legality of labeling.
- Consistency across the batch.

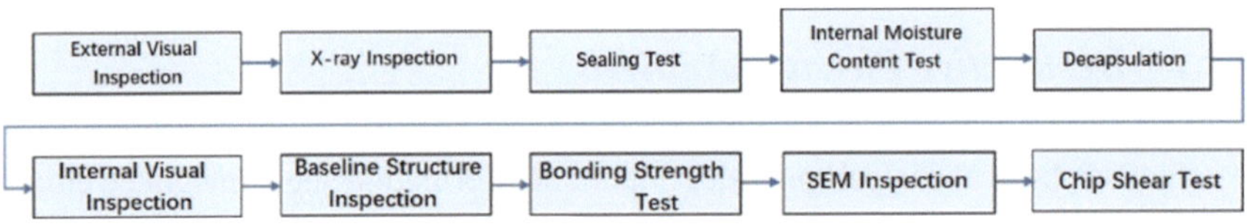

Fig. 9.1 Destructive physical analysis process for sealed packaging

Refer to GJB 548B—2005 Method 2009.1 for specific requirements and criteria.

(2) X-ray Inspection

The goal of X-ray inspection is to check for defects inside the packaging in a non-destructive manner. This includes the package body, chip, chip bonding, and lead wires. The main defects inspected are:

- Voids or foreign objects in the packaging.
- Voids in the chip bonding material.
- Misalignment of pins.
- Burrs on the lead frame.
- Poor bonding layout.
- Wire bonding collapse or lead breakage.
- Incorrect chip placement.

The inspection criteria and requirements can be found in GJB 548B—2005 Method 2012. Since X-rays can damage sensitive components, when using real-time X-ray inspection for integrated circuit screening, the total dose rate of the equipment should be estimated.

(3) Sealing Test

The sealing test aims to check the airtightness of the package. It mainly includes fine leak and coarse leak tests. Common leak detection methods are helium mass spectrometry and fluorine oil bubble methods. The helium method detects small gaps, while the fluorine oil method detects larger leaks.

(4) Internal Moisture Content Test.

The inspection criteria and requirements are based on GJB 548B—2005 Method 1014.

This test is used for hermetically sealed packages. Its purpose is to measure the moisture content in the gas inside metal or ceramic packages.

The method for the internal moisture content test is based on GJB 548B—2005 Method 1018.

(5) Internal Visual Inspection

Internal visual inspection checks defects related to chip dicing, chip mounting, and lead bonding. The inspection begins with a low magnification microscope (30-60x), followed by a high magnification microscope (75-200x). This helps identify defects at the chip and assembly levels. It also checks the batch consistency of the chips. The main defects include:

- Pinholes, peeling, or cracks in the glass passivation layer (especially damage caused by the molding plastic filler).
- Voids, corrosion, peeling, or lifting in the metallization layer.
- Bonding lifting, displacement, or excessive deformation.
- Evaluation of the batch consistency of chips.

The inspection areas for this procedure are based on GJB 548B—2005 Method 2013. If necessary, further checks can be made using scanning electron microscopy (SEM) to inspect bonding, passivation, or intermetallic compounds (IMC).

(6) Baseline Structure Inspection

The baseline structure inspection ensures the product meets the requirements of the baseline design documentation. Any deviations should be recorded as defects.

(7) Bond Strength Test

The purpose of the bonding strength test is to check whether the bonding strength meets the requirements in the order documents.

The wire bonding pull test follows GJB 548B—2005 Method 2011 Condition D. It is important to note that since the bonding wires are not fully exposed during opening, the criteria in GJB 548B—2005 may not apply. Generally, the weakest part of the bond is the neck. Bonding strength is significantly influenced by the environment in which the device is used. Storing in thermal cycling or high temperature and high humidity can cause bonding strength degradation. In molding plastics with flame retardants, intermetallic compounds (IMC) at the gold-aluminum bond interface can speed up degradation. In some cases, it is recommended to perform bond strength tests after accelerated tests to ensure device quality and long-term reliability.

(8) SEM Inspection

SEM inspection uses the material properties of the sample's surface to create micro-images. It is mainly used to observe microstructures, locate failures, and analyze defects.

The key defects inspected during SEM include:

- Bonding defects: Bonding point depression, bond pull-off, visible IMC larger than 0.1 mil around the solder ball, and microstructural features at the fracture of the bonding wire, bonding layer, and chip interface.
- Glass passivation layer integrity: Delamination, pinholes, and cracks caused by molding plastic fillers or packaging mechanical stress.
- Chip metallization layer defects: Planarization quality, via quality, or step quality.

Inspection follows GJB 548B—2005 Method 2018.1.

(9) Chip Shear Test

The chip shear test checks the integrity of the materials and process steps used to mount semiconductor chips and passive devices onto substrates or packaging shells. It focuses on defects at the chip mounting area.

This test follows GJB 548B—2005 Method 2019. It measures the force applied to the chip during the shear test, observes the types of failure caused by the force, and examines the interface morphology of the remaining chip bonding materials and the metalized layer of the substrate or packaging shell.

(10) Additional Analysis

① Glass Transition Temperature Check

The glass transition temperature is a crucial indicator of the reliability of plastic-packaged devices. When the temperature exceeds the glass transition temperature, the diffusion rates of water vapor and ions increase. This accelerates the diffusion of moisture inside the packaging. Thermal Mechanical Analyzers (TMA) and Differential Scanning Calorimeters (DSC) are commonly used to measure the glass transition temperature of molding plastics.

② Pin Plating Composition Analysis and Solderability

Due to environmental regulations, lead-free processes are now commonly used in the civilian and commercial electronics fields. These devices have pins plated with tin. Lead-free devices can grow tin whiskers in natural environments. These whiskers can cause short circuits between densely packed pins, which is a significant safety risk. In critical applications, it is best to analyze the composition of the pin plating. This analysis helps assess solderability and identify potential risks, such as heat resistance during soldering.

Figure 9.2 shows the EDS composition analysis of two samples' pins. The left sample is a leaded device. The pins are first plated with palladium (Pd), which provides protection, and then plated with nickel. The composition is shown in Fig. 9.2a. The C and O elements are contaminants, while the Au element results from pre-treatment. The right sample is a lead-free device. The pins are plated with tin, and the composition is shown in Fig. 9.2b. Leaded devices are no longer allowed in civilian and commercial fields. Most new devices on the market are lead-free. However, due to the stricter conditions of the lead-free process and the risk of tin whiskers, military devices still often use leaded processes. Therefore, it is essential to perform plating composition analysis before use to distinguish and prevent potential issues.

③ Water Vapor Diffusion Coefficient and Thermal Expansion Coefficient of Plastic-Packaged Circuits

The water vapor diffusion coefficient is determined by measuring the mass change of molding plastics over time. This is done under specific baking conditions. The thermal expansion coefficient reflects how material size changes with temperature. Measuring this coefficient helps us understand the thermomechanical properties of integrated circuits. Mismatched thermal expansion coefficients in plastic packaging

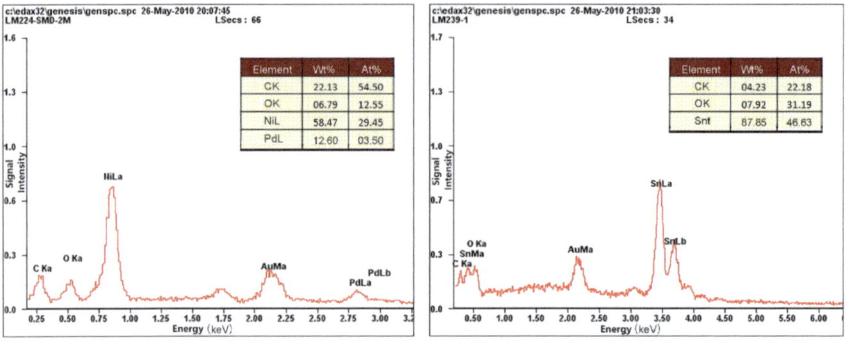

(a) EDS Composition Analysis of Nickel-Plated Pin (b) EDS Composition Analysis of Tin-Plated Pin

Fig. 9.2 EDS composition analysis of pins from two samples

can lead to serious thermomechanical fatigue problems. To minimize mechanical stress, the thermal expansion coefficient of the molding plastic should match that of the other materials used in the packaging. Therefore, testing the thermal expansion coefficient provides important information on material degradation. This information is crucial for assessing the reliability of the molding plastics.

④ SEM Examination of Packaging Filler Particles

To prevent damage to the chip passivation layer and metallization layer wiring, it is essential to avoid using filler particles that are too large or sharp. Such particles can cause damage under external mechanical stress. In critical applications, destructive physical analysis can supplement the examination of particle size and shape in the metallization layer. Typically, quartz sand particles in fillers are spherical and have a diameter of about 20 μm. However, the quality of fillers can vary significantly between manufacturers. Some manufacturers use fillers with larger, sharper particles, which can pose quality risks.

The following sections present several cases of destructive physical analysis. These cases illustrate how such analysis ensures the quality and reliability of integrated circuit packaging.

Case 1: **Destructive Physical Analysis of a SiP Module**

A SiP module underwent destructive physical analysis. First, a microscope was used for an external visual inspection of the module. This check assessed the surface for defects. The inspection revealed clear surface markings, and the bottom pin had detached. No other abnormalities were observed. Next, the SiP module was encapsulated and ground from the front to the storage chip layer. The storage devices were then subjected to destructive physical analysis, following the prescribed standards. This included X-ray and C-SAM checks. The results indicated delamination at the lead frame of the Pin 22 external bond point. Additionally, the solder joint of Pin 22 showed significant abnormalities compared to the other joints, which posed a risk of open circuits. A typical inspection result is shown in Fig. 9.3.

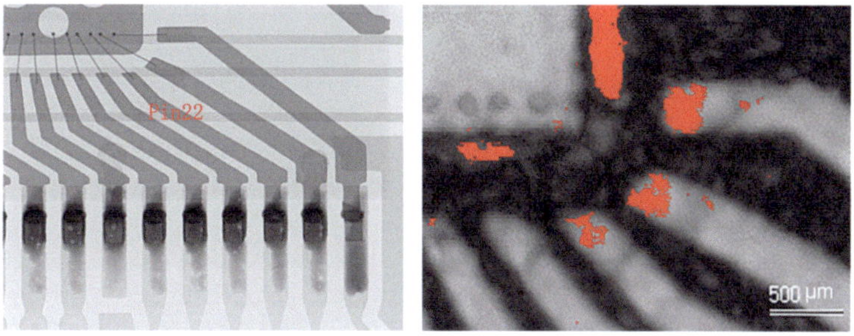

(a) X-ray morphology of the bond wire and pin at Pin22 of the storage device

(b) Front C-SAM morphology of the storage device at Pin22

Fig. 9.3 Typical inspection results

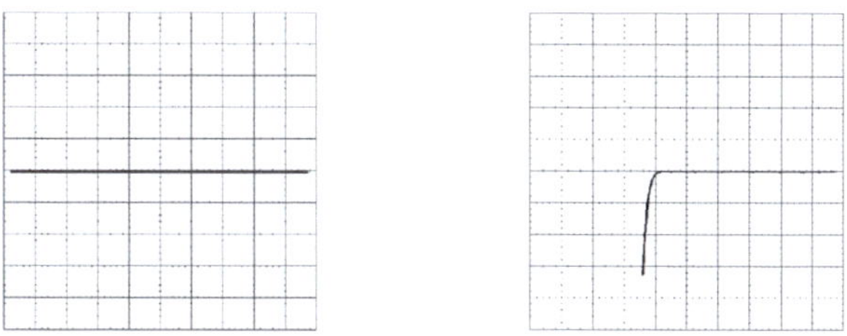

(a) I-V characteristic curve between Pin22 and GND of the storage device

(b) I-V characteristic curve between Pin21 and GND of the storage device

Fig. 9.4 Test results

After the SiP module was powered on, a functional anomaly was observed in the storage device. Using techniques such as planar grinding and laser opening, the bond wire and external bond point wire frame at Pin22 of the storage device were exposed. Testing revealed an open circuit between Pin22 and GND. The test results are shown in Fig. 9.4.

To examine the morphology of the solder joint at Pin22 of the storage device, encapsulation and sectioning were performed. Internal visual inspection found a significant gap between the Pin22 pin and the pad. The gap was filled with molding plastic (see Fig. 9.5). The solder connection area between the pin and pad was small, and a fracture open circuit was observed. This does not meet the internal visual inspection standards for destructive physical analysis.

Case 2: SEM examination of particles in packaging filling materials.

The morphology of the packaging filling material affects the quality of the plastic encapsulation. The particle analysis of the packaging filling material in the sample is shown in Fig. 9.6. Figure 9.6a shows the typical morphology of the filling material in an operational amplifier produced by Company A. The quartz sand particles are spherical, with a diameter of 20–30 μm and smooth edges. However, the devices produced by Company B have much larger quartz sand particles, as shown in Fig. 9.6b. The

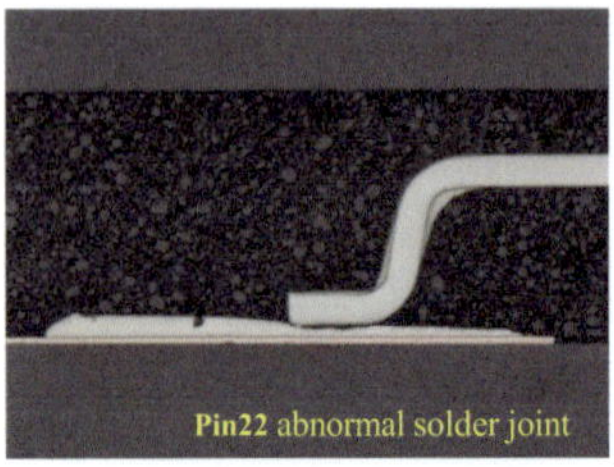

(a) Overall morphology of the solder joint cross-section at Pin22 of the storage device

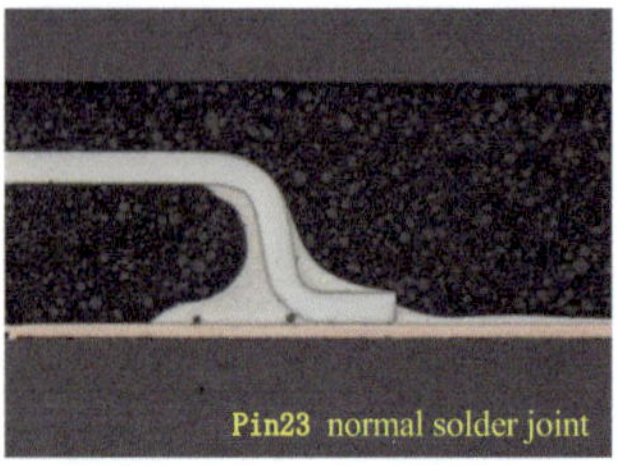

(b) Overall morphology of the solder joint cross-section at Pin23 of the storage device

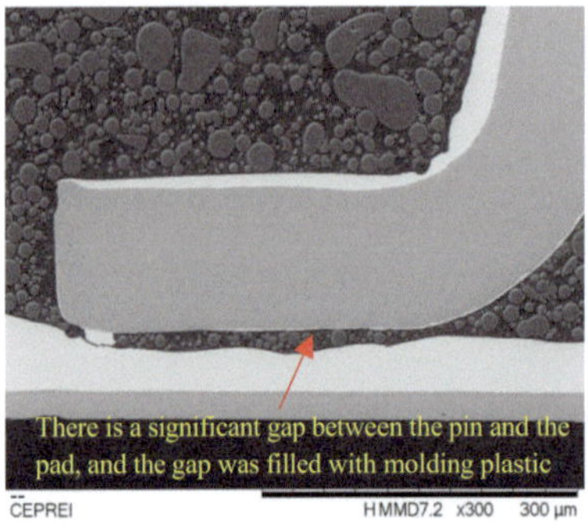

(c) SEM morphology of the solder joint cross-section at Pin22 of the storage device

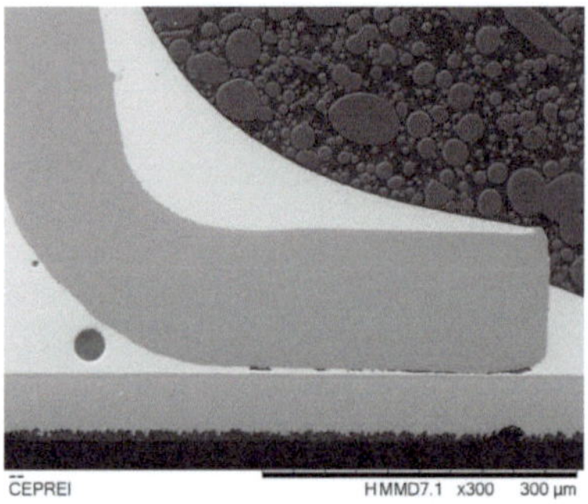

(d) SEM morphology of the solder joint cross-section at Pin23 of the storage device

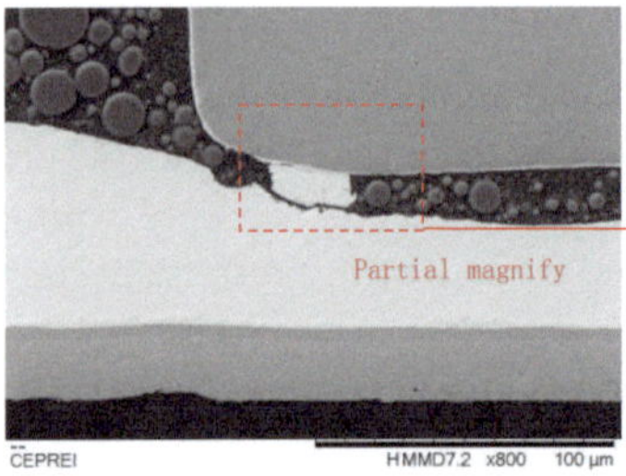

(e) Magnified SEM morphology of the solder joint cross-section at Pin22 of the storage device

(f) Partial magnified SEM morphology of the solder joint cross-section at Pin22 of the storage device

Fig. 9.5 SEM morphology of the solder joints in the storage device of SiP

particles are also sharp and could potentially impact the chip passivation layer and metallization under external mechanical stress. Figure 9.6c shows the SEM inspection results of the filling material particles in devices produced by Company C. The filling material particles are sharp, with diameters greater than 50 μm. Such sharp quartz sand particles can easily damage the passivation layer and metallized wiring of the integrated circuit. In contrast, the filling material particles in Company A's

(a) Particles of filling material from Company A

(b) Particles of filling material from Company B

(c) Particles of filling material from Company C

Fig. 9.6 Particle analysis of packaging filling materials in the sample

products are more reliable and advantageous compared to those of Companies B and C. Integrated circuits from Companies B and C have a potential risk of chip damage due to the larger filling material particles.

9.2.2 Structural Analysis

In the 1990s, the international aerospace community proposed a new method for evaluating the reliability of integrated circuits used in aerospace. This method is called structural analysis. Structural analysis mainly assesses the design, structure, and process rationality of integrated circuit products. By conducting in-depth structural analysis of the packaging, we can identify potential defects in the packaging structure that may cause failure or reliability problems. It can also determine whether the manufacturing process affects the reliability of the packaging. Therefore, structural analysis is a key method to ensure packaging quality and reliability.

Currently, there are no standardized guidelines for structural analysis, either domestically or internationally. Structural analysis for integrated circuit packaging can refer to GJB 1420B-2011 and GJB 2440A-2006. However, these standards do not address the impact of packaging processes, such as chip bonding, wire bonding, and sealing, on reliability. This makes them incomplete for structural analysis. At present, structural analysis mainly relies on personalized methods and experience. By referring to the failure analysis of similar packaging structures, common failure issues are included in the analysis plan. Test methods are then developed, test items are selected, and the packaging materials, structure, and processes are analyzed.

The general method for structural analysis is shown in Fig. 9.7. It mainly includes the following steps.

(1) Determine the packaging type

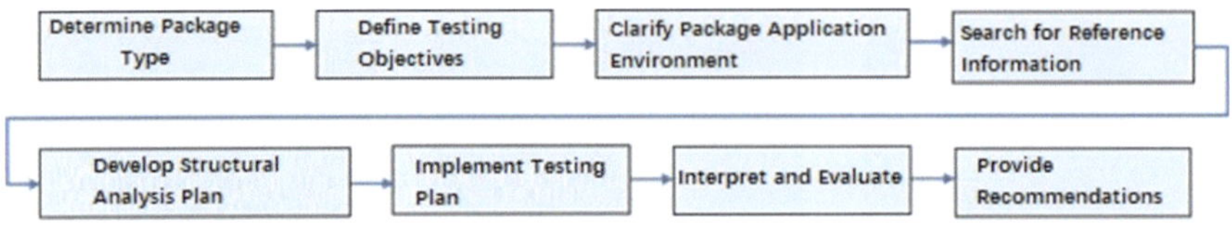

Fig. 9.7 General method for structural analysis

Conduct an initial analysis of the packaging structure type (e.g., plastic packaging, hermetic packaging, wire bonding, flip-chip soldering, stacking). For each packaging structure, determine its basic design and structural form.

(2) Determine the test objective

The test objective sets the direction and focus of the structural analysis. Clearly defining the test objective enables a more targeted structural analysis plan.

(3) Clarify the packaging application environment

The packaging's application environment greatly impacts its reliability. By understanding the application environment, potential issues can be considered. This allows the test plan to be designed accordingly. For instance, packaging used in marine environments should focus on corrosion resistance.

(4) Search for reference information

Look for structural analysis data from similar packaging types. Understand their structural features, which can serve as references. Summarize the issues encountered during destructive physical analysis and usage of similar packaging. Focus on analyzing relevant issues for the packaging structure under review.

(5) Develop a structural analysis plan

When creating the structural analysis plan, consider packaging type, test objectives, application environment, and reference information. The test items should cover the materials, structure, and process elements of the packaging.

Test items generally include non-destructive tests such as X-ray inspection, external visual inspection, label strength testing, Scanning Acoustic Microscope testing, SEM inspection, energy spectrum analysis, hermeticity testing, and internal visual inspection. Destructive tests include bond strength testing, chip shear force testing, and Focused Ion Beam (FIB) analysis.

(6) Implement the test plan

Follow the test plan's procedures and requirements. If the test plan does not achieve its objective during the testing process, modify the test items and order based on analysis.

(7) Judgment and evaluation

After completing the tests and inspections, evaluate the results. Structural analysis evaluation includes the following:

① Evaluate the rationality of the structural design.
② Check the process design and quality.
③ Analyze potential failure risks.
④ Identify prohibited or restricted processes and materials.

The conclusion of the structural analysis should provide an evaluation of the packaging's design, reliability, and failure risks. For structures with failure risks, provide specific analysis recommendations. For the overall packaging structure, conclusions such as usable, restricted, or prohibited should be given.

(8) Provide suggestions

After completing the structural analysis, provide targeted suggestions to both the development and usage parties based on the reliability issues and failure information obtained. For the identified issues, suggest specific improvements in design, structure, and processes to the development party. Offer advice to the user on the application environment, risk warnings, and mitigation plans. For structures with potential failure risks, propose follow-up testing and validation procedures.

Here is an example of how a structural analysis plan for integrated circuit packaging is formulated.

At the beginning of the structural analysis, the plan is mainly based on the test items. If the sequence of test items is not properly arranged, it can significantly affect the comprehensiveness and accuracy of the structural analysis. To avoid interference between different test items, the hierarchical nature of the packaging structure should be considered when developing the test plan. The analysis should proceed from external packaging to internal packaging, and finally, chip assembly.

For example, in the case of a radio frequency (RF) front-end module packaging, the structural analysis should address three areas: external packaging, internal packaging, and chip assembly. The test sequence should follow a logical progression: external packaging analysis, then internal packaging analysis, and finally chip assembly analysis. The RF front-end module packaging structural analysis content and key methods are shown in Fig. 9.8.

Additionally, the packaging structure can be broken down further into its smallest units. Typically, industry-recognized structures are used as the smallest units.

The structural unit decomposition diagram of a ceramic QFN package is shown in Fig. 9.9. Based on the decomposition diagram, and considering the unique characteristics and manufacturing processes of each unit, such as packaging materials, interconnection processes, and internal structures, appropriate test items are selected for the structural analysis of each unit. This ultimately leads to a comprehensive evaluation conclusion of the ceramic QFN package's structure.

9.2.3 Counterfeit and Refurbishment Analysis

1. Key Points for Inspecting Counterfeit, Refurbished, and Mixed-Batch Integrated Circuits

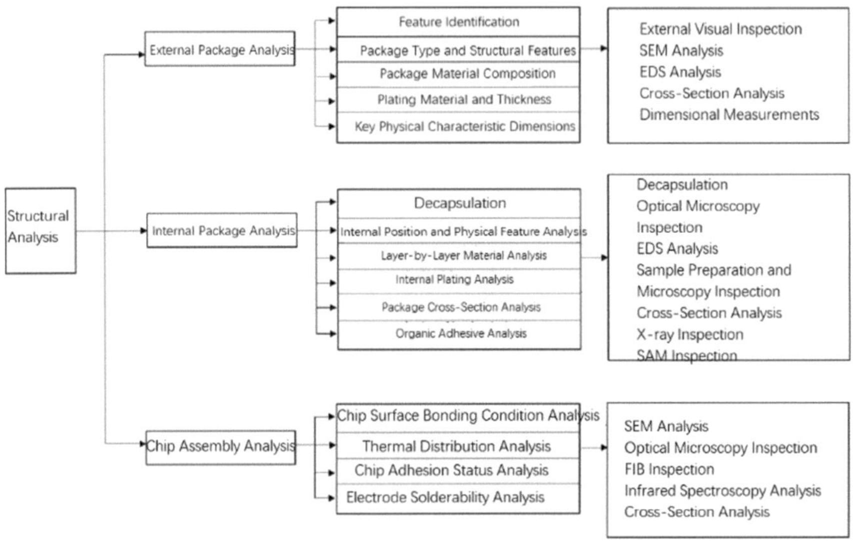

Fig. 9.8 Structural analysis content and main methods for RF front-end module packaging

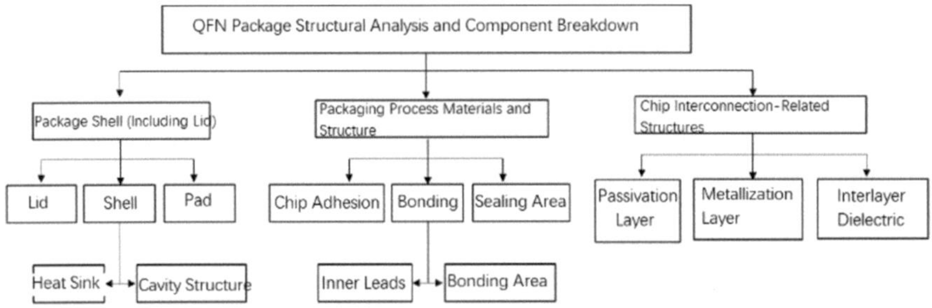

Fig. 9.9 Structural unit decomposition diagram of a ceramic QFN package

Counterfeit, refurbished, and mixed-batch integrated circuits pose significant potential risks to high-reliability applications. They also greatly hinder the development of domestic, self-developed new integrated circuits. Thus, both technical and management measures must be enforced. Refurbished integrated circuits usually have the following characteristics:

① Refurbished products have been on the market for a long time and are produced in large quantities

Refurbished integrated circuits often come from discarded electronic products. These circuits are typically sold after being on the market for a certain period. To ensure better sales, refurbished integrated circuits are often general-purpose, widely used, and easy to sell. Examples include memory chips, single-chip power supplies, communication interfaces, general-purpose analog circuits, and general logic arrays.

② Refurbished products differ from conventional integrated circuit processes

The standard process for plastic-sealed integrated circuits is: injection molding → labeling → lead plating → tie-bar cutting → testing → warehousing. However, during refurbishment, paint is applied, and tie-bar areas are re-plated, which differs from conventional processes.

③ Refurbishment affects the reliability of integrated circuits

The grinding process during refurbishment places pressure on the internal chip, which damages the chip's surface. This damage is particularly noticeable in thin-packaged integrated circuits, where surface grinding is obvious.

To strengthen the inspection of counterfeit, refurbished, and mixed-batch integrated circuits, the following key points are provided based on engineering experience:

(1) External Visual Inspection

Considering the characteristics of refurbished integrated circuits, the nature of the refurbishment, and the differences between conventional processes, it is not difficult to spot refurbished products. Recently introduced integrated circuits are usually not refurbished. Specialized integrated circuits are typically not refurbished. However, the following points should be noted: Integrated circuits with painted surfaces and labels on the painted area; Integrated circuits that have been ground, painted, and labeled on the painted surface; Integrated circuit leads with tie-bar areas coated with solder (this may also occur due to long storage, where soldering is done to improve solderability; it should be judged with other information).

The key points for external visual inspection are as follows:

- Check if the chip markings have been ground or if the original laser markings remain.
- Examine whether the chip edges are sharp and if the boundary lines are parallel.
- Use chemical paint removal to check for grinding. After paint removal, observe for quartz sand grinding.
- Identify the differences between the painted surface and other areas.
- Check the depth of marking holes to determine if they have been ground.
- Distinguish between paint applied after molding and the natural flow pattern of spray-painted surfaces.
- Check if the lead cross-sections have solder residue.
- Examine whether the natural fracture areas of the plastic are painted.
- Look for cracks, stains, or other defects on the package.

A typical appearance of a refurbished integrated circuit is shown in Fig. 9.10.

(2) X-ray Inspection

The key points for X-ray inspection are as follows:

- Check if the lead frame shape is the same.

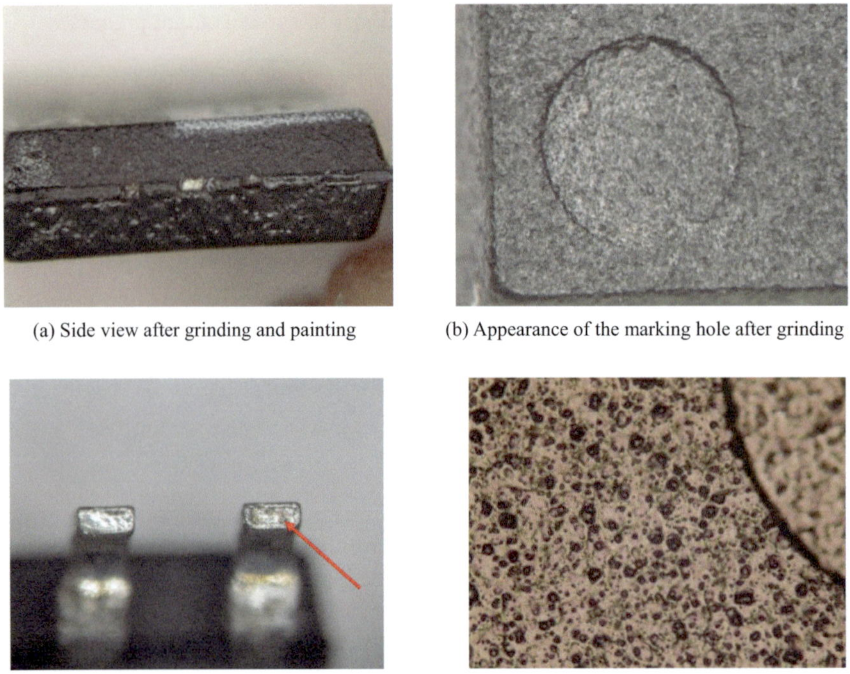

(a) Side view after grinding and painting

(b) Appearance of the marking hole after grinding

(c) Lead cross-section with solder residue

(d) Surface of plastic-sealed package after grinding and painting

Fig. 9.10 Typical appearance of refurbished integrated circuits

- Examine if the lead layout matches. Look for bond offset, bridging, or uneven soldering.
- Check if the chip size is consistent.

An X-ray image of a counterfeit integrated circuit is shown in Fig. 9.11, where the leads are disconnected.

(3) C-SAM Inspection

Key inspection points include:

- Checking for delamination at the chip/molding compound, lead frame/molding compound, and chip/substrate interfaces.
- Detecting cracks in the encapsulation material.
- Identifying fractures in the chip.

Figure 9.12 shows typical C-SAM defects in counterfeit integrated circuits.

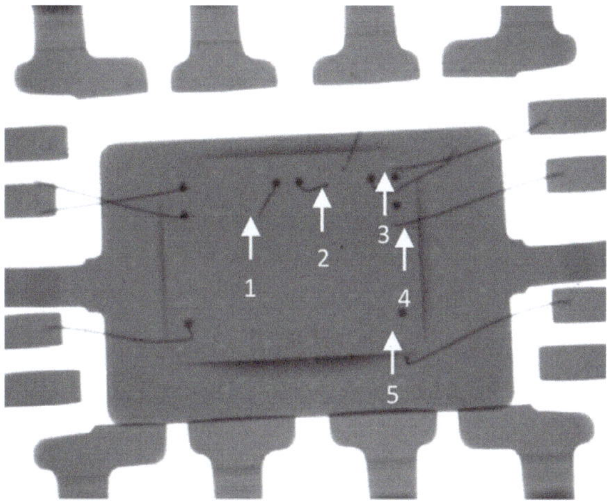

Fig. 9.11 X-ray image of a counterfeit integrated circuit

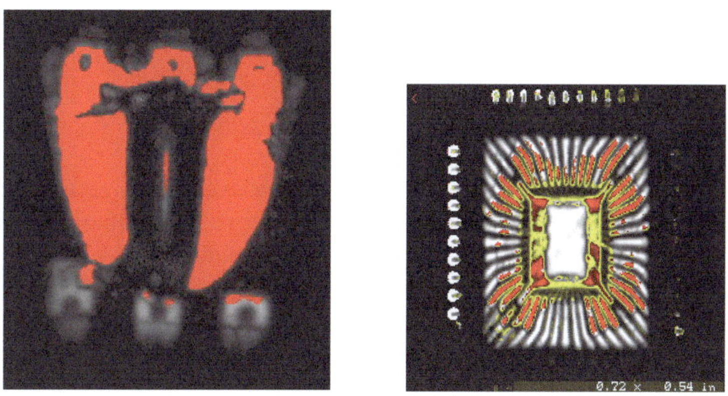

(a) Cracks in the molding compound

(b) Delamination between the lead frame and the molding compound

Fig. 9.12 Typical C-SAM defects in counterfeit integrated circuits

(4) Internal Visual Inspection

Key inspection points include:

- Verifying if the chip layout matches the package. It should be identical to genuine chips from authorized sources. The version must also be consistent.
- Checking for corrosion, overcurrent damage, electrostatic discharge marks, or overvoltage breakdown on the chip pads and metal traces.
- Inspecting chip bonding for damage, wire detachment, or other defects.
- Examining the passivation layer for cracks, punctures, or other irregularities.

2. Reliability Issues of Counterfeit, Refurbished, and Mixed-Batch Products

China's electronics industry has a weak foundation. Some high-end components, such as CPUs, high-capacity memory, DSPs, and power RF devices, rely on imports. However, procurement channels are difficult to control. Many chip distributors, driven by high profits, sell substandard products. They pass off refurbished or even counterfeit chips as new ones. In recent years, the global economy has undergone major adjustments. Many semiconductor manufacturers have cut production. This has led to chip shortages. As the economy recovers, demand for industrial-grade microcircuits has surged. The supply shortage has worsened the spread of counterfeit, refurbished, and mixed-batch products. Even the U.S. military has raised concerns about counterfeit chips appearing in military equipment.

Refurbished products have been used before, but their usage history is unclear. Their lifespan is significantly reduced. They may also have aging and corrosion issues. For example, destructive physical analysis revealed that a supposedly unused chip had been exposed to high current (see Fig. 9.13). This suggests it was a salvaged component. The refurbishment process itself can cause problems. Chips may suffer electrostatic stress damage. Their solderability may also be affected. These defects create serious risks for final products. Counterfeit chips make circuit design and debugging difficult. They delay the design process and introduce unpredictable hazards. Mixed-batch products pose another risk. They may contain low-quality chips or have inconsistent performance. This makes reliability and maintenance more difficult.

Counterfeit, refurbished, and mixed-batch chips are serious quality concerns. They must be identified and screened before installation. Destructive physical analysis and specialized counterfeit detection methods can help eliminate them.

The following case studies discuss the risks and hazards of these chips in electronic equipment.

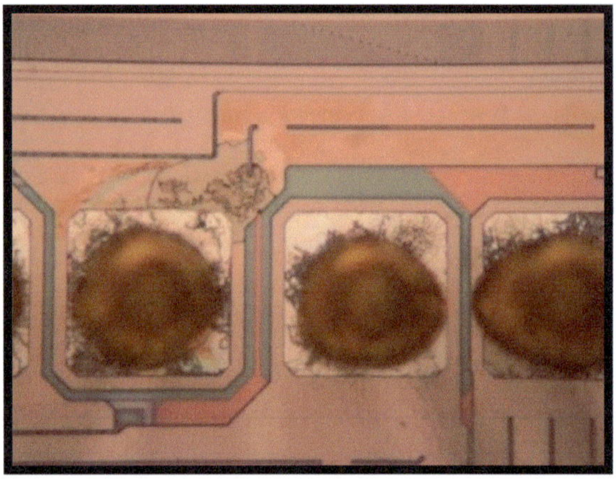

Fig. 9.13 A supposedly unused integrated circuit that had experienced high current exposure

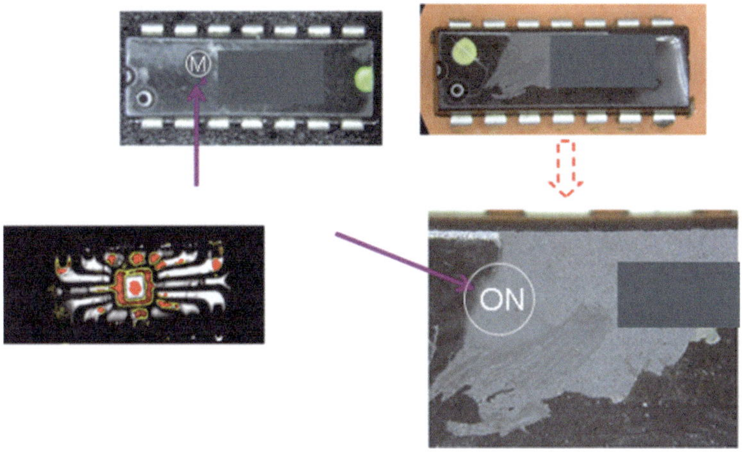

Fig. 9.14 Two different logos found on a failed integrated circuit after installation

Case 1: **A Failed Plastic-Packaged Integrated Circuit Identified as a Refurbished Product with Two Different Logos**

A failed integrated circuit was found with severe delamination in its lead frame. The surface showed the "M" logo of MOTOROLA. After scraping off the top coating, another logo, "ON" (ON Semiconductor), appeared (see Fig. 9.14). This confirmed that the sample was a refurbished product.

Case 2: **A Power Control Chip Identified as Refurbished Before Installation, Showing Passivation Layer Damage and Metal Corrosion**

A power control chip was used in a DC-DC power module. During production testing and customer use, failures occurred. The defective samples could not output PWM signals properly. The top and bottom edges were not parallel to the plastic packaging, as shown in Fig. 9.15.

An external visual inspection was conducted. The surface markings on all samples were identical. However, some samples had uneven surfaces. The edges of the same sample showed shape variations. Some samples had visible grinding marks, as shown in Fig. 9.16. Based on these findings, refurbishment was suspected.

The *I-V* characteristics of both functional and failed samples were tested. Compared to the functional samples, the failed samples showed abnormal port characteristics. The comparison results are shown in Fig. 9.17.

A C-SAM inspection was conducted on all samples. Delamination was found at the interface between the molded plastic and the lead frame. Some samples also had delamination at the molded plastic and chip interface. The typical C-SAM images of the failed samples are shown in Fig. 9.18. These images reveal delamination at the molded plastic/chip interface, the molded plastic/substrate edge, and the molded plastic/lead frame interface. The typical C-SAM images of the functional samples are

Fig. 9.15 Misalignment between the top and bottom edges of the failed sample and the plastic packaging

shown in Fig. 9.19. These samples also showed delamination at the molded plastic/substrate edge and the molded plastic/lead frame interface.

Both failed and functional samples were decapsulated under the same conditions. Two failed samples had different chip surface colors. Three failed samples and the functional samples showed varying degrees of metal corrosion. The corrosion of the aluminum metallization layer at the edge of the failed samples is shown in Fig. 9.20. The unused samples also had aluminum metallization layer corrosion and passivation layer fractures, as shown in Fig. 9.21.

Based on external visual inspection, C-SAM results, and internal examination, the failed and unused samples were analyzed. Their appearance, chip markings, corrosion, and passivation layer fractures confirm that these samples are refurbished products.

Case 3: **A Power Control Chip Identified as Refurbished/Counterfeit Before Installation, Showing Passivation Layer Damage and Metallization Layer Corrosion, with Reversed Enable Level**

During installation and debugging, a unit found that the integrated circuit had no functional output. A failure analysis was conducted. First, a visual inspection was performed. All samples had a coating on the surface. The coating covered the identification hole, making the country of origin marking "MALAY" appear blurred (see Fig. 9.22). When the coating was lightly scraped with a blade, it was removed, exposing the molded plastic underneath. The exposed plastic was darker than the coated surface (see Fig. 9.23). Under a high-magnification microscope, the coating did not fully cover the sample's side. A clear boundary was visible.

A C-SAM inspection was performed on the failed samples. Delamination was observed in the substrates and chips. The C-SAM images of the failed sample substrates are shown in Fig. 9.24. Delamination was present at the chip/molded plastic interface and the chip/substrate interface.

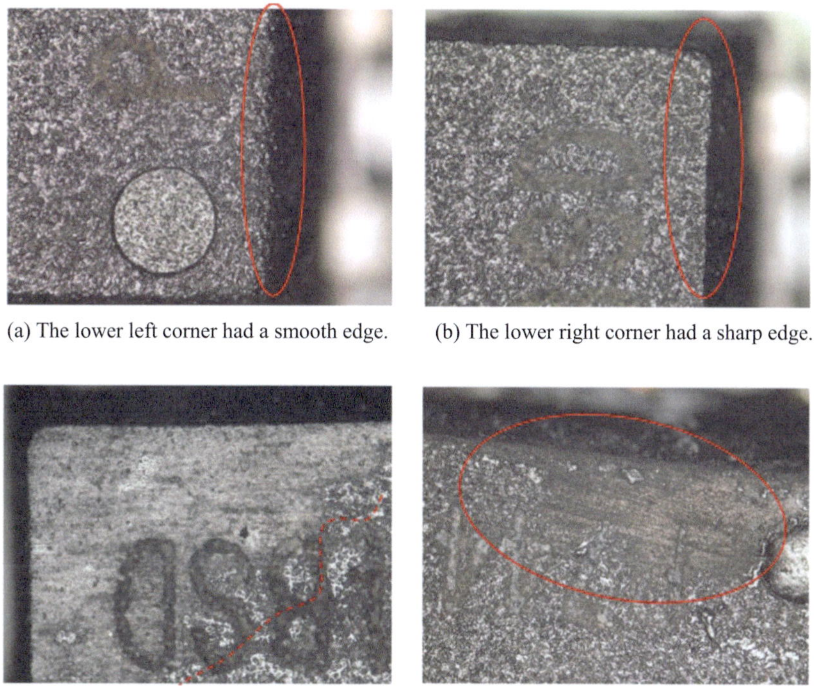

(a) The lower left corner had a smooth edge.

(b) The lower right corner had a sharp edge.

(c) After cleaning with alcohol, the molded plastic surface edge and central area looked different.

(d) After cleaning with alcohol, grinding marks became visible.

Fig. 9.16 Grinding marks observed on the surface of the failed sample during external visual inspection

It was found during opening that the chip marking on the sample did not match the model on the packaging. A review of the datasheets for two power chip models from the manufacturer revealed different enable level requirements. One model required a low-level enable, while the other needed a high-level enable. The enable port was tested in the circuit. It was found that both failed samples had a high-level enable, with an output voltage of around 1.13 V. When the enable port was grounded, there was no output. Chemical opening showed that the internal chip did not match the packaging marking.

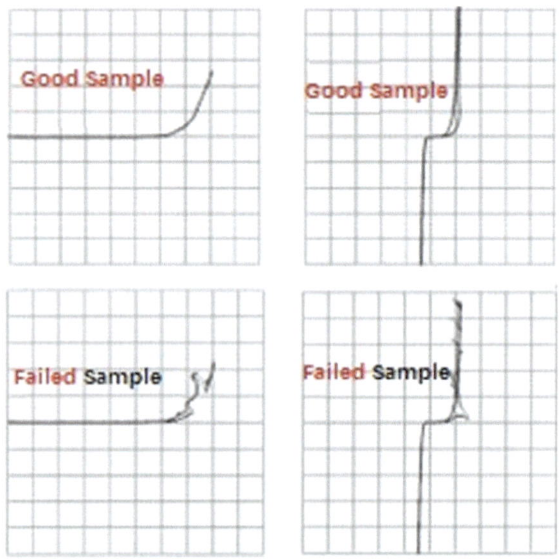

Fig. 9.17 Comparison of port characteristics between failed and good samples

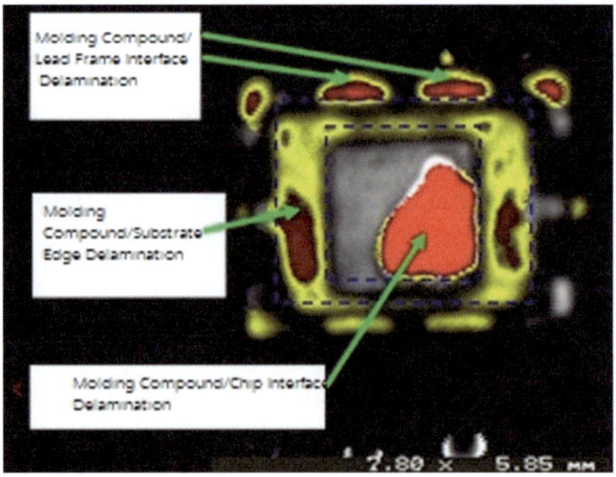

Fig. 9.18 Typical C-SAM images of failed samples

Therefore, the sample exhibited signs of being refurbished and counterfeit. The surface of the sample had a coating, and most of the failed samples had a chip model different from the surface marking. Due to the different pin port enable levels, the designed circuit could not function correctly.

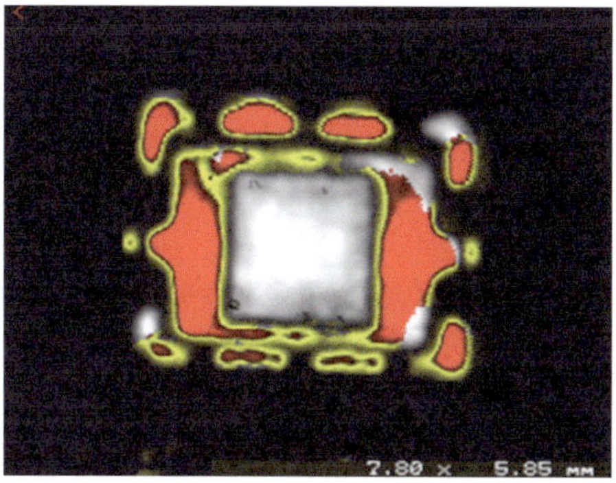

Fig. 9.19 Typical C-SAM images of functional samples

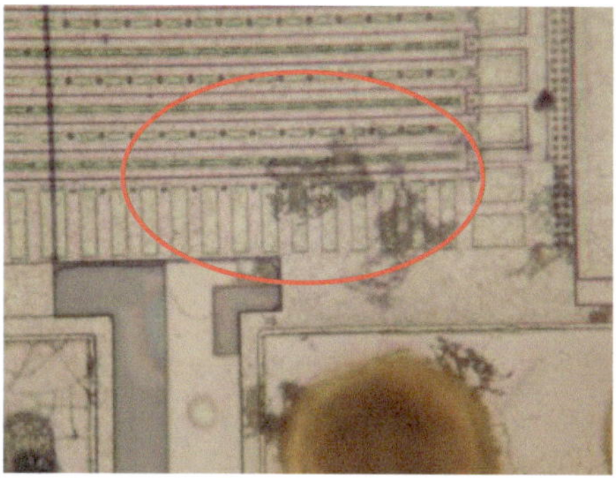

Fig. 9.20 Corrosion of the aluminum metallization layer in failed samples

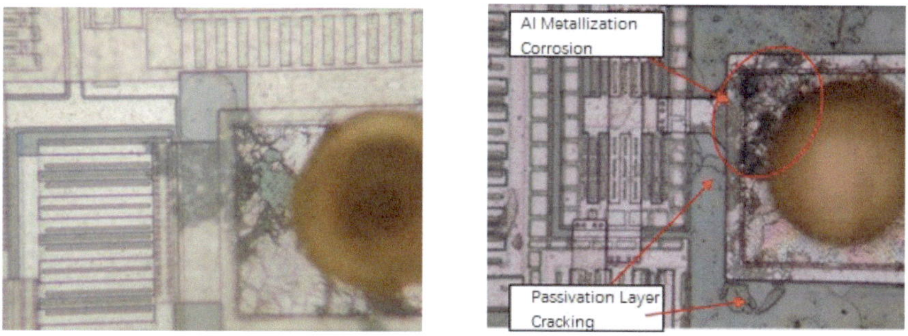

Fig. 9.21 Corrosion of the aluminum metallization layer and passivation layer fractures in unused samples

9.3 Accelerated Life Testing Evaluation

9.3.1 Accelerated Life Testing Method for Integrated Circuit Products

With the advancement of microelectronics, high-reliability integrated circuits are increasingly common. Under normal operating temperature conditions, the product's lifespan can reach millions of hours. This leads to the problem that, after long-term testing, no failures are observed, making it difficult to assess the reliability of integrated circuits effectively. In such cases, increasing the test temperature, while keeping failure mechanisms unchanged, can induce more failures. This is

Fig. 9.22 Blurred country of origin marking "MALAY" on the sample

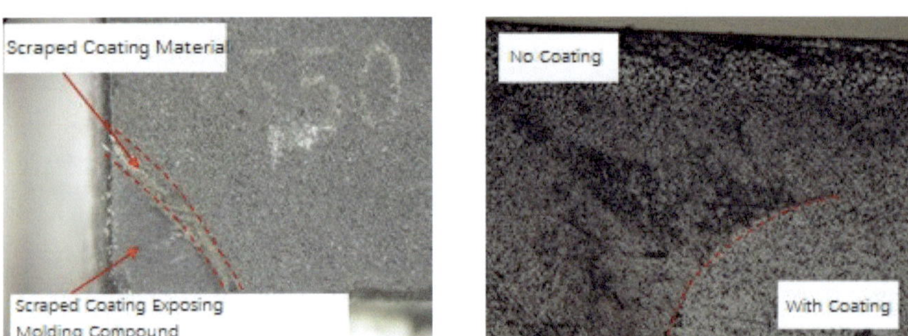

(a) Coating Removed

(b) Clear Boundary Between Coated and Uncoated Areas

Fig. 9.23 Surface morphology of the failed sample

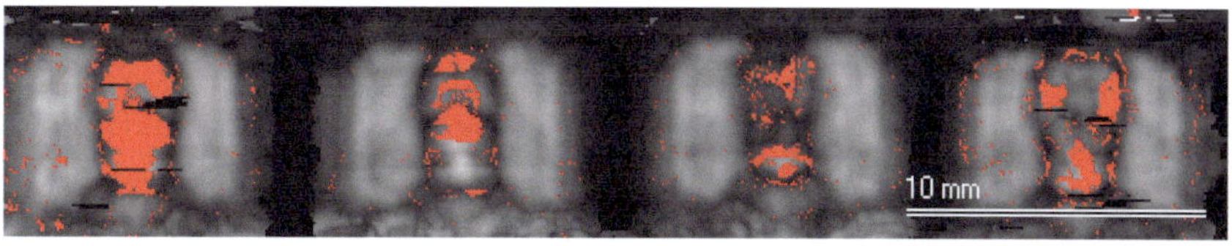

(a) Delamination at the Chip/Molded Plastic Interface

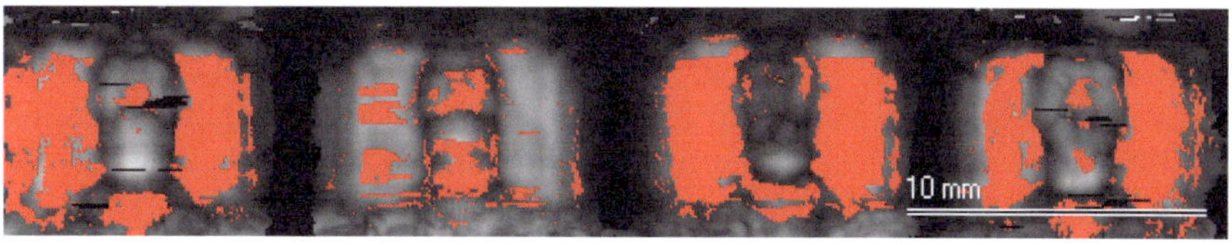

(b) Delamination at the Chip/Substrate Interface

Fig. 9.24 C-SAM images of the failed sample substrate

very helpful for estimating the high-temperature reliability of the samples. This type of test is known as accelerated life testing. Accelerated life testing for integrated circuit packaging can refer to the methods for testing integrated circuit products.

The goals of accelerated life testing for integrated circuits are as follows:

(1) Reducing testing costs and time: Accelerated life testing speeds up fault exposure by increasing operational or environmental stress, avoiding the high costs of large sample sizes and long testing periods.
(2) Obtaining reliable data: By using statistical methods and extrapolation, useful reliability data can be acquired, such as failure distribution, mean life, product characteristics over time, and reliability indicators under normal stress levels.
(3) Failure identification: Accelerated life testing can reveal failure information such as failure modes, failure models, and failure mechanisms.
(4) Evaluating structure, materials, and processes: It helps assess and improve the quality of the integrated circuit product.
(5) Using harsher testing conditions: This helps detect potential defects in products.
(6) Determining the maximum stress limits: Accelerated life testing helps determine the maximum stress levels that integrated circuit products can safely withstand.

1. Accelerated Life Testing Design

Accelerated life testing includes constant stress accelerated life testing, step stress accelerated life testing, and progressive stress accelerated life testing. Among these, constant stress accelerated life testing is the most developed. The error in the test results is small, and it is widely applied. This section will explain the design method for constant stress accelerated life testing.

First, a set of accelerated stress levels is chosen, such as S_1, S_2, …, S_x. All these levels must be higher than the normal stress level S_0. Then, the samples are divided into K groups. Each group undergoes accelerated life testing at a different stress level.

The test continues until a certain number of samples in each group fail. The value of K determines the precision and reliability of the acceleration factor extraction. A larger K value means more stress levels and more precise extraction. However, with more stress levels, the sample size and cost also increase. There is a trade-off between precision and cost.

(1) Selection of Stress Levels

The chosen stress levels must ensure that the failure modes and mechanisms at different stress levels are the same as those at normal stress levels. This allows the use of accelerated equations to extrapolate the product's reliability at normal operating conditions. If the failure mechanisms differ significantly, extrapolation will be difficult, and accelerated life testing cannot be conducted. If there is only one stress level in the accelerated life testing, there should be at least 4 stress levels. As more types of stress are added, the number of stress levels must also increase to get an accurate acceleration factor.

(2) Sample Size Selection

Each stress level must have at least 5 samples. Products are less likely to fail under low stress levels, so more samples should be used. At high stress levels, products fail more easily, so fewer samples are needed.

(3) Failure Criteria Determination

The failure of samples should be determined according to the product specifications. The failure criteria must be clear, and the exact failure time of each sample must be recorded.

(4) Test Termination

It is best to stop the test when all the samples have failed. This improves the statistical accuracy. However, for many products, achieving full failure would lead to long test times and higher costs. Therefore, it is recommended that more than half of the samples fail at each stress level. If this is not possible, at least 30% of the samples should fail. If there are only 5 samples at a given stress level, at least 3 should fail. Otherwise, the statistical error will be large, and reliable test data cannot be obtained.

2. Accelerated Life Testing Analysis

(1) Lifetime Distribution

After completing the accelerated life testing, statistical analysis is performed on the sample quantity and failure time data. The distribution of sample lifetimes is then used to extract the characteristic lifetime under different accelerated stress levels. In accelerated life tests of integrated circuits, the common lifetime distributions are exponential distribution, Weibull distribution, and log-normal distribution.

① Exponential Distribution

Exponential distribution is the most common lifetime distribution. The distribution and density functions are as follows:

$$F(t) = 1 - \exp\left(-\frac{t}{\theta}\right), t \geq 0 \tag{9.1}$$

$$f(t) = \frac{1}{\theta}\exp\left(-\frac{t}{\theta}\right), t \geq 0 \tag{9.2}$$

In this equation, θ represents both the average lifetime and the characteristic lifetime. When the product's lifetime follows an exponential distribution, the reciprocal of the average lifetime, $1/\theta$, equals its failure rate λ. According to the product's failure pattern, after removing infant mortality failures, the remaining products' lifetimes can be considered to follow an exponential distribution before entering the wear-out failure phase.

② Weibull Distribution

Weibull distribution is commonly used in reliability analysis. Many electronic components and devices have lifetimes that follow a Weibull distribution. By adjusting the parameters of the Weibull distribution, the entire product lifecycle can be modeled. Studies show that components, devices, and equipment that experience a local failure leading to a global failure can be approximated by a Weibull distribution.

The distribution and density functions for Weibull distribution are as follows:

$$F(t) = 1 - \exp\left[-\left(\frac{t}{\alpha}\right)^{\beta}\right], t \geq 0 \tag{9.3}$$

$$f(t) = \frac{\beta}{\alpha}\left(\frac{t}{\alpha}\right)^{\beta-1}\exp\left[-\left(\frac{t}{\alpha}\right)^{\beta}\right], t \geq 0 \tag{9.4}$$

The Weibull distribution has two parameters: α and β. The characteristic lifetime corresponds to the failure time at 63.212% of the failure probability. It can be denoted simply as $\alpha = t_{63}$.

③ Log-Normal Distribution

Log-normal distribution is also a common lifetime distribution in reliability analysis. The lifetime of insulators, diodes, and other components often follows a log-normal distribution. The density and distribution functions for log-normal distribution are as follows:

$$f(t) = \frac{1}{\sigma t\sqrt{2\pi}}\exp\left[-\frac{[\ln(t) - \mu]^2}{2\sigma^2}\right], t > 0 \tag{9.5}$$

$$F(t) = \Phi\left(\frac{\ln(t) - \mu}{\sigma}\right), t > 0 \tag{9.6}$$

In the equation, $\Phi(\cdot)$ represents the standard normal distribution function. The log-normal distribution has two parameters: μ (the logarithmic mean) and σ (the logarithmic standard deviation).

(2) Acceleration Model

The method used to evaluate product lifetime in accelerated life tests involves obtaining the product's lifetime under high-stress conditions. A relationship between stress levels and product lifetime is then established to predict the lifetime of the product under normal working conditions. For most products, the lifetime is shorter under high-stress levels and longer under low-stress levels. However, there are exceptions for some products. Focusing on individual product lifetimes makes it difficult to establish a clear relationship between lifetime and stress levels. Therefore, we typically analyze the overall data from multiple products. By using parameters such as characteristic lifetime, mean lifetime, and median lifetime, we establish the relationship between lifetime and stress level. This relationship is called the acceleration model.

The relationship between lifetime and stress level is usually nonlinear. To create a linear relationship, transformations such as logarithmic or reciprocal transformations are often applied. This allows for linear fitting.

(3) Acceleration Factor

The acceleration factor is a key parameter in accelerated life testing. It is the ratio of a specific lifetime characteristic at normal stress to the corresponding lifetime characteristic at the accelerated stress level. The acceleration factor is a dimensionless number. It reflects the accelerating effect of a given stress level and depends on the stress level. When the acceleration factor approaches 1, it indicates that the accelerated stress has no effect. The method to calculate the acceleration factor varies based on the lifetime distribution. For commonly used acceleration models, there are specific methods to calculate the acceleration factor. The acceleration factor can be used in product reliability assessment, quality identification, and reliability acceptance.

9.3.2 Commonly Used Acceleration and Failure Models in Packaging

1. Arrhenius Model

The Arrhenius model is the most widely used high-temperature durability acceleration model [9]. Many packaging structures, under high-temperature stress, can be modeled using the Arrhenius model. For example, this model is used for the degradation of solder balls in Ball Grid Array (BGA) packaging under high-temperature storage, as well as the high-temperature degradation of gold-tin bonding [10, 11]. As the temperature increases, the energy of atoms and electrons in the material increases. This accelerates chemical reactions inside the packaging structure and causes premature failure. The expression for the Arrhenius model is as follows:

$$\xi = A\exp\left(-\frac{E_a}{K_B T}\right) \tag{9.7}$$

In the equation, ξ represents a certain lifetime characteristic. K_B is the Boltzmann constant. A is a material-related coefficient. E_a is the activation energy. T is the absolute temperature. The Arrhenius model shows that as the temperature increases, the product's lifetime shortens exponentially. Taking the logarithm of both sides of the equation gives:

$$\ln\xi = a - b/T \tag{9.8}$$

In this equation, $a = \ln A$, and $b = E_a/K_B$. Therefore, the logarithm of the lifetime has a linear relationship with the reciprocal of the temperature. By fitting the data, the parameters a and b of the acceleration model can be obtained. This allows for the prediction of the product's lifetime under different operating stresses.

2. Coffin-Manson Model

Thermal fatigue failure caused by thermal cycling in packaging is a common failure mode. It often leads to cracks, delamination, and even open circuits. The Coffin-Manson model is used to describe failure caused by thermal fatigue under thermal cycling stress [12, 13]. The low-cycle fatigue life (N_f) of the material is related to the plastic strain range. This relationship is expressed by the Coffin-Manson equation.

$$N_f = \frac{1}{2}\left(\frac{\Delta\gamma_\tau}{2\varepsilon_f}\right)^{\frac{1}{c}} \tag{9.9}$$

In this equation, N_f represents the fatigue life; $\Delta\gamma_\tau$ is the equivalent shear strain amplitude; ε_f is the fatigue ductility coefficient; c is a parameter related to the thermal cycle profile, which can be determined using Eq. (9.10).

$$c = -0.442 - 0.0006T_{sj} + 0.0174\ln(1 + f) \tag{9.10}$$

In this equation, T_{sj} is the average cycle temperature (in °C); f is the cycle frequency (in cycles/day).

3. Black Model

The Black model is a classic model for evaluating electromigration. Electromigration is a phenomenon that occurs in metal interconnects under the influence of current and temperature stress. Electrons exchange momentum with the metal lattice. As a result, metal atoms migrate in the direction of electron flow, creating voids in their original positions. Simultaneously, metal atoms accumulate, forming hill-like protrusions [14]. Voids can cause open or broken leads, while the accumulation of metal atoms can lead to short circuits between multilayer wiring. This affects the chip's normal operation. Electromigration failure typically occurs slowly and may take months or even years to manifest. Therefore, accelerated stress tests using high temperatures and high currents are applied to assess electromigration life. Electromigration life is related to temperature and current density and follows the Black equation, expressed as:

$$\mathrm{MTTF} = A\frac{1}{J^n}\exp\left(\frac{E_a}{kT}\right)$$

In this equation, MTTF is the mean time to failure; A is a constant; J is the current density; n is the current density exponent; E_a is the activation energy; T is the temperature; k is the Boltzmann constant.

4. Lawson Model and Peck Model

The Lawson and Peck models are used to evaluate the accelerated life of electronic components under damp heat conditions. Environmental humidity often causes corrosion, coating peeling, and other issues in electronic components. Humidity can also penetrate the packaging to the internal chip, leading to chip corrosion and failure. This shortens the component's lifespan. Typically, humidity and temperature are combined as accelerated variables in life tests. The Lawson model is a common method to assess corrosion of active components caused by humidity. It is a comprehensive model that combines both temperature and humidity. The equation is:

$$K = A\exp\left(-\frac{E_a}{kT}\right)\exp\left(bH^2\right)$$

In this equation, K is the corrosion rate, k is the Boltzmann constant, A and b are constants, Ea is the activation energy, and H is the relative humidity.

The Peck model is another common damp heat acceleration model. It has better adaptability than the Lawson model. The Peck model accounts for both temperature and humidity. It is an extension of the Arrhenius model. This empirical model is based on many different life studies. The equation is:

$$t = A(\mathrm{RH})^{-n}\exp\left(\frac{E_a}{kT}\right)$$

In this equation, t is the failure time, A is a constant depending on material, process, and conditions, RH is the relative humidity, n is a constant, Ea is the activation energy, k is the Boltzmann constant, and T is the temperature.

In accelerated failure physical models for integrated circuit packaging, aside from the five models mentioned above, there are also inverse power law models for voltage, current, and power as acceleration stresses. The Eyring model is used for voltage and temperature acceleration. These models are not discussed here.

9.3.3 Accelerated Life Test Case for Integrated Circuit Packaging

In recent years, with the rapid development and wide application of advanced packaging technologies such as fan-out and 3D packaging, copper pillar bumps in high-density interconnections have been widely adopted. They are favored for their fine pitch, high density, short interconnection lines, and low signal transmission loss [15–17]. As the feature size and pitch of copper pillar bumps decrease, the current density and temperature in the bumps increase significantly under the same transmission current. This leads to severe electromigration reliability issues [18, 19].

The method for conducting accelerated life tests for copper pillar bumps due to electromigration is as follows. At least three sets of current stresses and three sets of temperature stresses are selected. These stresses are used to perform accelerated life tests. The chosen stress levels must ensure that the failure modes and mechanisms of the copper pillar bumps under various stress levels match those under normal stress levels. The resistance of the copper pillar bump interconnections is measured over time for each stress condition. The main focus is on observing the formation of cracks, hillocks, voids, and IMC in the copper pillar bumps [20].

A schematic diagram of the copper pillar bump structure is shown in Fig. 9.25. The copper pillar bump has a diameter of about 50 μm, a height of 55 μm, and a pitch of 80 μm between the bumps. The top of the copper pillar bump is a Sn1.8Ag solder ball, with a height of 23 μm. A 2 μm nickel layer is plated between the copper pillar bump and the solder. The substrate has a diameter of 66 μm, and the opening diameter of the polyimide (PI) on the substrate is 36 μm with a thickness of 5 μm. The copper interconnection between the copper pillar bumps is 136 μm long, 36 μm wide, and about 4 μm thick.

Figure 9.26 shows the sample and structural diagram of high-density copper pillar bump interconnections. A daisy-chain structure is designed. Nine sets of thermo-electric coupled accelerated stresses are applied to the product. These include three current density stress levels: 2×10^4 A/cm^2, 2.5×10^4 A/cm^2, and 3×10^4 A/cm^2. Additionally, three high-temperature stress levels: 100, 125, and 150 °C, are used for accelerated life testing. The resistance of the copper pillar bump interconnections under each stress level is measured over time. Figure 9.27 shows the curves of copper pillar bump interconnection resistance as a function of time under different temperatures and current densities.

The reliability model for copper pillar bump interconnections under thermoelectric combined stresses is constructed using the Black equation, which describes

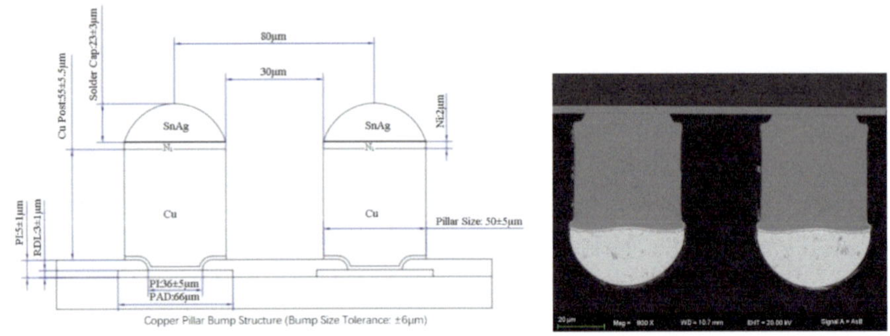

Fig. 9.25 Schematic diagram of copper pillar bump structure

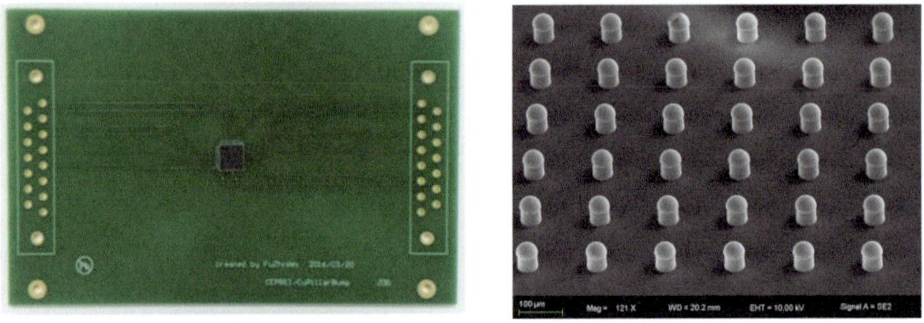

(a) Sample Diagram (b) Structural Diagram

Fig. 9.26 Sample and structural diagram of high-density copper pillar bump interconnections

atomic mass transport. The Black equation is:

$$\text{MTTF} = A\frac{1}{(j)^n}\exp\frac{E_\text{a}}{kT} \tag{9.11}$$

In this equation, MTTF is the mean time to failure, j is the current density, $k = 8.616 \times 10^{-5}$ is the Boltzmann constant, T is the interconnection wire temperature, E_a is the activation energy, n is the current density exponent, and A is the line width constant.

First, the MTTF data from the thermoelectric reliability tests are analyzed. The current density exponent and activation energy of the copper pillar bump interconnections are determined. Then, a copper pillar bump interconnection electromigration reliability model is established. Figures 9.28 and 9.29 show the relationships between MTTF of copper pillar bump interconnections due to electromigration, current density, and temperature. By fitting the data, the current density exponent and activation energy are extracted. The electromigration reliability model for copper pillar bump interconnections is obtained.

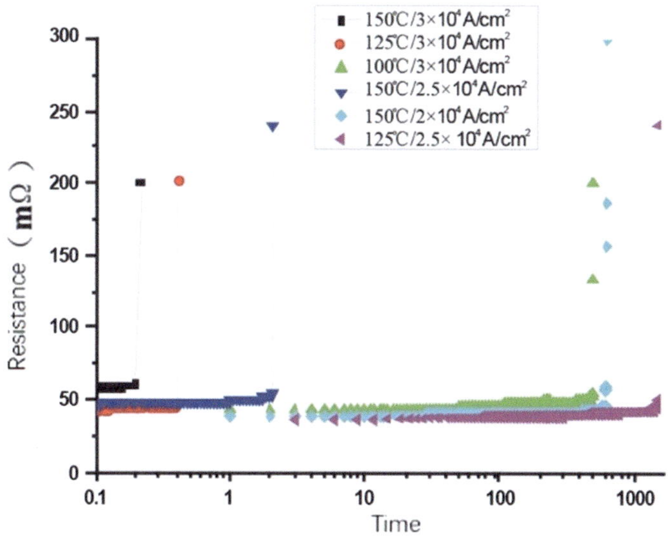

Fig. 9.27 Curves of copper pillar bump interconnection resistance over time at different temperatures and current densities

$$\mathrm{MTTF} = A\frac{1}{(j)^{17.75}}\exp\frac{0.51}{kT} \tag{9.12}$$

In the equation, A is a constant with a value of 5.76×10^9.

The failure threshold is set when the copper pillar bump interconnection resistance increases by 20% compared to its initial value. Statistical analysis is conducted on the copper pillar bump interconnection resistance data collected during the tests. The lifetime distribution curve of copper pillar bump interconnections under thermoelectric combined stress, fitted using the Weibull distribution, is shown in Fig. 9.30. From Fig. 9.30, it can be seen that the lifetime of the copper pillar bump interconnections under thermoelectric combined stress follows the Weibull distribution closely. The shape parameter is 7.78. Based on this distribution curve, it is concluded that the failure of the copper pillar bump interconnections under thermoelectric combined

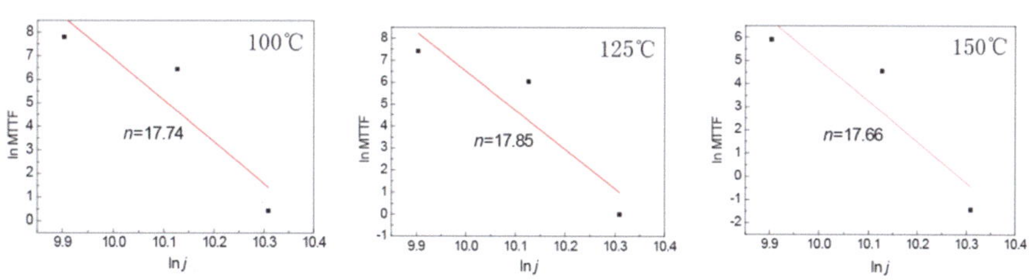

Fig. 9.28 The relationship between copper pillar bump interconnection electromigration MTTF and current density

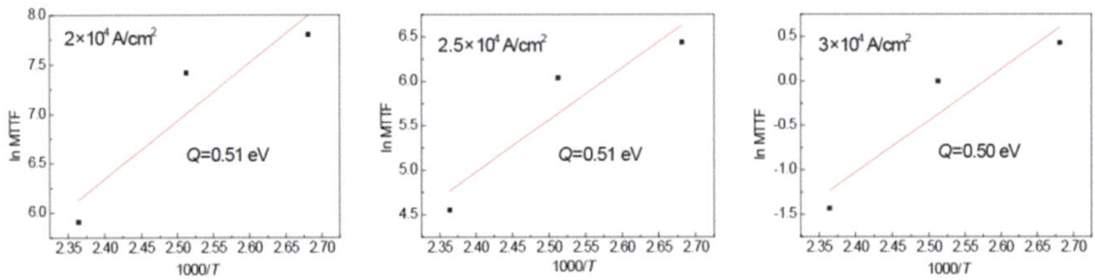

Fig. 9.29 The relationship between copper pillar bump interconnection electromigration MTTF and temperature

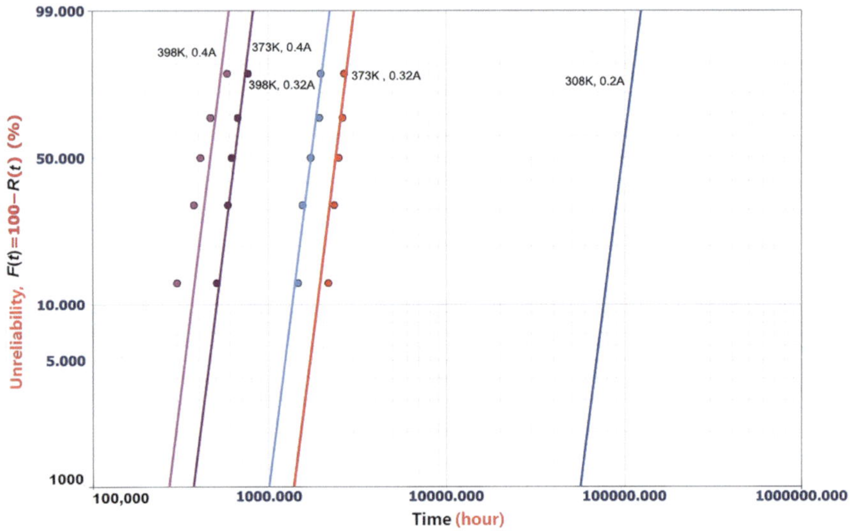

Fig. 9.30 The lifetime distribution curve of copper pillar bump interconnections under thermo-electric combined stress

stress follows the cumulative damage failure characteristic. Further analysis of the lifetime distribution curve allows the relationship between acceleration factor and acceleration stress to be established. This makes it possible to extrapolate the average lifetime of the copper pillar bump interconnection under conditions of 308 K temperature and 0.2A current stress. The estimated average lifetime is about 10.9 years [21].

References

1. Yu SY (2012) Studies on the thermal cycling reliability of BGA system-in-package (SiP) with an embedded die. IEEE Trans Compon, Packag, Manuf Technol 2(4):625–633
2. Driel W, Gils M, Silfhout R et al (2005) Prediction of delamination related IC & packaging reliability problems. Microelectron Reliab 45(9–11):1633–1638
3. Pham V L, Xu J, Pan K et al (2020) Investigation of underfilling BGAs packages-thermal fatigue. In: 2020 IEEE 70th electronic components and technology conference (ECTC), IEEE, pp 2252–2258
4. Ma S, Liu Y, Zheng F et al (2020) Development and reliability study of 3D WLCSP for automotive CMOS image sensor using TSV technology. In: 2020 IEEE 70th electronic components and technology conference (ECTC), IEEE, pp 461–466
5. Mattila TT, Marjamaki P, Kivilahti JK (2006) Reliability of CSP interconnections under mechanical shock loading conditions. IEEE Trans Compon Packag Technol 29(4):787–795
6. Tanaka N, Kitano M (1999) Evaluating IC-package interface delamination by considering moisture-induced molding-compound swelling. IEEE Trans Compon Packag Technol 22(3):426–432
7. Noma H, Okamoto K, Toriyama K et al (2015) HAST failure investigation on ultra-high density lines for 2.1 D packages. In: 2015 international conference on electronics packaging and iMAPS All Asia conference (ICEP-IAAC), pp 161–165
8. Feng W, Zhou J, Fu X et al (2009) THB reliability research for fine pitch substrate. In: 2009 16th IEEE international symposium on the physical and failure analysis of integrated circuits, pp 219–223
9. Keith J (1984) The development of the Arrhenius equation. J Chem Educ 61(6):494
10. Lee CB, Yoon JW, Suh SJ et al (2003) Intermetallic compound layer formation between Sn-3.5 mass %Ag BGA solder ball and (Cu, immersion Au/electroless Ni-P/Cu) substrate. J Mater Sci Mater Electron 14(8):487–493
11. Sharif A, Chan YC (2004) Dissolution kinetics of BGA Sn-Pb and Sn-Ag solders with Cu substrates during reflow. Mater Sci Eng B 106(2):126–131
12. Gektin V, Bar-Cohen A (1997) Coffin-Manson fatigue model of underfilled flip-chips. IEEE Trans Compon Packag Manuf Technol Part A 20(3):317–326
13. Ladani LJ (2010) Numerical analysis of thermo-mechanical reliability of through silicon vias (TSVs) and solder interconnects in 3-dimensional integrated circuits. Microelectron Eng 87(2):208–215
14. Chen KC, Liao CN (2007) Direct observation of electromigration-induced surface atomic steps in Cu lines by in situ transmission electron microscopy. Appl Phys Lett 90(20):192
15. Gerber M, Beddingfield C, O'connor S et al (2011) Next generation fine pitch Cu Pillar technology—Enabling next generation silicon nodes. In: 2011 IEEE 61st electronic components and technology conference (ECTC). IEEE, pp 612–618
16. Chang YJ, Hsieh YS, Chen KN (2014) Submicron Cu/Sn bonding technology with transient Ni diffusion buffer layer for 3DIC application. IEEE Electron Device Lett 35(11):1118–1120
17. Chen NC, Hsieh TH, Jinn J et al (2016) A novel system in package with fan-out WLP for high speed SERDES application. In: 2016 IEEE 66th electronic components and technology conference (ECTC), IEEE, pp 1495–1501
18. Islam N, Kim G, Kim KO (2014) Electromigration for advanced Cu interconnect and the challenges with reduced pitch bumps. In: 2014 IEEE 64th electronic components and technology conference (ECTC), IEEE, pp 50–55
19. Kuo KH, Mao C, Wang K et al (2015) The impact and performance of electromigration on fine pitch Cu pillar with different bump structure for flip chip packaging. In: 2015 IEEE 65th electronic components and technology conference (ECTC), IEEE, pp 626–631

20. Chen S, Zhao N, Qiao YY et al (2019) Growth behavior and orientation evolution of Cu_6Sn_5 grains in micro interconnect during isothermal reflow. In: 2019 IEEE 69th electronic components and technology conference (ECTC), IEEE, pp 1629–1634
21. Bin Z, Yun H, Yunfei E et al (2018) Interfacial behavior and failure mechanism of Cu/Ni/SnAg1.8/Cu flip-chip copper pillar bumps under thermal and electrical stress. Acta Phys Sin 67(2):028101

Chapter 10
Reliability of Integrated Circuit Board-Level Assembly

Electronic manufacturing can be divided into four levels: semiconductor manufacturing (Level 0), PCB/integrated circuit packaging/passive components/process materials manufacturing (Level 1), board-level assembly (Level 2), and system-level assembly (Level 3). The integrated circuit assembly discussed in this chapter belongs to Level 2 of electronic manufacturing (board-level assembly), which primarily refers to the process of soldering packaged integrated circuits onto a PCB to achieve thermal, electrical, and mechanical connections between the integrated circuit and the PCB through board-level assembly solder joints.

The reliability of integrated circuit board-level assembly is a crucial foundation and assurance for the overall reliability of electronic systems. This chapter discusses the impact of assembly processes, structures, materials, and environmental stresses on the reliability of board-level assembly solder joints in integrated circuits. It provides relevant reliability evaluation test methods and, in conjunction with typical failure cases of board-level solder joints in integrated circuits, explores failure control methods for board-level solder joint interconnections.

10.1 Board-Level Assembly Processes and Reliability

The functionality of integrated circuits relies on the board-level solder joint interconnections, and assembly processes span the entire life cycle of electronic products. Accompanying different service environments, these processes ultimately manifest as different failure modes that expose the impact of processes on solder joints. Therefore, the reliability of electronic systems largely depends on the reliability of integrated circuit assembly processes. The board-level assembly processes for integrated circuits have evolved over a long period, leading to a mature manufacturing industry chain. A wealth of relevant monographs and literature is also available for reference. Therefore, this section will not focus on the discussion of assembly process

B. Zhou et al., *Reliability Technology for Integrated Circuit Packaging*,
https://doi.org/10.1007/978-981-95-3885-0_10

technologies but will briefly introduce the evolution of integrated circuit board-level assembly processes and discuss the main reliability issues that may arise in the context of Surface-Mounted Technology (SMT).

10.1.1 The Evolution of Board-Level Assembly Processes

With the development of electronic packaging technology, the board-level assembly processes of integrated circuits have also been continuously improved, experiencing transitions through different stages such as manual soldering, through-hole insertion, and surface mounting. Before the 1960s, integrated circuits had very simple structures and were generally manually soldered to the bottom of a box or an installation board, as a mature packaging form had not yet been developed. By the 1970s, packaging technologies such as Single In-line Package (SIP), Dual In-line Package (DIP), and Pin Grid Array (PGA) began to be applied to the board-level assembly of integrated circuits. These types of packaging featured through-hole insertion leads, which utilized through-hole insertion technology to mount components onto the PCB. This method was simple and convenient to operate, and the corresponding wave soldering technology gradually became widely used in the industry during this period. In the 1980s and 1990s, driven by the demand for multi functionality, high density, and miniaturization of integrated circuits, packaging technologies developed rapidly. A variety of packaging forms such as Small Outline Package (SOP), Quad Flat Package (QFP), Ball Grid Array (BGA), and Chip Scale Package (CSP) emerged and were widely applied. The board-level assembly of integrated circuits entered the surface mounting era, with reflow soldering as the key process becoming the mainstream technology. From the early twenty-first century to the present, to meet the challenges of the physical limits of semiconductor manufacturing processes, integrated circuit packaging has evolved from 2 to 3D. System-in-Package (SiP) and 3D wafer-level packaging, which include chip stacking and Through Silicon via (TSV), have become important directions for the development of integrated circuit packaging. The innovations of these advanced packaging technologies are mainly focused on the internal structure of the package, while the external part still uses the same solder ball array leads as BGA packaging. Therefore, reflow soldering is still the main process used for surface mounting in board-level interconnections.

The development of integrated circuit board-level assembly processes is closely linked to changes in packaging forms. Figure 10.1 shows the development roadmap of electronic assembly processes. Before the 1960s, integrated circuits did not have a mature packaging form, and the assembly process had not yet entered the PCB era. Traditional assembly methods such as manual soldering and mechanical fastening were mostly used. Subsequently, with the widespread application of PCBs, board-level assembly entered the modern assembly process technology stage, evolving from through-hole insertion to surface mounting and composite assembly. Through-hole insertion components generally used wave soldering technology, while surface

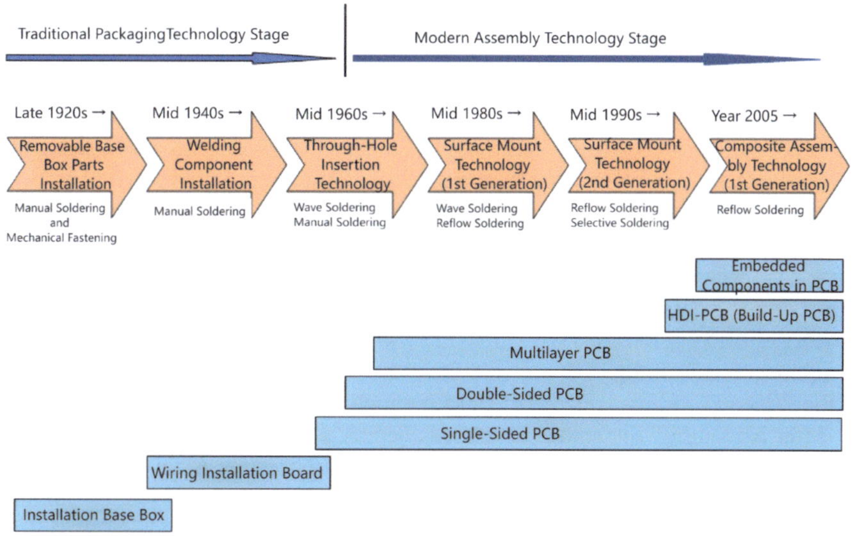

Fig. 10.1 Development route of electronic assembly processes [2]

mounting and composite assembly components mainly used reflow soldering technology. At present, surface mounting technology with reflow soldering as the key technology still occupies a dominant position. With the diversification of packaging forms, reflow soldering assembly is often combined with wave soldering (for some through-hole lead integrated circuits) and manual soldering (for products with special installation positions and low quantities). In addition, some advanced micro-assembly technologies have been applied to a certain extent in integrated circuit board-level assembly, such as Chip-on-Board (CoB) assembly, which directly installs bare chips onto the PCB through wire-bonding.

10.1.2 Mainstream SMT Technologies

SMT refers to an electronic assembly technology that uses automated equipment to directly mount or solder surface-mounted, miniaturized, leadless or short-lead components onto a PCB [3]. Currently, the mainstream SMT technologies include reflow soldering and wave soldering, with reflow soldering being the most commonly used and wave soldering typically used for the installation of through-hole lead components and PCB through-holes. The following sections will provide a brief introduction to these two mainstream SMT technologies.

1. Wave Soldering Assembly Technology

Wave soldering is mainly used for through-hole insertion, mixed assembly of surface mounting and through-hole insertion, and other similar processes. Integrated circuits using wave soldering mainly include SIP, DIP, and PGA packaged components with through-hole leads. Before wave soldering assembly, processes such as dispensing (or printing), component placement, adhesive curing, and component insertion must be carried out. The wave soldering assembly process consists of four main steps: flux application, preheating, soldering, and cooling [4]. During wave soldering, the components to be soldered first pass through a flux foaming (or spraying) trough, where a thin layer of flux is evenly applied to the bottom surface of the PCB and the component leads. After the flux application, the components enter the preheating zone for preheating. The high-temperature environment causes the solvent in the flux to evaporate and the activators in the flux to decompose and activate. The preheating process effectively removes oxidation and contamination from the solder pads and component leads on the PCB and alleviates the potential for thermal stress damage to the PCB and components due to sudden temperature increases during soldering.

After preheating, the soldering process begins. First, the bottom surface of the PCB passes through the first solder wave (a turbulent/vibrating wave, also known as the λ-wave), which covers the component leads and the solder pads on the bottom surface of the PCB with solder. The molten solder wets and spreads on the metal surface (which has been cleaned by the flux). Then, the bottom surface of the PCB passes through the second solder wave (a smooth wave, also known as the Ψ-wave), which separates the bridges between the leads.

Finally, the solder joints are solidified through natural cooling, completing the soldering process.

2. Reflow Soldering Assembly Technology

Before reflow soldering assembly, solder paste is applied to the solder pads on the PCB using methods such as printing or dispensing. Components are then placed in the designated positions using a pick-and-place machine. The solder paste melts and flows during the reflow heating process, creating connections. The typical reflow soldering assembly process includes four main steps: solder paste application, component placement, soldering, and inspection. The typical reflow soldering assembly process flow is shown in Fig. 10.2.

(1) Solder paste application

The process of applying solder paste evenly to the surface of a PCB is known as solder paste application or solder paste printing. This includes methods such as screen or stencil printing, dispensing, and roller coating, with screen or stencil printing being

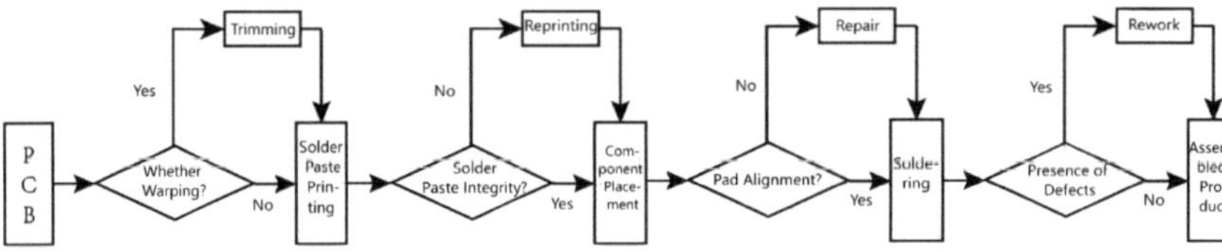

Fig. 10.2 Typical reflow soldering assembly process flow

the most common. Stencil printing involves placing a metal foil stencil on the PCB, aligning it with the solder pads. After the stencil pattern matches the solder pads on the PCB, solder paste is applied across the entire stencil using a squeegee. Once the PCB is separated from the stencil, the solder paste application is complete. Stencil printing offers advantages such as fast assembly speed and the ability for mass production. It also allows for precise control of pattern alignment and solder paste quantity, making it suitable for fine-pitch interconnections. However, this method requires a high degree of surface flatness for the PCB to be assembled.

(2) Component placement

The placement process involves using a pick-and-place machine or manual methods to attach components to the PCB surface where solder paste has been applied. This process consists of two steps: picking and placing. Pick-and-place machines can be categorized into two types: one where the placement head is fixed, and the component supply worktable and PCB move. This type of placement can cause component position shifts and does not allow for the replenishment of components on the supply worktable during the placement process. The other type has a fixed component supply worktable and PCB, with a movable placement head. This type of pick-and-place machine features a flexible placement system with a rotating head equipped with multiple vacuum nozzles. Components can be replenished during the placement process, and since the PCB is fixed, placed components will not shift.

(3) Soldering

Based on the set soldering profile, the components to be soldered undergo preheating, peak temperature soldering, and cooling. This process is known as reflow soldering. Depending on the heating method, reflow soldering can be divided into infrared reflow soldering, vapor phase reflow soldering, forced hot air convection reflow soldering, and laser reflow soldering. Among these, infrared reflow soldering and forced hot air convection reflow soldering, or a combination of these two heating methods, are widely used. With the application and promotion of lead-free assembly, higher demands are placed on reflow soldering. For example, lead-free assembly processes require more uniform soldering temperatures and sufficient wetting times. A combination of infrared radiation and forced hot air convection can avoid significant temperature differences in soldering, making it the most common heating method in lead-free reflow assembly processes.

(4) Inspection

Surface mounting processes require continuous process inspection, which mainly includes automated optical inspection (AOI), ultrasonic testing, X-ray inspection, and functional testing. Before solder paste application, the PCB should be inspected to ensure that the warpage is within acceptable limits. After solder paste application, the quality of the application should be checked to determine the position of the solder paste, the height of the solder paste, and any solder paste bridging. After soldering, a variety of technical means should be used to inspect and analyze the quality of the solder joints.

10.1.3 Process Factors Affecting the Reliability of Board-Level Assembly

The board-level assembly process of integrated circuits involves multiple key processes, and the quality control of these processes directly affects the quality and reliability of the board-level assembly. This section will briefly discuss the impact of process control in surface mounting on the quality and reliability of solder joints, focusing on key processes in surface mounting.

1. Process Factors Affecting the Reliability of Wave Soldering Assembly

The wave soldering assembly process includes flux application, preheating, soldering, and cooling. If the flux application is insufficient or uneven, it will not achieve good soldering effects. Excessive flux application can lead to too many residues, so there are inspection and control requirements for the amount and uniformity of flux application.

The preheating process has strict requirements for preheating temperature and time. By optimizing the preheating temperature and time, the activity of the flux can be maximized and thermal damage to the circuit board and components can be reduced. Insufficient preheating temperature or time will not fully decompose the activators in the flux, resulting in weakened soldering performance. On the other hand, excessive preheating temperature or time can cause the flux to decompose excessively, reducing its activity and potentially causing warpage and thermal stress damage to the circuit board and components.

Key process parameters for wave soldering include soldering temperature, conveyor belt speed, and conveyor belt angle. An appropriate soldering temperature ensures good wetting of the solder. Therefore, it is necessary to regularly calibrate the temperature of the soldering furnace. The conveyor belt speed should not be too fast to ensure adequate wetting of the soldering area. If the conveyor belt angle is too large, the solder joints will become too thin, and if the angle is too small, there will be too much solder.

In addition, attention should be paid to nitrogen inerting during the soldering process and solder pad dissolution issues. Wave soldering is currently mainly used for soldering through-hole lead integrated circuits, and the resulting solder joints are through-hole insertion solder joints. Common process defects include pinholes/voids, solder beads, bridging, and solder joint tinning. Subsequent sections will provide a detailed description of the reliability of through-hole insertion solder joints.

2. Process Factors Affecting the Reliability of Reflow Soldering Assembly

Quality control of solder paste is crucial for the reliability of the printing process in reflow soldering. Specific control measures include managing the metal particle content, viscosity, and conformability of the solder paste. The solder paste should be stored in a low-temperature environment before printing and used directly after removal, without extended dwell time at ambient conditions.

The two main factors affecting solder paste printing quality are the printing stencil and printing parameters. Printing stencils are typically made from materials with low

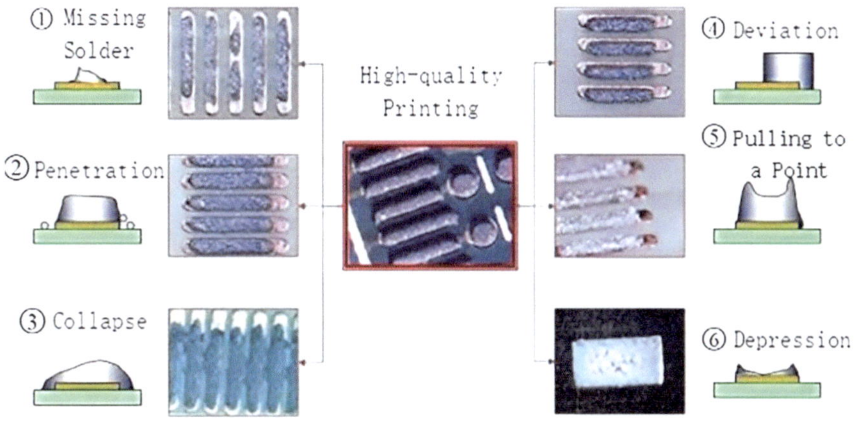

Fig. 10.3 Defect types in solder paste printing [5]

friction coefficients, such as copper or stainless steel, to facilitate the demolding and formation of solder paste and prevent solder paste bridging. The thickness of the printing stencil needs to be strictly controlled. A stencil that is too thin can lead to insufficient solder paste, deformed solder joints, and poor mechanical performance. A stencil that is too thick poses a risk of solder paste bridging. Printing parameters are optimized by controlling the speed and pressure of the squeegee to ensure the solder paste forms properly. Additionally, the flatness of the circuit board significantly impacts printing quality. An uneven circuit board can cause contact issues between the printing stencil and the circuit board. When the solder paste is squeezed, it may be pushed out of the solder pad or stick to the bottom surface of the stencil, resulting in solder paste collapse. Defect types in solder paste printing are shown in Fig. 10.3, including solder starvation, penetration, collapse, misalignment, bridging, and depression. These defects can be eliminated by optimizing the printing stencil and printing parameters.

In the reflow soldering assembly process, the setting and control of the reflow soldering temperature profile are the core factors determining soldering quality [6]. The reflow soldering temperature profile consists of preheating, activation, reflow, and cooling zones, as shown in Fig. 10.4.

The preheating zone initially raises the circuit board temperature from room temperature to the activation temperature, with a heating rate generally of 2–5 °C/s. An excessively high heating rate in the preheating zone can cause thermal damage to components, while a low heating rate may not sufficiently raise the circuit board to the activation temperature.

The activation zone allows the circuit board to stabilize at a certain temperature, enabling the flux to fully activate and volatile substances to evaporate from the solder paste, reducing the porosity of the solder joints. The reflow zone, also known as the peak temperature zone or final heating zone, raises the soldering temperature from the activation temperature to the peak temperature. The peak soldering temperature

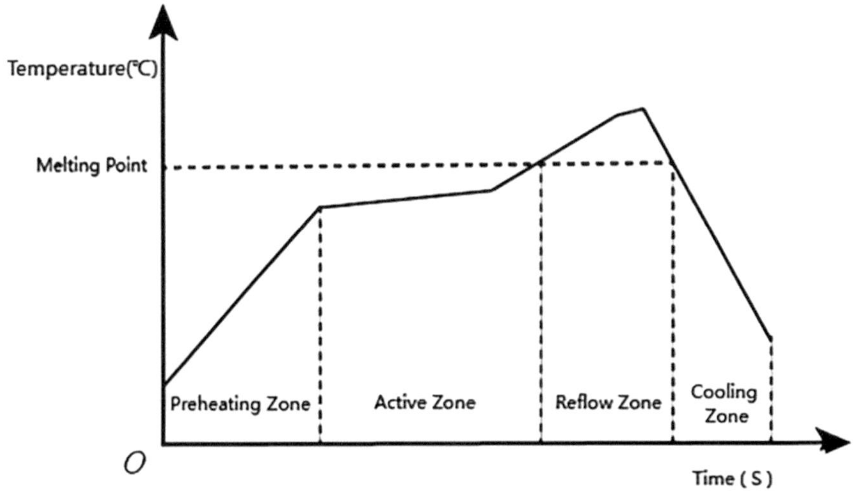

Fig. 10.4 Reflow soldering temperature profile [6]

should be higher than the melting point of the solder, while the activation temperature is generally lower than the solder's melting point. The heating rate in the reflow zone should not exceed 2–5 °C/s, and the peak temperature should not exceed the recommended value. Otherwise, it may cause excessive warping, delamination, or burning of the circuit board, as well as thermal damage to components. The cooling zone should ideally have a temperature profile that mirrors the reflow zone, with a cooling rate generally controlled at 4 °C/s. The cooling rate should not be too high to avoid thermal-shock stress.

With the continuous development of electronic packaging technology, a variety of board-level interconnection structures have emerged, involving different compositions such as leaded solder, lead-free solder, and mixed solder. The control effect of the reflow soldering process will ultimately be reflected in the quality and reliability of the board-level assembly solder joints. Subsequent sections will discuss the reliability of different structures and materials of board-level solder joints and further explore the impact of reflow soldering assembly processes on solder joint quality and reliability.

10.2 Board-Level Solder Joint Structure and Reliability

The structure of board-level assembly solder joints is related to the packaging form and the shape of the component leads. Different lead shapes determine the corresponding PCB pad shapes, which in turn affect the structure and morphology of the board-level assembly solder joints. This section will discuss the relationship between board-level solder joint structures and reliability based on different integrated circuit packaging forms.

10.2.1 Analysis of Solder Joint Structure Features

Early packaging forms such as SIP, DIP, and PGA used through-hole leads, which were soldered into the through-holes of the PCB to form through-hole insertion solder joints. With the development of packaging technology, various interconnection structures such as J-shaped solder joints, leadless interconnections, L-shaped lead solder joints, and ball-grid array (BGA and CGA solder joints) solder joints have emerged. Among them, L-shaped lead solder joints are mainly used in common packaging forms such as SOP and QFP, while BGA solder joints, known for their fine-pitch capacity and high lead density, are widely used in 2.5D/3D packaging board-level interconnections.

1. Through-hole Insertion Solder Joint Structure

The cross-sectional structure of through-hole insertion solder joints is shown in Fig. 10.5. The component leads are placed in the center of the plated through-hole (PTH) on the PCB, and the solder wets the surface plating of the leads and the inner wall plating of the PTH to form an effective solder joint. Common packaging forms that use this type of solder joint include SIP, DIP, and PGA.

2. L-shaped Lead Solder Joint Structure

The cross-sectional structure of L-shaped lead solder joints is shown in Fig. 10.6. Since the shape of L-shaped lead solder joints resembles a bird's wing, they are also known as wing-shaped lead solder joints and are commonly used in SOP, QFP, and other packaging interconnections [7]. During assembly, the leads are placed on the corresponding pads on the PCB, and the solder wets the surface plating of the leads and the pads to form the solder joint. The L-shaped leads are located on the outer

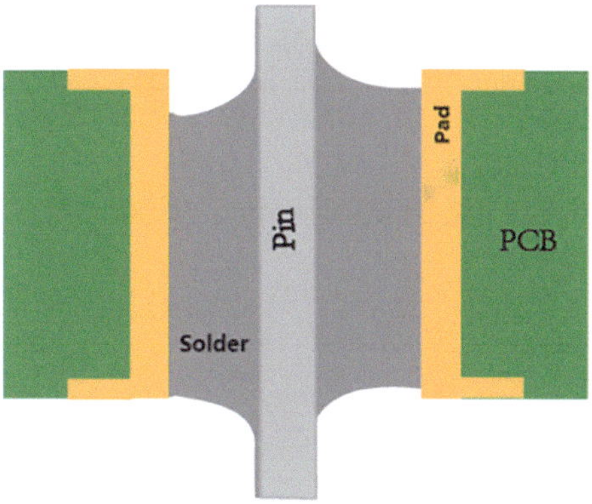

Fig. 10.5 Cross-sectional structure of through-hole insertion solder joints

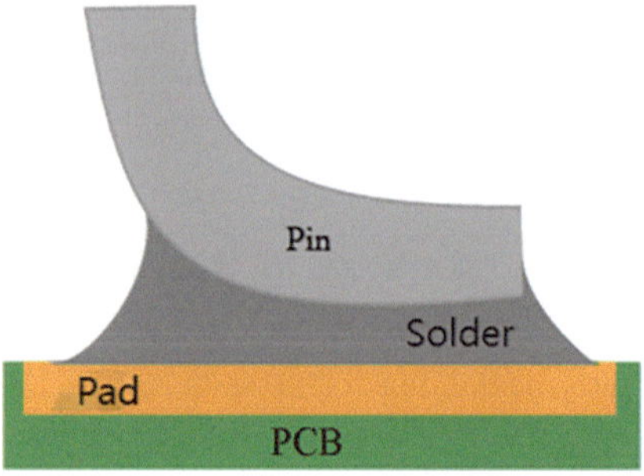

Fig. 10.6 Cross-sectional Structure of L-shaped lead solder joints

side of the package structure, so the solder joints are not covered by the component. The quality of the solder joint interconnection can be easily checked through visual inspection or low-magnification microscopy.

3. BGA Solder Joint Structure

The structure of BGA solder joints is shown in Fig. 10.7. BGA solder joints are commonly used in CSP, flip-chip (FC), multi-chip modules (MCM), SiP, and other packaging interconnections. After the substrate is ball-plated, the device is aligned with the PCB printed with solder paste. The solder wets the surface plating of the PCB pads to form BGA solder joints. Since BGA solder joints are located under the device, the inspection of these solder joints is more challenging and usually requires X-ray or some destructive techniques.

4. CGA Solder Joint Structure

The CGA solder joint structure is an improvement based on the BGA solder joint structure [8]. Compared to the ball structure, the column structure has better creep resistance. CGA solder joints are commonly used for interconnections between ceramic package shells/bases and PCBs, and they offer high-density packaging, superior thermal performance, and fatigue resistance, making them preferred in military and aerospace applications. The structure of CGA solder joints is shown in Fig. 10.8.

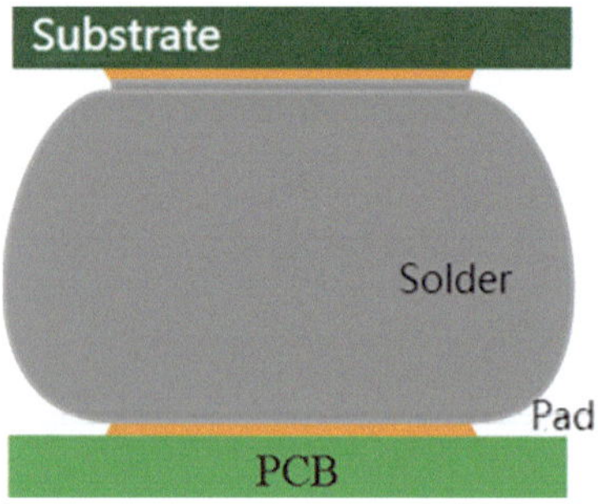

Fig. 10.7 BGA solder joint structure

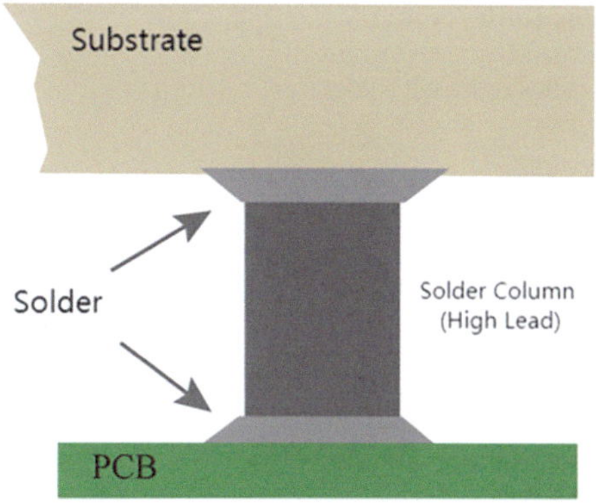

Fig. 10.8 CGA solder joint structure

5. Other Solder Joint Structure

In addition to the above structures, there are also J-shaped lead solder joints, leadless soldering, etc., which are used in various packaging forms such as QFN, LCCC, and SOJ. J-shaped lead solder joints form effective solder joints through the good wetting of solder on the surface plating of the leads and the pads. The area occupied by J-shaped lead solder joints on the circuit board is relatively small, and the inspection of these solder joints is quite challenging. The structure of J-shaped lead solder joints is shown in Fig. 10.9.

QFN, LCCC, and other packages typically use leadless soldering. The peripheral conductive pads of the package serve as electrical leads, and the solder wets the pads of the device and the PCB to form effective solder joints. The solder joints of leadless soldering are located under the package body, and the solder joint thickness

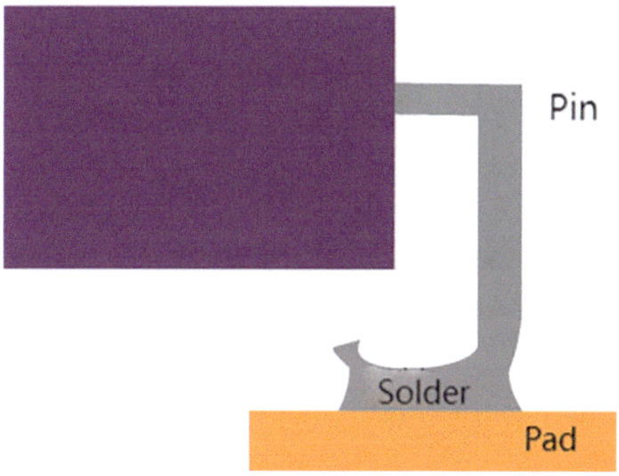

Fig. 10.9 J-shaped lead solder joint structure

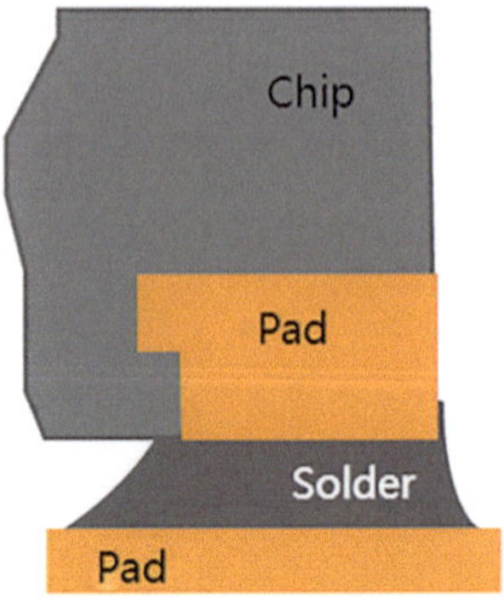

Fig. 10.10 Typical cross-sectional structure of leadless package solder joints

is relatively thin, making inspection difficult. The typical cross-sectional structure of leadless package solder joints is shown in Fig. 10.10.

10.2.2 Solder Joint Structure and Reliability

1. Reliability of Through-hole Insertion Solder Joint Structure

    (1) Formation of through-hole insertion solder joints

Through-hole insertion generally uses wave soldering to solder. After the through-hole is coated with flux to remove the oxide layer, the solder climbs up the through-hole based on the wetting action of the base metal and capillary phenomenon to form a solder joint. The force diagram for through-hole wave soldering fill is shown in Fig. 10.11. The solder climbs up the through-hole under the combined action of three forces: gravity (G), the static pressure (P_Y) formed by the PCB being immersed in molten solder, and the additional pressure (P_A) formed by surface tension [9]. Among them, P_A is the capillary force, which is the main force for the solder to climb up, while the static pressure (upward pressure) P_Y of the solder has a relatively smaller impact.

P_A is formed by surface tension and is the pressure that the solder surface interface has more than a flat interface. Its equation is:

$$P_A = 2\sigma/r \tag{10.1}$$

In the equation, r is the radius of curvature of the solder surface, and σ is the surface tension of the solder surface. Equation (10.1) shows that the capillary force P_A is directly proportional to the surface tension σ of the solder surface and inversely proportional to the radius of curvature r of the solder surface. The analysis diagram for the fill height of through-hole wave soldering is shown in Fig. 10.12. The final fill height of the through-hole can be expressed by Eq. (10.2).

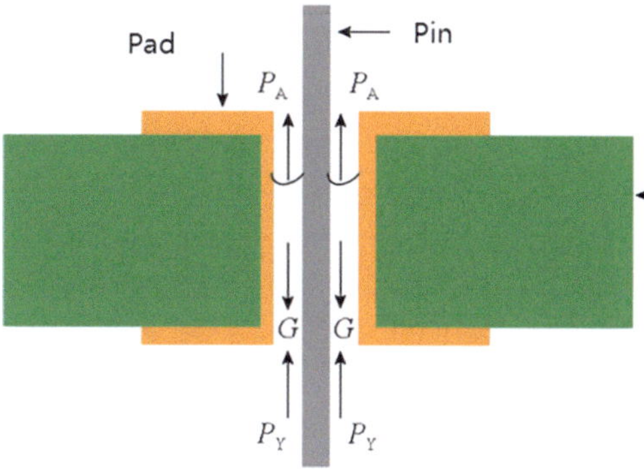

Fig. 10.11 Force diagram for through-hole wave soldering fill [9]

$$h = 2(\sigma_{sg} - \sigma_{sl})/\rho \tag{10.2}$$

In the equation, ρ, σ_{sl} and σ_{sg} are the density of the liquid solder, the solid–liquid interfacial tension, and the solid–gas interfacial tension, respectively; h is the maximum height of the liquid surface climb. The climb speed of the solder surface can be expressed by Eq. (10.3).

$$v = d\sigma_1 \cos\theta/(4\eta y) \tag{10.3}$$

In the equation, y is the climb height of the solder surface at any given moment; θ is the contact angle of the liquid surface; d is the diameter of the through-hole; η is the viscosity of the liquid solder; σ_1 is the surface tension of the liquid solder; v is the climb speed of the solder surface.

The final fill height h of the through-hole is inversely proportional to the gap distance between the PCB surface through-hole and the component lead, that is, the smaller the gap, the greater the final fill height of the through-hole. From Eq. (10.2), it can be seen that increasing the solid–gas interfacial tension σ_{sg} or reducing the

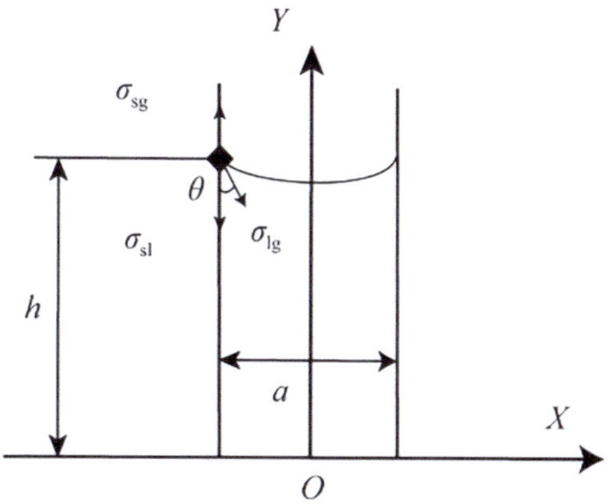

Fig. 10.12 Analysis diagram for the fill height of through-hole wave soldering [9]

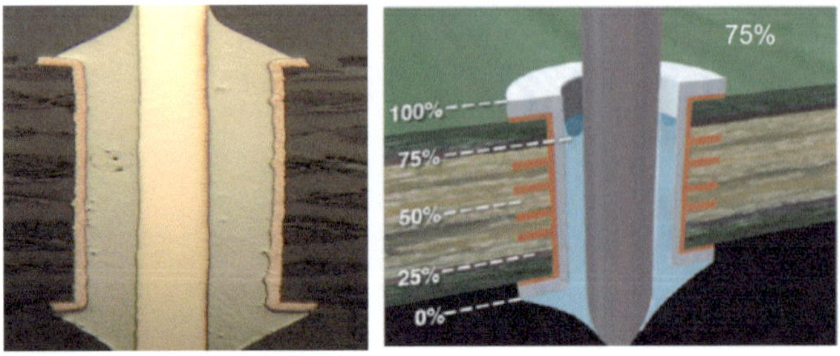

(a) Ideal filling (b) Example of fill height

Fig. 10.13 Fill height of through-hole wave soldering

solid–liquid interfacial tension σ_{sl} can increase the final fill height of the through-hole. The inner wall of the through-hole should be kept clean. Residues on the inner wall of the through-hole can lead to a decrease in the solid–gas interfacial tension σ_{sg}, resulting in insufficient fill height. The fill height of the through-hole is a key factor affecting the solder joint fill performance and directly impacts the reliability of the solder joint connection.

(2) Reliability of through-hole insertion solder joint morphology

The fullness of the through-hole insertion solder joint morphology and the adequacy of through-hole filling are key to the reliability of the solder joint connection, especially for thicker components with strong heat absorption capabilities. Solder joints with insufficient filling not only have low mechanical strength, but also poor thermal fatigue performance of the interconnection structure. According to the requirements of IPC-A-610E 7.3.5.1, the solder fill height in the through-hole after soldering needs to be greater than 75%. The fill height of through-hole wave soldering is shown in Fig. 10.13.

The main factors affecting the fill height of through-hole wave soldering include the solderability of the component, the type of flux, soldering process parameters, and the matching degree of the through-hole diameter. The solderability of the component determines the wettability of the soldering interface, thereby affecting the fill height of the through-hole. The type of flux determines the ability of the flux to remove the oxide film from the solderable end, affecting the wettability of the solderable termination surface to the solder. Other soldering process parameters, such as the inclination angle of the transport track, chain speed, flux spraying effect, preheating zone temperature rise curve, soldering temperature and time, and wave height, all have significant impacts on the solder fill in the PCB through-hole. The smaller the gap between the PCB and the through-hole lead, the greater the fill height h of the through-hole. However, if the gap between the PCB and the through-hole lead is too small, it will increase the difficulty of the insertion process. Therefore, the through-hole diameter ratio design of the PCB is crucial for whether the solder can fill the gap completely during wave soldering.

2. Reliability of L-shaped Lead Solder Joints

(1) Morphology of L-shaped lead solder joints

Under known conditions such as the volume and properties of the solder, the morphology of the L-shaped lead solder joint can be predicted based on the minimum energy principle and the theory of liquid solder wetting. The principle of minimum energy here refers to the fact that when the liquid system, which includes surface potential energy, gravitational potential energy, and potential energy of external forces, reaches a balanced state, the system's energy is minimized. The total energy E of the solder joint system consists of surface potential energy E_S, gravitational potential energy E_G, and potential energy of external forces E_f, as shown in Eq. (10.4).

$$E = E_S + E_G + E_f \tag{10.4}$$

In the equation, the gravitational potential energy E_G is

$$E_G = \int\int\int_V \rho gz dV \tag{10.5}$$

The potential energy of external force E_f is

$$E_f = Fh \tag{10.6}$$

In Eqs. (10.5) and (10.6), V is the volume of the solder, z is the vertical coordinate, ρ is the density of the solder, g is the acceleration due to gravity, h is the height of the solder joint, and F is the force on each solder joint. The surface potential energy E_S of the system consists of the surface potential energy E_{Sf} of the free liquid surface and the interfacial energy E_{Si} between the solder and the pad and lead surfaces, that is,

$$E_S = E_{Sf} + E_{Si} = \int\int_{S_{lg}} Td \int\int_{S_{sl}} - \cos \mathrm{d}_{lg_{lg}} \tag{10.7}$$

In the equation, T_{lg} is the surface tension of the molten solder, S_{lg} is the total area of the free liquid surface, S_{sl} is the total area of the solid–liquid interface, and θ is the wetting angle. Assuming the volume of the solder joint is V_0, the volume constraint condition is

$$\int\int\int_V dV - V_0 = 0 \tag{10.8}$$

Introducing the function I:

$$I = \int\int_{S_{lg}} Td \int\int_{S_{sl}} \cos \mathrm{d} \int\int\int_V d\ (\int\int\int_V d o_{0_{lg_{lg}}} \tag{10.9}$$

In the equation, λ is the Lagrange multiplier. When the integrand satisfies the Euler–Lagrange equation, the extremum of the functional can be obtained, thereby predicting the shape of the solder joint [10]. In the L-shaped lead solder joint, the factors that mainly affect the reliability of the interconnect structure include the thickness of the solder joint after soldering, the height of the lead, and the contact length between the lead and the substrate.

(2) Reliability of L-shaped lead solder joint structure

The center region of the L-shaped lead solder joint that contacts the pad is the most stress-concentrated area, and the stress size is affected by the combined effects of the thickness of the solder joint, the height of the lead, and the contact length between the lead and the substrate. This area is the weakest part of the L-shaped lead solder joint. For L-shaped lead solder joints, the greater the thickness of the solder joint, the greater the stiffness of the interconnection structure, and the less buffering effect on displacement, which leads to greater stress transmitted to the solder joint structure and an increase in the maximum stress value of the solder joint structure.

The height of the L-shaped lead solder joint also affects the maximum stress value of the solder joint structure. Figure 10.14 shows the relationship between the height of the L-shaped lead and the maximum stress value. As the height of the lead increases, the maximum stress value of the L-shaped lead solder joint structure first increases and then decreases. Figure 10.14 shows that when the height of the lead reaches 0.7 mm, the stress value of the L-shaped lead solder joint structure reaches its maximum. The design of L-shaped lead solder joints must fully consider factors such as the size of the device assembly and the stress-bearing capacity of the solder joint structure. In addition, as the contact length between the lead and the substrate increases, the maximum stress value in the solder joint decreases rapidly. The L-shaped lead solder joint structure serves as the main mechanical support for the package. Appropriately increasing the contact length between the lead and the substrate can effectively mitigate the stress in the solder joint structure.

3. Reliability of BGA Solder Joint Structure

(1) Morphology analysis of BGA solder joints

The morphology of BGA solder joints is mainly determined by parameters such as the angle of the PCB, the volume of the solder joint, and the height of the solder joint. The geometric relationship between the angle α of the solder joint with the PCB, the height H of the solder joint, the volume V of the solder joint, and the radius R_0 of the pad is shown in Fig. 10.15 [12].

Assuming that the horizontal deviation distance of the center of the circular arc from the center of the solder joint is L, the geometric relationship between the angle α of the solder joint with the PCB, the height H of the solder joint, the volume V of the solder joint, and the radius R_0 of the pad can be expressed as

$$V = 2\pi \int_0^{\frac{H}{2}} R(z)^2 dz - 2\pi R_0^2 \tau \tag{10.10}$$

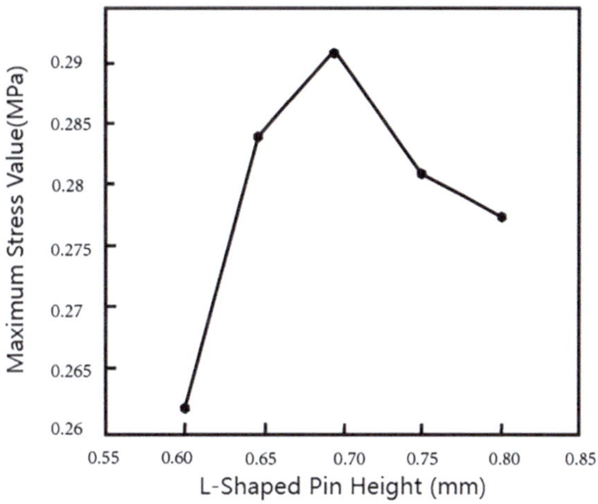

Fig. 10.14 Relationship between the height of the L-shaped lead and the maximum stress value [11]

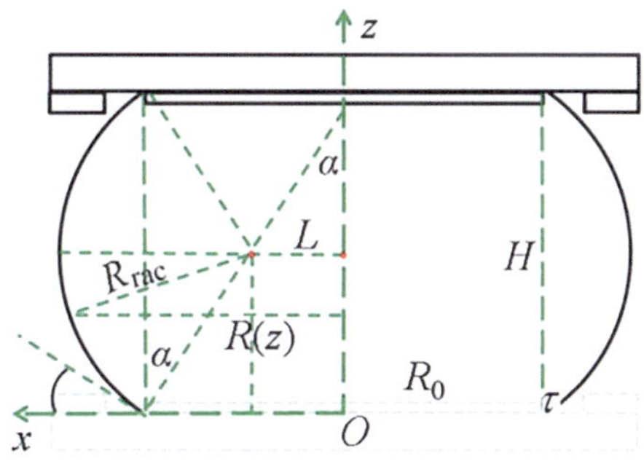

Fig. 10.15 Geometric structure analysis of BGA solder joints

In the equation,

$$R(z) = \sqrt{R_{\text{rac}}^2 - (\frac{H}{2} - z)^2} + L \tag{10.11}$$

$$R_{\text{rac}} = \frac{H}{2\cos\alpha} \tag{10.12}$$

After simplification and calculation, and consulting the integral table, we can obtain

$$V = \pi H\left(R_{\text{rac}}^2 + L^2\right) - \frac{\pi H^3}{12} - 2\pi R_0^2 \tau + \pi HL\sqrt{R_{\text{rac}}^2 - \frac{H^2}{4}} + 2\pi LR_{\text{rac}}^2 \arcsin\frac{H}{2R_{\text{rac}}} \tag{10.13}$$

In addition, according to the principle of similar triangles, we can get

$$L = R_0 - \frac{H}{2}\tan\alpha \tag{10.14}$$

Substituting Eq. (10.14) into Eq. (10.13), we can obtain

$$\begin{aligned}V = {} & \frac{\pi H^3}{4\cos^2\alpha} - \frac{\pi H^3}{12} + \pi H\left(R_0 - \frac{H}{2}\tan\alpha\right)\sqrt{\frac{H^2}{4\cos^2\alpha} - \frac{H^2}{4}} \\ & + 2\pi\left(R_0 - \frac{H}{2}\tan\alpha\right)\frac{H^2}{4\cos^2\alpha}\arcsin(\cos\alpha) \\ & + \pi H(R_0 - \frac{H}{2}\tan\alpha)^2 - 2\pi R_0^2\tau\end{aligned} \tag{10.15}$$

From Eq. (10.15), it can be seen that by determining the angle $\boldsymbol{\alpha}$ of the solder joint with the PCB, the height H of the BGA solder joint can be solved, thereby determining the morphology of the solder joint. In actual service, BGA solder joints with different heights, radii, and angles will exhibit different fatigue lives. Unreasonable morphological design can also easily lead to reliability problems such as uneven height, unsatisfactory solder joint shape, and short circuits during the assembly process.

(2) Morphology and reliability of BGA solder joints

The assembly process of BGA solder joints is prone to reliability issues such as voids, uneven height, unsatisfactory solder joint shape, and short circuits. For example, as the size of integrated circuit packages increases, warpage deformation is likely to occur between the substrate and the PCB during the soldering process. The molten solder joints will be stretched or compressed into different shapes due to the deformation, and when the solder joints solidify, they will present different shapes, forming residual stresses inside the solder joint and package structure, causing phenomena such as uneven height of the BGA solder joint array. The thermal deformation of the package during the reflow soldering process mainly includes concave and convex deformation, as shown in Fig. 10.16.

The difference in coefficient of thermal expansion (CTE) of different materials is an important reason for the warpage deformation of the package during the soldering process. During the soldering process, the compressive strength of the BGA solder joint is very weak. The package will act on the solder joint under the combined effects of thermal stress and gravity, causing the solder joint to have inconsistent height after solidification. The main reason for the unsatisfactory shape of the BGA solder joint is the solder starvation caused by the capillary effect during the assembly

Fig. 10.16 Schematic diagram of thermal deformation structure

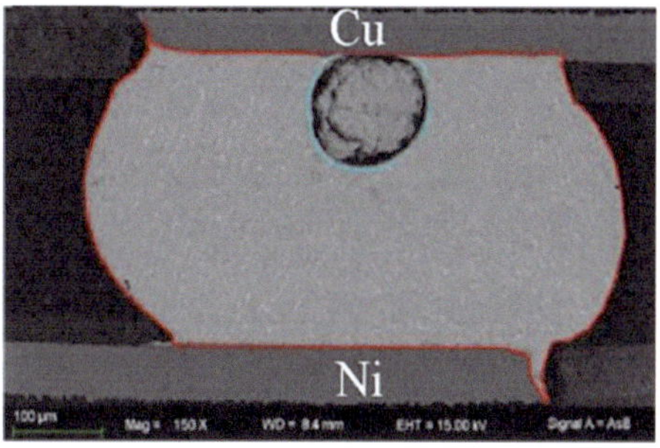

Fig. 10.17 Internal voids of BGA solder joints

process. If there are through-holes on the BGA pad, the solder can easily flow into the through-hole due to the capillary effect. Solder starvation can also be caused by misalignment of the component placement/solder paste, and the lack of solder mask isolation around the pad and through-hole. In addition, if the coplanarity of the device and PCB is poor, unsatisfactory solder joints are also likely to occur. Therefore, the main measures to solve the problem of unsatisfactory BGA solder joints are to print sufficient solder paste, use solder mask to cover the through-holes, and replace the through-holes with micro through-holes.

As the pitch of BGA solder joint interconnections continues to shrink, solder joint bridging and short circuits have become common defects in the assembly process. To address the bridging phenomenon of BGA solder joints, methods such as improving the solderability of the packaging components and strictly controlling the amount of solder paste printed are mainly used. During the soldering process, the gases absorbed inside the solder joint are decomposed, volatilized, and released. If there are still gases trapped in the surface layer of the solder after solidification, voids will form in the solder joint. The internal voids of BGA solder joints are shown in Fig. 10.17.

The formation of voids is affected by the reflow soldering temperature profile. More voids are likely to form under elevated peak temperature conditions. Under lower peak temperature conditions, the solvent is less likely to volatilize, reducing the viscosity of the flux residue, making it easier for the residue to escape from the solder, and thus reducing the formation of voids in the solder joint. In addition, factors such as the soldering gas atmosphere, solder solvent boiling point, pad design, and surface treatment methods will all affect the formation of voids in BGA solder joints.

Voids not only affect the mechanical properties of the interconnection solder joints, such as reducing the strength, ductility, and shortening the fatigue life of the solder joints, but also form hot spots in the solder joint structure, reducing the service reliability of the assembly interconnection. IBM has pointed out that the void ratio in BGA solder joints should not exceed 15%, and when the void ratio exceeds 20%, it will significantly affect the interconnection reliability. The acceptance standards for voids in solder joints in different positions according to IPC-7095A standard are shown in Table 10.1. Since different levels of electronic products have different requirements for their own quality and reliability, the acceptance standards for voids in different positions are also different.

3. Reliability of CGA Solder Column Structures

(1) Morphology of CGA solder columns

CGA solder column structures can be mainly divided into cast solder columns, copper tape-wrapped solder columns, copper-plated solder columns, and micro-coil solder columns. The cast solder column was developed by IBM in the late 1970s. This type of solder column uses high-lead-content Sn-Pb solder. The surface of the column is smooth, it has a high melting point, and it is cost-effective. Compared with BGA solder joint structures, cast solder columns have a significant improvement in thermal shock resistance, but their mechanical properties are relatively poor.

In the early 1980s, to enhance the thermal shock resistance of solder columns, Raychem proposed the copper tape-wrapped solder column structure. The main body of the copper tape-wrapped solder column still uses high-lead-content Sn-Pb solder, but copper tape is used to wrap the outer side of the column. Due to the excellent thermal conductivity of copper, the thermal shock resistance of copper tape-wrapped solder columns is better. The structure of copper tape-wrapped solder columns is shown in Fig. 10.18.

In 2000, IBM Corporation developed the copper-plated solder column structure. By using high-temperature-resistant solder, the oxidation of the solder column during

Table 10.1 Acceptance criteria for voids in electronic products soldering according to IPC-7095A standard [13]

Voids location	Class I products	Class II products	Class III products
Internal of BGA solder Joint	36% of solder joint cross-sectional area or 60% of solder joint diameter	20% of solder joint cross-sectional area or 45% of solder joint diameter	9% of solder joint cross-sectional area or 30% of solder joint diameter
Interface of BGA solder joint	25% of solder joint cross-sectional area or 50% of solder joint diameter	12% of solder joint cross-sectional area or 35% of solder joint diameter	4% of solder joint cross-sectional area or 20% of solder joint diameter

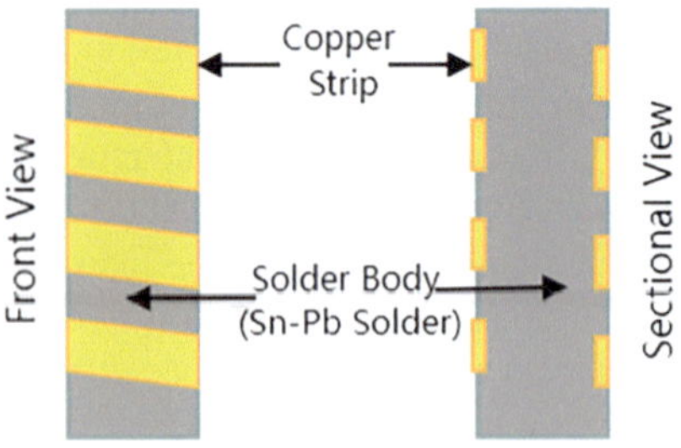

Fig. 10.18 Structure of copper tape-wrapped solder columns

the soldering process is prevented. This also mitigates issues such as the reduction in fatigue resistance caused by surface roughening of the solder column. For example, when soldering a solder column with 90Pb10Sn solder, the soldering temperature can reach 320 °C. Ordinary cast solder columns and copper-wrapped solder columns would melt under such conditions. The interior of the copper-plated solder column is a high-temperature-resistant solder wire, which is enveloped by a copper layer. An external tin plating is then applied over the copper layer, making it difficult for the solder column to melt. The tin plating can cover defects on the surface of the copper layer, enhancing the smoothness of the solder column surface and preventing oxidation of the copper layer. This helps maintain good electrical conductivity [17]. The tin plating also serves as a barrier layer to meet the EU RoHS standard's requirements for lead content. The copper-plated solder column has excellent thermal conductivity, facilitating heat dissipation from the integrated circuit. Compared to the copper-wrapped solder column, the copper-plated solder column has a shorter thermal conduction path. The structure of the copper-plated solder column is shown in Fig. 10.19.

In 2012, NASA adopted beryllium copper alloy to change the solder column structure to a micro-coil interconnection. The micro-coil ends are connected by solder paste to form a micro-coil solder column structure, as shown in Fig. 10.20 [18]. The micro-coil solder column structure allows more physical space between the PCB and the device substrate, thus having better creep resistance. Combined with the excellent anti-vibration performance of the coil structure, micro-coil solder columns are expected to enhance the reliability of integrated circuit assemblies in harsh service environments. However, this structure significantly reduces the thermal conductivity of the interconnection. Studies have shown that the thermal conductivity of micro-coil solder columns is only about half that of BGA solder joint structures.

(2) Reliability of CGA solder joint structures.

The column interconnection configuration of CGA solder joints effectively mitigates the mismatch of thermal expansion coefficients between the substrate and the circuit board. However, there are reliability issues in the assembly process of solder

Fig. 10.19 Copper-plated solder column structure

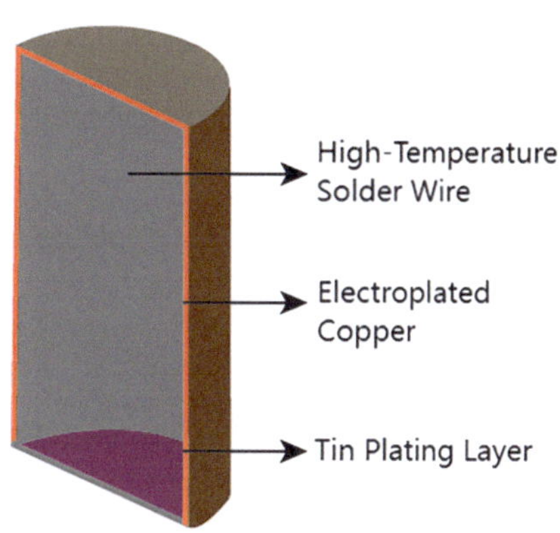

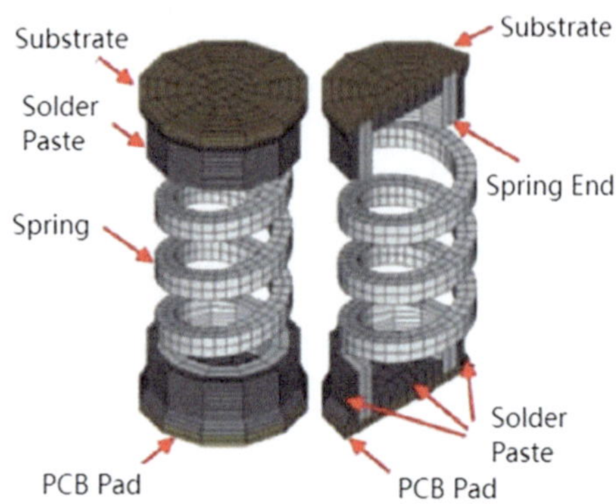

Fig. 10.20 Micro-coil solder column structure

columns, such as voids in the solder columns, offset of the solder columns, and virtual soldering at the connection of the solder columns. When high-temperature soldering is performed, the flux in the solder will volatilize or release gases. An inappropriate process profile can cause gases to fail to escape in time, forming voids. Voids in CGA devices can be inspected and identified using X-ray/SEM. The presence of voids in the solder columns directly affects the service reliability of CGA interconnections and can even reduce the application level of the device. Figure 10.21 shows the interfacial voids of CGA solder columns.

Solder column offset (see Fig. 10.22) poses more severe impacts on the reliability of CGA interconnections than solder column voids and is more difficult to repair [19]. To address the phenomena of solder column offset and bending, it is generally specified that the tilt of the solder column before device assembly should not exceed 5°, and after assembly is completed, the tilt of the solder column must not exceed 10°. CGA solder column offset can be categorized into pre-soldering offset and post-soldering offset. Post-soldering offset is mainly due to poor solderability of the solder columns, and rapid temperature changes during high-temperature melting and low-temperature cooling processes cause deformation. Therefore, ensuring the verticality

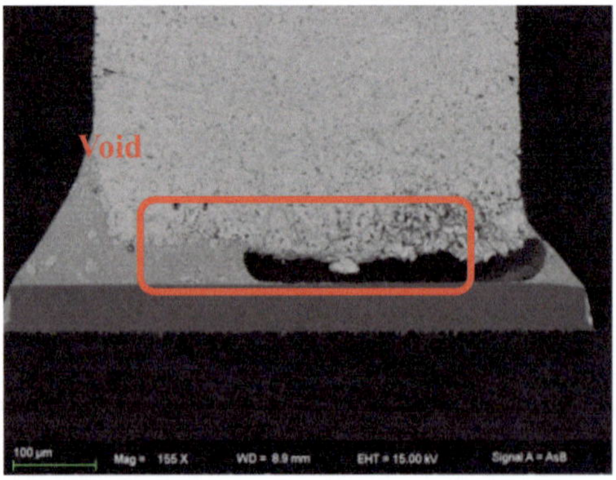

Fig. 10.21 Interface voids of CGA solder column

Fig. 10.22 CGA solder column offset [20]

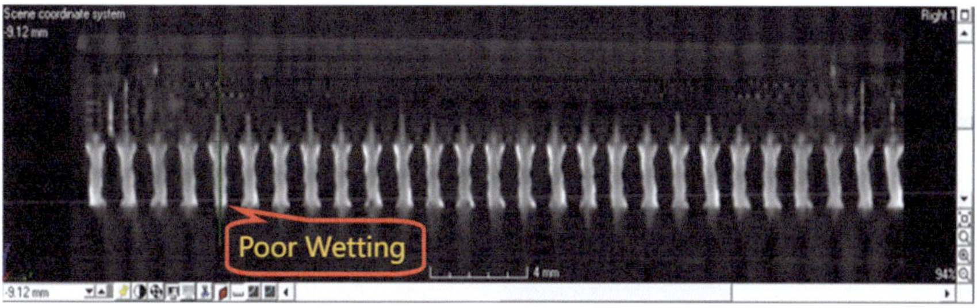

Fig. 10.23 Poor wetting of the CGA solder column

and solderability of the solder columns during soldering is an important measure to suppress deformation and offset.

Virtual soldering defects in CGA solder columns are also very common in the assembly process. The inspection and identification of virtual soldering defects in CGA solder columns are more challenging, specifically manifested as poor wetting of the solder columns (see Fig. 10.23), uneven edges leading to poor contact of the interconnection, etc.

10.3 Board-Level Solder Joints and Reliability

10.3.1 Reliability of Leaded Solder Assembly

Tin-lead solder has the advantages of good wettability, low cost, mature process, and low soldering temperature. Therefore, solder joints assembled with this material exhibit excellent interconnection performance and are highly favored in some high-reliability fields such as medical, aerospace, and military devices.

1. Impact of Reflow Soldering on Leaded Assembly

During the reflow soldering process, molten SnPb solder reacts with the PCB pads and component leads to achieve electrical and mechanical connections between the assembly board and the components. SnPb solder reacts with the pads or component leads to form intermetallic compounds (IMCs) at the interface.

Common IMCs in SnPb solder include Cu_3Sn near the pad end and Cu_6Sn_5 near the solder joint end. Generally, in the leaded solder assembly process, Cu_6Sn_5 is the main IMC. IMCs are brittle structures that reduce the toughness and low-cycle fatigue resistance of the solder joint, leading to interconnection failure between the solder joint and the circuit board [21]. The growth of IMCs during the assembly process is affected by surface mount technology parameters such as the reflow soldering temperature profile, solder paste viscosity, and conveyor belt speed. For example, reducing the conveyor belt speed prolongs the heating time of the SnPb solder, leading to excessive interfacial reactions and the formation of rich Cu_6Sn_5 and Cu_3Sn compound layers.

The growth of IMCs can be effectively retarded by controlling the liquid phase time and the peak temperature of the reflow soldering process, thereby improving the mechanical properties of the board-level assembly solder joints. To ensure good metallurgical bonding between the solder and the pads and component leads, it is recommended that the peak temperature of the reflow soldering be 30–40 °C higher than the melting point of SnPb solder. The liquid phase time of the solder should be controlled between 30 and 90 s. This not only ensures sufficient diffusion and reaction time for the solder but also prevents over-depletion of the pads and leads [22].

In addition, the temperature change rate during the reflow soldering process also affects the reliability of SnPb solder assembly. In particular, the cooling rate in the cooling zone has a more significant impact on the solder joints. For example, properly controlling the cooling rate can alleviate thermal stress during the soldering process and can also affect grain growth within the solder joint, thereby altering the mechanical strength and reliability of the solder joint. The bonding strength between the solder joints and the pads of 30 samples under cooling rates of 1.05 °C/s, 0.65 °C/s, and 0.45 °C/s is shown in Fig. 10.24.

The data on the bond strength between SnPb solder joints and pads in Fig. 10.24 shows that when the cooling rate is 1.05 °C/s, the bonding strength between SnPb solder joints and pads fluctuates between 8 and 10 N, with most samples having a bonding strength between 8.5 and 9.5 N, and an average value of 9.02 N. When the cooling rate is 0.65 °C/s, the bonding strength between SnPb solder joints and pads fluctuates between 7.5 and 9.0 N, with most samples exhibiting a bonding strength of around 8N, and an average value of 8.27 N. When the cooling rate is 0.45 °C/s, the bonding strength between SnPb solder joints and pads ranges from 6.5 to 8.5 N, with an average value of 7.52 N. Therefore, as the cooling rate increases, the bonding strength between SnPb solder joints and pads gradually increases.

Fig. 10.24 Bonding strength between SnPb solder joints and pads for 30 samples [23]

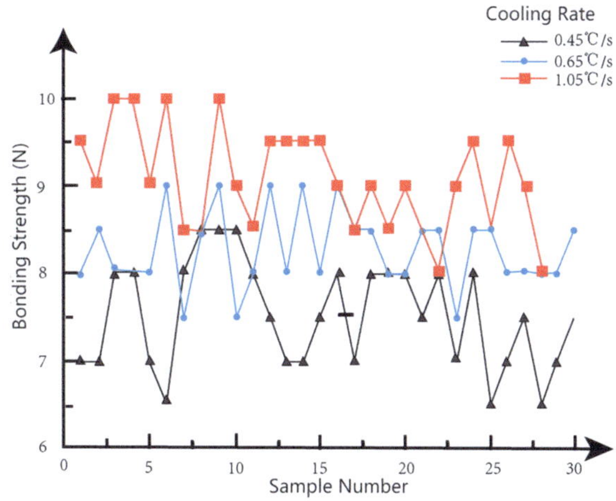

2. Reliability Issues in SnPb Assembly

(1) Pb segregation in SnPb solder joints

During the SnPb solder joint assembly process, Sn atoms in the solder diffuse into the base material (such as Cu pads), while Pb atoms do not diffuse significantly and remain at the interface, forming Pb segregation. The phenomenon of Pb segregation in SnPb solder joints is shown in Fig. 10.25.

When selective diffusion occurs, Sn atoms near the Cu pad diffuse into the Cu pad, and Pb atoms retard the diffusion of Sn atoms farther from the Cu pad. After a certain period, a Pb-rich layer will form near the Cu pad [24]. This phenomenon is known as Pb segregation. Segregation tends to form low-melting-point brittle phases in the solder joint. The mechanical properties of the interface near this layer are poor, and even under low stress, it can become a starting point for failure. The main methods to suppress segregation are to control the soldering temperature reasonably to avoid overheating during soldering; to control the soldering time reasonably to

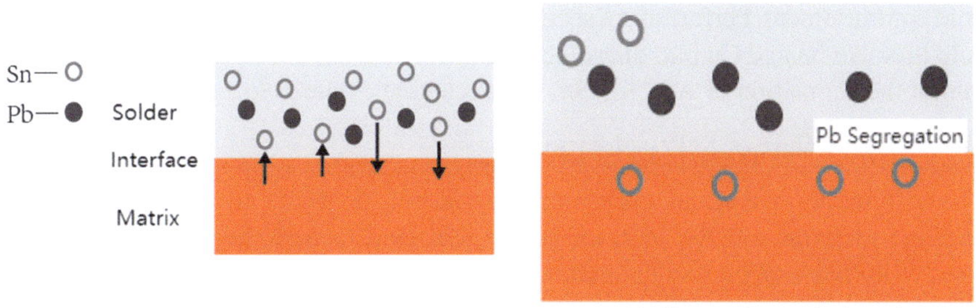

(a) Diffusion of Sn into the base material (b) Pb remains at the interface

Fig. 10.25 Phenomenon of Pb segregation in SnPb solder joints [2]

avoid prolonged heating of the solder joint; and to take measures to prevent Pb contamination when using lead-free soldering.

(2) Gold Embrittlement in SnPb Solder Joints

In aerospace product applications, the embrittlement of solder joints caused by the reaction of SnPb solder with gold-plated pads, terminals, or leads (gold embrittlement) is a significant reliability hazard [25]. Gold embrittlement in solder joints leads to open circuit failures and has become a critical common issue that must be solved in the application of aerospace products [26].

During the reflow soldering process, when the soldering time is insufficient, Au dissolves into the solder but cannot diffuse uniformly, resulting in local high-concentration Au layers that reduce the mechanical strength of the solder joint. As the Au content in SnPb solder joints increases, the mechanical properties (tensile strength, shear strength, and elongation) of the solder joints vary. When the Au content is below its solubility in SnPb solder, the mechanical properties of the solder joint increase with increasing Au content. However, when the Au content reaches its maximum solubility, the mechanical strength of the SnPb solder joint decreases with increasing Au content, while the elongation increases with increasing Au content and reaches a peak at an Au content of 3 wt%, rapidly decreasing at an Au content of 6 wt%. As early as the early 1960s, Foster et al. pointed out that when the Au content in SnPb solder exceeds 3 wt%, the brittleness of the solder joint will increase significantly. This finding has been followed within the industry. However, controlling the Au content at 3 wt% is difficult in actual engineering, and with the continuous reduction in solder joint size, the standard for triggering gold embrittlement in solder joints still needs further research.

As the service time of board-level assembly solder joints extends, a deposit of Au-Ni-Sn multinary compounds will continuously form at the pad interface, and the bonding strength between the solder and the Ni-Au surface will continuously decrease. To suppress the formation of $Au_xNi_{1-x}Sn_4$ phases, adding Ni or Ag to the solder matrix can be considered. Adding a small amount of Ni to the solder allows Ni atoms to react with Sn to form Ni_3Sn_4 particles within the solder joint, thereby reducing the deposition of $Au_xNi_{1-x}Sn_4$ phases [29]. Similarly, adding Cu to the solder can effectively inhibit the deposition of multinary compounds and prevent gold embrittlement. Performing a secondary reflow soldering at high temperature can help dissolve Ni and Cu into the solder matrix, which has been proven to effectively inhibit the deposition of $Au_xNi_{1-x}Sn_4$ phases and alleviate gold embrittlement.

(3) Fatigue failure of SnPb solder joints

The difference in thermal expansion coefficients between the integrated circuit package material and the PCB assembly will lead to periodic cyclic alternating stresses during temperature fluctuations, ultimately resulting in crack formation and failure due to fatigue. SnPb solder joints exhibit certain softening characteristics during temperature cycling service, that is, the peak stress in the solder joint will decrease with increasing temperature cycle counts [31]. Figure 10.26 shows the

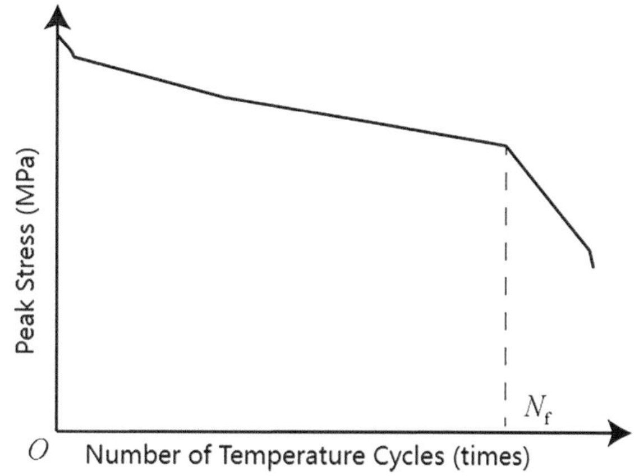

Fig. 10.26 Change curve of the peak stress of Sn63Pb37 solder [31]

change curve of the peak stress of Sn63Pb37 solder with different temperature cycle counts.

In Fig. 10.26, the peak stress of Sn63Pb37 solder during temperature cycling conditions can be divided into three stages: transient stage, steady-state stage, and subsequent failure stage. The first stage is the transient stage, which is the shortest in duration, generally not exceeding 100 temperature cycles. The peak stress in the SnPb solder joint rapidly decreases from its initial state. The second stage is the steady-state stage, which lasts the longest. The peak stress in the SnPb solder joint slowly decreases at a nearly constant rate with increasing temperature cycle counts. The third stage is the failure stage, during which the peak stress in the SnPb solder joint decreases at an accelerating rate, leading to the failure of the SnPb solder joint.

10.3.2 Reliability of Lead-Free Solder Assembly

The toxicity of Pb elements and their compounds in leaded assembly poses serious hazards to human health and the environment. During the "Eleventh Five-Year Plan" period in China, industrial-grade electronic components have basically achieved lead-free status, and imported industrial-grade key electronic components from abroad have fully realized lead-free status. However, the assembly of lead-free solders, due to differences in composition and soldering processes, has introduced new reliability issues into the board-level assembly of integrated circuits.

1. Factors Affecting the Reliability of Lead-free Solder Assembly

(1) Impact of lead-free solder alloy

Currently, the mainstream lead-free solder alloy is still the tin-silver-copper (Sn-Ag-Cu, SAC) series, with a melting point generally ranging from 217 to 227 °C. Lead-free solder has a higher melting temperature during the soldering process, and the solder contains a large amount of Sn, which is prone to oxidation of both the solder and the pad and excessive growth of IMCs. The thickness of IMCs will continuously increase with service time, which is one of the key factor affecting the reliability of solder joint assembly.

Compared to Sn37Pb solder, which has a melting point of 183 °C, the mainstream lead-free solder alloy has a melting point approximately 34 °C higher. Therefore, the shift from leaded to lead-free assembly is not just a simple material replacement; changes in the mechanical and electrical properties of the solder alloy also need to be thoroughly analyzed. Changes in the solder alloy affect multiple aspects such as melting temperature, wettability, adhesion rate between the solder and the pad, manufacturing cost, process compatibility, and material recyclability during the manufacturing and assembly processes.

(2) Impact of SMT process flow

In response to the characteristics of lead-free solder, appropriate adjustments need to be made to the SMT process flow. For example, lead-free solder has poor self-correction ability and requires higher precision in solder paste printing and component placement. Therefore, the reflow soldering temperature profile should be redefined, and nitrogen atmosphere control should be used to improve wettability. Table 10.2 provides a comparison of soldering temperature process parameters for Sn37Pb and SAC assemblies.

Table 10.2 Comparison of soldering temperature process parameters for Sn37Pb assembly and SAC assembly

Process parameter		Solder type	
Process stage	Parameter	SAC	Sn37Pb
Heating	Temperature(°C)	20–100	20–100
	Time(s)	100–200	60–90
	Process window(s)	Slow heating	–
Preheating	Temperature(°C)	100–150	100–150
	Time(s)	40–70	60–90
Rapid heating	Temperature(°C)	150–217	150–183
	Time(s)	50–70	30–60
	Process window(s)	20	30
Reflow soldering (PCB/FR4 max high-temperature resistance 240 °C)	Peak temperature(°C)	235–245	210–230
	Temperature(°C)	$240 - 235 = 5$	$240 - 210 = 30$
	Time(s)	50–60	60–90
	Process window(s)	10	30

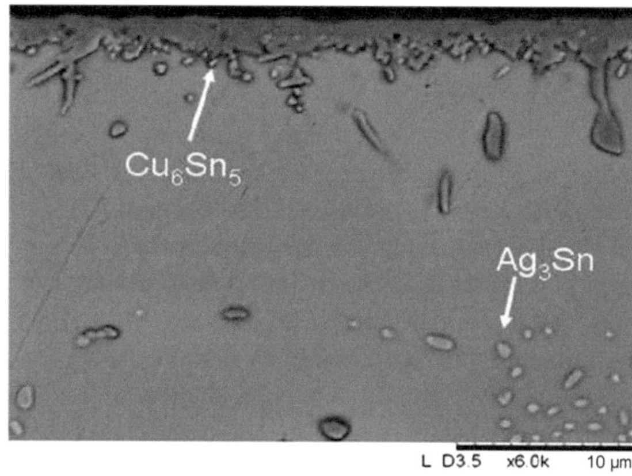

Fig. 10.27 Microstructure of SAC105 solder [33]

2. Microstructure of Lead-free solder

(1) Microstructure within solder joints

The microstructure of SnPb solder mainly consists of Sn-rich and Pb-rich phases. For the SAC series of lead-free solders, the microstructure mainly includes Cu-Sn and Ag-Sn phases [32]. Figure 10.27 shows the microstructure of SAC105 solder.

In Fig. 10.27, the SAC105 solder mainly contains Cu_6Sn_5 and Ag_3Sn IMCs. Cu_6Sn_5 is distributed in a layered form at the interface of the solder joint ends, while Ag_3Sn mainly exists in the form of particles within the solder joint. High-temperature aging, temperature cycling, or current stress will accelerate the formation and growth of Cu_6Sn_5 and Ag_3Sn in the solder joint. The lead-free solder also forms Cu_3Sn phase over time. There are many types of lead-free solders, including Sn-Ag, Sn-Ag-Cu, Sn-Ag-Bi, and Sn-Ag-Bi-In. The composition of the solder determines the microstructure of the solder joint, which in turn directly determines the failure mode of the solder joint. A stable microstructure has better fatigue resistance and can effectively alleviate stress concentration during plastic deformation.

(2) Growth of IMC

Excessive thickness of IMCs deteriorates the solder joint structure and reduces the mechanical and electrical properties of the interconnection. Under high-temperature stress conditions, the growth of IMC thickness over time can be characterized by the Dybkov equation [13].

$$Y = \sqrt{Dt} + Y_0 \tag{10.16}$$

In the equation, D is the thermal diffusion coefficient, t is the time under temperature loading, and Y_0 is the initial thickness of the IMC. Under temperature stress alone, the growth of IMCs is mainly influenced by atomic thermal diffusion, and the change in thickness of the IMCs is related to the square root of the time under temperature loading. Under coupled thermal and electrical conditions, the growth of IMCs is also affected by electromigration effects. Studies have shown that when the current density in the solder joint reaches 1000 A/cm^2, the IMC at the anode of the solder joint grows faster, while the IMC at the cathode gradually stabilizes. The migration law of metal atoms in lead-free solder under coupled thermal and electrical conditions is still complex and requires further research.

3. Process Defects in Lead-free Assembly

Common process defects in lead-free assembly include microsegregation, lifting of pad edges, via pad delamination, and package warpage. These defects occur relatively infrequently in SnPb solder and have not attracted widespread attention.

(1) Large-grain IMC

The transition of IMCs in SnAg lead-free solder from eutectic to compound composition will lead to significant performance changes. For example, as the Ag content in lead-free solder increases from zero, the tensile strength of the solder joint increases, while the elongation decreases. However, when the Ag content exceeds 3.5 wt%, the tensile strength will significantly decrease [34]. In addition to the formation of fine Ag_3Sn particles, large plate-like Ag_3Sn particles, known as primary crystals, is the solid state organization precipitated during the solidification of the solder, will also form. These large Ag3Sn primary crystals (coarse and large IMCs) reduce the strength of the solder joint and severely degrade its creep, fatigue, and impact resistance. For SAC305 solder, the initial crystalline IMC is elongated rod-like Cu_6Sn_5, and no obvious Ag_3Sn primary crystals are found. By reducing the Ag content in the solder alloy, especially when the Ag content is below 3.2 wt%, the formation of large-grain IMCs can be effectively inhibited. Additionally, increasing the cooling rate and reducing the undercooling are effective methods to suppress the formation of large-grain IMCs.

(2) Pad edge lifting

Pad edge lifting is a common defect in lead-free wave soldering, especially on metalized through-hole substrates. The main reason for pad edge lifting is the significant difference in thermal expansion coefficients between the substrate and the solder, Cu pad, and other materials. For example, the thermal expansion coefficient of the thickness direction of FR4 PCB material is more than ten times that of Sn. If there is a liquid phase at the interface, thermal contraction occurs, causing the corner to lift from the substrate, and this lifting is irreversible.

Bi alloy elements are also a cause of pad edge lifting. As dendrites grow, Bi elements segregate in the liquid phase, and the concentration of Bi increases at the front end of the dendrites near the Cu pad interface, slowing down dendrite growth. Since heat is transferred from the inside of the through-hole to the Cu pad, the part of the solder near the pad interface accumulates more heat and solidifies more slowly. When solidification occurs from the top of the corner, various stresses (solidification shrinkage, thermal shrinkage, and substrate thermal shrinkage) are generated, causing the corner to delaminate from the pad. Therefore, Bi and other solute elements promote delayed solidification and are a cause of pad edge lifting. To improve pad edge lifting, single-sided substrates with low thermal shrinkage can be selected, and solder alloys containing Bi, In, and other solute elements should not be used. The mechanism model of pad edge lifting in lead-free solder alloys is shown in Fig. 10.28.

(3) Solidification cracks

Solidification cracks form on the fillet surface of the solder during the solidification process of the solder. Since the solidification of solder on the substrate cannot be completely uniform, this process is controlled by the direction and speed of solidification. Under the application of temperature cycling and other loading conditions, solidification cracks do not cause cracking due to stress concentration. The main method to suppress solidification cracks in lead-free solder is to select solder alloys close to the eutectic composition or those that do not show significant lifting, such as Sn4.0Ag0.9Cu alloys.

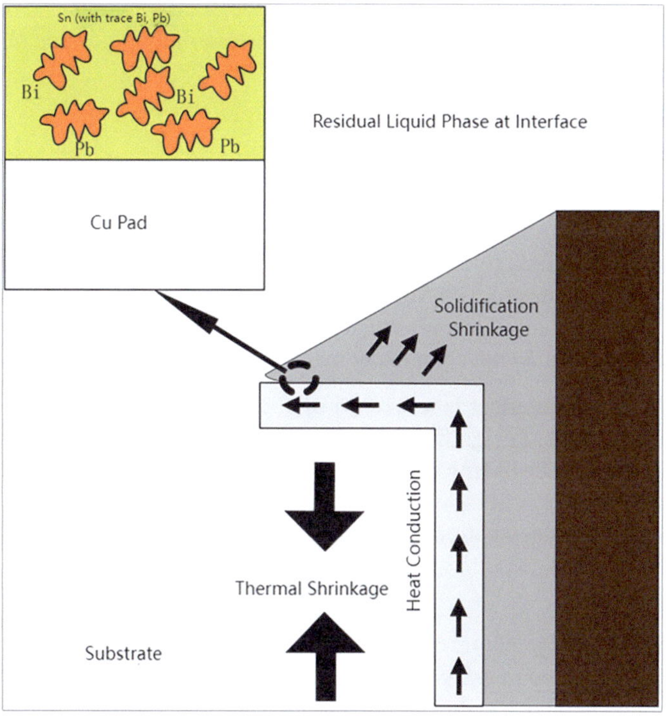

Fig. 10.28 Mechanism model of warpage in lead-free solder alloy

(4) Pad delamination

Pad delamination mainly refers to the separation between the substrate and the pad. When the solder joint solidifies, the substrate begins to cool and gradually returns to its original flat shape. The residual stress accumulated in the solder joint during the cooling process remains unreleased, forming residual stress. The residual stress is concentrated at the interface between the pad and the substrate. When the thermal adhesion force between the pad and the substrate is less than the cohesive force of the solder, it can cause pad lifting or surface cracking of the solder. Selecting a Cu substrate with high adhesion can effectively alleviate the occurrence of pad delamination during high-temperature soldering. Other effective inhibition methods include increasing the strength of the substrate itself and reasonably selecting solder mask defined (SMD) structures.

10.3.3 Reliability of Mixed Solder Assembly

Although the reliability of lead-free interconnections has gradually improved and been proven, the complete transition from leaded to lead-free manufacturing still requires a certain period of transition. Therefore, to meet the special requirements of high-reliability electronic products, the aerospace, transportation, and other high-reliability fields still commonly use mixed solder assembly processes of lead-free and leaded solders.

1. Compatibility of Mixed Solder Assembly

Mixed solder assembly of leaded and lead-free materials can be divided into two main types: one is the combination of lead-free SAC solder joints and SnPb solder paste, known as backward compatibility; the other is the combination of SnPb solder joints and lead-free SAC solder paste, known as forward compatibility. When SAC solder joints are combined with SnPb solder paste, due to the difference in melting points, the following consequences may occur during the reflow soldering process if the SAC solder joints do not completely melt [16].

- The self-correction effect is weakened or absent, leading to local open circuits, which are particularly significant for fine-pitch lead components.
- The collapse of the solder joint is insufficient, causing severe coplanarity issues for the device, which in turn leads to solder joint open circuits.
- The solder joint does not collapse, and there is minimal mixing of solder and solder paste alloys, resulting in an uneven microstructure of the solder joint and internal stress.

When SAC solder joints are combined with SnPb solder paste, the reliability does not decrease compared to using SnPb solder joints/SnPb solder paste combinations when the temperature exceeds 225 °C and the SAC solder joints completely melt. However, when SnPb solder joints are combined with SAC solder paste, the leaded

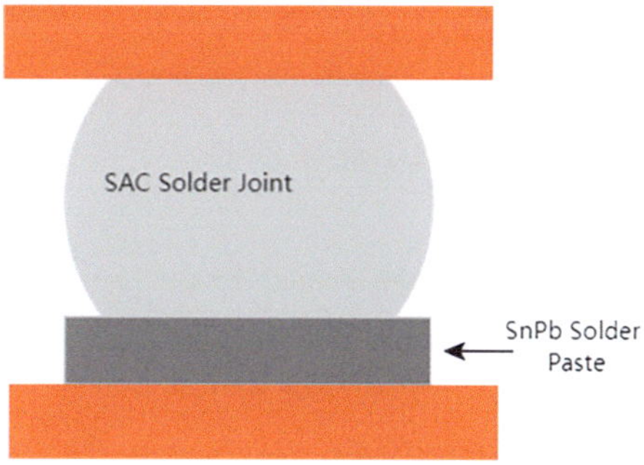

Fig. 10.29 Model of SAC solder joints combined with SnPb solder paste

solder joints will melt first, covering the surface of the pad and device leads. The volatile products of the flux are not easily completely expelled, which can easily lead to void formation. Therefore, in mixed assembly processes, the melting point of the solder joint must not be lower than that of the solder paste. Currently, the most widely used mixed assembly in the electronics industry is the backward compatibility combination of SAC solder joints and Sn37Pb solder paste. The model of SAC solder joints combined with SnPb solder paste is shown in Fig. 10.29.

2. Reflow Soldering of Mixed Solder Joints

When SAC solder joints are soldered with SnPb solder paste, some companies choose to use the reflow soldering temperature profile for SnPb leaded soldering or the reflow soldering temperature profile for SAC lead-free soldering. Adopting the pure leaded soldering temperature profile for SnPb will result in the SnPb solder paste on the pad melting while the SAC solder joint has not yet melted. The incomplete melting of the SAC solder joint leads to poor self-correction effects, which in turn affects the precision of component placement and creates potential open circuit hazards in the solder joints. In addition, some mixed assembly processes can cause microstructural segregation and degradation of interfacial bonding, as well as an increase in voids.

Pure lead-free SAC solder joints require higher temperatures during the reflow soldering process, which are more susceptible to defects in device packaging due to moisture and thermal stress. Therefore, it is not appropriate to directly use the pure lead-free reflow soldering temperature profile for mixed solder joints. The reflow soldering temperature profile is an important link in ensuring the quality and process reliability of interconnections. The reflow soldering temperature test profile for backward compatibility (SAC solder joint/SnPb solder paste) is shown in Fig. 10.30.

When performing mixed solder joint assembly based on Sn37Pb solder paste and backward compatibility, the melting point of the SAC alloy is 217 °C, while the peak temperature for the reflow soldering of Sn37Pb solder paste is between 205 and 225 °C. When the reflow soldering temperature is high enough (greater than 225 °C), the SAC solder joint and SnPb solder paste can be well fused, and the self-correction effect will improve the quality of the mixed assembly. Research has shown that when SAC lead-free solder joints are combined with SnPb leaded solder paste, considering

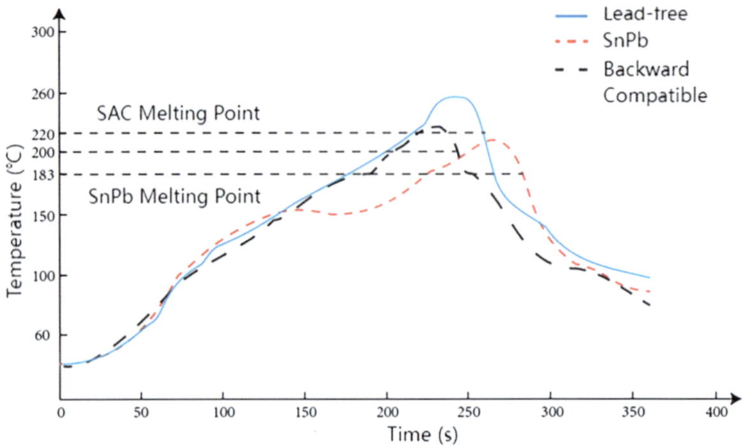

Fig. 10.30 Reflow soldering temperature test profile for backward compatibility (SAC solder joint/SnPb solder paste)

both the device's temperature tolerance and process reliability, it is recommended that the peak soldering temperature range be 225–235 °C. When both the solder end and the solder joint use high-melting-point soluble materials (such as Sn, SAC), the peak temperature of the soldering profile can be increased to 230 °C (for solder ends coated with SAC) or 235 °C (for solder ends coated with Sn). When the solder end coating is made of soluble materials (such as Ag, Pd, Au) and the solder joint material is SAC, the peak temperature of the leaded soldering process profile needs to be increased to above 217 °C.

3. Impact of Pb Content on Solder Joint Solidification Behavior

The content of Pb affects the melting and solidification curves of mixed solder joints. Figure 10.31 shows the melting and solidification curves of solder joints with different Pb contents. When Sn3Ag0.5Cu is mixed with a certain amount of Pb, both the melting and solidification processes of the solder joint change significantly. The initial melting temperature of the lead-free solder decreases with increasing Pb content, and the melting behavior of the solder joint gradually transitions from being controlled by Sn3Ag0.5Cu to being controlled by Sn37Pb solder. The solidification curve also changes with increasing Pb content.

Figure 10.32 shows the microstructure of Ni/SAC305/Cu dual-interface mixed solder joints. When the Pb content is 4.67 wt%, both $(Cu,Ni)_6Sn_5$ and $(Ni,Cu)_3Sn_4$ IMCs coexist in the solder joint. As the Pb content continues to increase, the initial solidification temperature of the solder joint decreases, and Ag_3Sn and β-Sn overlap with the rapid precipitation of Pb elements, resulting in a single exothermic peak. When the Pb content exceeds 26.49 wt%, there is insufficient Ag and Cu to convert all $(Ni,Cu)_3Sn_4$ into $(Cu,Ni)_6Sn_5$ alloy, and $(Ni,Cu)_3Sn_4$ is formed near the Ni layer side of the welding interface. The solidification behavior of the solder joint begins to be dominated by Sn37Pb.

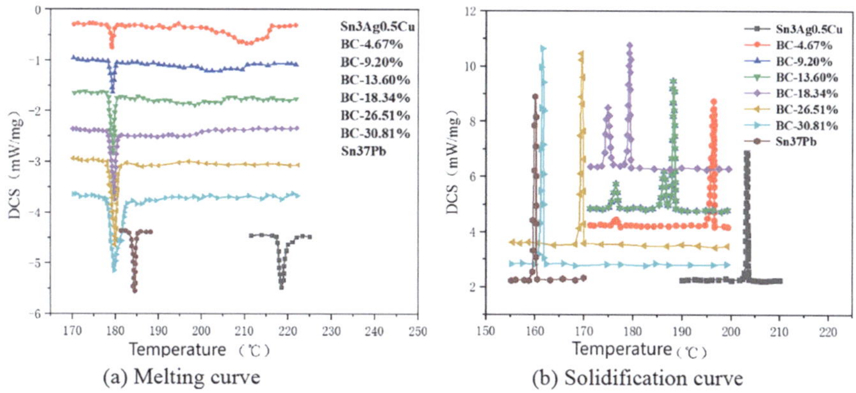

Fig. 10.31 Melting and solidification curves of solder joints with different Pb contents [36]

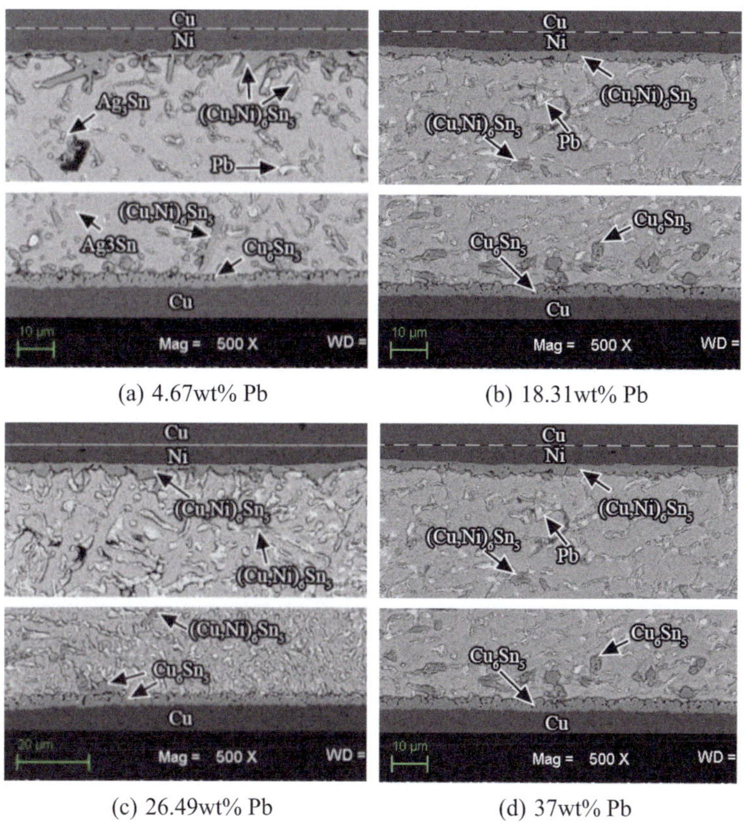

Fig. 10.32 Microstructure of Ni/SAC305/Cu dual-interface mixed assembly solder joint

4. Optimization Analysis of Solder Joints Based on Pb Content

The content of Pb affects the melting and solidification behavior of mixed solder joints, thereby influencing the microstructure of the solder joints and further altering their thermal and mechanical properties [37]. Figure 10.33 shows the microstructure of Ni/SAC305/Cu dual-interface mixed solder joints after high-temperature aging. After high-temperature aging, the needle-like IMC-(Cu, Ni)$_6$Sn$_5$ on the upper side of the welding interface becomes coarse, and the Cu_6Sn_5 at the lower side interface shows a layer-like distribution. With the addition of Pb (4.67 wt%), the Pb elements change from elongated to spherical particles, and the originally needle-like (Cu, Ni)$_6$Sn$_5$ on the upper side gradually transforms into a scallop shape during high-temperature aging. As the Pb content further increases, the IMC morphology on the upper side of the interface completely changes to a scallop shape (Cu, Ni)$_6$Sn$_5$, while the lower side remains in a layered structure of Cu6Sn5. When the Pb content exceeds 26.49 wt%, the compound composition of the welding interface changes from (Cu, Ni)$_6$Sn$_5$ to (Ni, Cu)$_3$Sn$_4$.

The microstructure of mixed solder joints is similar to that of SnPb solder joints. The IMCs in the welding interface become coarse and the IMC thickness increases during high-temperature aging, but the specific microstructural characteristics vary with different Pb contents. As the high-temperature process continues, Cu atoms from the pad continuously diffuse into the solder joint, forming Cu_6Sn_5 type IMCs,

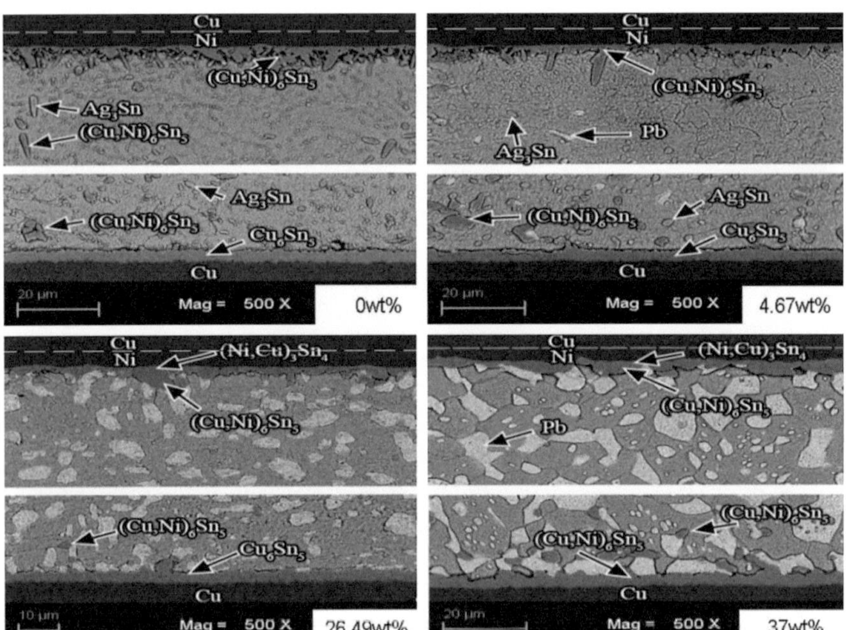

Fig. 10.33 Microstructure of Ni/SAC305/Cu dual-interface mixed solder joints after high-temperature aging

and Ag_3Sn particles in the solder joint gradually grow larger. When the Pb content exceeds 18.31 wt%, the solidification behavior of the mixed solder joint begins to be dominated by Sn37Pb, and β-Sn grains can observe Pb-rich phases and SnPb eutectic phases around them. When the Pb content increases to more than 26.49 wt%, the welding interface near the pad side of the upper interface forms Ni_3Sn_4 type IMCs.

Figure 10.34 shows the shear strength curve of solder joints after high-temperature service with different Pb contents. The introduction of Pb elements in mixed solder joints enhances the strength of the solder joints. However, when the Pb content exceeds 26.49 wt%, the shear strength of the solder joints begins to gradually decrease, but the overall shear strength is still higher than that of Sn37Pb solder joints. Although the shear strength of the solder joints fluctuates to some extent with the change in Pb content, the introduction of a small amount of Pb element has a more obvious effect on strengthening the shear strength of the solder joints. After 1000 h of high-temperature aging at 125 °C, the shear strength of the solder joints decreases to some extent, but the overall trend of the shear strength of the solder joints with increasing Pb content remains parabolic and does not change.

When the Pb content is high, β-Sn grains are surrounded by a network structure formed by discrete α-Pb compositions, which facilitates the plastic deformation of β-Sn grains, directly manifested as a decrease in shear strength. After the solder joint has been in service for 1000 h at a high temperature of 125 °C, the β-Sn grains in the solder joint matrix undergo significant coarsening, and the reduction in grain boundary area decreases the bonding force between β-Sn grains. At the same time, the growth of dispersed phases (such as Ag_3Sn, Cu_6Sn_5) reduces their dispersion strengthening effect, ultimately leading to a sharp decrease in the shear strength of the solder joint after high-temperature aging.

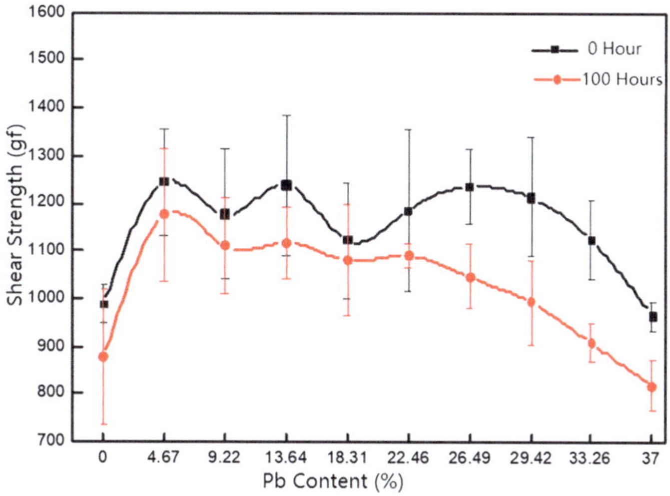

Fig. 10.34 Shear strength curve of solder joints after high-temperature service with different Pb contents

The following summary of the effects of Pb content on mixed assembly solder joints is divided into three aspects: solder joint microstructure distribution, interfacial IMC characteristics, and mechanical properties. For solder joints in the as-soldered state, the main effects are as follows: the introduction of Pb can refine the microstructure of the solder joint; the changes in interfacial IMC during high-temperature aging tests vary with the Pb content; trace amounts of Pb in the mixed solder can enhance the shear strength of the solder joint; when the Pb content is between 4.67 and 26.49 wt%, the solder joint exhibits relatively better shear strength, and the relationship between shear strength and Pb content is parabolic.

The introduction of Pb reduces the initial melting temperature of the solder joint (178 ± 2 °C). When the Pb content is between 4.67 and 18.31 wt%, a low-melting eutectic phase of Sn–Pb-Ag appears during solidification, and the content of this low-melting eutectic phase increases with the increase in Pb content. The presence of the low-melting eutectic phase can easily lead to thermal stress failure of the solder joint, so it needs to be controlled reasonably, and the safe operating temperature range for the product should be specified.

When the Pb content reaches 4.67 wt%, the shear strength of the solder joint in the as-soldered state reaches its peak. When the Pb content varies within the normal mixed assembly process range (4.67–18.31 wt%), the shear strength remains relatively stable and is significantly higher than that of pure lead and lead-free solder joints. After 1000 h of high-temperature aging at 125 °C, the shear strength of the solder joints generally decreases. However, the solder joint with 4.67 wt% Pb content exhibits the least reduction in shear strength. Therefore, by controlling the Pb content of the mixed assembly solder joint to around 4.67 wt% through pad and stencil aperture design, not only can higher shear strength be achieved, but also the reduction in shear strength during high-temperature aging can be minimized. This is crucial for ensuring the reliability of mixed assembly solder joints.

10.4 Environmental Stresses and Board-Level Assembly Reliability

10.4.1 Thermal Stress and Board-Level Assembly Reliability

1. Creep Behavior of BGA solder joints

Under high-temperature thermal stress conditions, BGA solder joints are prone to severe microstructural degradation and creep deformation. The creep performance of solder joints is influenced by parameters such as solder composition, pad and plating materials, and the geometry of the solder joints. Changes in solder composition and microstructure can lead to differences in creep deformation rates of up to 2–3 orders of magnitude. Figure 10.35 shows the creep curves of Cu/SAC/Cu solder joints with different heights in BGA structures (high-temperature aging at 125 °C/1000 h).

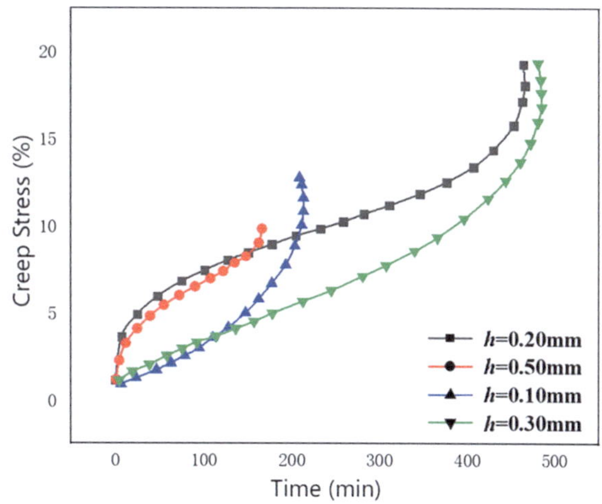

Fig. 10.35 Creep curves of Cu/SAC/Cu solder joints with different heights in BGA structures

The curve shown in Fig. 10.35 includes three stages of creep: initial creep stage, steady-state creep stage, and accelerated creep stage [38]. The total time of these three stages can be considered as the creep life of the solder joint. The creep life increases with the height of the solder joint and shows a parabolic relationship with the height of the solder joint. When the height of the solder joint is 0.30 mm, the creep life reaches its maximum value, and then decreases with increasing height. The creep life of Cu/SAC/Cu structured BGA solder joints is very sensitive to changes in solder joint height.

The steady-state creep stage is the second stage of the creep curve, which lasts the longest time, and the creep life of the solder joint is mainly determined by this stage. In the engineering field, the steady-state creep stage is commonly used as one of the key parameters to assess the reliability of interconnection solder joints. The steady-state creep strain rate is a function related to stress and temperature, which can be expressed by the following equation.

$$\dot{\varepsilon} = A\left(\frac{Gb}{RT}\right)\left(\frac{\sigma}{G}\right)^n \exp\left(\frac{-Q_c}{RT}\right) \tag{10.17}$$

In the equation, $\dot{\varepsilon}$ is the minimum steady-state creep strain rate, Q_c is the creep activation energy, σ is the applied stress, n is the creep stress index, G is the shear modulus related to temperature, b is the Burgers vector, R is the gas constant, T is the thermodynamic temperature, and A is a constant.

Figure 10.36 shows the creep fracture morphology of Cu/SAC/Cu solder joints with different heights after high-temperature aging at 125 °C/1000 h. The fracture location of the solder joint is mainly concentrated in the center of the solder joint and the solder matrix near the interface. As the height of the solder joint increases, the fracture location of the solder joint shifts from the center region to the interface. Long-term high-temperature exposure has no obvious effect on the creep fracture

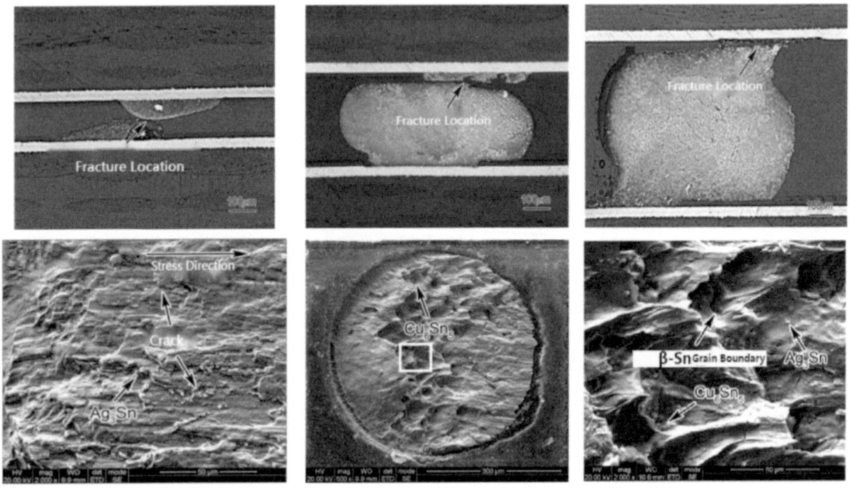

(a) At the height of 0.10mm (b) At the height of 0.30mm (c) At the height of 0.50mm

Fig. 10.36 Creep fracture morphology of Cu/SAC/Cu solder joints with different heights

location of the solder joint. The fracture surface of the 0.10 mm height solder joint has many flat step-like slip surfaces, and no obvious β-Sn grain deformation is observed. The creep fracture mode is mainly intergranular fracture. When the height of the solder joint is 0.30 mm, the fracture surface of the solder joint is relatively smooth, and no obvious grain boundaries are observed. The grain deformation direction is consistent with the shear load direction. The fracture surface of the 0.50 mm height solder joint is located in the solder matrix near the interface, showing typical ductile fracture characteristics. There are many micro-pits on the creep fracture surface, and Cu_6Sn_5 particles are visible at the bottom, accounting for about 50–60% of the entire pad area. Under shear stress, there is an obvious stress and strain concentration phenomenon at the end of the solder joint and the solder matrix near the interface, which increases with the height of the solder joint.

2. Temperature Cycling Fatigue

Temperature cycling testing exposes board-level assembly devices to alternating thermal stresses over a short period to assess their tolerance to temperature changes. The difference in thermal expansion coefficients between integrated circuit packages and PCB materials creates mechanical stresses and relative displacements at the material junctions. As the number of thermal cycles increases, stress relaxation occurs in the board-level assembly solder joints, plastic strain accumulates, and cracks form and propagate within the solder joints [39].

Figure 10.37 shows the cracking failure of mixed solder joints under temperature cycling stress. Cracks mainly extend along the interface between the ductile Pb-rich region and the brittle solder region. Under alternating thermal stress, Pb particles in the solder joint gradually accumulate, forming larger white Pb-rich regions. The

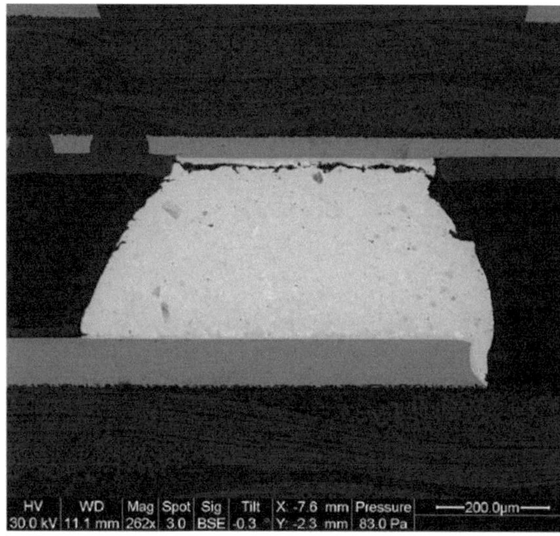

Fig. 10.37 Cracking failure of mixed solder joints under temperature cycling stress [40]

boundary between the ductile Pb phase and the brittle solder matrix is prone to grain boundary slip under cyclic stress, leading to solder joint cracking failure.

The thermal fatigue life of solder joints under temperature cycling stress can be assessed using life prediction models based on plastic deformation, creep deformation, fracture parameters, and energy. Among these, life prediction models based on plastic deformation mainly consider time-independent plastic effects, assuming that the accumulation of plastic strain under temperature cycling stress causes fatigue damage or failure. The Coffin-Manson model, Engelmaier model, and Soloman model are commonly used for life prediction based on plastic strain. These models include the empirical relationship between the plastic shear strain and the number of cycles to failure for each cycle.

The plastic strain of the solder joint can be obtained through theoretical calculations, numerical simulations, and experiments. Equation (10.18) shows the Engelmaier fatigue model, where the number of cycles to failure is determined by the total strain and the modified fatigue ductility index c.

$$N_f = \frac{1}{2}\left(\frac{\Delta\gamma}{2_{\varepsilon_f}}\right)^{\frac{1}{c}} \tag{10.18}$$

In the equation, N_f is the number of cycles to failure of the solder joint, and $\Delta\gamma$ is the total plastic shear strain amplitude.

Creep is generally considered to be the result of grain boundary slip or dislocation glide in the matrix. The Kench-Fox model applies the theory of matrix dislocation creep to life prediction of solder joints, as shown in Eq. (10.19).

$$N_f = \frac{C}{\Delta \gamma_{\mathrm{mc}}} \tag{10.19}$$

In the equation, N_f is the number of cycles to failure of the solder joint, $\Delta \gamma_{\mathrm{mc}}$ is the creep strain amplitude, and C s a constant, which is related to the microstructure of the solder.

Based on fracture mechanics principle, the fracture process of solder joints can be divided into crack initiation and crack propagation stages. Life prediction models based on fracture parameters can effectively illustrate the initiation and propagation of cracks. Equation (10.20) shows the stress intensity factor model based on fracture parameters.

$$\Delta K_{\mathrm{eff}} = \sqrt{\Delta K_I^2 + \Delta K_{\mathrm{II}}^2} \tag{10.20}$$

In the equation, ΔK_{eff} is the effective stress intensity factor amplitude of the solder joint crack, and ΔK_I and ΔK_{II} are the stress intensity factors for mode I and mode II cracks, respectively.

Life prediction models based on energy consider the hysteresis effect. To perform the analysis using this model, the cycle of crack occurrence must first be determined, followed by calculating the crack propagation rate based on fracture mechanics theory to estimate the number of cycles to failure. The sum of the cycle of crack occurrence and the estimated number of cycles to failure gives the total cycle life.

10.4.2 Mechanical Stress and Board-Level Assembly Reliability

1. Reliability Under Random Vibration Stress

(1) Modal analysis

Random vibration refers to vibration that cannot be represented by a deterministic function and can only be characterized by probabilistic or statistical methods. Reliability analysis of board-level assembly under random vibration stress first requires an analysis of the vibration characteristics of the structure under vibration stress, that is, modal analysis. The modes of a structure are determined by the form of the structure and the characteristics of the material itself and are independent of external loading conditions. Modal analysis methods include experimental modal analysis and theoretical modal analysis. Theoretical modal analysis is based on linear vibration theory and is used to analyze the relationship between inputs, responses, and systems, establishing a theoretical model for the system. This theoretical modal analysis method has been incorporated into some finite element simulation software, and modal parameters can be obtained through modeling, adding constraints and loads, and post-processing. Figure 10.38 shows and an example of the test results of the first

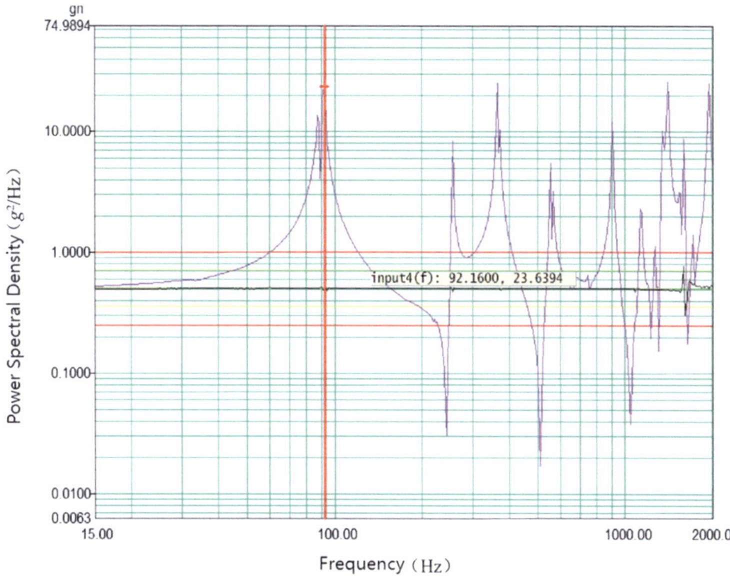

Fig. 10.38 Example of the first order natural frequency test result

order natural frequency (approximately 92 Hz) of the printed circuit board assembly (PCBA) obtained using a frequency sweep test.

(2) Random vibration testing

In random vibration testing, the PCBA component can be fixed to a vibration table using fixtures, and the acceleration sensor can be glued on the vibration table and the PCB test board. The random vibration excitation conditions are shown in Fig. 10.39, where the maximum power spectral density (PSD) extracted from the vibration microenvironment of an atmospheric data computer system is used as the excitation. The frequency range is 15–2000 Hz, and the root mean square acceleration is 8.299 Grms. PSD represents the variance of the excitation or response with frequency, and the area under the PSD curve is equal to the variance of the excitation or response.

The random vibration test process generally uses circuit disconnection or a 20% increase in resistance as the criterion to confirm sample failure [41]. Figure 10.40 shows the micrograph of the failure of the board-level assembly solder joint after random vibration testing.

Under random vibration conditions, the board-level assembly solder joints mainly exhibit three failure modes: ① fatigue cracking of the solder joint near the PCB pad; ② fatigue of the solder joint near the BGA device interface; ③ device substrate pad is stripped from inside the laminate. During random vibration, the bending deformation of the PCBA component causes the solder joints to withstand alternating tensile and compressive stresses, leading to solder joint failure. When the PCB bends towards the component side, the BGA solder joints are subjected to compressive stress from

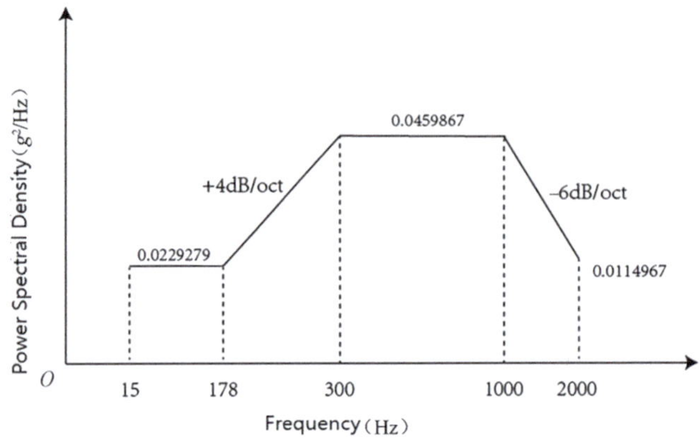

Fig. 10.39 Random vibration excitation condition

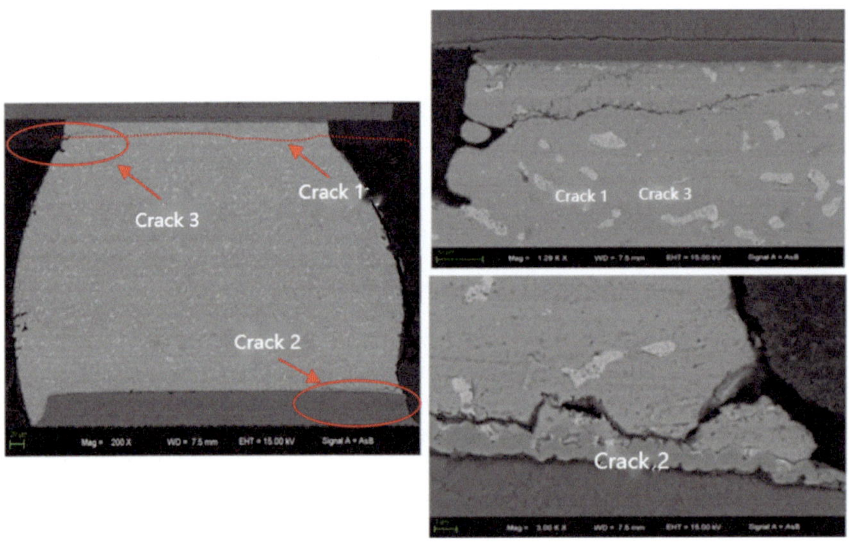

Fig. 10.40 Micrograph of the failure of the board-level assembly solder joint after random vibration testing

both the substrate and the PCB, causing the cracks to close. Conversely, when the cracks are subjected to tensile stress, they open up. As the solder joint structure continuously deteriorates, cracks propagate, and internal damage accumulates, the solder joint eventually fails due to disconnection.

(3) Physics models of failure under random vibration

① Manson high-cycle fatigue model

In the Manson-Coffin equation for high-cycle fatigue life, the total strain-life curve is obtained by superimposing the elastic strain and plastic strain life curves of the structure, as shown in the following equation.

$$\frac{\Delta\varepsilon_t}{2} = \frac{\Delta\varepsilon_{el}}{2} + \frac{\Delta\varepsilon_{pl}}{2} = \frac{\sigma_f'}{E}\left(2N_f\right)^b + \varepsilon_f'\left(2N_f\right)^c \tag{10.21}$$

In the equation, $\Delta\varepsilon_t$ is the total strain range, $\Delta\varepsilon_{\mathrm{el}}$ is the elastic strain range, $\Delta\varepsilon_{\mathrm{pl}}$ is the plastic strain range, $\sigma_f\prime$ is the fatigue strength coefficient, E is the material's elastic modulus, b is the fatigue strength index, $\varepsilon_f\prime$ is the fatigue plasticity coefficient c is the fatigue plasticity index, and N_f is the number of fatigue cycles.

② Miner linear fatigue damage accumulation model

One of the commonly used fatigue damage theories in engineering applications is the linear fatigue damage accumulation theory. This theory states that under cyclic loading conditions, fatigue damage caused by different stresses is independent and can be linearly superimposed. When the cumulative damage reaches a critical threshold, the sample is considered to have failed [17]. Miner's theory is a typical linear fatigue damage accumulation theory. Assuming that the sample is subjected to m constant amplitude alternating stresses S_1, S_2,..., S_m, the total fatigue damage of the sample can be expressed as shown in Eq. (10.22).

$$D = \sum_{i=1}^{m} \frac{n_i}{N_i} \tag{10.22}$$

In equation, n_i is the actual number of cycles the sample experiences under stress S_i, N_i is the number of cycles to failure of the sample under stress S_i, D is the total fatigue damage on the sample, and when the total fatigue damage D equals 1, the sample is considered to have failed. All damages are linearly superimposed to form the damage accumulation in Miner's theory, and this calculation method is simple and convenient, so it is often used in practical engineering.

③ Steinberg high-cycle fatigue model

The aviation system often uses the Steinberg model to predict the life of board-level assembly solder joints. It is believed that the cyclic life of solder joints in board-level assembly under high-cycle fatigue conditions can be expressed by the Basquin equation.

$$\sigma N^b = C \tag{10.23}$$

In the equation, N is the cycle life, σ is the stress amplitude, b is the fatigue strength index, and C is a constant coefficient. This type of failure can be described as a straight line on the double logarithmic coordinate axis with stress and failure cycle number as coordinates, where the stress of the solder joint is related to the displacement of the PCB. Based on this assumption, the high-cycle fatigue relationship is

$$Z_1(N_1)^b = Z_2(N_2)^b \tag{10.24}$$

In the equation, Z_1 and Z_2 are are displacements, and N_1 and N_2 are lifetimes. By converting Eq. (10.24), Eq. (10.25) is obtained.

$$N_2 = N_1(\frac{Z_1}{Z_2})^{\frac{1}{b}} \tag{10.25}$$

In the Steinberg equation, N_1 and Z_1 represent the lifetime and displacement, respectively, and the lifetime N_2 can be calculated through them.

2. Reliability Under Drop Shock

The reliability under drop shock is mainly divided into product-level, board-level, and joint-level. Among them, board-level drop shock targets the interconnection between integrated circuits and PCB assemblies, focusing on reliability issues within secondary packaging structures. Board-level drop shock reliability analysis is closer to the actual drop shock environment and has lower test costs, so it receives more attention. Figure 10.41 shows an example of the bending phenomenon of a PCB during drop shock. Since integrated circuits and PCBs have different elastic moduli, they will produce different degrees of bending during drop shock, thereby causing the board-level assembly solder joints to withstand different types of stresses. These stresses mainly include tensile and compressive stresses caused by different longitudinal displacements, shear stresses caused by different lateral displacements, and torsional stresses caused by the combined action of tensile and compressive stresses and shear stresses. Among them, tensile and compressive stresses are the main stresses causing solder joint failure. The repeated bending up and down during drop shock is the key reason for the crack failure of solder joints.

The main failure modes of board-level assembly solder joints under drop shock include IMC fracture, mixed fracture of matrix and IMC [42], and Cu pad spalling, etc. Since tensile and compressive stresses are mainly concentrated at the corners of the peripheral solder joints, and Cu_6Sn_5 in IMC is a hard-brittle structure, IMC solder joint failure is the main type under drop shock conditions. The failure process of board-level assembly solder joints under drop shock is difficult to observe directly. The resistance testing method can be used to monitor the failure process of the PCB. Figure 10.42 shows the resistance changes during the drop shock failure process of solder joints.

During drop shock, the failure process of board-level assembly solder joints can be divided into three stages: the first stage is the initiation of solder joint cracks, with

Fig. 10.41 Example of the bending phenomenon of a PCB during drop shock

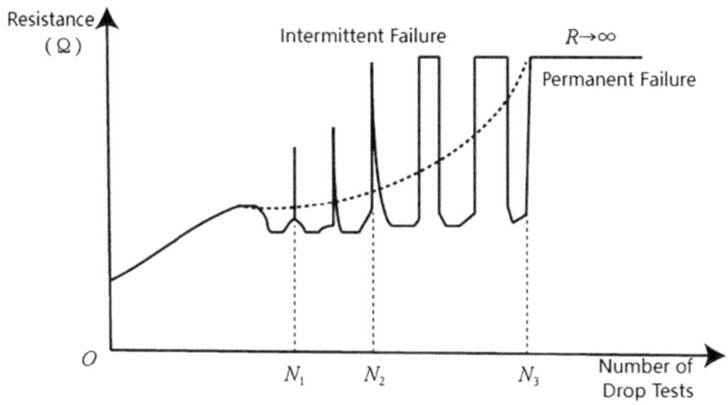

Fig. 10.42 Resistance changes during the drop shock failure process of solder joints

a slow increase in the resistance of the interconnection structure; the second stage is the continuous propagation of cracks in the solder joints, with a further increase in solder joint resistance, entering the rapid failure stage; the third stage is the complete disconnection of the interconnection structure, with the resistance jumps to infinity. The drop shock failure process is similar to the metal material fatigue failure process, so some related research has analyzed the failure process of board-level assembly solder joints from a fatigue perspective.

The drop shock acceleration and the average drop shock life of the product are believed to follow an exponential relationship, as shown in Eq. (10.26).

$$N = C_1 A^{C_2} \tag{10.26}$$

In the equation, N is the average drop shock life, A is the peak acceleration of the drop shock, and C_1 and C_2 are related parameters.

10.4.3 Current Stress and Board-Level Assembly Reliability

With the miniaturization of board-level assembly solder joints, the current density carried by the solder joints increases rapidly. Current stress has become an important factor affecting the reliability of board-level assembly solder joints. When the current density exceeds 10^3 A/cm^2, the electromigration effect will significantly impact the growth and evolution of IMCs in the solder joints. In addition, process defects in the solder joints can cause local current crowding in the solder joints, generating significant Joule heating effects, thereby reducing the reliability of board-level assembly solder joint interconnections. Under current stress, the evolution laws of IMCs in solder joints are related to the grain orientation, grain boundary type, and other factors in the solder.

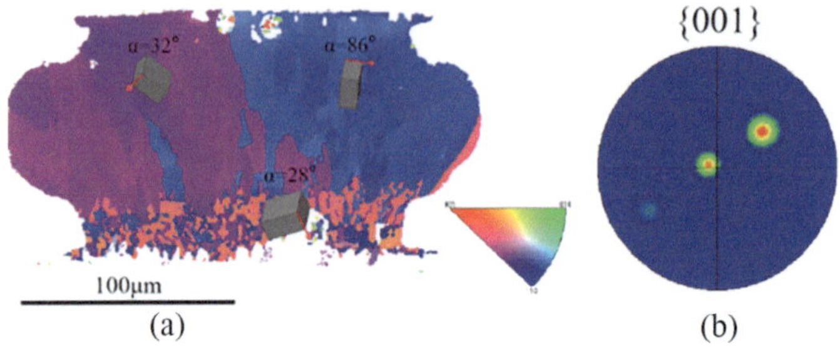

Fig. 10.43 Electron backscatter diffraction image of a BGA solder joint

1. Influence of Grain Orientation on IMC Migration

Figure 10.43 shows the electron backscatter diffraction image of a BGA solder joint. The solder joint structure contains three grains with α angles of 32°, 28°, and 86°, respectively. Figure 10.43b shows the pole figure of the BGA solder joint on the {001} plane. The pole figure shows three clusters in the plane, representing the three grains.

Figure 10.44a–c are the SEM images of the solder joint before reflow soldering and after electromigration testing, respectively. Figure 10.44a shows that the initial state of the BGA solder joint forms a good metallurgical bond between the solder joint and the pad, and a thin layer of Cu_6Sn_5 compound is formed at the interface. The lower interface has a thicker IMC layer due to the double reflow soldering process. Figure 10.44b and c show the SEM images of the solder joint after 100–200 h of service under a current density of 1.5×10^4 A/cm^2 at 25 °C, respectively. In Fig. 10.44b, after 100 h of current stress, a small amount of Cu_6Sn_5 precipitates on the left side of the solder joint, while no precipitation is observed on the right side. In Fig. 10.44c, after 200 h of current stress, the precipitation of Cu_6Sn_5 on the left side of the solder joint increases significantly, while there is still no obvious change on the right side. The precipitation of Cu_6Sn_5 on the surface of the solder joint is mainly due to the migration of Cu atoms from the cathode to the anode under the action of electron wind. When Cu atoms reach the maximum solubility in the solder matrix during migration, they react with Sn to form Cu_6Sn_5. The precipitation of Cu_6Sn_5 in the solder joint shown in Fig. 10.44 exhibits obvious clustering. Combining the analysis of Fig. 10.43, it can be seen that Cu_6Sn_5 only forms in the grain with an α angle of 32°, indicating that when the angle between the grain orientation and the current direction is small, it is conducive to the migration of Cu atoms; conversely, when the angle is large, the migration of Cu atoms is more difficult [43].

2. Influence of Grain Boundaries on IMC Migration

The grain boundaries within the solder have a significant impact on the electromigration of the solder joint. Figure 10.45 shows the micrograph of a BGA solder joint after 200 h of current stress at 25 °C and 1.5 × 10^4 A/cm^2. After 200 h of current stress, Cu_6Sn_5 in the solder joint has migrated significantly. Combining the analysis of grain orientation and grain boundary images, Cu_6Sn_5 mainly migrates along the grain boundaries. Detection and analysis have found that the solder joint structure contains a large-angle grain with an α angle between 54° and 67°, which is not conducive to the diffusion of Cu atoms within the grain. Therefore, small-angle grains play an important role in the migration of IMCs.

Under the same test conditions, no obvious electromigration was found in the solder joint structure shown in Fig. 10.46. The electron backscatter diffraction map in Fig. 10.46a shows that the solder joint mainly consists of three large-angle grains. The α angles of these grains are relatively large, at 62°, 63.5°, and 86°, which effectively suppress the migration of Cu atoms. The grain boundaries within the solder joint are primarily large-angle grain boundaries, with an orientation difference of approximately 55° to 65°, making it difficult for IMCs to diffuse along these boundaries. The large-angle grain boundaries in the solder joint shown in Fig. 10.46 are

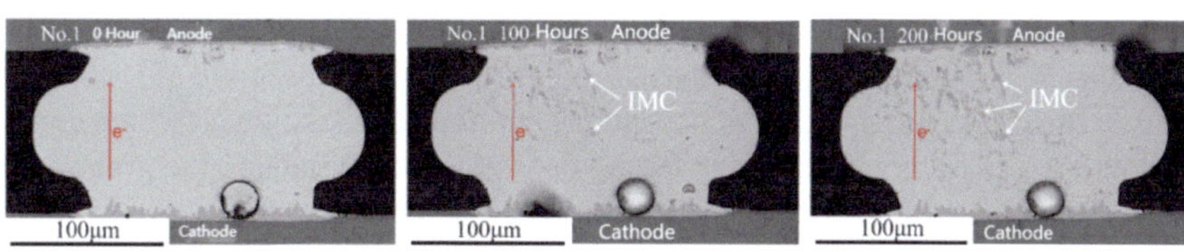

(a) Initial state (b) 100 hours of current stress (c) 200 hours of current stress

Fig. 10.44 SEM images of BGA solder joint

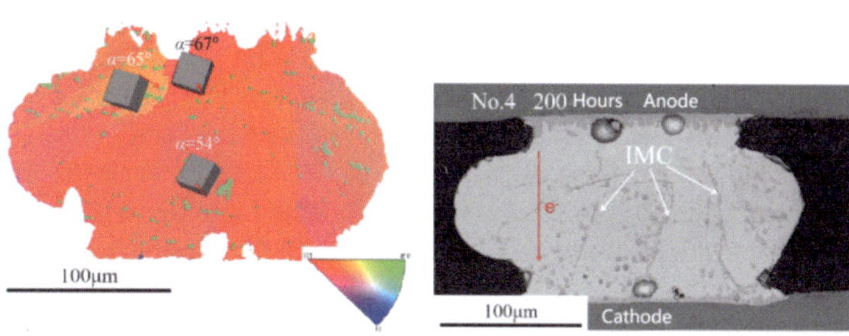

(a) Electron backscatter diffraction image (b) SEM image under 200 hours of current stress

Fig. 10.45 Micrograph 1 of BGA solder joints

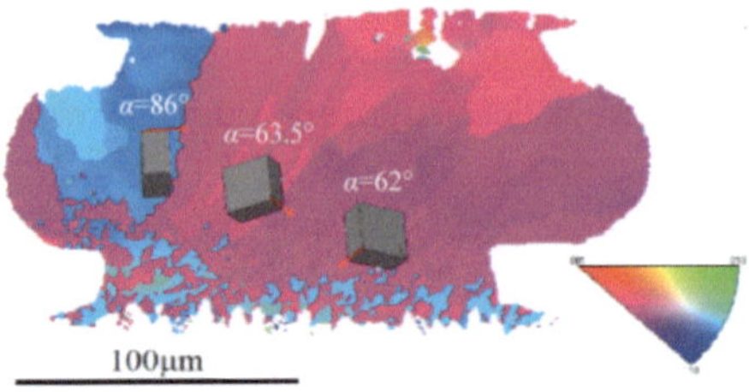

(a) Electron backscatter diffraction image

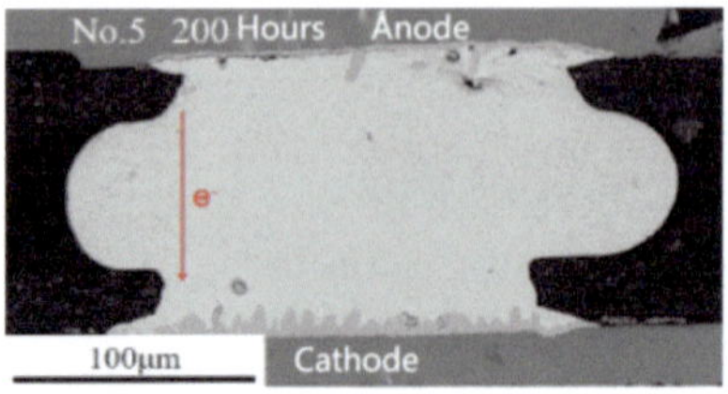

(b)SEM image under 200 hours of current stress

Fig. 10.46 Micrograph 2 of BGA solder joints

mainly cyclic twin boundaries. These boundaries are coherent and have fewer dislocations and defects. In contrast, ordinary grain boundaries contain a large number of dislocations and defects, which facilitate atomic diffusion. According to relevant studies, the diffusion rate of atoms along grain boundaries is three orders of magnitude higher than that within the grain body [44].

3. Grain Rotation Induced by Electromigration

The mechanism of grain rotation in solder is related to grain orientation. Due to the anisotropy of Sn, the diffusion flux of vacancies also exhibits anisotropy [45]. The difference in orientation of two adjacent grains leads to the difference in vacancy diffusion flux, forming a vacancy concentration difference in the grain, which corresponds to the stress gradient, under the action of the stress gradient, the grains will rotate. Figure 10.47 shows the principle diagram of grain rotation.

Assuming that the *a*-axis (length) of grain I is parallel to the electron flow direction, and the *c*-axis (height) of grain II is parallel to the electron flow direction, the electron flow direction is from right to left, and the vacancy flow direction is the same as the electron flow direction, so the vacancy flow direction is from right to left, as shown in Fig. 10.47a. Since the vacancy diffusion flux in grain I is larger, a large number of vacancies diffuse from grain I to the interface between grains I and II (interface I). Grain II has its *c*-axis parallel to the electron flow direction, so only a small part of the vacancies diffuse from interface I to the interface between grains II and III (interface II), resulting in a large accumulation of vacancies at interface I. However, since the *a*-axis of grain III is parallel to the electron flow direction, a large number of vacancies diffuse away from interface II through grain III, causing a sharp decrease in vacancies at interface II. Finally, as shown in Fig. 10.47c, there is a significant vacancy concentration gradient on both sides of grain II, forming a stress gradient that causes grain II to rotate, with the side near interface I sinking and the side near interface II rising.

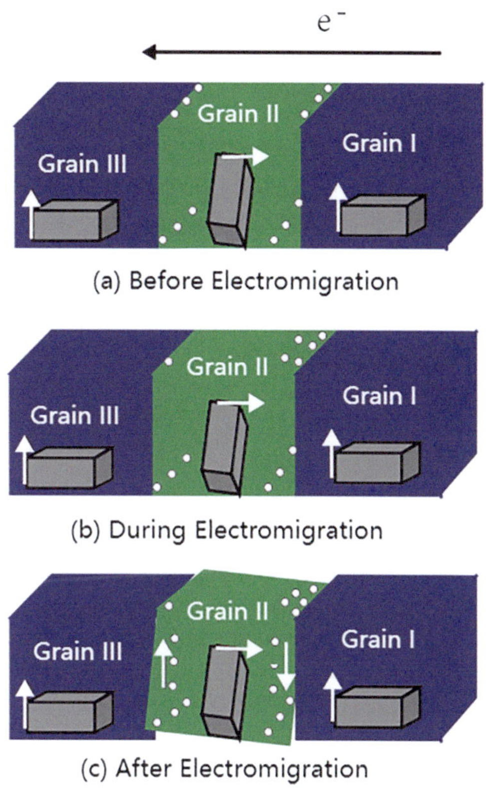

Fig. 10.47 Principle diagram of grain rotation [46]

10.4.4 Coupled Stresses and Board-Level Assembly Reliability

1. Electromigration Reliability Under Ultra-low Temperature Stress

(1) Brittle-to-ductile transition phenomenon at low temperature

When the ambient temperature is below 13 °C, pure Sn solder undergoes allotropy transformation, and the body-centered tetragonal structure of white β-Sn will transform into the cubic crystal structure of gray α-Sn, resulting in a brittle-to-ductile transition phenomenon. The allotropy transformation includes two stages: nucleation and growth. The incubation cycle of nucleation can last from several months to several years. The volume increase of Sn during allotropy transformation (about 27%) can cause local fractures in solder joints. Adding elements such as Pb, Sb, and Bi to pure Sn solder can lower the brittle-to-ductile transition temperature of the solder. For example, the brittle-to-ductile transition temperature of Sn37Pb solder is around −100 °C, and when the mass fraction of Pb exceeds 90%, it is difficult to observe the low-temperature phase transition phenomenon of Sn-based solder at temperatures above − 200 °C.

Compared with traditional Sn37Pb, the SAC series of lead-free solders is more prone to brittle-to-ductile transition at low temperatures. For example, the brittle-to-ductile transition temperature of Sn0.5Cu and Sn0.5Cu(Ni) alloys is around −125 °C, while the transition temperature of Sn-Ag and SAC alloys is between −78 and −45 °C, and increases with increasing Ag content. The brittle-to-ductile transition temperature of Sn1.0Ag0.1Cu alloy can even rise to −10 °C. The influence of different elements on Sn phase transformation is quite different. For example, soluble substances such as Bi, even at a content of only 0.0035 wt%, can effectively inhibit the β-phase to α-phase transition; non-soluble substances, such as Zn, Al, Mg, and Mn, can significantly lower the β-phase to α-phase transition temperature. Elements such as Cu, Fe, and Ni have little effect on inhibiting the β-phase to α-phase transition.

Low-temperature environments have a significant impact on the mechanical properties of solder joints, such as the increase in tensile strength of solder with decreasing temperature [47]. However, after undergoing a brittle-to-ductile transition at low temperatures, Sn-based solder joints are prone to cracking. The tensile strength of solder alloy increases with decreasing temperature in a parabolic trend. The tensile strength first increases with decreasing temperature, reaches a maximum around −150 °C, and then decreases rapidly. Figure 10.48 shows the effect of temperature on solder joint shear strength. As the temperature decreases, the shear strength of solder joints also increases. For example, at −150 °C, when the Pb content is 22.46 wt%, the shear strength of the solder joint is about 103 MPa, which is about 85% higher than that at room temperature. Ultra-low temperature environments increase the shear strength of solder joints. It is believed that ultra-low temperature stress weakens the thermal activation process of dislocation motion in the material, increases the resistance of dislocations in crystal motion, and macroscopically manifests as an increase in material shear strength.

(2) Solder joint electromigration under ultra-low temperature conditions

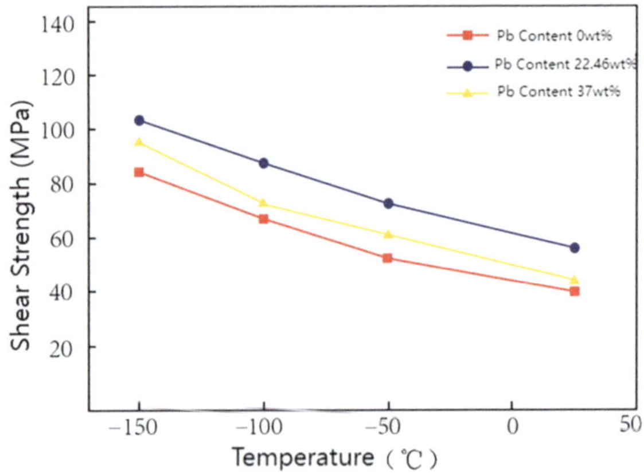

Fig. 10.48 Effect of temperature on solder joint shear strength

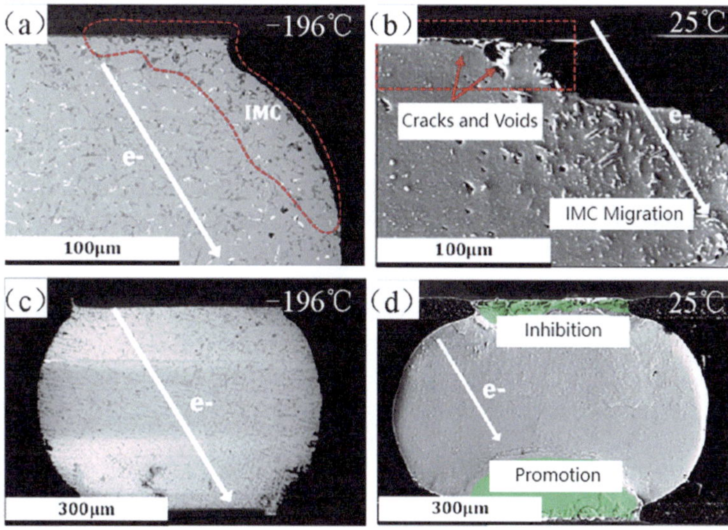

Fig. 10.49 SEM image of sac305 solder joint after electromigration testing

Under ultra-low temperature environment, the diffusion rate of atoms in solder joints decreases sharply, and the influence of Joule heating effect is weakened, significantly extending the electromigration life of solder joints. Under room temperature conditions, current stress causes atoms at the cathode of the solder joint to migrate to the anode, forming voids and hillocks. Under ultra-low temperature conditions, the combined effect of high-density electron flow impact and Joule heating is greatly reduced, and the microstructural characteristics of electromigration degradation in solder joints are different. The electromigration effect is significantly inhibited, and the migration of IMCs is weakened. Figure 10.49 shows the SEM image of SAC305 solder joint after electromigration testing.

Under 25 °C conditions, the IMC at the cathode of the solder joint dissolves and migrates to the anode, and the thickness of the IMC at the anode increases and forms a protrusion. Under −196 °C conditions, the IMC only migrates slightly in the electron flow direction and no cracks and holes are formed at the interface. Vacancies in the solder matrix continuously undergo thermal vibrations, and the higher the temperature, the larger the amplitude of the vibrations and the higher the probability of vacancy formation, resulting in a higher vacancy concentration in the solder matrix. However, under ultra-low temperature conditions, the amplitude of atomic vibrations decreases, and a greater driving force is required for atoms to migrate. Therefore, under the same current density conditions, the higher the temperature of the solder joint, the greater the migration flux of electromigration. In addition, changes in solder joint temperature can also change the main elements that migrate in the solder joint, thereby affecting the degradation laws of electromigration in solder joints.

2. Electromigration Reliability Under High-temperature Stress

(1) Failure of solder joints under thermoelectric coupled stress

Figure 10.50 shows the electromigration segregation morphology of mixed solder joints under conditions of 125 °C ambient temperature and 1.8×10^4 A/cm^2 current density. Pb in the mixed solder joint diffuses to the anode and mainly aggregates in the upper right corner of the solder joint where the current density is relatively low. As the electromigration time extends, the segregation becomes more severe.

High temperatures increase the atomic diffusion rate in solder joints, making the electromigration effect more pronounced and accelerating the reaction between solder and pad, resulting in severe corrosion of Cu pads while the Ni plating remains relatively intact. Figure 10.51 shows the corrosion of Cu pads at different electromigration times under conditions of 135 °C ambient temperature and 1.5×10^4 A/cm^2 current density. As the current stress test time increases, the corrosion of Cu pads by solder becomes more severe.

In the initial stage of electromigration, the Cu pad does not react with the solder due to the protection of the Ni plating, and the solder slowly corrodes the Ni coating. Once the solder breaks through the Ni plating, it will quickly react with the Cu pad. The failure modes of mixed SAC-SnPb solder joints under electromigration include cathodic layer voids, continuous or abnormal growth of local IMCs, and erosion of Cu wiring at electron injection ports. The failure process is often influenced by factors

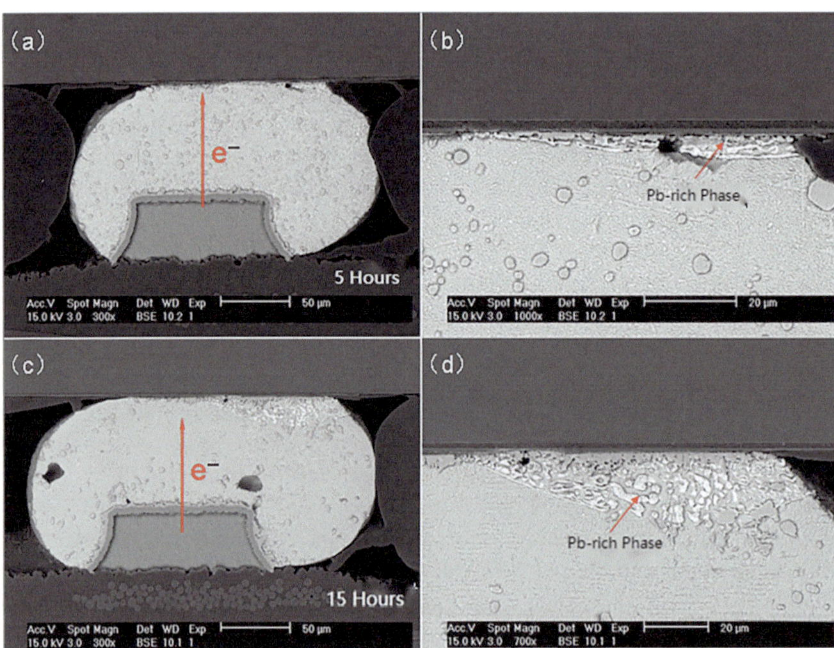

Fig. 10.50 Electromigration segregation morphology of mixed solder joints

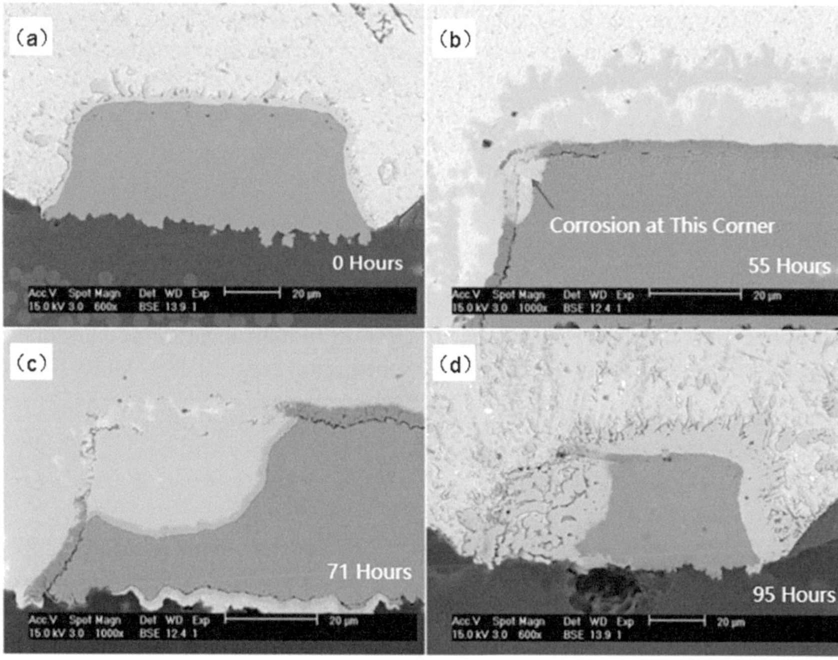

Fig. 10.51 Corrosion of Cu pads at different electromigration times

such as current density, ambient temperature, test time, and solder joint structure, but the essential mechanism is the directional migration of atoms.

(2) Reliability model under thermoelectric coupled stress

The electromigration effect in solder joints is a mass transport process under the combined action of thermal and electrical stresses. Its reliability model can be established by referring to the Black equation. Using solder joint temperature T and current density j as variables, the relationship between failure time and temperature and current density can be obtained to determine the electromigration reliability model of mixed solder joints, as shown in Eq. (10.27).

$$\mathrm{MTTF} = A\frac{1}{(j)^n}\exp\frac{Q}{kT} \tag{10.27}$$

In the equation, MTTF is the mean time to failure, j is the current density, k is the Boltzmann constant, T is the temperature, and Q is the electromigration activation energy. This equation indicates that the geometry of the solder joint, material, current density, ambient temperature, and other factors all affect the electromigration effect in solder joints. To establish an accurate electromigration reliability model for solder joints, these factors must be fully considered.

According to the Black equation, under the same temperature conditions, the logarithm of current density $\ln j$ and the logarithm of mean time to failure ln MTTF show a linear relationship, as shown in Eq. (10.28).

$$\ln MTTF = -n \cdot \ln j + B \tag{10.28}$$

$$B = -\ln A + \frac{Q}{kT} \tag{10.29}$$

In the equation, B is the constant. From Eq. (10.28), it can be seen that the slope of the linear fit of failure life data in the ln MTTF-ln j double logarithmic coordinates is the absolute value of the current index n. Substituting Eqs. (10.28) into(10.27), we get

$$\ln MTTF = \ln A - n \cdot \ln j + \frac{Q}{kT} \tag{10.30}$$

From Eq. (10.30), it can be seen that ln MTTF and $1/T$ show a linear relationship. The linear relationship between the absolute temperature inverse $1/T$ and the logarithm of mean time to failure ln MTTF can be fitted, and the slope of the fitted line is the electromigration activation energy of the solder joint.

In the Black equation, the variables j and T represent the average current density and ambient temperature, respectively. In actual solder joints, electromigration failures typically occur in high-current density areas, and the Joule heating effect often causes the working temperature of the solder joint to be much higher than the actual ambient temperature. Therefore, it is necessary to modify the established reliability model based on the Joule heating effect and current crowding effect to obtain the accurate relationship between mean time to failure and current density and temperature.

10.5 Reliability Evaluation of Board-Level Assembly

10.5.1 Reliability Evaluation Methods for Board-Level Assembly

Reliability testing of board-level assembly is mainly used to evaluate the reliability of solder joints in integrated circuit board-level assembly. It can effectively expose potential defects in the board-level assembly process, facilitating timely correction and the proposal of improvement measures. This section will mainly discuss the reliability evaluation test methods, standards, and failure criteria for solder joints in board-level assembly of integrated circuits, without involving specific device-level reliability test methods.

1. Reliability Test Standards for Board-level Assembly Solder Joints

The content of reliability testing for board-level assembly is determined based on the actual application service environment and existing failure modes. Transportation, storage, and service conditions are key factors that determine the service stress and failure modes of products. After clarifying the specific failure modes corresponding to the stress conditions, relevant reliability test projects can be arranged in a targeted manner. Table 10.3 shows the relationship between failure modes of board-level assembly solder joints, environmental stresses, and reliability tests.

There are relatively few international standards specifically for the reliability evaluation of board-level assembly solder joints. The main ones include IPC-9701 *Performance Test Methods and Qualification Requirements for Surface Mount Solder Attachments* and IPC-SM-785 *Guidelines for Accelerated Reliability Testing of Surface Mount Attachments*, both of which reference standards such as JESD22-A104 and IPC-TM-650.

Drop shock environmental adaptability is a common test item for portable mobile electronic products, with board-level assembly solder joints being the focus of attention in these tests. Specific reference standards include JESD22-B110 *Subassembly Mechanical Shock* and JESD22-B111 *Board Level Drop Test Method of Components for Handheld Electronic Products*. For electronic components that operate long-term in high-temperature environments, it is necessary to consider the growth of IMCs in

Table 10.3 Relationship between failure modes of board-level assembly solder joints, environmental stresses, and reliability tests

Failure mode	Stress category	Possible environmental stress (specified conditions)	Test project and method
Thermal fatigue fracture Creep fracture	Thermal stress	Environmental temperature changes Temperature changes due to switching on/off	Temperature cycling
		Temperature changes due to on−site transfer	Temperature shock
		Thermal stress during storage	High−temperature storage
Electrochemical migration Corrosion Degradation of insulation performance	Chemical Electrochemical stress	High−temperature and high−humidity working environment	Damp heat test powered test
			Damp heat test
			HAST
Static fracture Vibration fracture Creep fracture	Mechanical stress	Drop	Mechanical drop
		Mobile use	Random vibration
		Improper operation	Three−point bending

board-level assembly solder joints caused by thermal stress. High-temperature test evaluation can refer to the standard JESD22-A103 *High Temperature Storage Life*.

European and American countries are leading in the field of board-level assembly reliability evaluation technology, so the industry mainly relies on and refers to IPC, JEDEC, and other related standards to conduct qualification tests. There are very few authoritative domestic standards in China. A limited number of enterprises will refer to standards such as GB/T 2423 to conduct qualification tests based on the actual application conditions of their products. For high-reliability military products, the main references are GJB 150A—2009 *Military Equipment Laboratory Environmental Test Methods* and GJB 548B—2005 *Microelectronic Devices Test Methods and Procedures*. When referring to standards for testing, the test conditions should be determined in combination with the actual application environment and user requirements. If comparable test results are required, the referenced standards should remain consistent.

2. Failure Criteria and Analysis Methods

The reliability evaluation of board-level assembly solder joints requires clear failure criteria, and the definition of these criteria should fully consider the failure modes and mechanisms of the solder joints under test stress. The IPC-9701 standard defines the failure criterion for solder joints as an increase in resistance value exceeding 20% of the initial value, or the detection of no less than 10 events with a resistance value exceeding 1000 Ω within 1 μs. The IPC-A-610 standard considers solder joint cracks reaching 25% as unacceptable. For corrosion failures, when there is a visible discoloration corrosion phenomenon and the insulation resistance value decreases by more than 10% of the initial value, it is considered a failure.

Reliability tests for board-level assembly solder joints, such as thermal shock, temperature cycling, and random vibration, typically use failure criteria based on resistance values or crack conditions, in line with IPC standards. These tests usually result in fatigue and overload cracking failures, which are indicated by a sustained increase in solder joint resistance. For high-temperature and high-humidity tests, corrosion failure is often used as the criterion. Solder joint failures in these tests are characterized not only by color changes but also by a decrease in insulation performance between solder joints, which can be detected through visual inspection and insulation resistance testing. For electronic products with special applications and requirements, failure criteria need to be defined separately based on actual conditions.

Different failure criteria correspond to different detection and analysis techniques. For example, for the failure criterion of a 20% increase in resistance value, a resistance testing system is needed for monitoring. For the failure criterion of detecting no less than 10 events with a resistance value exceeding 1000 Ω within 1 μs, an event recorder can be used. For the failure criterion of cracks reaching 25%, a combination of X-ray detection, dyeing and penetration testing, metallographic sectioning, and SEM observation methods is required. For solder joint corrosion failures, the appearance morphology should be observed with an optical microscope, and the insulation performance should be tested with an insulation resistance tester. If necessary, the microstructure and composition changes of the solder joint should also be analyzed. The core function of board-level assembly solder joints is to provide electrical interconnection between integrated circuits and PCBs. Most solder joint degradations in reliability tests are manifested as crack propagation and increased resistance values. Therefore, the failure criteria defined in IPC-9701 and IPC-A-610 standards are highly recognized in the industry.

3. Reliability Test Methods for Board-level Assembly Solder Joints

Common reliability test methods for board-level assembly solder joints involve thermal, mechanical, and chemical aspects. This section will mainly introduce common evaluation methods such as temperature cycling/temperature shock testing, vibration testing, drop testing, and high-temperature testing. When conducting reliability testing for specific products, the test content and test evaluation of board-level assembly solder joints requires clear failure.

(1) Temperature cycling/ temperature shock testing

Temperature cycling/ temperature shock testing is mainly used to evaluate the thermal fatigue failure of board-level assembly. Key parameters for temperature cycling testing include the rate of temperature change, the range of temperature change, the number of test cycles, and the dwell time at high and low temperatures. Temperature shock testing has a higher rate of temperature change and requires a specified conversion time between high and low temperatures, but it does not require a controllable rate of temperature change. Temperature cycling testing represents environments with gradual temperature changes, with a temperature change rate generally less than or equal to 20 °C/min. Temperature shock testing represents environments with sudden temperature changes and is more suitable for evaluating extreme environmental applicability.

To evaluate the thermal fatigue failure of board-level assembly solder joints using temperature cycling testing, the IPC-9701 standard can be referred to. Table 10.4 lists the requirements and recommended conditions for temperature cycling testing. The preferred temperature range for the test is 0–100 °C. Under the premise of ensuring that the failure mechanism of the test is consistent with the actual application, the temperature range of the test can be appropriately expanded. For example, for PCBs with high T_g properties, the temperature range of the test can be expanded to -55–125 °C. The test is usually stopped when 50% or 63.2% of the samples fail, or the test time is determined based on the actual reliability of the solder joints. The failure life of the samples is recorded using a resistance monitoring system or an event recorder. The test samples are usually test structures that are consistent with the material, structure, and assembly process of the actual product.

(2) Vibration testing

Laboratory environments often use vibration testing to simulate and evaluate the reliability of board-level assembly solder joints under different vibration stress environments. Vibration stress is one of the key factors causing failure of board-level assembly solder joints. Products such as automotive, airborne, and satellite-borne

Table 10.4 The requirements and recommended conditions for temperature cycling testing

Items		Test conditions	Notes
Temperature cycle (TC)	TC1	0 ~ 100 °C	Prefferd
	TC2	-25 ~ 100 °C	–
	TC3	-40 ~ 126 °C	–
	TC4	-55 ~ 125 °C	–
	TC5	-55 ~ 100 °C	–
Test time Number of test cycles (NTC)	NTC-A	50% (best 63.2%) of solder joints fail or 200 cycles	According to failure time or number of cycles
	NTC-B	500 cycles	
	NTC-C	1000 cycles (preferred for TC2, TC3, TC4)	
	NTC-D	3000 cycles	
	NTC-E	6000 cycles (preferred for TC1)	
Low−temperature dwell time		10 min	Generally set based on one cycle per hour
Low−temperature tolerance		$+0/-10$ °C($+0/-5$ °C)	
High−temperature dwell time		10 min	
High−temperature tolerance		$+10/-0$ °C($+5/-0$ °C)	
Temperature change rate		$\leq$ 20 °C/min	
Number of samples		33	Consider return for repair, add 10 to the number of samples
Circuit board thickness		2.35 mm	Or actual assembly board thickness
Items		Test conditions	Notes
Packaging situation		Daisy chain structure or actual assembly structure	–
Monitoring method		Continuous resistance monitoring/ event monitoring	Solder joint failure time monitoring

electronic components are subject to different frequency ranges and levels of vibration stress due to the trajectory, speed, and environmental resistance of the product's movement. Long-term vibration stress can lead to the accumulation of fatigue damage and may directly cause solder joint over-stress cracking. When there are process or design defects in board-level assembly solder joints, vibration testing can quickly exacerbate defects and solder joint failures, thereby exposing reliability issues.

Vibration testing mainly includes random vibration testing, scanning vibration testing, sinusoidal vibration testing, and composite vibration testing, among which random vibration testing and sinusoidal vibration testing are more common. Sinusoidal vibration testing simulates the vibration caused by rotation, pulses, and oscillations in actual application scenarios and is divided into constant-frequency testing and sweep-frequency testing. The stress intensity of sinusoidal vibration testing depends on the frequency range, amplitude value, and test duration. Random vibration testing is mainly used to simulate the transportation environment of products to evaluate the vibration resistance of the structure. The stress intensity of random vibration testing depends on the frequency range, Grms (root mean square acceleration), test duration, and axis. The installation of test samples in vibration testing must consider the consistency with the actual application scenario. For example, board-level components should be fixed as much as possible according to the actual equipment installation situation to evaluate the true vibration resistance of the assembly solder joints.

Vibration testing can refer to international standards such as IEC 68-2-34 and IEC 68-2-35, and the commonly used domestic standard is GB/T 2423. These standards have clear regulations and requirements for the specific test procedures and conditions of vibration testing.

(3) Drop testing

Drop testing is mainly used to evaluate the anti-drop impact performance of test objects after free falling from a certain height. Board-level assembly solder joints are prone to cracking failures under drop impact, so portable electronic products generally require relevant anti-drop impact performance tests.

Drop testing has requirements for the material and hardness of the test bench surface, the direction of the drop, the height, and the number of drops. For small-volume, low-mass samples, drop testing should be performed on a hard wooden floor, while large-volume, high-mass samples should be tested on a hard surface such as reinforced concrete. The IEC standard recommends a drop height of 1–1.2 m for samples with a mass not exceeding 20 kg, and the drop direction should be performed with each of the six faces and four edges and corners facing down. After the test, the damage to the solder joints can be assessed through visual inspection or resistance testing.

Drop testing can refer to international standards such as IEC 68-2-32 Basic Environmental Testing Procedure and JESD22-B111 *Board Level Drop Test Method of Components for Handheld Electronic Products*, and the commonly used domestic standard is GB/T 2423.15—2008 *Environmental Testing for Electrical and Electronic Products Part 2: Test Methods Test Ga and Guidelines Steady-State Acceleration*. These standards have clear regulations and requirements for the specific test procedures, test conditions, and failure monitoring methods of drop testing.

(4) High-temperature testing

High-temperature testing is mainly used to evaluate the adaptability of electronic products to high-temperature stress. High-temperature stress can accelerate the growth of IMCs in solder joints. IMCs are high-resistance brittle structures that not only reduce the structural strength and electrical transmission performance of solder joint interconnections but also cause the formation of Kirkendall voids at the welding interface, thereby affecting the reliability of board-level assembly solder joints.

Standards such as JESD22-A103 *High Temperature Storage Life* and GB/T 2423 can provide references for high-temperature testing of board-level assembly solder joints. The key elements of high-temperature test conditions need to be reasonably determined, namely the test temperature and duration. If the test temperature is too low, the test duration will be greatly prolonged, and if the test temperature is too high, it may introduce new failure modes and mechanisms. For example, when the test temperature exceeds the T_g temperature of the PCB substrate, it will cause PCB deformation, thereby exerting greater mechanical stress on the solder joints and possibly introducing new failure mechanisms.

The duration of high-temperature testing for board-level assembly solder joints is generally long, and it is recommended to use online testing methods to continuously monitor the degradation of solder joints, so as to find the failure of solder joints in time. In addition, metallographic sectioning and SEM image analysis can be performed at fixed intervals to obtain the changes in the microstructure of the solder joints and the growth of IMCs.

10.5.2 Reliability Evaluation and Failure Cases of Board-Level Assembly

1. Optimization Design Case of Mixed Solder Joint Process

(1) Analysis of soldering quality before process optimization

A case where a company experienced solder joint cracking failures near the edges of components during BGA solder joint mixed assembly is analyzed. Through reasonable process optimization, the quality and reliability of board-level assembly processes can be improved. Figure 10.52 shows the metallographic micrograph of the failed BGA solder joint. It can be seen from Fig. 10.52 that Pb element diffuses evenly in the solder joint, but there is a noticeable Pb element enrichment on the side near the component. The solder joint cracking is concentrated between the solder and the IMC on the BGA component pad. There are voids inside the solder joint, all located on the side near the PCB pad. The inside of the voids in the solder joint is black, indicating the presence of non-volatile substances inside the voids.

The solder joint shown in Fig. 10.52 was analyzed by SEM to obtain the microstructure of the failed BGA solder joint, as shown in Fig. 10.53. The IMC between the PCB pad and the solder is uniform and continuous, with a thickness of about 1–2 μm. The thickness of the IMC between the solder and the component lead (or BGA component pad) is about 0.8–1.3 μm. There is a noticeable Pb element enrichment on the side of the BGA component pad, and a large amount of Pb elements were detected at the interface of the BGA component side in the cracked state.

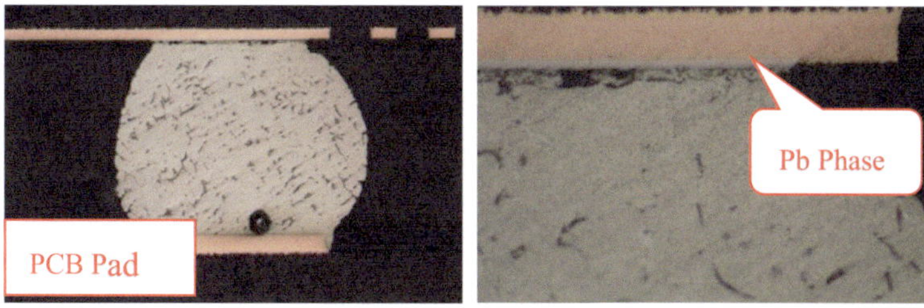

Fig. 10.52 Metallographic micrographic of failed BGA solder joints

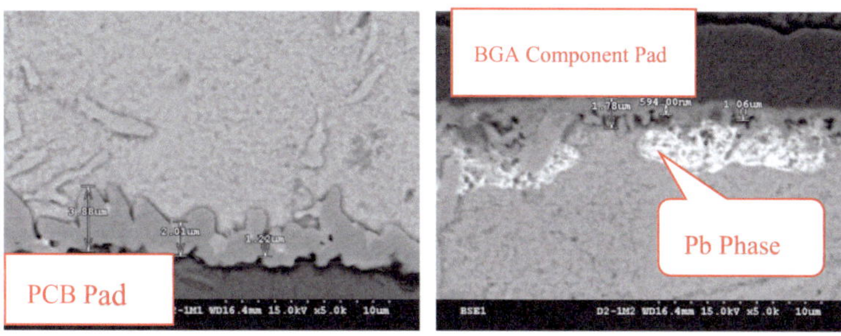

Fig. 10.53 Microstructure morphology of failed BGA solder join

Based on Figs. 10.52 and 10.53, it can be inferred that the solder and pad on the PCB have fully alloyed after soldering (Pb elements have diffused to the component side). During the cooling process, since the PCB side has many through-holes, its heat dissipation is faster than that of the component side. Therefore, the solder on the side near the PCB will solidify first. When there is obvious warpage deformation at the component edge, the temporarily unsolidified solder joint on the component side will be subjected to contraction stress and separate.

Figure 10.54 shows the soldering process curve. It can be seen from the parameter confirmation that the temperature curve of the soldering shown in the figure is the test temperature of the bare board, not the test temperature of the typical solder joint during the soldering process of the PCBA component. Therefore, there is a serious error in this soldering temperature test method.

(2) Specific measures for process optimization

In response to the soldering quality issues of PCBA components made with the original process, optimizations were made in key process material pre-treatment and control, as well as reflow soldering processes.

Key process material pre-treatment and control: In addition to the conventional baking treatment of PCB materials, BGA components were baked at 125 °C for 8 h to remove moisture.

First soldering process optimization: The reflow soldering temperature curve should measure the temperature of the BGA solder joint, not the temperature of

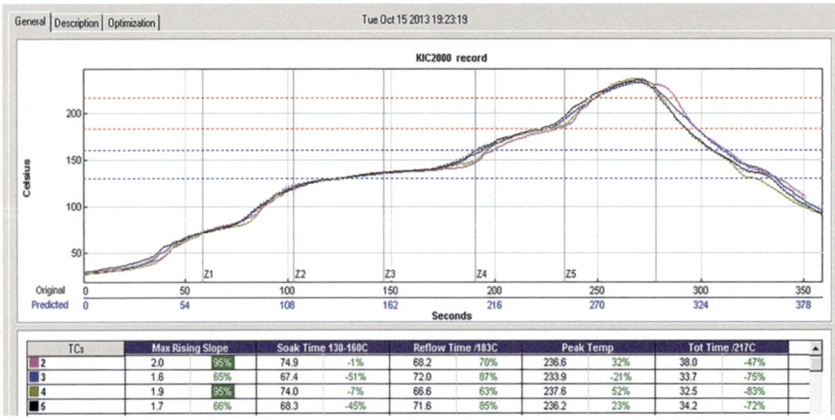

Fig. 10.54 Soldering process curve

the bare board. Therefore, a high-lead solder wire can be used to fix the thermocouple to calibrate the temperature of the BGA solder joint during soldering. Based on the existing equipment, the heating zone temperature parameters and chain speed were optimized, considering the heat requirements of the solder joint during reflow soldering.

Second soldering process optimization: Multiple PCBA components were made using different process parameters, and the optimized soldering quality was analyzed. It was found that after reflow soldering of BGA solder joints, the soldering effect was good. However, after wave soldering, the solder joint will have cracking defects and Pb element enrichment was observed at the cracking locations. By monitoring the temperature of the solder joints during wave soldering, it was found that the temperature reached as high as 177 °C. This temperature can cause partial remelting of areas with a quaternary eutectic structure of (Sn) + (Pb) + Ag_3Sn + Cu_6Sn_5 (with a melting point as low as 176 °C), and combined with the warpage deformation of the BGA, it leads to solder joint cracking. The wave soldering process curve was optimized using the same method as the first soldering process optimization.

(3) Soldering quality analysis after process optimization.

Based on the first and second soldering process optimizations, the final optimized PCBA soldering quality analysis and process parameter confirmation were carried out. Table 10.5 shows the soldering quality statistics after process optimization.

The data in Table 10.5 shows that, except for the slight Pb element enrichment at the solder joint interface on the component side of 4# samples, the BGA solder joints of 1#, 2#, and 3# samples have good solder joint collapse, uniform Pb element diffusion in the solder joints. The solder joints after soldering have some small voids, but the void rate is within the standard limits. Figure 10.55 shows the metallographic results of BGA solder joints after soldering process optimization.

The comparison of metallographic images before and after process optimization shows that after process optimization, the solder joints have good solder joint

Table 10.5 Statistical results of soldering quality after process optimization

PCBA no	Quality evaluation
1#	Good solder joint collapse, uniform Pb element diffusion, small voids present in solder joints but within standard limits
2#	Good solder joint collapse, uniform Pb element diffusion, no noticeable Pb element enrichment at component side solder joint interface
3#	Good solder joint collapse, uniform Pb element diffusion, small voids present in solder joints but within standard limits
4#	Good solder joint collapse, uniform Pb element diffusion, but slight Pb element enrichment at component side solder joint interface for a few solder joints

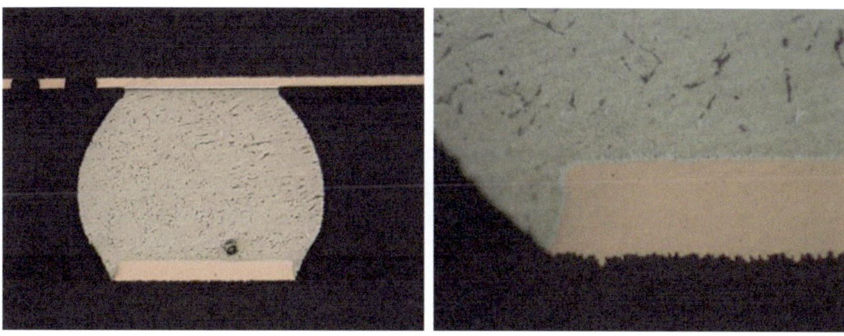

Fig. 10.55 Metallographic results of BGA solder joints after soldering process optimizaiton

collapse, good bonding with the component and PCB pad, uniform Pb element diffusion in the solder joints, and the cracking phenomena, void defects, and Pb element enrichment in the solder joints have all been improved and alleviated.

2. Case of Predicting Board-level Assembly Solder Joint Thermal Fatigue Life

Finite element simulation was used to obtain the stress–strain response of board-level assembly solder joints. A board-level test component based on daisy chain monitoring circuit was designed, and the actual life of mixed solder joints was obtained through thermal fatigue accelerated life testing. The fatigue toughness coefficient for thermal fatigue failure of mixed solder joints was calculated based on the Engelmaier equation and the life test results of mixed solder joints, and a thermal fatigue life prediction model for mixed solder joints was established.

(1) Sample design and simulation

The daisy chain test structure sample was designed based on the material, size, and other parameters of the actual device assembly solder joints, as shown in the circuit layout and physical image of the daisy chain sample in Fig. 10.56. The diameter of the BGA solder joint is approximately 0.5 mm, the height is 0.35 mm, and the pad diameter is 0.35 mm.

Fig. 10.56 Circuit layout and physical image of the daisy chain sample

A finite element simulation model of BGA solder joints was established, which mainly includes structures such as the silicon chip, PCB, plastic encapsulation layer, solder mask layer, substrate, and pads, as shown in Fig. 10.57. Based on the symmetry of the BGA solder joint arrangement and the PCB, a 3D solid model of the middle position BGA solder joint and its corresponding PCB was established without affecting the calculation accuracy.

(2) Stress loading of temperature cycling testing

An event monitor (STD128) was used to continuously monitor the state of the solder joints in the daisy chain circuit during temperature cycling testing. According to the IPC-SM-785 standard, the failure criterion for solder joints was defined as the event monitor continuously detecting 10 or more instances of instantaneous resistance values not less than 300Ω. Thirty groups of daisy chain samples were selected for the test, and a total of 3000 temperature cycles were performed.

The temperature cycling range was set to −55~125 °C for both simulation and testing. The time to rise from low temperature to high temperature was 18 min, the high-temperature dwell time was 12 min, the time to drop from high temperature to low temperature was 18 min, and the low-temperature dwell time was 12 min. The total time for one cycle was 60 min. The temperature cycling load is shown in Fig. 10.58.

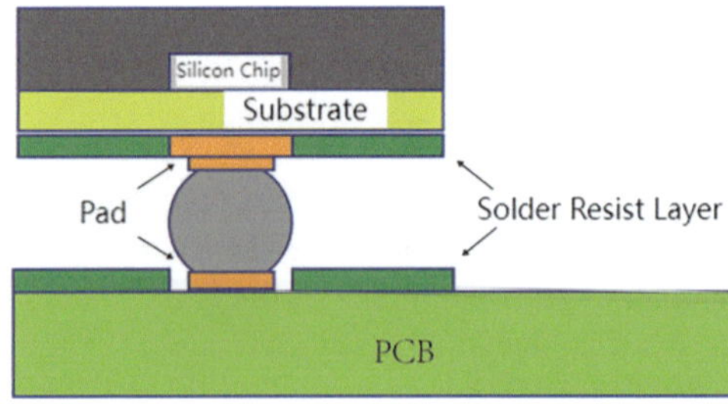

Fig. 10.57 Finite element simulation model of BGA solder joints

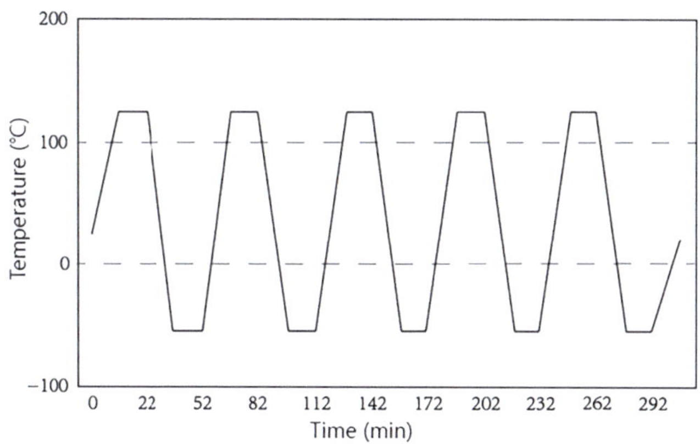

Fig. 10.58 Temperature cycling load

(3) Establishment of thermal fatigue life prediction model

The stress and strain distribution in the solder joints obtained from temperature cycling simulation is shown in Fig. 10.59. The maximum stress–strain point in the solder joint is located at the part connected to the chip-side pad of the outer diagonal solder joint. The simulation results of temperature cycling show that the maximum equivalent plastic strain point appears at the interface between the solder joint and the pad.

The variation of the maximum equivalent stress and strain over time in the solder joint was extracted from the simulation results. Figure 10.60 shows the equivalent strain-equivalent stress curve, from which the cycle plastic strain range of the critical solder joint was calculated to be 0.0203.

The Engelmaier equation and the life test results of mixed solder joints were combined to calculate the fatigue toughness coefficient for thermal fatigue failure of mixed solder joints, and a thermal fatigue life prediction model for mixed solder

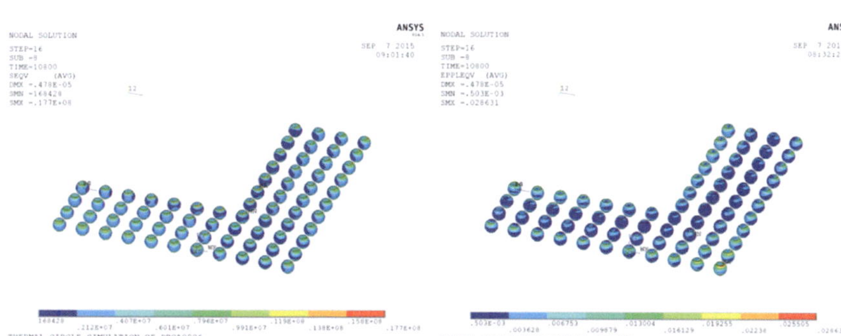

(a) Stress distribution map (b) Strain distribution map

Fig. 10.59 Stress and strain distribution

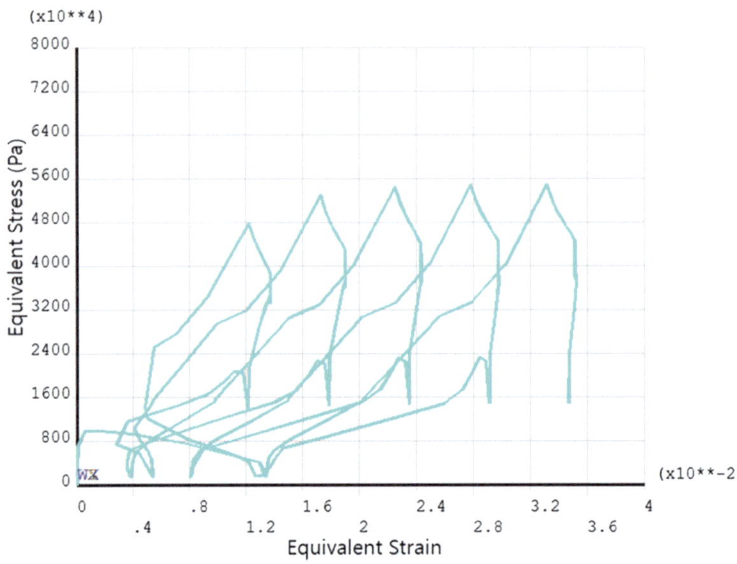

Fig. 10.60 Equivalent strain-equivalent stress curve

joints was established. After 3000 temperature cycles, 29 groups of mixed solder joints successively failed with open circuit. The two-parameter Weibull distribution curve of the life of mixed solder joints is shown in Fig. 10.61. The shape parameter β was calculated to be 3.614, the characteristic life was 2440 cycles, and the average life was 2204 cycles.

Figure 10.62 shows the cracking failure of mixed solder joints under temperature cycling stress. The cracks mainly extend along the interface between the ductile Pb-rich region and the brittle solder region. Under temperature cycling stress, Pb particles in the solder joint gradually accumulate, forming larger white Pb-rich regions. The boundary between the ductile Pb phase and the brittle solder matrix is prone to grain boundary slip under temperature cycling conditions, which can lead to solder joint cracking.

In combination with IPC-SM-785 standard, the solder joint repair life failure model is derived as follows:

$$N_f(x\%) = \frac{1}{2}\left(\frac{\Delta\gamma_p}{2\varepsilon_f}\right)^{\frac{1}{c}}\left[\frac{\ln(1-0.01x)}{\ln 0.5}\right]^{\frac{1}{\beta}} \tag{10.31}$$

In the equation, $\Delta\gamma_p$ is the cyclic plastic strain range, obtained from temperature cycling simulations, and is taken as $\Delta\gamma_p = 0.0203$; the shape parameter β is 3.614; ε_f is the thermal fatigue toughness coefficient to be determined; $N_f(x\%)$ is the number of cycles corresponding to a failure probability of $x\%$; c is the fatigue toughness index, expressed as:

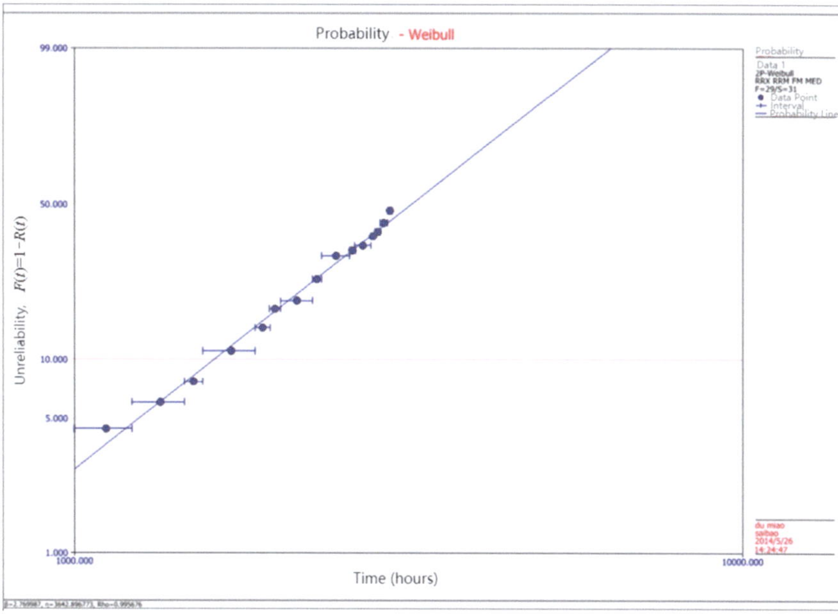

Fig. 10.61 Two-parameter Weibull distribution curve of the life of mixed solder joints

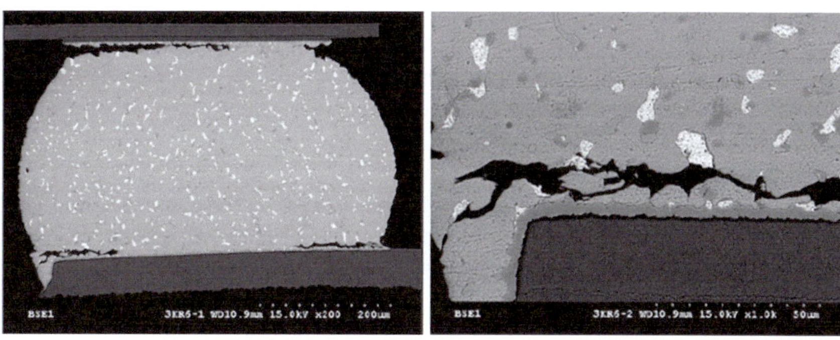

Fig. 10.62 Cracking failure of mixed solder joints under temperature cycling stress

$$c = -0.442 - 6 \times 10^{-4} T_{SJ} + 1.74 \times 10^{-2} \ln(1 + \frac{360}{t_D}) \quad (10.32)$$

In the equation, T_{SJ} is the average temperature during temperature cycling; t_D is the hold time per half-cycle in minutes. Referring to Fig. 10.58, T_{SJ} and t_D are determined to be 35 °C and 15 min, respectively, leading to $c = -0.407$. Based on experimental data, the thermal fatigue toughness coefficient ε_f for the mixed assembly solder joint is found to be 0.308. Consequently, the thermal fatigue life prediction model for the mixed assembly solder joint is:

$$N_f(x\%) = \frac{1}{2}\left(\frac{\Delta\gamma_p}{0.616}\right)^{\frac{1}{-0.407}}\left[\frac{\ln(1-0.01x)}{\ln 0.5}\right]^{0.2767} \tag{10.33}$$

3. Failure Analysis Case of BGA Solder Joints

For a certain type of mobile phone frequently experiencing failures such as inability to turn on, sudden shutdown, and automatic power off phenomenon, it was initially suspected that the BGA components on the mobile phone motherboard had solder joint failures, and a failure analysis was conducted on the PCBA component to determine the specific cause of failure.

X-ray transmission was used to inspect the good and failed PCBA components. The X-ray transmission results of the solder joints are shown in Fig. 10.63. The inspection results showed that, in addition to some solder joints having voids, no obvious solder joint missing, bridging, or obvious virtual soldering welding abnormalities were found in the designated BGA solder joints of the good and failed samples.

Dyeing tests were performed on one failed PCBA assembly with BGA packages, and the results are shown in Fig. 10.64. They indicate partial local cracking and complete cracking in some BGA solder joints, with cracks occurring in the array corners and peripheral regions.

A cross-sectional metallographic section analysis was performed on 1pcs failed PCBA sample's BGA device solder joint. The cross-sectional observation results showed that cracking occurred between the solder and the PCB pad, as well as in some of the PCB material under the pad. A metallographic section analysis was also performed on the same position solder joint of a good sample, and cracking between the solder joint and the PCB pad was also found. The metallographic slice of the solder joint of a good PCBA component is shown in Fig. 10.65.

The cross section of failed cracking solder joint was analyzed using SEM. The SEM image showed that a clear IMC was formed between the solder and the PCB pad. There was a P-rich layer between the Ni plating of the PCB pad and the IMC, and the cracking location of the solder joint was between the IMC and the P-rich layer on the surface of the PCB pad. The local IMC in the solder joint was relatively

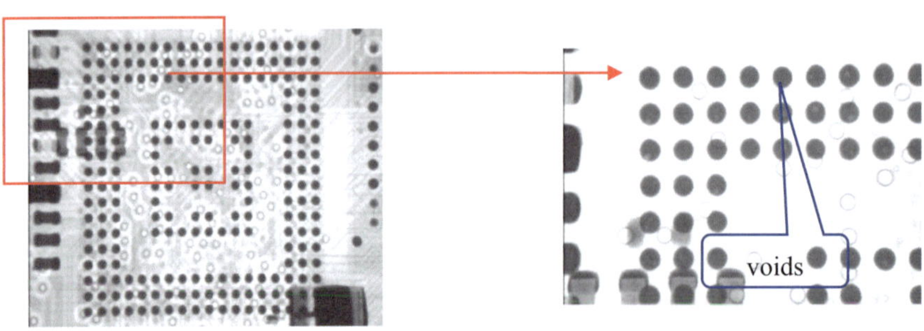

Fig. 10.63 X-ray transmission results of solder joints

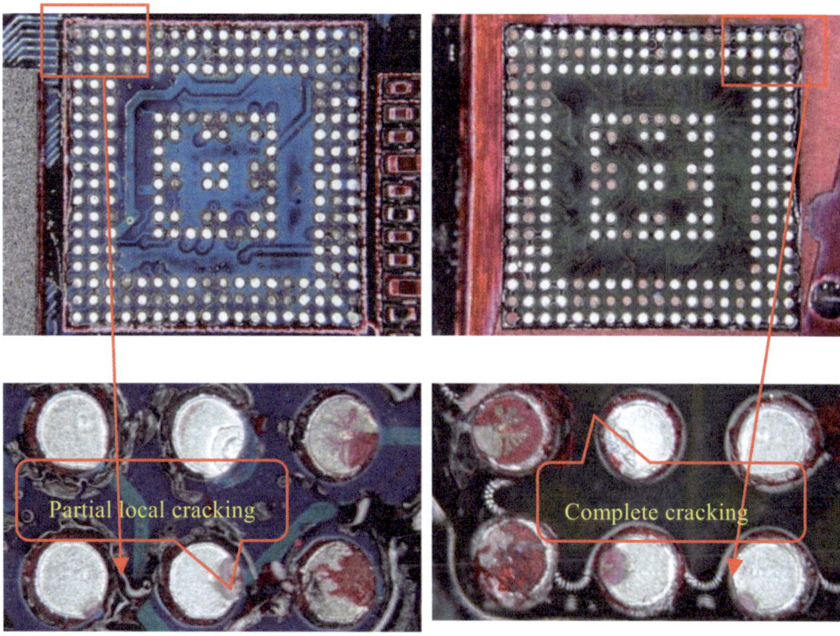

Fig. 10.64 Dyeing test results

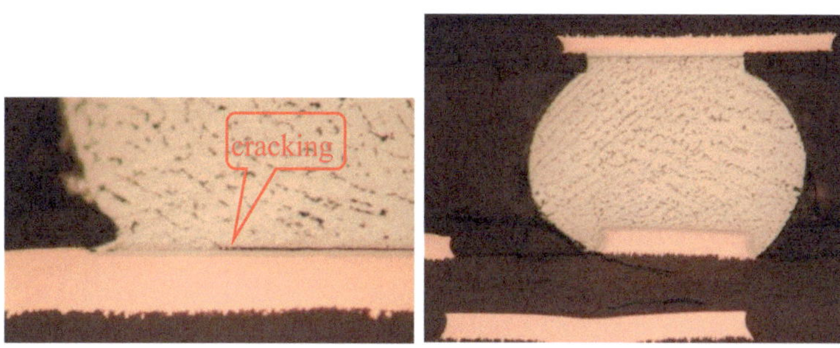

Fig. 10.65 Metallographic section of a good solder joint in a PCBA component

thick, with a thickness exceeding 6 μm. The SEM image of the failed solder joint is shown in Fig. 10.66.

The comprehensive analysis indicates that the functional failure of the PCBA component is due to the complete cracking of some corner BGA solder joints. The thick IMC in the solder joint leads to the formation of the P-rich layer, which in turn reduces the mechanical strength of the solder joint. The corner BGA solder joints are subjected to greater stress under working conditions and fail due to cracking when they cannot withstand the stress. Therefore, it is recommended to optimize the

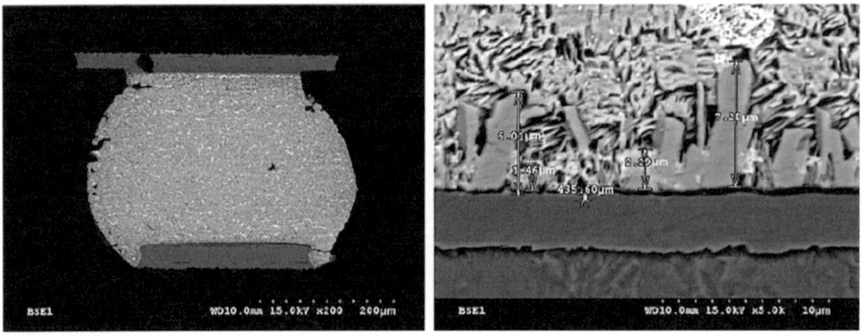

Fig. 10.66 SEM image of the failed solder joint

soldering process to control the influence of the P-rich layer, and if necessary, take measures (such as using underfill) to relieve the stress on the solder joints.

References

1. Wenli W, Yanshu Y (2011) Reliability of electronic assembly processes. Publishing House of Electronics Industry, Beijing
2. Rongrong F (2012) Reliability of modern electronic soldering processes. Publishing House of Electronics Industry, Beijing
3. Zhaohua W (2002) Fundamentals of surface mount technology. Beijing: National Defense Industry Press
4. Sitek J, Bukat K (2005) Influence of flux activity on process parameters and solder joints in lead-free wave soldering. In: International spring seminar on electronics technology: meeting the challenges of electronics technology progress, pp 100–105
5. Xianbing L (2014) Solder paste printing defects and quality improvement. In: Proceedings of the 2014 China high-end SMT academic conference, pp 229–235
6. Xuexia Y, Yu Z, Xuefeng S (2016) Experimental study on the influence of reflow soldering temperature profile on solder joint shape. Electron Compon Mater 35(2):70–72
7. Hu J, Jing B, Huang Y et al (2020) A health indicator for interconnect structure of QFP package under vibration and steady temperature. IEEE Access (99):1–1
8. Zhu Y, Liu H, Chang S (2020) Research on the board-level reliability of ceramic column grid array packaging component. In: 2020 21st international conference on electronic packaging technology (ICEPT), IEEE, pp 1–4
9. Daojun L, Guanghui H, Yabing Z (2015) Reliability technology and case study of electronic assembly processes. Beijing: Publishing House of Electronics Industry
10. Chiang KN, Chen WL (1998) Electronic packaging reflow shape prediction for the solder mask defined ball grid array. J Electron Packag 120(2):175–178
11. Yuxiu W, Songbai X, Yong H (2005) Influence of lead dimensions on the reliability of CPGA wing lead solder joints. Weld J 26(10):105–108
12. Xunping L (2012) Microstructure formation and evolution of Cu(Ni)/Sn-3.0Ag-0.5Cu/Ni(Cu) microsolder joints in BGA structure and shear fracture behavior of dimensional effects. Guangdong: South China University of Technology
13. Design and assembly process implementation for BGAs: IPC-7095

14. Ghaffarian R (2016) Effect of column properties and CGA assembly reliability by testing and analysis. Therm Thermomechanical Phenom Electron Syst, IEEE, pp 268–277
15. Tong L, Jiang C, Ao G (2018) Research on the board level reliability of high density CBGA and CCGA under thermal cycling. In: 2018 19th international conference on electronic packaging technology (ICEPT), IEEE, pp 1382–1386
16. Ghaffarian R (2006) CCGA packages for space applications. Microelectron Reliab 46(12):2006–2024
17. Shouwei L, Chongchong M, Dandan Y (2016) Development status and challenges of CCGA solder columns. Electron Packag 16(10):6–10
18. Wei Z, Shouhong S, Hui S (2011) Reliability assembly and mechanical reinforcement process of CCGA devices. Electron Process Technol 32(6):349–352
19. Rongzheng D, Yibo Y, Bo C et al (2012) Reliability study of CBGA and CCGA ball and column soldering and rework. Electron Packag 12(12):9–13
20. Yinglei C (2010) Reliability analysis and structural design of high-density large-sized CCGA secondary packaging. Harbin: Harbin Institute of Technology
21. Fu ZW, Zhou B, Yao RH et al (2017) Electromigration effect on kinetics of Cu-Sn intermetallic compound growth in lead-free solder joint. IEEE Trans Device Mater Reliab 17(4):773–779
22. Feifei T, Hao T, Ming Z (2019) Reliability study of eutectic tin-lead solder and thin gold solder joints. Solid-State Electron Res Prog 39(01):72–76
23. Yanpeng Z, Yanfu T, Wei W (2017) Influence of cooling rate on microstructure and mechanical properties of SnPb solder joints. Electron Process Technol 38(004):197–199
24. Cao H, Zhang Z, Yu Y (2018) Research on evolution behaviors of SnPb solder under high current stressing. In: 2018 19th international conference on electronic packaging technology (ICEPT), pp 891–894
25. Du X, Li J, Wang K et al (2020) Reliability evaluation of typical gold-containing solder joints for products. In: 2020 21st international conference on electronic packaging technology (ICEPT), pp 1–5
26. Xing X, Gaiqing C, Mingsheng C (2016) Reliability study of eutectic Tin-Lead solder joints in deep cold environment. Electron Process Technol 37(06):323–326
27. Foster FG (1962) Embrittlement of solder by gold from plated surfaces. Am Soc Test Mater 1:13
28. Xiaoming W, Yanping F (2013) Research on failure mechanism of eutectic Tin-Lead solder joints with gold-plated layers. Spacecr Eng 22(02):108–112
29. Gu X, Huang L, Wang G (2018) Investigation of ENEPIG/Sn-10Sb solder joint. In: 2018 19th international conference on electronic packaging technology (ICEPT), pp 731–733
30. Li Y, Wang H, Li Y et al (2020) Simulation and thermal fatigue analysis for board level BGA connection of HTCC packaging. In: 2020 21st international conference on electronic packaging technology (ICEPT), pp 1–5
31. Stolkarts V, Keer LM, Fine ME (1999) Damage evolution governed by microcrack nucleation with application to the fatigue of 63Sn–37Pb solder. J Mech Phys Solids 47(12):2451–2468
32. Wu J, Alam MS, Hassan KR et al (2020) Investigation and comparison of aging effects in SAC+ X solders exposed to high temperatures. In: 2020 IEEE 70th electronic components and technology conference (ECTC), pp 492–503
33. Burke C, Punch J (2017) A comparison of the creep behavior of joint-scale SAC105 and SAC305 solder alloys. IEEE Trans Compon Packag Manuf Technol 4(3):516–527
34. Cao H, Gan G, Tian M et al (2020) Effect of Ag content on microstructure and fatigue crack of SAC lead-free solder balls under rapid thermal fatigue. In: 2020 21st international conference on electronic packaging technology (ICEPT), pp 1–6
35. Hariharan G, Bhat C, Chaware R et al (2017) Thermo-mechanical reliability prediction for copper pillar 3D IC devices. In: 2017 IEEE 19th electronics packaging technology conference (EPTC), pp 1–7
36. Bin Z (2018) Research on failure mechanism and reliability of micro bump interconnections under thermoelectric stress. South China University of Technology, Guangdong

37. Ma B, Zhou B, Wang S et al (2017) Mechanical properties and microstructure of mixed SnAgCu-SnPb solder joints at cryogenic temperature. In: 2017 18th international conference on electronic packaging technology (ICEPT), pp 1582–1585
38. Le W-K, Zhu Z-W, Cao S et al (2018) Shear creep and fracture behavior of micro-scale BGA structured Cu/Sn–3.0 Ag–0.5 Cu/Cu joints under electro-thermo-mechanical coupled loads. In: 2018 19th international conference on electronic packaging technology (ICEPT), pp 620–624
39. Chen J, Wan N, Li J et al (2018) Effect of rapid inducted heating on the microstructure of solder joint in IC. In: 2018 19th international conference on electronic packaging technology (ICEPT), pp 815–818
40. George E, Das D, Osterman M et al (2011) Thermal cycling reliability of lead-free solders (SAC305 and Sn3.5Ag) for high-temperature applications. IEEE Trans Device Mater Reliab 11(2):328–338
41. Zhao Z, Hu C, Yin F (2018) Failure analysis for vibration stress on ball grid array solder joints. In: 2018 19th international conference on electronic packaging technology (ICEPT), pp 486–490
42. Chen Z, Che F, Ding MZ et al (2017) Drop impact reliability test and failure analysis for large size high density FOWLP package on package. In: 2017 IEEE 67th electronic components and technology conference (ECTC), pp 1196–1203
43. Sun H, Huang M (2017) Dominant effect of Sn grain orientation on electromigration-induced failure mechanism of Sn-3.0 Ag-0.5 Cu flip chip solder interconnects. In: 2017 18th international conference on electronic packaging technology (ICEPT), pp 1296–1299
44. Li S, Basaran C (2010) Effective diffusivity of lead-free solder alloys. Comput Mater Sci 47(1):71–78
45. Huang ML, Kuang JM, Sun HY (2019) Electromigration-induced β-Sn grain rotation in lead-free flip chip solder bumps. In: 2019 IEEE 69th electronic components and technology conference (ECTC), pp 2036–2041
46. Wu AT, Gusak AM, Tu KN et al (2005) Electromigration-induced grain rotation in anisotropic conducting beta tin. Appl Phys Lett 86(24):241902
47. Wang X, Liu X, Ding Y et al (2020) Study on the low temperature reliability of leaded solder. In: 2020 21st international conference on electronic packaging technology (ICEPT), pp 1–5

MIX
Papier aus verantwortungsvollen Quellen
Paper from responsible sources
FSC® C105338

If you have any concerns about our products,
you can contact us on
ProductSafety@springernature.com

In case Publisher is established outside the EU,
the EU authorized representative is:
Springer Nature Customer Service Center GmbH
Europaplatz 3, 69115 Heidelberg, Germany

Printed by Libri Plureos GmbH
in Hamburg, Germany